SCALING ANALYSIS IN MODELING TRANSPORT AND REACTION PROCESSES

The Wiley Bicentennial–Knowledge for Generations

Each generation has its unique needs and aspirations. When Charles Wiley first opened his small printing shop in lower Manhattan in 1807, it was a generation of boundless potential searching for an identity. And we were there, helping to define a new American literary tradition. Over half a century later, in the midst of the Second Industrial Revolution, it was a generation focused on building the future. Once again, we were there, supplying the critical scientific, technical, and engineering knowledge that helped frame the world. Throughout the 20th Century, and into the new millennium, nations began to reach out beyond their own borders and a new international community was born. Wiley was there, expanding its operations around the world to enable a global exchange of ideas, opinions, and know-how.

For 200 years, Wiley has been an integral part of each generation's journey, enabling the flow of information and understanding necessary to meet their needs and fulfill their aspirations. Today, bold new technologies are changing the way we live and learn. Wiley will be there, providing you the must-have knowledge you need to imagine new worlds, new possibilities, and new opportunities.

Generations come and go, but you can always count on Wiley to provide you the knowledge you need, when and where you need it!

William J. Pesce
President and Chief Executive Officer

Peter Booth Wiley
Chairman of the Board

SCALING ANALYSIS IN MODELING TRANSPORT AND REACTION PROCESSES

A Systematic Approach to Model Building and the Art of Approximation

William B. Krantz
Isaac M. Meyer Chair Professor
National University of Singapore
Rieveschl Ohio Eminent Scholar and Professor Emeritus
University of Cincinnati
President's Teaching Scholar and Professor Emeritus
University of Colorado

AIChE®

BICENTENNIAL 1807 WILEY 2007 BICENTENNIAL

WILEY-INTERSCIENCE

A JOHN WILEY & SONS, INC., PUBLICATION

Cover: Smaller and Smaller, a 38 cm × 38 cm wood engraving by Maurits Cornelis Escher (1898–1972), is a beautiful marriage between art and mathematics that employs hyperbolic tiling of two-dimensional space. It provides an appropriate introduction to the subject of this book—whereas the reptilian creatures become progressively smaller toward the center of the figure, they are the same size in scaled space based on a characteristic length for their geometry.

Published by John Wiley & Sons, Inc., Hoboken, New Jersey.
Published simultaneously in Canada.

Wiley Bicentennial Logo: Richard J. Pacifico

Library of Congress Cataloging-in-Publication Data:

Krantz, William B., 1939–
Scaling analysis in modeling transport and reaction processes : a systematic approach to model building and the art of approximation / William B. Krantz.
p. cm.
Includes bibliographical references and index.
ISBN 978-0-471-77261-3 (cloth)
1. Transport theory—Mathematics. 2. Chemical reactions—Mathematical models. 3. Scaling laws (Statistical physics) 4. Approximation theory. 5. Mensuration. I. Title.
QC175.25.S8K73 2007
530.13′8–dc22
2006030750

Printed in the United States of America.

10 9 8 7 6 5 4 3 2 1

To my wife, June, for her continued love, support, patience, and understanding throughout the years of our marriage and particularly during the time that I was writing this book!

CONTENTS

PREFACE

How big is big? How small is small? How wide is wide? How tall is tall?[1]

Scaling analysis as defined in this book involves a systematic method for nondimensionalizing the dependent and independent variables as well as their derivatives in a set of describing equations for a physical process. The unique aspect of this nondimensionalization is that it is done to ensure that the variables and their derivatives are bounded of order one; this implies that the magnitude of the dimensionless variables and their derivatives can range between zero and more-or-less 1. When this order-of-one scaling is done, the magnitude of the resulting dimensionless groups permits assessing the relative importance of the various terms in the describing equations; this in turn has many applications. The magnitude of the dimensionless groups appearing in the resulting dimensionless describing equations can be used to assess possible simplifying approximations. Order-of-one scaling analysis results in the minimum parametric representation of the describing equations. As such, this systematic method of scaling offers many advantages relative to dimensional analysis using the Pi theorem, which does not necessarily result in the minimum number of dimensionless groups. A particular advantage of scaling analysis is that it permits assessing the usefulness of a process or technology without the need for prior bench- or pilot-scale data. It also provides a template for the design of experiments to explore a new process or to validate a mathematical model.

The motivation for developing the approach to scaling analysis presented in this book extends back over 40 years, to when the author was a graduate student in the Department of Chemical Engineering at the University of California at Berkeley. The author had difficulty in grasping constructs such as hydrodynamic boundary-layer theory that were introduced using rather intuitive arguments such as those found in Schlichting's classic book.[2] The author strongly believed that boundary-layer theory as well as approximations such as creeping and lubrication flows, film theory, penetration theory, quasi-steady-state, and so on, which were introduced using intuitive arguments, could in fact be developed systematically. During this time the author rather serendipitously became aware of the work of Hellums and Churchill, who used a type of scaling analysis for systematically arriving at the

[1]Anonymous—attributed to an inquisitive young child!
[2]H. Schlichting, *Boundary Layer Theory*, McGraw-Hill, New York, 1960.

form of dimensionless variables that permitted similarity solutions to partial differential equations.[3] The book by Hansen, who used the Lie group of uniform magnifications and contractions to explore the spectrum of differential equations and associated boundary and initial conditions that would permit similarity solutions, was invaluable for establishing the mathematical basis for scaling analysis.[4] The pioneering books of van Dyke[5] and Nayfeh,[6] which focused on the use of perturbation expansions for solving differential equations, were particularly helpful in developing the concepts of ordering and multiple scales. Development of the microscale–macroscale modeling concept for modeling heterogeneous systems was strongly influenced by the book *Mass Transfer with Chemical Reaction* by Astarita.[7] Since the author was a product of the "transport phenomena" generation of engineering students, many of the examples and problems in this book were inspired by his perceived need to justify the assumptions underlying material presented in the classic textbook *Transport Phenomena* by Bird, Stewart, and Lightfoot.[8] The confluence of these influences led the author to develop the approach to scaling analysis presented in this book. Whereas this systematic method for scaling analysis borrows liberally from these prior mathematical developments, the author believes this to be an original contribution that he proffers to the community of scholars in science and engineering as both a teaching and a research tool.

When the author began his academic career in the Department of Chemical Engineering at the University of Colorado in 1968, he introduced the use of scaling analysis in the courses he taught involving fluid dynamics, heat transfer, mass transfer, and reactor design. The use of scaling analysis was very well received by the many engineering students who passed through the courses the author taught during his 38 years in academia. In particular, it helped his students by providing a systematic method for understanding subtle concepts such as creeping and boundary-layer flows in fluid dynamics, the Boussinesq approximation for thermally driven free convection, Taylor dispersion in mass transfer, and the various reaction regimes in mass transfer with chemical reaction. Scaling analysis was also helpful to students because it provided a unified approach to teaching transport and reaction processes. For example, scaling analysis provides a systematic method of illustrating the analogous roles played by the Reynolds number in fluid dynamics and the Peclet number in heat or mass transfer, or the Biot number in heat transfer and the Damköhler number in mass transfer with chemical reaction. The effective use of scaling analysis as a pedagogical tool in his courses contributed in no small way to the author being recognized by many teaching awards, including lifetime designation as a President's Teaching Scholar of the University of Colorado and several awards from the American Society for Engineering Education. The

[3] J. D. Hellums and S. W. Churchill, *A.I.Ch.E.J.*, **10**, 110 (1964).

[4] A. G. Hansen, *Similarity Analysis of Boundary Value Problems in Engineering*, Prentice-Hall, Englewood Cliffs, NJ 1964.

[5] M. Van Dyke, *Perturbation Methods in Fluid Mechanics*, Parabolic Press, Stanford, CA, 1975.

[6] A. H. Nayfeh, *Perturbation Methods*, Wiley, New York, 1973.

[7] G. Astarita, *Mass Transfer with Chemical Reaction*, Elsevier, New York, 1967.

[8] R. B. Bird, W. E. Stewart, and E. N. Lightfoot, *Transport Phenomena*, Wiley, New York, 1960.

favorable response the author received from his students to the use of scaling analysis in his courses led him to present and publish several papers on this subject. This national as well as international exposure for this systematic approach to scaling analysis catalyzed a response from the academic community whose encouragement motivated the author to write this book.

Whereas scaling analysis clearly is invaluable as a pedagogical tool, it also has application in research and development. For example, scaling analysis allows one to assess the value of a new process by providing order-of-magnitude estimates of the anticipated performance. It also can be used to establish the process parameters in the design of both numerical and laboratory experiments to explore new technologies. For this reason, timely examples drawn from the author's experience are included that effectively illustrate how scaling analysis was used to design a novel membrane–lung oxygenator,[9] to assess the use of pulsed pressure-swing adsorption in producing oxygen from air,[10] to develop a model for polymeric membrane fabrication,[11] and to explore the potential of a novel process for producing hydrogen from methane using solar energy.[12] Hence, the book has been written to serve as both a textbook and as a reference book for researchers.

The book includes 62 examples that are worked in some detail to illustrate the scaling method as well as 165 unworked problems that can be assigned when the book is used as a textbook. Many of these problems are open ended; as such, they provide excellent material to stimulate creative thinking for students. The author has used selected chapters of this book complemented either with an appropriate textbook or his lecture notes to teach courses in transport phenomena, fluid dynamics, heat transfer, mass transfer, and reactor design. Whether one intends to use this as a reference book or a textbook, it is necessary to read Chapters 1 and 2, which provide an overview of the systematic approach to scaling analysis. A course in fluid dynamics can easily cover Chapters 1 through 3, including working nearly all the problems at the end of the Chapter 3. A course in heat transfer would cover Chapters 1, 2, and 4, and a course in mass transfer would cover Chapters 1, 2, and 5. For both courses, parts of Chapter 3 would be required to consider convective heat or mass transfer. A course in mass transfer with chemical reaction would cover Chapters 1, 2, and 6 as well as the parts of Chapters 3 and 5 needed to consider convective mass transfer. A course in modeling transport and reaction processes necessarily would involve all the chapters. The author used a draft version of this book to teach a course on process modeling to graduate students whose pre- or corequisites included at least a graduate-level course in transport phenomena; this one-semester course covered all the chapters, at least in part.

The author conscientiously tried to ferret out the errors in the book. but, a few more were found with each rereading. Unfortunately, perfection is a quality

[9]R. R. Bilodeau, R. J. Elgas, W. B. Krantz, and M. E. Voorhees, U.S. patent 5,626,759, issued May 6, 1997.

[10]E. M. Kopaygorodsky, V. V. Guliants, and W. B. Krantz, *A.I.Ch.E. J.*, **50**(5), 953 (2004).

[11]D. Li, A. R. Greenberg, W. B. Krantz, and R. L. Sani, *J. Membrane Sci.*, **279**, 50 (2006).

[12]J. K. Dahl, A. W. Weimer, and W. B. Krantz, *Int. J. Hydrogen Energy*, **29**, 57 (2004).

accorded only to the gods! The author would greatly appreciate receiving corrections and suggestions for improving the book. In particular, he welcomes contributions of new examples and problems that will possibly be included in subsequent editions of the book, with due credit given to contributors.

ACKNOWLEDGMENTS

Preparing these acknowledgments made the author reflect on the wonderful people who made this book possible. We are influenced by many people in our lives. However, a few people really make a difference in what we become and what we accomplish. Educators should take some encouragement from this first acknowledgment since it indicates that students remember at least some of what you say to them—in this case the student carried the sage advice of his teacher given to him over 50 years ago! The late Dom Wulstan Mork, the author's high school English teacher, conveyed to his class that every person can write one good book. Although the author has edited research monographs, this is his first solo effort to write a book that more or less represents his own work. Hence, the author sincerely hopes that this effort will be his "good book" and hopefully, not his last!

The author first pursued a degree in chemistry at Saint Joseph's College in Indiana. The small liberal arts college environment helped to develop the author as a "whole man" by cultivating his interests in music, philosophy, languages, mathematics, and science. His life would have taken a much different path had it not been for the guidance of the late Brother John Marling at Saint Joe's, who encouraged him to pursue engineering studies at the University of Illinois.

At the University of Illinois the author was extremely fortunate to take an undergraduate course from Professor John Quinn, who used the first printing of the first edition of *Transport Phenomena* by Bird, Stewart, and Lightfoot as the textbook. This superbly written book stimulated the author to begin asking the questions regarding how the underlying assumptions in the many models used to describe transport phenomena might be rigorously justified—the seed for writing this book had been planted! The author would be remiss if he did not acknowledge the profound influence that Professor Max Peters had on his career. Max had an inimitable way of giving advice that he must have picked up while serving in the 10th Mountain Division of the U.S. Army during World War II! Despite the author's intention of taking a job in industry after receiving his B.S. degree in chemical engineering, Max "issued orders" for him to apply to graduate school, which he dutifully obeyed!

The University of California at Berkeley provided an unbelievably fertile environment for graduate studies in the early 1960s. The undercurrent of thought at Berkeley was to question and challenge irrespective of whether the subject involved science or societal issues. Professor John Prausnitz, who was the author's academic advisor, was particularly influential in stimulating the author to think more creatively. Indeed, "thinking outside the box" did not come naturally to a young man who had spent four years in the highly disciplined environment of a military high school and an all-men's Catholic college isolated deep in the cornfields of Indiana—you did what you were told to do and did not question the reason for doing it! Serving as a teaching assistant for several courses at Berkeley gave the author the opportunity to discover something that he might not otherwise have known: namely that his vocation in life was to be an educator.

Once having decided to enter academia, the decision as to where to apply for a faculty position was easy. Professor Max Peters had moved on from Illinois to become Dean of Engineering at the University of Colorado. The author wanted to work for the man who was so influential in charting the course of his professional life. The author's academic career began with some very good advice from the late Professor R. Curtis (Curt) Johnson, then chairman of the Department of Chemical Engineering at Colorado, who told him to first become a good teacher and then focus on building a credible research program—uncommon advice for young educators today. The author's senior faculty mentor at Colorado was Professor Klaus Timmerhaus, who provided invaluable guidance for his career development. Without Klaus's continued encouragement and support, it is doubtful that this book would ever have been written.

The author would be an ingrate if he did not acknowledge the many students at the University of Colorado, the University of Cincinnati, and most recently at the National University of Singapore who provided useful feedback on the scaling analysis method that he inculcated into his courses. The author sincerely wishes that he could acknowledge all these students by name. If they had not responded favorably to the use of scaling analysis in his courses, there would have been no impetus to write the book.

This book is significantly improved as a result of the opportunity provided to the author by Professor Raj Rajagopalan, head of the Department of Chemical and Biomolecular Engineering at the National University of Singapore. Raj offered a visiting faculty appointment to the author and also encouraged him to teach a graduate-level course on scaling analysis using a draft copy of the book. This convinced the author that the book could be covered in sufficient depth in a one-semester course.

Finally, and most appropriately, the author would like to acknowledge the staff of his publisher, John Wiley & Sons. The directives provided by the Wiley staff made preparing this book much easier. In particular, the author acknowledges Bob Esposito, Executive Editor for Wiley's Scientific, Technical, and Medical Division, with whom he worked most closely in preparing the manuscript for publication.

1 Introduction

Through and through the world is infested with quantity: To talk sense is to talk quantities. It is no use saying the nation is large . . . How large?

It is no use saying the radium is scarce . . . How scarce?

You cannot evade quantity. You may fly to poetry and music, and quantity and number will face you in your rhythms and your octaves.[1]

1.1 MOTIVATION FOR USING SCALING ANALYSIS

This book is directed to a broad spectrum of readers since modeling transport and reaction processes is common to many fields of pure and applied science. The book should be useful to educators who are seeking effective pedagogical tools for introducing their students to an ever-expanding body of knowledge in the field of transport phenomena and reactor design. It should also be of value to engineers and scientists who need to apply and develop mathematical models for transport and reaction processes. It will be helpful to students who are seeking ways to better understand the broad range of subjects encompassed by transport and reaction processes.

As defined in this book, the subject of scaling analysis, deals with a systematic method for nondimensionalizing a system of describing equations for transport or reaction processes. The resulting dimensionless system of equations represents the minimum parametric representation of the process. By this we mean that the solution for any quantity that can be obtained from these equations will be at most a function of the dimensionless independent variables and the dimensionless groups generated by the scaling process. For example, scaling a heat-conduction process will lead to a set of dimensionless equations whose solution for the dimensionless temperature will be a function of the dimensionless spatial and temporal

[1]Alfred North Whitehead (1861–1947), in *The World of Mathematics*, J. R. Newman, ed., Simon & Schuster, New York, 1956.

Scaling Analysis in Modeling Transport and Reaction Processes: A Systematic Approach to Model Building and the Art of Approximation, By William B. Krantz

independent variables and dimensionless parameters such as the Prandtl number ($C_p\mu/k$, in which μ is the shear viscosity, C_p the heat capacity, and k the thermal conductivity). Quantities that are obtained by evaluating the solution to the dimensionless equations at fixed values of the spatial and temporal variables or by integrating a dimensionless dependent variable over the spatial or temporal domain will be functions of a reduced set of dimensionless spatial or temporal variables and the relevant dimensionless groups. In some cases the dimensionless dependent variable of interest might be a function of only the dimensionless groups. For example, in a steady-state heat-conduction process, the dimensionless heat-transfer coefficient (Nusselt number) will be a function of the relevant dimensionless groups, such as the Prandtl number and geometric aspect ratios. This minimum parametric representation of a transport or reaction process is useful since it identifies the dimensionless variables and groups that can be used to correlate data from either laboratory or numerical experiments (i.e., computer simulations). The resulting dimensionless groups can also be used for scale-up or scale-down analyses by invoking the principles of geometric and dynamic similarity.

There is no unique set of dimensionless dependent and independent variables and associated dimensionless groups for a system of equations describing a transport or reaction process. For any system of describing equations, one set of dimensionless dependent and independent variables and corresponding dimensionless groups can always be obtained from any other set. However, one can scale a system of describing equations in a unique way to ensure that the relevant dependent and independent variables and their derivatives are bounded of order one. By this we mean that the magnitude of the particular dimensionless variable or its derivative is bounded between zero and more-or-less 1. For those familiar with formal ordering arguments, we are bounding our variables to be *little oh* of 1 [i.e., $o(1)$] as opposed to *big oh* of 1 [i.e., $O(1)$], which means that the quantity is essentially 1. Note that by *of order one* we do not mean exactly 1. In $o(1)$ scaling, one can say, for example, that $0.8 \cong 1$ or $3 \cong 1$; that is, the quantity is well within an order of magnitude of 1. In this book this special application of scaling that leads to unique dimensionless variables and groups is referred to as $o(1)$ *scaling*.

The nondimensionalization associated with $o(1)$ scaling is indeed unique. However, arriving at this unique scaling often involves a process of trial and error. That is, one has to assume that a particular transport or reaction process is dominated by some mechanism(s) (e.g., heat conduction in a particular direction for a multidimensional heat-transfer process) and then has to nondimensionalize the describing equations by comparing the other terms to the one that embodies this mechanism. After obtaining a system of dimensionless describing equations, one evaluates the resulting dimensionless groups for the relevant geometric and physical parameters of interest. If all the dimensionless groups are bounded of order one [i.e., $o(1)$], the original assumption as to the controlling mechanism(s) was correct. However, if any of the dimensionless groups is much larger than 1, it can indicate that the scaling was not correct for the geometric and physical parameters of interest or that there is a region of influence or boundary layer in which a temporal or spatial derivative becomes very large. In either case one has to repeat the scaling analysis with a different set of assumptions as to the controlling mechanism(s). The possibility also exists that proper scaling will yield a dimensionless group that is much larger than 1, which multiplies some grouping that involves the difference between two dimensionless quantities each of which is bounded of o(1). In this case the large dimensionless group implies that the grouping it multiplies is much less than 1. We will see that scaling analysis is forgiving in that, when done correctly, all terms in the relevant equations will be bounded of order one; that is, the product of any dimensionless group and the grouping of dimensionless dependent and/or independent variables that it multiplies is o(1).

The utility of o(1) scaling is that when all the relevant dependent and independent variables and their derivatives are bounded of order one in the resulting dimensionless describing equations, one can assess the importance of various terms on the basis of the values of the dimensionless groups that multiply them. If all the dimensionless dependent variables and their derivatives and the independent variables are bounded of o(1), the dimensionless groups should also be bounded between 0 and 1. Hence, if a dimensionless group is of order 0.01 or less, the term that it multiplies can be ignored in developing a model for the particular transport or reaction process while incurring only a very small ($\sim 1\%$) error. Hence, by using o(1) scaling, one can appropriately simplify the describing equations for a transport or reaction process. For example, the equations of motion can be nondimensionalized appropriately using o(1) scaling to determine the condition required to neglect the inertia terms; that is, a very small Reynolds number, which is the familiar creeping-flow approximation.

The trial-and-error process involved in arriving at the proper o(1) scaling is of particular value in designing experiments. In the absence of solving model equations, o(1) scaling permits determining the values of the geometric and process parameters that are required to achieve certain experimental conditions. For example, o(1) scaling permits determining the adsorbent bed properties required to ensure that an adsorption process is controlled by equilibrium considerations rather than intraparticle diffusion.

Scaling analysis is also useful for developing perturbation expansion solutions to the describing equations. Scaling will identify dimensionless parameters whose limiting values (i.e., very large or very small) permit making certain approximations in solving the describing equations. For example, when the Reynolds number is very small, one can develop an analytical solution for the flow around a sphere falling at its terminal velocity in a Newtonian fluid with constant physical properties; the result is the familiar Stokes flow solution for creeping flow over a sphere. However, one can account for the neglected inertia terms in the equations of motion by considering a perturbation expansion solution to the describing equations in terms of the small Reynolds number. The zeroth-order term in this perturbation expansion corresponds to the Stokes solution for creeping flow. The first-order term that accounts for some effects of the inertia terms was first worked out by Proudman and Pearson.[2] Perturbation solutions that are well behaved in the limit

[2] I. Proudman and J. R. A. Pearson, *J. Fluid Mech.*, **2**, 237 (1957).

of the perturbation parameter becoming very small or very large are referred to as *regular perturbation expansions*. Perturbation expansions that are not well behaved in the limit of a perturbation parameter becoming very small or very large are referred to as *singular perturbation expansions*. An example of the latter is very high Reynolds number flows. If one tries to solve the equations of motion in the limit of very large Reynolds numbers by attempting a perturbation expansion in the (small) reciprocal Reynolds number, one cannot properly account for the neglected viscous terms. This is a direct consequence of the reduction in the order of the describing equations when one develops the zeroth-order solution in the reciprocal Reynolds number. To solve singular perturbation expansion problems, one needs to use the method of multiple scales, whereby different scales are used in the inner region, the outer region, and the overlap region between them. Scaling analysis is an invaluable tool for determining when perturbation solutions are possible and in determining the proper scales for the various regions. This book complements classical references on perturbation expansion methods.[3,4]

For the same reason that scaling analysis is useful in determining the scales and expansion parameters in perturbation analyses, it is useful in assessing potential problems that can occur in solving a system of describing equations numerically. That is, when certain dimensionless groups become very small or very large, problems can be encountered in solving the resulting system of describing equations numerically. For example, when the Reynolds number becomes very large, the viscous effects will be confined to a very thin region in the vicinity of the solid boundaries. If one uses a coarse mesh or does not employ a numerical routine with a remeshing capability, the numerical routine will not provide sufficient resolution in the vicinity of the solid boundaries and thereby either will not run or will provide erroneous results. Scaling analysis can be used to identify these *boundary-layer regions* so that a proper numerical method can be employed to solve the problem.

Scaling analysis is particularly useful to an educator who is faced with explaining seemingly unrelated topics such as creeping flows, boundary-layer flows, film theory, and penetration theory. Topics such as these often are developed in textbooks in a rather intuitive manner. Scaling analysis provides a systematic way to arrive at these model approximations that eliminates guesswork; that is, scaling analysis provides an invaluable pedagogical tool for teachers. Disparate topics in transport and reaction processes can be presented in a unified and integrated manner. For example, a *region of influence* in scaling provides a means for presenting a unified approach to boundary-layer theory in fluid dynamics, penetration theory in heat and mass transfer, and the wall region for confined porous media.

Scaling analysis also provides a very effective learning tool for students. Textbooks on transport and reaction processes generally justify simplifying assumptions leading to the creeping-flow, boundary-layer, penetration theory, and plug-flow reactor equations and others through ad hoc arguments rather than by a systematic approach such as that provided by scaling analysis. Hence, a student might

[3] M. Van Dyke, *Perturbation Methods in Fluid Mechanics*, Parabolic Press, Stanford, CA, 1975.
[4] A. H. Nayfeh, *Perturbation Methods*, Wiley, New York, 1973.

not see the interrelationship between the various approximations made in describing transport and reaction processes, such as the analogy between boundary-layer theory in fluid dynamics and penetration theory in heat or mass transfer. Moreover, the ad hoc approach to simplifying the equations describing transport and reaction processes does not provide students with a basis for simplifying more complex problems not described in textbooks.

1.2 ORGANIZATION OF THE BOOK

Scaling analysis is used by many pure and applied scientists at least in some form; for example, in dimensional analysis. Many textbooks use order-of-magnitude arguments to simplify the describing equations for transport and reaction processes. However, what is lacking is a systematic treatment of scaling analysis that can be used reliably without the need for the intuition that is either an inherent talent or has been learned through years of practical experience. Hence, in Chapter 2 we present scaling analysis in general terms as a series of steps to be followed. We distinguish between the steps used in scaling for the purpose of dimensional analysis, which leads to nonunique dimensionless groups, and those to be followed for the special case of o(1) scaling, which leads to a unique minimum parametric representation.

Since this is intended to serve as both a reference book and as a textbook for a course in mathematical modeling, the subject matter covered by Chapters 3 through 5 is organized according to the conventional topics in transport phenomena: fluid dynamics, heat transfer, and mass transfer. The rationale for this organization is that one needs to know how to scale the fluid dynamics to handle scaling of convective heat and mass transfer. The latter is a necessary precursor to treating the special topic of mass transfer with chemical reaction, which is covered in Chapter 6.

Chapter 7 is an integrating chapter in which we consider the application of scaling to process design, which can involve coupled fluid flow, heat and mass transfer, and chemical reactions. In particular, we illustrate how scaling can be used to assess a new process or to design experiments (e.g., the sizing of equipment) to ensure that desired conditions are met.

We presume that the reader has a basic knowledge of transport and reaction processes; the book is not intended to replace textbooks that treat these subjects in depth. A basic knowledge of the language of continuum mechanics (i.e., vector and tensor mathematics) is assumed. However, the appendices summarize useful background material relevant to modeling transport and reaction processes. Since there is no general agreement in the literature on the sign convention in the constitutive equations or surface forces in the equations of motion, the appendices include a brief review of the sign convection used in the book. The appendices also summarize the forms of the continuity, equations of motion for both conventional fluid flow and flow through porous media, and energy- and species-balance equations in generalized vector–tensor notation as well as in rectangular, cylindrical, and spherical coordinates. Useful integral relationships for scalars, vectors, and tensors are also included in the appendices.

The format in Chapters 3 through 5 is designed to illustrate the application of scaling analysis by means of problems drawn from fluid dynamics, heat transfer, and mass transfer. These problems are organized to illustrate how scaling can be used to develop basic concepts such as creeping flows, boundary-layer theory, film theory, and penetration theory. The format is to begin by indicating what the problem is supposed to demonstrate. For example, analysis of an impulsively oscillated plate is presented to illustrate both how to handle time scaling and to show what is meant by a region of influence. Several problems are illustrated in detail, followed by a comparable number of example problems that are outlined in less detail.

Chapter 6 is organized somewhat differently since it considers problems in mass transfer with chemical reaction that require scaling analysis on both the micro- and macroscales: for example, on the scale of a small adsorbent particle and on the much larger scale of the contacting device that contains these particles. Hence, after introducing the concepts of micro- and macroscaling, the problems in this chapter focus on the use of scaling to identify the various reaction regimes that can be encountered in mass transfer with chemical reaction.

Whereas scaling analysis is used in Chapters 3 through 6 to justify classical approximations made in fluid dynamics, heat and mass transfer, and mass transfer with chemical reaction, in Chapter 7 we use scaling analysis to design and assess novel technologies. The four examples considered in this chapter are considerably more complex since they involve coupled transport and in some cases chemical reaction as well. These examples were chosen because scaling analysis contributed significantly to the process design and technology development.

Chapters 3 through 7 end with a summary that emphasizes the principles of scaling analysis that were illustrated in the worked problems. Unworked practice problems included at the end of each chapter explore in more detail the examples considered in the chapter and apply scaling analysis to related problems.

2 Systematic Method for Scaling Analysis

At its best, physics eliminates complexity by revealing underlying simplicity...

The beauty of the Standard Model (of particle physics) is in its symmetry;

mathematicians describe its symmetries with objects known as Lie groups."[1]

2.1 INTRODUCTION

In this chapter, scaling analysis is presented as a stepwise procedure. The procedure differs depending on whether one seeks to obtain the minimum parametric representation for dimensional analysis or to do o(1) scaling to simplify a set of describing equations or to design an experiment. We begin by considering o(1) scaling since this is the primary focus of the book. Scaling as an alternative method for dimensional analysis is included for completeness at the end of the chapter. Implementation of the stepwise procedure for either scaling or dimensional analysis in modeling transport and reaction processes is the subject of subsequent chapters. We begin by providing a brief overview of the mathematical basis for scaling analysis.

2.2 MATHEMATICAL BASIS FOR SCALING ANALYSIS

Scaling analysis has its mathematical foundation in Lie group theory, specifically the continuous group of uniform magnifications and contractions.[2] The properties of the latter group are useful when considering the operations involved when

[1]C. Seife, Can the laws of physics be unified? *Science*, **309**, 82 (2005).
[2]A review of group theory is given in A. G. Hansen, *Similarity Analysis of Boundary Value Problems in Engineering*, Prentice Hall, Englewood Cliffs, NJ, 1964, Chap. 4.

Scaling Analysis in Modeling Transport and Reaction Processes: A Systematic Approach to Model Building and the Art of Approximation, By William B. Krantz

we change the units on the quantities that appear in dimensional equations. For example, when converting the length unit of centimeters to meters, all quantities expressed either totally or partially in terms of length units (heights, widths, velocities, accelerations, densities, etc.) experience a uniform magnification or contraction; that is, all heights become smaller when expressed in terms of meters rather than centimeters, whereas all densities become larger.

The connection between uniform magnifications and contractions and scaling analysis might not be clear in view of the fact that one is not changing units when one nondimensionalizes a system of equations. When one nondimensionalizes a quantity, it involves dividing the quantity by another quantity or combination of quantities that have the same units. Quantities are broadly classified as either primary or secondary. *Primary quantities* are measured in terms of units of their own kind: for example, a length quantity measured in terms of meters or a force quantity measured in terms of Newtons. *Secondary quantities* are measured in terms of the units used for primary quantities: for example, velocity measured in terms of a length divided by a time, or force measured in terms of kilograms multiplied by meters divided by seconds squared. Note that any secondary quantity can be converted to a primary quantity merely by measuring it in terms of units of its own kind. Indeed, in the preceding examples, force was considered as both a primary and a secondary quantity. However, the same could be done with a quantity such as velocity; for example, we could define 1 Vel to be the velocity associated with a moving a distance of 1 meter in 1 second. Scaling analysis is equivalent to considering every scaled quantity to be a primary quantity since when we nondimensionalize a quantity, we are dividing it by something that has the same units. Hence, the properties of the group of uniform magnifications and contractions also underlie the operations that we use in scaling analysis. More could be said and done with group theory in exploring the full implications of scaling analysis. However, this would not serve the purpose of this book, which is to show how scaling analysis can be used to model transport and reaction processes.

2.3 ORDER-OF-ONE SCALING ANALYSIS

The procedure that is involved in o(1) scaling analysis can be reduced to the following eight steps:

1. Write the dimensional describing equations and their initial, boundary, and auxiliary conditions appropriate to the transport or reaction process being considered.
2. Define unspecified scale factors for each dependent and independent variable as well as appropriate derivatives appearing explicitly in the describing equations and their initial, boundary, and auxiliary conditions.
3. Define unspecified reference factors for each dependent and independent variable that is not referenced to zero in the initial, boundary, and auxiliary conditions.

4. Form dimensionless variables by introducing the unspecified scale factors and reference factors for the dependent and independent variables and the appropriate derivatives.
5. Introduce these dimensionless variables into the describing equations and their initial, boundary, and auxiliary conditions.
6. Divide through by the dimensional coefficient of one term (preferably one that will be retained) in each of the describing equations and their initial, boundary, and auxiliary conditions.
7. Determine the scale and reference factors by ensuring that the principal terms in the describing equations and initial, boundary, and auxiliary conditions are o(1) (i.e., they are bounded between zero and of order one).
8. The preceding steps result in the minimum parametric representation of the problem (i.e., in terms of the minimum number of dimensionless groups); appropriate simplification of the describing equations may now be explored.

The dimensional describing equations involved in step 1 for transport and reaction processes are usually differential equations with prescribed initial and/or boundary conditions as well as auxiliary conditions to determine the location of moving boundaries or free surfaces. These describing equations incorporate any simplifications that one is certain are justified; for example, assuming an incompressible flow for a liquid. However, one cannot eliminate any of the terms whose magnitude scaling analysis is being used to assess; for example, the inertia terms when one is seeking to justify the creeping-flow approximation. In implementing this step, one must write down at least formally all the differential and algebraic equations necessary to solve the particular problem. For example, one might have an elliptic differential equation that requires a downstream boundary condition that is not known; indeed, one might be using scaling analysis to determine when the elliptic equation can be simplified to a parabolic equation that obviates the need to specify this problematic boundary condition. Nonetheless, one needs to specify this unknown boundary condition at least formally. One also needs to include appropriate equations of state, kinetic relationships, and so on, required to ensure that the problem is determined completely.

In step 2 one defines scale factors for each dependent and independent variable that appears explicitly in the describing equations and their initial, boundary, and auxiliary conditions. However, in addition, one might have to define scale factors for certain derivatives of the dependent variables that appear explicitly in the describing equations and their initial, boundary, and auxiliary conditions. One sees that this procedure in step 2 is a dramatic departure from that used in conventional dimensional analysis. The reason for introducing scale factors on derivatives as well as dependent and independent variables is to ensure that the resulting dimensionless derivatives are o(1). This is a critical step since one would like to have every relevant dimensionless variable as well as their derivatives be of o(1) so that the magnitude of the dimensionless groups multiplying the dimensionless variables and/or their derivatives indicates the relative importance of the particular term in the describing equations. Often, the derivatives will scale with the same scale factors as those used for the dependent and independent variables. This occurs when the particular dependent variable experiences its characteristic change over a distance or time that corresponds to the characteristic length or time. However, this is not always the proper way to scale derivatives. For example, in moving boundary problems, one usually does not scale the time derivative of the location of the moving boundary with the characteristic length scale divided by the characteristic time scale. The proper way to scale derivatives can best be illustrated by means of the problems discussed in subsequent chapters.

Step 3 introduces reference factors for any dependent or independent variable that is not naturally referenced to zero. In fluid dynamics, dependent variables such as velocities are often naturally referenced to zero because of the no-slip and impermeable boundary conditions at solid surfaces. However, this is generally not true for heat- and mass-transfer problems. For example, a one-dimensional heat-conduction problem might have boundary conditions that involve different constant temperatures at two planar surfaces. If one wants the dimensionless temperature to be bounded between zero and 1, it is clearly necessary to introduce a reference temperature, which scaling will naturally determine to be the lowest known temperature for the process. Note that reference factors are sometimes needed for independent variables as well. For example, in solving a fluid-flow problem in an annulus the zero for the radial coordinate should be referenced to the inner wall of the annulus, not to the axis of symmetry for the cylindrical coordinate system. Introducing a reference factor for variables not naturally referenced to zero is critical in achieving $\circ(1)$ scaling. If this is not done for a variable that is not naturally referenced to zero, the parametric representation of the problem will involve an additional unnecessary dimensionless group.

In step 4 we form dimensionless variables for all dependent and independent variables and their relevant derivatives. These are defined by dividing the dimensional value of the particular variable relative to the unspecified reference factor (for those variables not naturally referenced to zero) by the unspecified scale factor.

Step 5 involves using the chain rule of differentiation to recast the dimensional describing equations in terms of the dimensionless variables. This is generally quite straightforward since the scale and reference factors are considered to be constants in scaling analysis. In some problems involving a region of influence such as boundary-layer flows, the scale factor might be a function of one of the independent variables. However, in such cases we are considering "local scaling" at a fixed value of the independent variable. Hence, the scale factors involving the region of influence are still treated as constants in the change of variables involved in the nondimensionalization. This will become clearer when the example problems involving a variable region of influence (e.g., boundary-layer flows) are considered. However, to reference a spatial variable to zero in some problems, (e.g., when a moving boundary is involved), the new scaled spatial variable will be a function of time through a reference factor that defines the location of the boundary. In these cases the chain rule of differentiation must be applied with caution since time derivatives in a fixed reference frame do not transfer as simple time derivatives with respect to the moving reference frame. An additional term is generated that involves the velocity of the moving boundary. This situation arises in problems involving moving boundaries, due to phase transition, net mass addition or removal from a system, and dissolution or precipitation of a phase.

In step 6 we divide through by the dimensional coefficient of one term in each differential and algebraic equation involved in the describing equations for the particular transport or reaction process. These dimensional coefficients will consist of known parameters of the process, such as the density, viscosity, thermal conductivity, and so on, as well as the unspecified scale and reference factors used to nondimensionalize the variables in the describing equations. In implementing this step, one should try to divide through by the dimensional coefficient of a term that must be retained in each of the describing equations, to retain physical significance. For example, to satisfy certain boundary conditions, one might need to retain the highest-order spatial derivative in a coordinate direction. Another example is the force that causes flow in a fluid dynamics problem, such as the axial pressure gradient. However, in some cases one might not know which terms must be retained. In such cases one divides through by the dimensional coefficient of some chosen term. If, in fact, the term chosen is not a significant term for the particular conditions being considered, other terms in the same describing equation will be multiplied by dimensionless groups that are significantly greater than 1, indicating that the latter terms are the most important in the equation being considered.

Step 7 is the most subtle step in scaling analysis. In this step one determines the unspecified scale and reference factors by demanding that the dimensionless dependent and independent variables and their relevant derivatives in the describing equations be o(1). To accomplish this, one sets appropriate dimensionless groups containing the unspecified scale and reference factors equal to 1 (for scale factors) or zero (for reference factors). In some cases the scale factor for one dependent variable (e.g., a particular velocity component in a fluid dynamics problem) might be obtained by integrating the scale factor for another variable (e.g., the derivative of this same velocity component). The manner in which this step is implemented is best learned by example, which is why most of the book is devoted to applying scaling analysis to a variety of problems in fluid dynamics, heat transfer, mass transfer, and reaction processes.

Step 8 is the desired end result of the scaling analysis: the unique minimum parametric representation of the describing equations for the process that ensures o(1) scaling. Since all the dimensionless dependent and independent variables and their relevant derivatives are o(1), the magnitude of each term in the describing equations is determined by the magnitude of the dimensionless group that multiplies this term. One dimensionless term in each describing equation will have a coefficient of unity since in step 6 we divided through by the dimensional coefficient of one term in each of these equations. Hence, one is comparing to 1 the magnitude of each term in each describing equation. How one proceeds in this step depends on what information is being sought in the scaling analysis. If one is seeking to determine the conditions required to ignore a particular term or terms in the describing equations, one merely demands that the dimensionless coefficient of the term be much less than 1 [e.g., o(0.1) or o(0.01)]. If one is seeking to determine what approximations are allowed for a particular problem for which the process parameters are known, one evaluates all the dimensionless groups in the describing equations to assess their magnitude. If the scaling analysis is correct for the particular process conditions, the magnitudes of all the dimensionless groups must be o(1). If any of the dimensionless groups are much greater than o(1), one of the following is indicated: (1) the term containing this group should have been the one whose dimensional coefficient was divided through to form the dimensional groups in step 6; (2) there is a region of influence or boundary layer in which a temporal or spatial derivative becomes very large; or (3) the group of dimensionless dependent variables and/or their derivatives that the large dimensionless group multiplies is very small. In the first two situations one has to rescale the describing equations either by dividing through by the appropriate dimensional coefficient in each equation or by introducing a region of influence. The third situation will be discussed in more detail shortly. One sees that scaling analysis is "forgiving" in that if one makes an incorrect assumption, it will lead to an apparent contradiction which indicates that the scaling was incorrect. When one has arrived at the correct scaling indicated by having all the dimensionless terms bounded of o(1), one can determine allowable assumptions from the magnitude of those dimensionless groups that are o(0.1). For example, if the dimensionless group (i.e., Reynolds number) multiplying the inertia terms in the equations of motion is O(0.1), the error incurred in dropping these terms typically will be approximately 10%. If this dimensionless group is O(0.01), the error typically will be approximately 1%. This will be illustrated in several of the example problems by comparing the approximate solution justified by scaling analysis with the solution to the full set of describing equations.

A particular advantage of an o(1) scaling analysis is that it also yields the minimum parametric representation of the dimensionless describing equations. Moreover, if the scaling analysis has been done correctly, the dimensionless groups usually are of o(1) as well. However, in some cases a proper scaling analysis will yield dimensionless groups that are much larger than O(1). Since each term in the describing equations must be bounded of o(1), this can occur only when a large dimensionless group multiplies a dimensionless term that can become very small. For example, a dimensionless group that is a measure of the reaction rate might multiply a term of the form $1 - X_c$, where X_c is the fractional conversion of the reactant. A large dimensionless group in this case would mean that $1 - X_c$ is quite small; that is, the large dimensionless group would imply nearly complete conversion.

Although scaling analysis can be described in terms of the eight steps described above, its use can best be explained through detailed examples. This constitutes the subject matter in the remaining chapters. However, before discussing the application of o(1) scaling analysis to transport and reaction processes, it is useful to consider the special application of scaling analysis as an alternative to the Buckingham Pi theorem for dimensional analysis.

2.4 SCALING ALTERNATIVE FOR DIMENSIONAL ANALYSIS

In $\circ(1)$ scaling analysis we arrive at a unique minimum parametric representation that permits assessing the relative magnitudes of the various terms in the describing equations. However, in some cases we seek to obtain a minimum parametric representation of the describing equations that is optimal for correlating experimental or numerical data, extrapolating known empirical equations, or scaling-up or scaling-down some transport or reaction process; the latter procedure is usually referred to as *dimensional analysis*. The conventional procedure for dimensional analysis is to use the Pi theorem, which involves the following steps:

1. List all quantities on which the phenomenon depends.
2. Write the dimensional formula for each quantity.
3. Demand that these quantities be combined into a functional relation that remains true independent of the size of the units.

In step 3, one invokes the Pi theorem, which states that $n - m$ dimensionless groups are formed from n quantities expressed in terms of m units. A proof of the Pi theorem and discussion of the special case $n = m$ is given in Bridgman.[3] Unfortunately, using the Pi theorem approach is not always straightforward. For example, how do we select the quantities? When do we include dimensional constants such as g_c (the Newton's law constant) or R (the gas constant)? How are dimensionless quantities such as angles involved? How many units must be considered? For example, force can be considered to be a primary quantity expressed in units of its own kind (e.g., Newtons) or a secondary quantity expressed in terms of mass, length, and time (e.g., kg·m/s^2). This problem also arises with quantities involving energy or temperature units since both can be considered as either primary or secondary quantities. The Pi theorem also does not identify quantities that always appear in combination; for example, a problem might involve the kinematic viscosity ν, but the Pi theorem approach would introduce the shear viscosity μ and the mass density ρ (i.e., $\nu = \mu/\rho$) as separate quantities, thereby generating an additional dimensionless group that in fact is not needed. The aforementioned difficulties in using the Pi theorem approach can preclude obtaining the minimum parametric representation, as illustrated in Chapter 3.

Scaling analysis can be used to circumvent the difficulties encountered in using the Pi theorem for dimensional analysis. The $\circ(1)$ scaling analysis procedure outlined in Section 2.3 leads to the minimum parametric representation for a set of describing equations; that is, to identifying the minimum number of dimensionless groups required for dimensional analysis. However, carrying out an $\circ(1)$ scaling analysis can be somewhat complicated and time consuming. Moreover, the dimensionless groups obtained from an $\circ(1)$ scaling analysis often are not optimal for correlating experimental or numerical data, for extrapolating empirical correlations, or for scale-up or scale-down analyses. The scaling analysis approach to

[3]P. W. Bridgman, *Dimensional Analysis*, Yale University Press, New Haven, CT, 1922.

dimensional analysis illustrated in this section is much easier and quicker to implement. However, it does not provide as much information as o(1) scaling analysis for achieving the minimum parametric representation. In particular, it does not lead to groups whose magnitude can be used to assess the relative importance of particular terms in the describing equations. It also does not identify regions of influence or boundary layers, whose identification in some cases can reduce the number of dimensionless groups. In this section we outline the stepwise procedure for using scaling analysis for dimensional analysis, which includes systematic methods for reducing the number of dimensionless groups and for casting them into alternative forms that are optimal for correlating data.

The stepwise procedure in implementing scaling analysis for dimensional analysis consists of the following 11 steps:

1. Write the dimensional describing equations and their initial, boundary, and auxiliary conditions appropriate to the transport or reaction process being considered.
2. Define unspecified scale factors for each dependent variable and its derivatives and each independent variable appearing explicitly in the describing equations and their initial, boundary, and auxiliary conditions.
3. Define unspecified reference factors for each dependent and independent variable that is not referenced to zero in the initial, boundary, and auxiliary conditions.
4. Form dimensionless variables by introducing the unspecified scale factors and relevant reference factors for each dependent variable and its derivatives and each independent variable.
5. Introduce these dimensionless variables into the describing equations and their initial, boundary, and auxiliary conditions.
6. Divide through by the dimensional coefficient of one term in each of the describing equations and their initial, boundary, and auxiliary conditions.
7. Determine the scale and reference factors by setting dimensionless groups equal to 1 (for scale factors) or zero (for reference factors); this yields the minimum parametric representation in the form

$$f(\Pi_1, \Pi_2, \ldots, \Pi_k) = 0 \tag{2.4-1}$$

   where Π_i denotes a dimensionless group. These Π_i's include dimensionless groups formed from combinations of the physical and geometric quantities and any dimensionless independent variables.
8. The dimensionless groups in step 7 are not unique; it might be advantageous to isolate two or more dimensional quantities into one group (if possible) to determine their interdependence; this is done by forming a new group from the k dimensionless groups via the operation

$$\Pi_p = \phi \cdot \Pi_1^a, \Pi_2^b, \ldots, \Pi_k^j \tag{2.4-2}$$

where ϕ is a dimensionless constant and $a, b, \ldots, j$ are constants chosen to isolate the desired quantities into the new dimensionless group Π_p; one can then use this new group along with any $k-1$ of the original groups; however, this operation cannot result in eliminating a dimensional quantity from the analysis.

9. The number of groups can be reduced further when a Π_i is either very large or very small by expanding equation 2.4-1 in a Taylor series in the small (or reciprocal of a large) Π_i:

$$f(\Pi_1, \Pi_2, \ldots, \Pi_k) = f\bigg|_{\Pi_i=0} + \frac{\partial f}{\partial \Pi_i}\bigg|_{\Pi_i=0} \Pi_i + \mathrm{O}(\Pi_i^2) \qquad (2.4\text{-}3)$$

If equation 2.4-3 can be truncated at the first term, the minimum parametric representation will involve $k-1\Pi_i$'s.

10. Any dimensionless group that contains the sum or difference of two dimensional quantities γ and δ, either of which appears individually in any other dimensionless group, can be redefined to exclude this particular quantity; that is,

$$\text{if } \Pi_p \equiv \alpha^a\beta^b(\gamma-\delta)^c, \text{ it can be replaced by } \Pi_p' \equiv \alpha^a\beta^b\gamma^c \text{ when } \delta \text{ is contained in another of the dimensionless groups} \qquad (2.4\text{-}4)$$

11. Any dimensionless group that contains one or more of the dimensionless groups that appear in the dimensional analysis can be redefined to exclude these redundant dimensionless groups; that is,

If $\Pi_p = f(\Pi_1, \Pi_2, \ldots, \Pi_k)$ and Π_j contains one or more of the other Π_i's, it can be redefined to exclude the redundant Π_i's. (2.4-5)

Step 1 in the scaling analysis procedure for dimensional analysis is the same as that used for o(1) scaling. In dimensional analysis it is essential to begin by writing the appropriate equation for the quantity that one seeks to correlate. For example, one might be seeking to correlate the drag force on a particle immersed in a fluid; one must then write the appropriate integral equation for the total drag; the latter will, in turn, require the solution to the appropriate form of the equations of motion, which then must also enter into the dimensional analysis. It is important to emphasize in implementing step 1 that one must write all the algebraic and differential equations along with the appropriate initial, boundary, and auxiliary conditions needed to solve the problem. One must also use all the available information to simplify these describing equations appropriately; for example, one eliminates the inertia terms if it is a creeping flow. However, it is sufficient to write the appropriately simplified equations of motion, energy-conservation equation, or species balances in generalized vector–tensor notation; that is, it is not necessary to expand any of these equations in a particular coordinate system.

Steps 2 through 6 are similar to those used in o(1) scaling analysis with one exception. In step 2, one scales only the dependent and independent variables; one does not scale any of the derivatives. The reason for this is that we are merely seeking to achieve a minimum parametric representation. We are not trying to scale to ensure that all the dimensionless variables and their derivatives are o(1).

The Π_i's obtained in step 7 constitute dimensionless groups formed from the physical properties, geometric and process parameters, and dimensionless dependent and independent variables. Note that some of the dimensionless dependent and/or independent variables will not appear in the Π_i's if they are integrated out or evaluated at fixed spatial or temporal conditions. For example, one might seek a correlation for the total drag force at the surface of a sphere falling at its terminal velocity. The total drag force involves integrating the local shear stress and pressure over the surface of the sphere. Hence, although the dimensionless local shear stress and pressure depend on the dimensionless spatial coordinates, the total drag force on the sphere does not depend on these quantities, due to integration over the surface. This will become clearer when representative problems are considered in subsequent chapters.

Step 8 is the procedure whereby one obtains the dimensionless groups that are optimal for the desired correlation. This step states merely that one can form a new set of dimensionless groups by multiplying two or more of the groups obtained in step 7 raised to arbitrary powers and multiplied by arbitrary constants. One does this to isolate certain quantities into just one dimensionless group. The only precaution to be observed here is that the resulting dimensionless groups must be independent and equal in number to the original set of groups. In addition, this procedure cannot result in eliminating any quantity from the dimensional analysis. For example, one might want to correlate the average velocity for fully developed laminar flow in a smooth cylindrical tube as a function of the relevant parameters. Steps 1 through 7 will result in two dimensionless groups. One possible set of dimensionless groups is the conventional friction factor and the Reynolds number. However, both of these groups contain the average velocity. This implies that a correlation for the friction factor as a function of Reynolds number would require a trial-and-error solution to obtain the average velocity. However, step 8 indicates that one can multiply the friction factor by the square of the Reynolds number to obtain a new dimensionless group that is independent of the average velocity. One can then plot the Reynolds number as a function of this new dimensionless group to obtain a correlation that is optimal for determining the average velocity.

Step 9 provides the formalism necessary to reduce the number of dimensionless groups when one or more of the Π_i's is either very small or very large. If one can assume that the dimensionless correlation has continuous derivatives with respect to the particular small Π_i (or reciprocal large Π_i), the correlation can be expanded in a Taylor series in the small Π_i (or reciprocal large Π_i). For a sufficiently small Π_i this Taylor series can be truncated at the first term, thereby formally eliminating this small dimensionless group from the correlation. For example, the friction factor (dimensionless drag force) for flow over a sphere is a function of the Reynolds number. However, at very small reciprocal Reynolds numbers the friction factor becomes independent of the Reynolds number and approaches a constant value.

Step 10 is a consequence of the fact that the set of dimensionless groups involved in a minimum parametric representation of the describing equations is not unique. If two quantities appear in the describing equations only as a sum or difference, the number of dimensionless groups can be reduced by using the sum or difference as a single dimensional quantity. However, if either of the quantities appearing in a sum or difference appears individually in any other dimensionless group, there is no advantage to considering the sum or difference as a separate dimensional quantity. Step 10 is particularly useful when one is trying to isolate particular quantities into just one dimensionless group in order to correlate experimental or numerical data. For example, a dimensional analysis correlation for the heat-transfer coefficient might involve the temperature difference $T_w - T_\infty$, where T_w and T_∞ are the wall and bulk-fluid temperatures, respectively. However, if a heterogeneous chemical reaction is occurring at the wall, T_w might enter the dimensional analysis separately, due to the dependence of the reaction rate constant on temperature. In this case the quantity $T_w - T_\infty$ can be replaced by T_w and T_∞ as separate quantities. Doing this might be advantageous if one is trying to isolate T_w into just one dimensionless group to study its effect on the performance. Sums or differences of dimensional quantities are often encountered when one obtains the minimum parametric representation of a set of describing equations by invoking $\circ(1)$ scaling; for example, the characteristic temperature scale might be $T_w - T_\infty$. This temperature difference is appropriate for the scaling analysis to assess what approximations might be justified. However, it might be inconvenient for a dimensional analysis correlation if one is seeking to isolate T_w or T_∞ into a single dimensionless group.

Step 11 also follows from that fact that the set of dimensionless groups involved in a minimum parametric representation of the describing equations is not unique. If a particular dimensionless group contains one or more of the other of the dimensionless groups in the dimensional analysis, this group can be redefined to exclude these redundant dimensionless groups. For example, in convective heat-transfer problems the Reynolds number, Peclet number, and Prandtl number can arise. However, the Peclet number is the product of the Reynolds and Prandtl numbers. If the Reynolds and Prandtl numbers appear independently in the dimensional analysis, one can eliminate the Peclet number. Dimensionless groups containing other dimensionless groups are more often encountered when one obtains the minimum parametric representation of a set of describing equations by invoking $\circ(1)$ scaling. For example, describing equations for mass transfer with chemical reaction might involve a dimensionless group containing the characteristic reaction-rate parameter, which in turn is a function of a characteristic temperature ratio that appears as an independent dimensionless group. Incorporating the temperature ratio into the definition of the characteristic reaction-rate parameter is critical to carrying out accurate scaling analysis. However, it will be very inconvenient for a dimensional analysis correlation if one is seeking to isolate either of the temperatures in this ratio into just one dimensionless group.

2.5 SUMMARY

An eight-step procedure was outlined whereby all the dependent and independent variables and their derivatives are bounded to be o(1). This procedure results in the minimum parametric representation of the describing equations; that is, the dimensionless describing equations involve the minimum number of dimensionless groups. The o(1) scaling procedure permits assessing the relative importance of each term in the describing equations, which then suggests possible approximations that can be made. Scaling analysis is forgiving in that it indicates if variables and/or their derivatives have been scaled incorrectly. This is usually indicated by one or more terms that are not bounded of o(1).

Although o(1) scaling analysis always leads to the minimum parametric representation of the describing equations, implementing this procedure can be tedious and time consuming. When one just seeks to do dimensional analysis without assessing the relative importance of the various terms in the describing equations, one can implement the scaling approach to dimensional analysis in the absence of any effort to bound the variables and their derivatives to be o(1). The latter procedure is straightforward and quick to implement. However, the resulting dimensionless groups do not provide nearly as much information on the physics and chemistry of the transport and reaction processes as would be obtained from an o(1) scaling analysis. An 11-step procedure was outlined for implementing the scaling approach to dimensional analysis. However, only the first seven steps are essential in implementing this procedure. The remaining four steps involve manipulations that permit recasting the dimensionless groups into a form that is optimal for correlating data or process scale-up. Using either o(1) scaling analysis or the alternative simpler scaling analysis approach to dimensional analysis offers many advantages relative to using the Pi theorem. In particular, natural groupings of the variables are identified, and complications related to choosing the proper dimensions and the need to introduce dimensional constants are avoided.

3 Applications in Fluid Dynamics

To simplify the Navier–Stokes equations within the boundary layer, we can then utilize the fact that the thickness of this layer is very small compared to its length along the body.[1]

3.1 INTRODUCTION

In this chapter we consider the application of scaling analysis to fluid dynamics. This will serve not only to illustrate how scaling analysis is implemented but will also provide a systematic means for introducing somewhat abstract concepts in fluid dynamics, such as creeping, lubrication, boundary-layer, quasi-steady-state, quasi-parallel, incompressible, and other flows. The material in this chapter thus provides a useful supplement to a foundation course in fluid dynamics. No attempt is made in this chapter or elsewhere in the book to provide a detailed derivation of the describing equations that are used in scaling analysis. However, the reader is referred to the appendices, which summarize the continuity equations and equations of motion along with the corresponding forms of Newton's constitutive equation in generalized vector–tensor notation as well as in rectangular, cylindrical, and spherical coordinates. These equations serve as the starting point for each example problem.

In this chapter we use the two ordering symbols introduced in Chapter 2: o(1) and O(1). The symbol o(1) implies that the magnitude of the quantity can range between 0 and more-or-less 1. The symbol O(1) implies that the magnitude of the quantity is more-or-less 1; that is, it is never much less than 1. In the problems in Sections 3.2 through 3.10 on o(1) scaling and dimensional analysis, the steps involved in each of these scaling procedures are discussed in detail. In the subsequent example problems, less detail is given; however, the steps involved are noted parenthetically.

We begin by considering the use of o(1) scaling to simplify laminar flow problems. We then use o(1) scaling to justify classical approximations made in fluid

[1] V. G. Levich, *Physicochemical Hydrodynamics*, Prentice-Hall, Englewood Cliffs, NJ 1962, p. 14.

dynamics, specifically creeping, lubrication, boundary-layer, and quasi-steady-state flows. We then apply o(1) scaling to show how it can be used to justify ignoring end and sidewall effects. We also consider the application of o(1) scaling in simplifying more complex flows, such as those involving free surfaces, porous media, and compressible fluids. In Section 3.10 we consider the use of scaling in dimensional analysis. The methodology used in this chapter is to illustrate the use of scaling by considering detailed examples. Additional worked example problems followed by several unworked practice problems are included at the end of the chapter.

3.2 FULLY DEVELOPED LAMINAR FLOW

Our first example of scaling analysis will consider a straightforward flow problem for which an exact analytical solution is available. Of course, one would not need to scale a problem that can be solved exactly analytically. However, this will permit us to assess the error made when particular assumptions are invoked based on scaling analysis. This example illustrate, use of the o(1) scaling analysis procedure. It also illustrates *region of influence scaling*, whereby we seek to determine the thickness of a region in which some important effect is concentrated. Region of influence scaling is particularly important since it forms the basis of hydrodynamic boundary-layer theory, considered in Section 3.4, and penetration theory in heat and mass transfer, considered in Chapters 4 and 5, respectively. Finally, this problem is used to illustrate the *forgiving nature* of scaling analysis. By this we mean that if an incorrect assumption is made concerning scaling, proper analysis will indicate the contradiction when values of the physical and geometric properties are substituted into the relevant dimensionless groups that emanate from the scaling.

Consider the steady-state fully developed laminar flow of a viscous Newtonian fluid having constant physical properties between two infinitely wide parallel plates as shown in Figure 3.2-1. The lower plate is stationary and the upper plate moves at a constant velocity U_0. This flow is also subject to a constant axial pressure driving force $\Delta P \equiv P_0 - P_L > 0$ applied over the length L. Note that the conditions required to ensure that any of the aforementioned assumptions are reasonable could be assessed using scaling analysis; we merely invoke these assumptions so that we

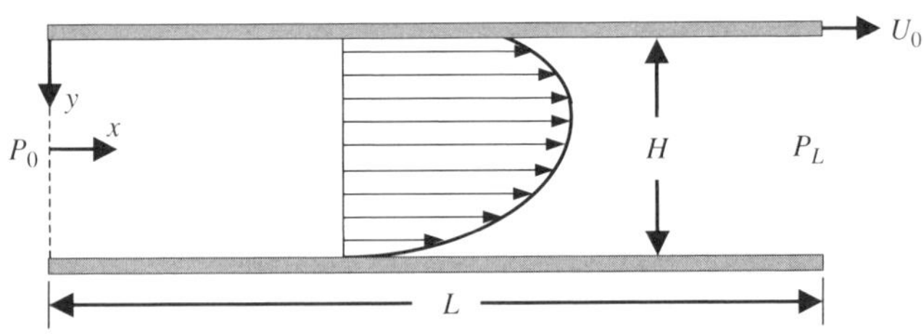

Figure 3.2-1 Steady-state fully developed laminar flow of a viscous Newtonian fluid that has constant physical properties between two infinitely wide parallel flat plates due to a pressure driving force $\Delta P \equiv P_0 - P_L$ applied over length L; the lower plate is stationary and the upper plate moves at constant velocity U_0.

can focus on assessing the applicability of just one assumption: when the effect of the upper plate velocity U_0 can be neglected. We invoke the stepwise o(1) scaling analysis procedure outlined in Chapter 2. In this first example of o(1) scaling analysis, we show all the steps in detail and provide a discussion of the rationale for each step.

Step 1 involves writing the describing equations, in this case the equations of motion and their boundary conditions simplified appropriately for this problem statement. Equations (D.1-10) and (D.1-11) in Appendix D simplify to the following for the flow conditions specified:

$$0 = -\frac{\partial P}{\partial x} + \mu \frac{d^2 u_x}{dy^2} \tag{3.2-1}$$

$$0 = -\frac{\partial P}{\partial y} + \rho g \tag{3.2-2}$$

$$u_x = U_0 \quad \text{at} \quad y = 0 \tag{3.2-3}$$

$$u_x = 0 \quad \text{at} \quad y = H \tag{3.2-4}$$

Equation (3.2-2) can be integrated and combined with equation (3.2-1) to obtain

$$0 = \frac{\Delta P}{L} + \mu \frac{d^2 u_x}{dy^2} \tag{3.2-5}$$

Step 2 involves introducing arbitrary scale factors for each dependent and independent variable. Step 3 is unnecessary in this problem since both the velocity and spatial coordinate are naturally referenced to zero. Step 4 involves defining the following dimensionless variables:

$$u_x^* \equiv \frac{u_x}{u_{xs}} \quad \text{and} \quad y^* \equiv \frac{y}{y_s} \tag{3.2-6}$$

One might reasonably ask why a separate scale factor is not introduced for the second derivative in equation (3.2-5). Indeed, one could introduce a scale factor for the second derivative. If this were done, one would find that there was no dimensionless group to determine the appropriate scale factor for the velocity. However, the latter could be obtained by integrating the scale for the second derivative of the velocity, in which case one would obtain the same scale for the velocity that will be obtained here by introducing its scale factor directly. Alternatively, one could introduce a scale factor for the first derivative of the velocity. Integrating the resulting scale factor again gives the same scale factor for the velocity within a multiplicative factor of O(1). This is explored further in Practice Problem 3.P.1.

In step 5 these dimensionless variables are substituted into the describing equations (3.2-3), (3.2-4), and (3.2-5):

$$0 = \frac{\Delta P}{L} + \frac{\mu u_{xs}}{y_s^2}\frac{d^2 u_x^*}{dy^{*2}} \tag{3.2-7}$$

$$u_{xs}u_x^* = U_0 \qquad \text{at} \quad y^* = 0 \tag{3.2-8}$$

$$u_{xs}u_x^* = 0 \qquad \text{at} \quad y_s y^* = H \tag{3.2-9}$$

Step 6 involves dividing through by the dimensional coefficient of the viscous term in equation (3.2-7) since this term must be retained in order to satisfy the two no-slip conditions at the solid boundaries. Similarly, in the two boundary conditions we divide through by the dimensional coefficient of the dimensionless dependent variable, which yields

$$0 = \frac{y_s^2 \Delta P}{\mu u_{xs} L} + \frac{d^2 u_x^*}{dy^{*2}} \tag{3.2-10}$$

$$u_x^* = \frac{U_0}{u_{xs}} \qquad \text{at} \quad y^* = 0 \tag{3.2-11}$$

$$u_x^* = 0 \qquad \text{at} \quad y^* = \frac{H}{y_s} \tag{3.2-12}$$

Step 7 involves determining the scale factors to ensure that the dimensionless term causing the flow (i.e., the pressure term) balances the term resisting the flow (i.e., the viscous term) and ensuring that the relevant dimensionless dependent and independent variables are o(1) for the region of the flow that is of interest, in this case all the fluid between the two flat plates. When considering any problem in fluid dynamics, it is important to ask the question "What causes the flow?" since in scaling, the term or terms causing the flow must balance the term or terms resisting the flow. The former might constitute a pressure gradient, a moving boundary, or body forces such as a gravitational, centrifugal, electric, or magnetic field; the latter might constitute viscous forces, inertia effects, pressure effects, or body forces (note that pressure and body forces can resist as well as cause flow). Balancing what causes the flow with the principal term(s) that resist flow generally determines one or more of the scales. Since we are scaling this problem for conditions such that the flow is caused principally by the pressure gradient, we ensure that the dimensionless viscous force and pressure terms balance by demanding that the dimensionless group in equation (3.2-10) be equal to 1; that is,

$$\frac{y_s^2 \Delta P}{\mu u_{xs} L} = 1 \tag{3.2-13}$$

Since our region of interest spans the entire fluid between the two flat plates, an appropriate scale for the spatial coordinate is obtained by demanding that the dimensionless group in equation (3.2-12) be equal to 1; that is,

$$\frac{H}{y_s} = 1 \Rightarrow y_s = H \tag{3.2-14}$$

Hence by combining equations (3.2-13) and (3.2-14), we obtain our velocity scale, which is given by

$$u_{xs} = \frac{H^2 \, \Delta P}{\mu L} \tag{3.2-15}$$

Note that this velocity scale is directly proportional to the maximum velocity for flow between two flat plates driven only by a pressure gradient. This scaling, ensures that the dimensionless velocity goes through a change of order one over a dimensionless distance of order one. Note that *a change of order one* implies that the dimensionless variable goes from its minimum value of zero to its maximum value, which has magnitude of order one. Note that in this case our dimensionless velocity will always be less than 1 since equation (3.2-15) overestimates the maximum velocity (by a factor of 8 when the motion of the upper plate can be neglected). In this case we know the exact value of the scale required to bound the dimensionless velocity between 0 and 1 since we can solve this particular problem analytically. However, in general we would not know any of the scales beforehand; indeed, determining these scales is one of the goals of the systematic scaling method. The dimensionless groups emanating from the scaling analysis that contain this velocity scale (Π_1 for the problem being considered here) could be eight times smaller (for this problem) or larger (if the reciprocal of Π_1 were used as the dimensionless group in this problem) than that obtained when the relevant dimensional physical and geometric properties are substituted to evaluate them. Hence, we see that the criteria expressed in terms of dimensionless groups that emanate from scaling analysis are generally within an order of magnitude. For this reason it is good practice to demand that the dimensionless groups emanating from a particular scaling analysis be at least two orders of magnitude less than 1, denoted by $\text{o}(0.01)$, to justify the particular assumption being considered.

Our dimensionless equations now become

$$0 = 1 + \frac{d^2 u_x^*}{dy^{*2}} \tag{3.2-16}$$

$$u_x^* = \frac{U_0 \, \mu L}{H^2 \, \Delta P} \quad \text{at} \quad y^* = 0 \tag{3.2-17}$$

$$u_x^* = 0 \quad \text{at} \quad y^* = 1 \tag{3.2-18}$$

Step 8 then involves using our scaled dimensionless describing equations to assess the criterion for which we can ignore the effect of the motion of the upper plate.

Equation (3.2-17) indicates that the motion of the upper plate will have an insignificant influence on the fluid flow if the following condition holds:

$$\frac{U_0\,\mu L}{H^2\,\Delta P} \equiv \Pi_1 \ll 1 \tag{3.2-19}$$

Note that the criterion that has emerged from our scaling analysis for ignoring the effect of the moving upper plate on the flow is in terms of a dimensionless group Π_1. The physical significance of this dimensionless group is that it is the ratio of the magnitude of the velocity of the upper plate to the magnitude of the characteristic velocity due to the applied pressure gradient. The dimensionless groups that emerge from scaling analysis will always have a physical significance that can be determined by examining how a particular group was formed: in this case, by dividing the velocity of the upper plate by the characteristic velocity determined by balancing the pressure and viscous forces in the equations of motion.

The question arises as to how small Π_1 has to be for the assumption of ignoring the upper plate motion to be reasonable. The answer to this question depends of course, on the error that one can tolerate in their answer. Since o(1) scaling involves order-of-magnitude analysis, one can project that if the dimensionless group in equation (3.2-17) is O(0.1), the error will be approximately 10 to 100%; if this group is O(0.01), the error will be approximately 1 to 10%. For example, the dimensionless velocity profile obtained from solving equations (3.2-16), (3.2-17), and (3.2-18) while retaining the effect of the moving upper plate is given by

$$u_x^* = \Pi_1 + \left(\frac{1}{2} - \Pi_1\right) y^* - \frac{1}{2} y^{*2} \tag{3.2-20}$$

The corresponding value of the dimensionless average velocity is given by integrating the foregoing velocity profile across the flow to obtain

$$\langle u_x^* \rangle = \frac{\Pi_1}{2} + \frac{1}{12} \tag{3.2-21}$$

One now can access the error in determining the dimensionless average velocity when the motion of the upper plate is ignored (i.e., when $\Pi_1 \ll 1$). For example, when $\Pi_1 = 0.1$, the error in determining the average velocity when the upper plate velocity is ignored is 38%. However, when $\Pi_1 = 0.01$, the error is reduced to 5.7%. This demonstrates clearly that proper scaling analysis provides results within an order of magnitude. If one wants to be certain that some assumption can be invoked with confidence, two orders of magnitude should be demanded for any "much less than" or "much greater than" condition.

Note, however, that the error encountered in making an approximation, such as assuming that $\Pi_1 \ll 1$, depends not only on the magnitude of Π_1 but also on the quantity that is being determined from the solution. The average velocity is an integral quantity whose value is not particularly sensitive to small errors in the velocity profile near the upper plate. However, if one is interested in determining the point or local velocity anywhere in the flowing fluid, the error is 100% at the upper moving plate irrespective of the value of Π_1 ! One might reasonably suspect that this large error in the point or local velocity is encountered only very near the upper moving plate. But the question arises: How near the upper plate is a significant error encountered in determining the point velocity? Scaling analysis can also address this question by considering *region-of-influence scaling* used to determine the thickness of a region within which some effect is important: in this case, the thickness of the region near the wall where the moving plate has a significant effect on determining the point velocity.

To carry out region of influence scaling, the unspecified length scale factor in fact becomes the thickness of the region of influence, which we denote by the symbol δ_m to emphasize its particular physical significance. By this we mean that the relevant dependent variable, in this case the velocity, is O(1) within this region. Let us rescale this problem to determine the magnitude of δ_m. Since we are considering the region near the upper wall, equations (3.2-10) through (3.2-12) remain the same. However, our velocity scale is no longer given by equation (3.2-15), which characterizes the velocity across the entire flow. Rather, in the region near the upper wall, the velocity scale is determined by the dimensionless group in the boundary condition at the upper wall:

$$u_x^* = \frac{U_0}{u_{xs}} = 1 \Rightarrow u_{xs} = U_0 \tag{3.2-22}$$

However, since the pressure force must still balance the viscous term near the upper wall, the dimensionless group in equation (3.2-13) must again be set equal to 1. When equation (3.2-22) is substituted into this dimensionless group, one obtains a measure of δ_m, the thickness of the region of influence within which one cannot ignore the influence of the moving plate on the point velocity:

$$\frac{\delta_m^2 \, \Delta P}{\mu U_0 L} = 1 \Rightarrow \delta_m^2 = \frac{\mu U_0 L}{\Delta P} \Rightarrow \frac{\delta_m^2}{H^2} = \frac{\mu U_0 L}{H^2 \, \Delta P} = \Pi_1 \tag{3.2-23}$$

One sees from equation (3.2-23) that $\sqrt{\Pi_1}$ is a measure of the fractional distance between the two plates, within which the effect of the moving plate on the point velocity is significant. Hence, if $\Pi_1 = \mathrm{O}(0.01)$, the moving plate has a significant effect on the local velocity across 10% of the distance between the two plates. This will seriously affect determining the point velocity through a significant portion of the flow but will have a relatively minor effect on integral quantities such as the average velocity.

One can conclude from this scaling analysis that the solution obtained for equations (3.2-1) through (3.2-4) when one assumes that $U_0 \cong 0$ will be reasonably accurate for determining integral quantities such as the average velocity when $\Pi_1 = \mathrm{o}(0.01)$ and that it will provide an accurate estimate for the local velocity when $\sqrt{\Pi_1} < y^* \leqslant 1$, where $y^* \equiv y/H$.

Before leaving this example it is instructive to see the forgiving nature of scaling analysis. For the purpose of illustration, let us assume that we set the dimensionless group in equation (3.2-22) incorrectly as our velocity scale; that is, we used the velocity of the upper plate as our velocity scale. Note that this is a gross underestimate of the proper scale for the dimensionless velocity when the effect of the motion of the upper plate has a negligible effect on the overall flow. However, if we had chosen U_0 as our velocity scale, equation (3.2-10) would have assumed the following dimensionless form:

$$0 = \frac{H^2 \Delta P}{\mu U_0 L} + \frac{d^2 u_x^*}{dy^{*2}} = \frac{1}{\Pi_1} + \frac{d^2 u_x^*}{dy^{*2}} \tag{3.2-24}$$

Note that if we have scaled the dimensionless velocity and spatial coordinate properly, the dimensionless second-order derivative in equation (3.2-24) should be o(1). We see from the above that if $\Pi_1 \ll 1$, as would be the case if the motion of the upper plate were negligible compared to the flow caused by the pressure gradient, equation (3.2-24) would be a statement that the sum of a very large term and a term of o(1) is equal to zero; this is clearly impossible. As a result of our incorrect scaling of the dimensionless velocity, we have encountered a contradiction. This indicates that we need to consider another scaling; this is, of course, the scaling wherein the velocity scale factor is given by equation (3.2-15). Hence, we see that scaling analysis is indeed forgiving in that improper scaling leads to a contradiction, which indicates that scaling needs to be repeated. When the proper scaling is found for the known physical and geometric properties of the problem, all terms will be bounded of o(1).

In carrying out this o(1) scaling analysis, we were seeking to determine the criterion for neglecting the motion of the upper plate on the fluid flow. We saw that the conditions required to assure minimal error in neglecting this term depended on the quantity that one sought to determine from the describing equations. In general, the criterion is less demanding for integral quantities such as the average velocity, volumetric flow rate, total drag force, and the like, than for quantities such as the velocity or shear stress at some point in the continuum. We could also have carried out o(1) scaling to determine when the pressure force could be neglected relative to the fluid motion caused by the moving boundary. This is left as an exercise in Practice Problem 3.P.2.

3.3 CREEPING- AND LUBRICATION-FLOW APPROXIMATIONS

Now that the procedure for o(1) scaling analysis has been illustrated in detail, we use this method to explore the various approximations made in classical fluid dynamics. We begin by using o(1) scaling analysis to explore the *creeping*- and *lubrication-flow approximations*. The latter is particularly important for flows involving very narrow gaps, such as journal bearings and fluid couplings. The problem that we consider is steady-state one-dimensional uniform or plug flow of a viscous Newtonian fluid having constant physical properties and constant velocity U_0 impinging on two nonparallel infinitely wide flat plates, as shown in Figure 3.3-1. This creates a developing flow with nonzero x- and y-velocity components.

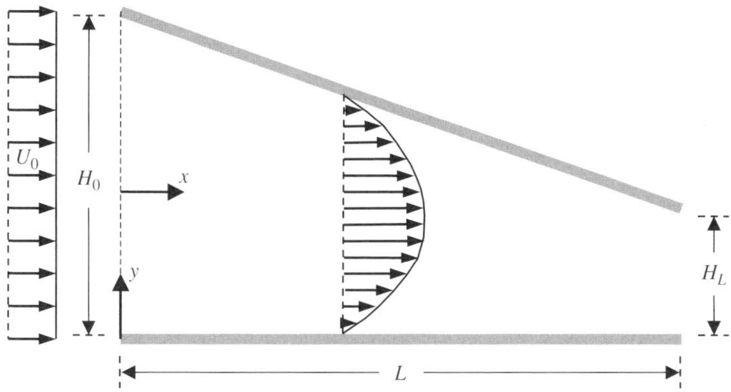

Figure 3.3-1 Steady-state developing laminar flow of a viscous Newtonian fluid that has constant physical properties between two infinitely wide nonparallel flat plates; only the local axial velocity profile is shown.

To appreciate the utility of scaling analysis, it is instructive to write the describing equations before we consider the approximations that we might use to simplify these equations. Indeed, one utility of scaling analysis is that it allows exploring the approximations that might be made to obtain a tractable solution to a problem. The continuity equation and equations of motion given by equations (C.1-1), (D.1-10), and (D.1-11) simplify to the following for the assumed flow conditions (step 1):

$$\rho u_x \frac{\partial u_x}{\partial x} + \rho u_y \frac{\partial u_x}{\partial y} = -\frac{\partial P}{\partial x} + \mu \left(\frac{\partial^2 u_x}{\partial x^2} + \frac{\partial^2 u_x}{\partial y^2} \right) \tag{3.3-1}$$

$$\rho u_x \frac{\partial u_y}{\partial x} + \rho u_y \frac{\partial u_y}{\partial y} = -\frac{\partial P}{\partial y} + \mu \left(\frac{\partial^2 u_y}{\partial x^2} + \frac{\partial^2 u_y}{\partial y^2} \right) - \rho g \tag{3.3-2}$$

$$\frac{\partial u_x}{\partial x} + \frac{\partial u_y}{\partial y} = 0 \tag{3.3-3}$$

The appropriate boundary conditions for this flow are given by

$$u_x = U_0, \quad u_y = 0 \quad \text{at} \quad x = 0 \tag{3.3-4}$$

$$u_x = f_1(y), \quad u_y = f_2(y) \quad \text{at} \quad x = L \tag{3.3-5}$$

$$u_x = 0, \quad u_y = 0 \quad \text{at} \quad y = 0 \tag{3.3-6}$$

$$u_x = 0, \quad u_y = 0 \quad \text{at} \quad y = H_0 - \frac{H_0 - H_L}{L} x \tag{3.3-7}$$

Equation (3.3-5) prescribes the downstream boundary conditions in terms of two functions, $f_1(y)$ and $f_2(y)$, which might be unknown. The tangential and normal velocity components to the sloped plate must be zero, due to the no-slip and impermeable wall boundary conditions. Equation (3.3-7) for the x- and y-velocity components follows directly from the aforementioned conditions. Note that we do not need to put any boundary conditions on the pressure since the average velocity U_0 is known, which permits determining the pressure driving force over the length L.

Note that this is a nontrivial problem to solve. As a result of the developing flow, there is strong bidirectional coupling between equations (3.3-1), (3.3-2), and (3.3-3) through the velocity components and pressure. Moreover, the presence of the inertia terms in equations (3.3-1) and (3.3-2) make both equations nonlinear. A further complexity is introduced by the elliptic nature of the describing equations. The presence of the second-order axial derivatives requires that downstream boundary conditions be specified. In many problems such as this, these downstream conditions are not known, which precludes solving the describing equations either analytically or numerically. Clearly, one would like to know how and when these describing equations might be simplified to permit a tractable solution. In particular, one would like to know when the inertia terms and axial viscous terms might be neglected. We use o(1) scaling to determine these conditions.

We begin by defining dimensionless variables involving unspecified scale factors (steps 2, 3, and 4):

$$u_x^* \equiv \frac{u_x}{u_{xs}}; \qquad u_y^* \equiv \frac{u_y}{u_{ys}}; \qquad P^* \equiv \frac{P}{P_s}; \qquad x^* \equiv \frac{x}{x_s}; \qquad y^* \equiv \frac{y}{y_s} \tag{3.3-8}$$

Note that we do not need to introduce any reference factors since all the dependent and independent variables are naturally referenced to zero.

We then introduce these dimensionless variables into the describing equations and divide through by the coefficient of one term in each of these equations that we believe should be retained (steps 5 and 6):

$$\frac{\rho u_{xs} y_s^2}{\mu x_s} u_x^* \frac{\partial u_x^*}{\partial x^*} + \frac{\rho u_{ys} y_s}{\mu} u_y^* \frac{\partial u_x^*}{\partial y^*} = -\frac{P_s y_s^2}{\mu u_{xs} x_s} \frac{\partial P^*}{\partial x^*} + \frac{y_s^2}{x_s^2} \frac{\partial^2 u_x^*}{\partial x^{*2}} + \frac{\partial^2 u_x^*}{\partial y^{*2}} \tag{3.3-9}$$

$$\frac{\rho u_{xs} u_{ys} y_s}{P_s x_s} u_x^* \frac{\partial u_y^*}{\partial x^*} + \frac{\rho u_{ys}^2}{P_s} u_y^* \frac{\partial u_y^*}{\partial y^*} = -\frac{\partial P^*}{\partial y^*} + \frac{\mu u_{ys} y_s}{P_s x_s^2} \frac{\partial^2 u_y^*}{\partial x^{*2}} + \frac{\mu u_{ys}}{P_s y_s} \frac{\partial^2 u_y^*}{\partial y^{*2}} - \frac{\rho g y_s}{P_s} \tag{3.3-10}$$

$$\frac{\partial u_x^*}{\partial x^*} + \frac{u_{ys} x_s}{u_{xs} y_s} \frac{\partial u_y^*}{\partial y^*} = 0 \tag{3.3-11}$$

$$u_x^* = \frac{U_0}{u_{xs}}, \quad u_y^* = 0 \quad \text{at} \quad x^* = 0 \tag{3.3-12}$$

$$u_x^* = f_1^*(y^*), \quad u_y^* = f_2^*(y^*) \quad \text{at} \quad x^* = \frac{L}{x_s} \tag{3.3-13}$$

$$u_x^* = 0, \quad u_y^* = 0 \quad \text{at} \quad y^* = 0 \tag{3.3-14}$$

$$u_x^* = 0, \quad u_y^* = 0 \quad \text{at} \quad y^* = \frac{H_0}{y_s} - \frac{(H_0 - H_L) x_s}{L y_s} x^* \tag{3.3-15}$$

Note that in implementing step 6 for equation (3.3-9), we have elected to divide through by the dimensional coefficient of the principal viscous term since it must be retained to satisfy the no-slip boundary conditions at the two flat plates. In equation (3.3-10) we have elected to divide through by the dimensional coefficient of the pressure term since it causes the flow in the y-direction.

Now let us proceed to determine the scale factors (step 7). The axial velocity scale is obtained from the dimensionless group in equation (3.3-12). We choose the dimensionless group in equation (3.3-15) that gives the larger scale for y_s since we want to bound y^* to always be less than or equal to 1. The dimensionless group in equation (3.3-13) provides the axial length scale. Since pressure causes flow in the axial direction, we set the coefficient of the dimensionless pressure term in equation (3.3-9) equal to 1 to determine the pressure scale. Finally, we set the dimensionless group in equation (3.3-11) equal to 1 since the two terms in the continuity equation have to balance for a developing flow. Hence, we obtain the following scale factors:

$$u_{xs} = U_0; \qquad u_{ys} = \frac{H_0}{L} U_0; \qquad P_s = \frac{\mu U_0 L}{H_0^2}; \quad x_s = L; \quad y_s = H_0 \quad (3.3\text{-}16)$$

Note that the scale for the y-component of velocity is dependent on the aspect ratio; for long closely spaced plates, the scale factor for u_y will be considerably smaller than that for u_x. This is perfectly reasonable since u_y arises from a need for the velocity profile to be rearranged to accommodate the change in spacing between the two plates. Less rearrangement is required for relatively closely spaced long plates. Note also that the scale factor for the pressure is a measure of the viscous drag stress multiplied by an aspect ratio. This also is reasonable since we have balanced the pressure force with the principal viscous stress, which indeed is τ_{yx}.

Substitution of the scale factors defined in equation (3.3-16) into equations (3.3-9) through (3.3-15) yields

$$\mathrm{Re}\frac{H_0}{L}u_x^*\frac{\partial u_x^*}{\partial x^*} + \mathrm{Re}\frac{H_0}{L}u_y^*\frac{\partial u_x^*}{\partial y^*} = -\frac{\partial P^*}{\partial x^*} + \frac{H_0^2}{L^2}\frac{\partial^2 u_x^*}{\partial x^{*2}} + \frac{\partial^2 u_x^*}{\partial y^{*2}} \quad (3.3\text{-}17)$$

$$\mathrm{Re}\frac{H_0^3}{L^3}u_x^*\frac{\partial u_y^*}{\partial x^*} + \mathrm{Re}\frac{H_0^3}{L^3}u_y^*\frac{\partial u_y^*}{\partial y^*} = -\frac{\partial P^*}{\partial y^*} + \frac{H_0^3}{L^3}\frac{\partial^2 u_y^*}{\partial x^{*2}} + \frac{H_0^2}{L^2}\frac{\partial^2 u_y^*}{\partial y^{*2}} - \frac{\mathrm{Re}}{\mathrm{Fr}}\frac{H_0}{L} \quad (3.3\text{-}18)$$

$$\frac{\partial u_x^*}{\partial x^*} + \frac{\partial u_y^*}{\partial y^*} = 0 \quad (3.3\text{-}19)$$

$$u_x^* = 1, \quad u_y^* = 0 \quad \text{at} \quad x^* = 0 \quad (3.3\text{-}20)$$

$$u_x^* = f_1^*(y^*), \quad u_y^* = f_2^*(y^*) \quad \text{at} \quad x^* = 1 \quad (3.3\text{-}21)$$

$$u_x^* = 0, \quad u_y^* = 0 \quad \text{at} \quad y^* = 0 \quad (3.3\text{-}22)$$

$$u_x^* = 0, \quad u_y^* = 0 \quad \text{at} \quad y^* = 1 - \left(1 - \frac{H_L}{H_0}\right)x^* \quad (3.3\text{-}23)$$

where $\text{Re} \equiv \rho U_0 H_0/\mu$ is the *Reynolds number*, a measure of the ratio of the kinetic energy per unit volume of the flow to the principal viscous stress, and $\text{Fr} \equiv U_0^2/gH_0$ is the *Froude number*, a measure of the ratio of the kinetic energy to the gravitational potential energy of the flow.

Now let us consider how we might be able to simplify the set of dimensionless equations above to obtain a tractable solution (step 8). If $\text{Re} \ll 1$, say $\text{Re} = \text{o}(0.01)$, we see that we can safely neglect the nonlinear inertia terms in equations (3.3-17) and (3.3-18) provided that $H_0/L = \text{o}(1)$. This simplification is referred to as the *creeping-flow* or *Stokes' flow approximation* or simply the *low Reynolds number approximation*. The creeping-flow approximation is very important for flow through porous media, microporous membranes, and packed beds, and for the flow of small particles such as dusts and mists. If, in addition, we can assume that the aspect ratio $H_0^2/L^2 \ll 1$, say $H_0^2/L^2 = \text{o}(0.01)$, we can neglect the second-order axial derivative in equation (3.3-17) as well as both viscous terms in equation (3.3-18). When both $\text{Re} \ll 1$ and $H_0^2/L^2 \ll 1$, it is referred to as the *lubrication-flow approximation*; that is, all lubrication flows are also creeping flows, but not all creeping flows are lubrication flows. Lubrication flows are very important in the design of journal bearings and fluid couplings as well as in the drainage of viscous films. Hence, in summary, the conditions for the applicability of the creeping-or Stokes' flow and lubrication-flow approximations are

$$\text{Re} \ll 1 \Rightarrow \text{creeping- or Stokes' flow approximation} \tag{3.3-24}$$

$$\text{Re} \ll 1 \quad \text{and} \quad \frac{H_0^2}{L^2} \ll 1 \Rightarrow \text{lubrication-flow approximation} \tag{3.3-25}$$

Note in the criteria for the lubrication-flow approximation that H_0 denotes a transverse length scale and L denotes a length scale in the principal direction of flow.

If we make the lubrication-flow approximation, our dimensionless describing equations for the flow shown in Figure 3.3-1 become

$$0 = -\frac{\partial P^*}{\partial x^*} + \frac{\partial^2 u_x^*}{\partial y^{*2}} \tag{3.3-26}$$

$$0 = -\frac{\partial P^*}{\partial y^*} \tag{3.3-27}$$

$$u_x^* = 1 \quad \text{at} \quad x^* = 0 \tag{3.3-28}$$

$$u_x^* = 0 \quad \text{at} \quad y^* = 1 - \left(1 - \frac{H_L}{H_0}\right) x^* \tag{3.3-29}$$

Equation (3.3-27) implies that there is a negligible pressure drop in the y-direction. Note that we ensured that the axial pressure gradient was o(1) because we bounded the pressure and axial distance scales to be O(1). However, we did not do anything in our scaling to ensure that the y-derivative of the pressure was O(1); indeed, equation (3.3-18) indicates that the transverse pressure gradient is $\text{O}(H_0^2/L^2)$, which is considerably less than 1 for this lubrication flow. The lubrication-flow approximation has permitted us to eliminate the strong coupling between the equations of motion and to convert our complex system of elliptic differential equations into a set of differential equations that obviates the need to satisfy the boundary conditions given by equation (3.3-21). Indeed, equations (3.3-26) through (3.3-29) can be solved analytically in closed form. That is, equation (3.3-27) implies that the axial pressure gradient is a function of only the axial coordinate x. This in turn implies that equation (3.3-26) can be integrated directly. Note that the dependence of u_x on x enters indirectly through the boundary condition given by equation (3.3-29). The axial pressure profile can be obtained from the axial velocity profile and the known average velocity U_0.

Before leaving this example it is important to realize the limitations implied by the creeping-and lubrication-flow approximations. Our o(1) scaling analysis indicated that the creeping-flow assumption is reasonable when Re $\ll 1$. However, inspection of equation (3.3-17) indicates that an additional condition required to ignore the inertia terms is that

$$\mathrm{Re}\frac{H_0}{L} \ll 1 \tag{3.3-30}$$

That is, it is not sufficient in this case that just the Reynolds number be very small; in addition, the aspect ratio cannot be too large. Note, however, that the length L was arbitrary in that L could denote any value of the axial coordinate in the principal direction of flow. This is the principle of *local scaling*, whereby we scale the problem for a fixed but arbitrary value of some coordinate, usually that in the principal direction of flow. Note that both the creeping- and lubrication-flow approximations break down near the leading edge of the two plates. The creeping-flow approximation breaks down when equation (3.3-30) is not satisfied. The lubrication-flow approximation breaks down when either equation (3.3-30) or (3.3-25) is not satisfied. This is explored in Practice Problem 3.P.5.

It is again instructional to illustrate the forgiving nature of scaling for this example. Let us assume that we incorrectly balanced the pressure term with the inertial terms in equation (3.3-9), which leads to $P_s = \rho u_{xs}^2$ as our (incorrect) pressure scale; note that this pressure scale is a measure of the kinetic energy per unit volume. If we use this pressure scale in equation (3.4-9), we obtain the following dimensionless x-component of the equations of motion:

$$\mathrm{Re}\frac{H_0}{L}u_x^*\frac{\partial u_x^*}{\partial x^*} + \mathrm{Re}\frac{H_0}{L}u_y^*\frac{\partial u_x^*}{\partial y^*} = -\mathrm{Re}\frac{H_0}{L}\frac{\partial P^*}{\partial x^*} + \frac{H_0^2}{L^2}\frac{\partial^2 u_x^*}{\partial x^{*2}} + \frac{\partial^2 u_x^*}{\partial y^{*2}} \tag{3.3-31}$$

If we now consider the lubrication-flow approximation, (i.e., Re $\ll 1$ and $H_0^2/L^2 \ll 1$), equation (3.3-31) simplifies to

$$0 = \frac{\partial^2 u_x^*}{\partial y^{*2}} \tag{3.3-32}$$

However, equation (3.3-32) is not a physically realistic result since there is no mechanism to cause the flow; that is, equation (3.3-32) states that there is no force to counteract the principal viscous stress. This indicates that the scaling was improper and that new scales need to be determined to achieve a proper balance between the relevant terms in the describing equations. Of course, we know that this error was due to using an improper pressure scale; the pressure force must balance the principal viscous term, not the inertia terms that drop out in the creeping-or lubrication-flow limit.

3.4 BOUNDARY-LAYER-FLOW APPROXIMATION

The next example is the complement of the creeping-flow approximation considered in Section 3.3: the *boundary-layer-flow approximation*, which is applicable in the limit of large Reynolds numbers. In this introductory example we consider the classical problem of a uniform plug flow of a viscous Newtonian liquid having constant physical properties intercepting a stationary semi-infinitely long infinitely wide horizontal flat plate, as shown in Figure 3.4-1. Boundary-layer flows are also examples of region of influence scaling, for which we use scaling to determine the thickness of a region wherein some effect is confined, in this case the effect of the flat plate that is propagated into the fluid by the action of viscosity. This example also illustrates the principle of *local scaling*, in which we carry out the scaling at some arbitrary but fixed value of one of the spatial coordinates.

The traditional approach to introducing hydrodynamic boundary-layer theory is to begin by assuming the existence of the boundary layer. This can be very confusing, especially for students, since the boundary layer is an abstract concept. Here we arrive at the need to define some region of influence (e.g., the boundary layer) by being faced with a paradox due to incorrect scaling. That is, we are going to scale this problem initially without assuming the existence of a boundary layer. This will lead to a contradiction since the scaling analysis was improper. However, by virtue of the forgiving nature of scaling analysis, we naturally arrive at the concept of a hydrodynamic boundary layer without the need to introduce the boundary layer initially.

In view of the preceding discussion, the continuity equation and equations of motion given by equations (C.1-1), (D.1-10), and (D.1-11) simplify to the following for the assumed flow conditions (step 1):

$$\rho u_x \frac{\partial u_x}{\partial x} + \rho u_y \frac{\partial u_x}{\partial y} = -\frac{\partial P}{\partial x} + \mu \left(\frac{\partial^2 u_x}{\partial x^2} + \frac{\partial^2 u_x}{\partial y^2} \right) \tag{3.4-1}$$

$$\rho u_x \frac{\partial u_y}{\partial x} + \rho u_y \frac{\partial u_y}{\partial y} = -\frac{\partial P}{\partial y} + \mu \left(\frac{\partial^2 u_y}{\partial x^2} + \frac{\partial^2 u_y}{\partial y^2} \right) - \rho g \tag{3.4-2}$$

$$\frac{\partial u_x}{\partial x} + \frac{\partial u_y}{\partial y} = 0 \tag{3.4-3}$$

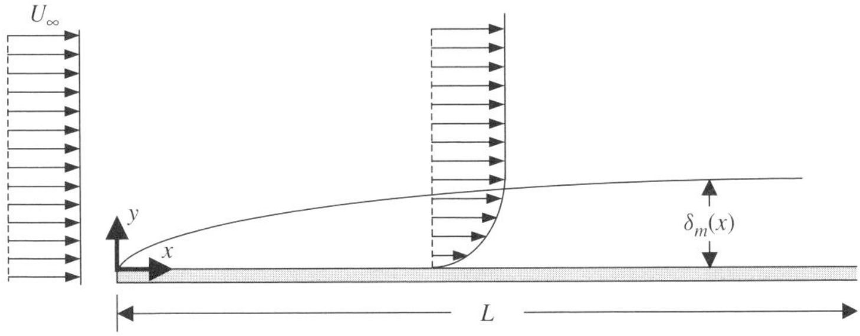

Figure 3.4-1 Uniform plug flow of velocity U_∞ for a viscous Newtonian fluid that has constant physical properties, intercepting a stationary semi-infinitely long infinitely wide horizontal flat plate.

The corresponding boundary conditions for this flow are given by

$$u_x = U_\infty, \quad u_y = 0 \quad \text{at} \quad x = 0 \tag{3.4-4}$$

$$u_x = f_1(y), \quad u_y = f_2(y) \quad \text{at} \quad x = L \tag{3.4-5}$$

$$u_x = 0, \quad u_y = 0 \quad \text{at} \quad y = 0 \tag{3.4-6}$$

$$u_x = U_\infty, \quad u_y = 0 \quad \text{at} \quad y = \infty \tag{3.4-7}$$

where $f_1(y)$ and $f_2(y)$ are unspecified functions. Equations (3.4-1) through (3.4-3), as well as the boundary conditions given by equations (3.4-4), (3.4-5), and (3.4-6), are identical to equations (3.3-1) through (3.3-3) for the lubrication-flow problem considered in Section 3.3. This is, of course, because both the present example and that considered in Section 3.3 are developing flows. However, the similarity ends here because here we are going to consider the limit of a very large Reynolds number; we thus consider the limit at the other end of the Reynolds number spectrum. Note also that the remaining boundary conditions differ in the two problems. Equation (3.4-7) states that the axial velocity becomes equal to the initial plug-flow velocity and that the transverse velocity becomes zero infinitely far above the flat plate.

Keep in mind that at least initially, we are going to scale this problem incorrectly to prove the point that scaling analysis can be used to arrive at the boundary-layer approximation systematically. We begin by defining dimensionless dependent and independent variables (steps 2, 3, and 4):

$$u_x^* \equiv \frac{u_x}{u_{xs}}; \qquad u_y^* \equiv \frac{u_y}{u_{ys}}; \qquad P^* \equiv \frac{P}{P_s}; \qquad x^* \equiv \frac{x}{x_s}; \qquad y^* \equiv \frac{y}{y_s} \tag{3.4-8}$$

Introduce these dimensionless variables into the describing equations and divide through by the dimensional coefficient of one term in each equation that should be retained to maintain physical significance (steps 5 and 6):

$$u_x^* \frac{\partial u_x^*}{\partial x^*} + \frac{u_{ys} x_s}{u_{xs} y_s} u_y^* \frac{\partial u_x^*}{\partial y^*} = -\frac{P_s}{\rho u_{xs}^2} \frac{\partial P^*}{\partial x^*} + \frac{\mu}{\rho u_{xs} x_s} \frac{\partial^2 u_x^*}{\partial x^{*2}} + \frac{\mu x_s}{\rho u_{xs} y_s^2} \frac{\partial^2 u_x^*}{\partial y^{*2}} \tag{3.4-9}$$

$$u_x^* \frac{\partial u_y^*}{\partial x^*} + \frac{u_{ys} x_s}{u_{xs} y_s} u_y^* \frac{\partial u_y^*}{\partial y^*} = -\frac{P_s x_s}{\rho u_{xs} u_{ys} y_s} \frac{\partial P^*}{\partial y^*} + \frac{\mu}{\rho u_{xs} x_s} \frac{\partial^2 u_y^*}{\partial x^{*2}} + \frac{\mu x_s}{\rho u_{xs} y_s^2} \frac{\partial^2 u_y^*}{\partial y^{*2}} - \frac{g x_s}{u_{xs} u_{ys}} \tag{3.4-10}$$

$$\frac{\partial u_x^*}{\partial x^*} + \frac{u_{ys} x_s}{u_{xs} y_s} \frac{\partial u_y^*}{\partial y^*} = 0 \tag{3.4-11}$$

$$u_x^* = \frac{U_\infty}{u_{xs}}, \quad u_y^* = 0 \quad \text{at} \quad x^* = 0 \tag{3.4-12}$$

$$u_x^* = f_1^*(y^*), \quad u_y^* = f_2^*(y^*) \quad \text{at} \quad x^* = \frac{L}{x_s} \tag{3.4-13}$$

$$u_x^* = 0, \quad u_y^* = 0 \quad \text{at} \quad y^* = 0 \tag{3.4-14}$$

$$u_x^* = \frac{U_\infty}{u_{xs}}, \quad u_y^* = 0 \quad \text{at} \quad y^* = \infty \tag{3.4-15}$$

Note that we have divided equations (3.4-9) and (3.4-10) through by the dimensional coefficient of the axial inertia term since we are considering a large Reynolds number flow for which the inertia terms must be retained.

The dimensionless groups in equations (3.4-11), (3.4-12) or (3.4-15), and (3.4-13) are set equal to 1 to determine the following scales (step 7):

$$u_{xs} = U_\infty; \qquad x_s = L; \qquad u_{ys} = \frac{y_s}{L} U_\infty \tag{3.4-16}$$

Moreover, since pressure causes the flow in the y-direction, we set the dimensionless coefficient of the pressure term in equation (3.4-10) equal to 1 to determine the pressure scale:

$$P_s = \rho U_\infty^2 \frac{y_s^2}{L^2} \tag{3.4-17}$$

The immediate problem we see is that there is no dimensionless group to determine y_s. In view of this, let us set the transverse length scale equal to the axial length scale; that is, $y_s = x_s = L$.

Now let us substitute these scales into equations (3.4-9) through (3.4-15) to obtain

$$u_x^* \frac{\partial u_x^*}{\partial x^*} + u_y^* \frac{\partial u_x^*}{\partial y^*} = -\frac{\partial P^*}{\partial x^*} + \frac{1}{\mathrm{Re}_L} \frac{\partial^2 u_x^*}{\partial x^{*2}} + \frac{1}{\mathrm{Re}_L} \frac{\partial^2 u_x^*}{\partial y^{*2}} \tag{3.4-18}$$

$$u_x^* \frac{\partial u_y^*}{\partial x^*} + u_y^* \frac{\partial u_y^*}{\partial y^*} = -\frac{\partial P^*}{\partial y^*} + \frac{1}{\mathrm{Re}_L} \frac{\partial^2 u_y^*}{\partial x^{*2}} + \frac{1}{\mathrm{Re}_L} \frac{\partial^2 u_y^*}{\partial y^{*2}} - \frac{1}{\mathrm{Fr}} \tag{3.4-19}$$

$$\frac{\partial u_x^*}{\partial x^*} + \frac{\partial u_y^*}{\partial y^*} = 0 \tag{3.4-20}$$

$$u_x^* = 1, \quad u_y^* = 0 \quad \text{at} \quad x^* = 0 \tag{3.4-21}$$

$$u_x^* = f_1^*(y^*), \quad u_y^* = f_2^*(y^*) \quad \text{at} \quad x^* = 1 \tag{3.4-22}$$

$$u_x^* = 0, \quad u_y^* = 0 \quad \text{at} \quad y^* = 0 \tag{3.4-23}$$

$$u_x^* = 1, \quad u_y^* = 0 \quad \text{at} \quad y^* = \infty \tag{3.4-24}$$

where $\text{Re}_L \equiv U_\infty \rho L/\mu$ is the Reynolds number based on the local axial distance as the characteristic length and $\text{Fr} \equiv U_\infty^2/gL$ is the Froude number.

Now consider the limit of very large Re_L (Reynolds number) (step 8). One sees that the principal viscous term (i.e., the second-order derivative with respect to y) drops out in both equations (3.4-18) and (3.4-19); this means that it is not possible to satisfy both boundary conditions given by equations (3.4-23) and (3.4-24). This, indeed, is a contradiction since if equation (3.4-24) is not satisfied, there is no mechanism to cause the flow; that is, in this case it is the free stream velocity that "pulls" along the fluid whose motion is being impeded by the presence of the flat plate. However, if equation (3.4-23) is not satisfied, the no-slip condition will be violated at the surface of the flat plate. What we have arrived at is *d'Alembert's paradox*[2]; that is, in the limit of large Reynolds numbers, the equations of motion appear unable to admit any restraining drag force since in this limit the inertia terms overwhelm the viscous terms. The conclusion we must come to here is that there must be some region of influence near the flat plate within which the effects of viscosity are important regardless of how large the Reynolds number is. We seek to use o(1) scaling to determine the thickness of this region and to arrive at a minimum parametric representation of the describing equations that circumvents d'Alembert's paradox.

The contradiction that we encountered in the above incorrect scaling arose because we arbitrarily chose $y_s = L$. This scale implies that the velocity goes from a minimum value of 0 to a maximum value of U_∞ over a length that goes from a minimum value of 0 to a maximum value of L. However, since L can be quite large, this scaling implies that the second derivative of u_x with respect to y could be grossly underestimated. For a large Reynolds number flow for which the action of viscosity is confined to the vicinity of the boundaries, the transverse length scale in general should be considerably smaller than L. Let us refer to this region of influence for the effect of the viscosity by the symbol δ_m; that is, we say that $y_s = \delta_m$.

Now let us rescale equations (3.4-1) through (3.4-7) and again introduce the dimensionless variables defined by equation (3.4-8) with the proviso that we replace

[2]Jean Le Rond d'Alembert (1717–1783) studied experimentally the drag force on a sphere in a flowing fluid. He expected that the force would approach zero as the viscosity of the fluid approached zero. However, the drag force observed converged on a nonzero value as the viscosity became very small. The disappearance of the viscous drag force for very high Reynolds number flows is known as *d'Alembert's paradox*.

y_s by δ_m. We again obtain the scale factors given in equations (3.4-16) and (3.4-17), where we have replaced y_s everywhere by δ_m. However, in view of the fact that the principal viscous term in equation (3.4-9) has to be important at least within some small region in the vicinity of the flat plate, we set the dimensionless group in front of this term equal to 1 to ensure that this term is of the same size as the inertia terms that are being retained. This yields the following equation for the thickness of the region of influence or hydrodynamic boundary layer:

$$\delta_m^2 = \frac{\mu L}{\rho U_\infty} \Rightarrow \delta_m = \frac{L}{\sqrt{\mathrm{Re}_L}} \tag{3.4-25}$$

where $\mathrm{Re}_L \equiv L\rho U_\infty/\mu$ is the Reynolds number based on the arbitrary downstream length along the plate. We see that the boundary layer becomes thinner for larger Reynolds numbers, but also becomes thicker with increasing distance L along the flat plate. Note that L is arbitrary in that it can be any fixed value of the axial length coordinate; that is, our scaling was done for an arbitrary length L of a semi-infinite flat plate, which is what is meant by the concept of *local scaling*. The general behavior of $\delta_m(x)$ is shown in Figure 3.4-1. It should not be surprising that the above estimate of boundary-layer thickness is within a multiplicative constant of $\mathrm{O}(1)$ of the value obtained via analytical solutions to the boundary-layer equations.[3]

If we now rewrite our dimensionless describing equations in terms of the scales defined by equations (3.4-16), (3.4-17), and (3.4-25), we obtain

$$u_x^* \frac{\partial u_x^*}{\partial x^*} + u_y^* \frac{\partial u_x^*}{\partial y^*} = -\frac{1}{\mathrm{Re}_L} \frac{\partial P^*}{\partial x^*} + \frac{1}{\mathrm{Re}_L} \frac{\partial^2 u_x^*}{\partial x^{*2}} + \frac{\partial^2 u_x^*}{\partial y^{*2}} \tag{3.4-26}$$

$$u_x^* \frac{\partial u_y^*}{\partial x^*} + u_y^* \frac{\partial u_y^*}{\partial y^*} = -\frac{\partial P^*}{\partial y^*} + \frac{1}{\mathrm{Re}_L} \frac{\partial^2 u_y^*}{\partial x^{*2}} + \frac{\partial^2 u_y^*}{\partial y^{*2}} - \frac{\sqrt{\mathrm{Re}_L}}{\mathrm{Fr}} \tag{3.4-27}$$

$$\frac{\partial u_x^*}{\partial x^*} + \frac{\partial u_y^*}{\partial y^*} = 0 \tag{3.4-28}$$

$$u_x^* = 1, \quad u_y^* = 0 \quad \text{at} \quad x^* = 0 \tag{3.4-29}$$

$$u_x^* = f_1^*\left(y^*\right), \quad u_y^* = f_2^*\left(y^*\right) \quad \text{at} \quad x^* = 1 \tag{3.4-30}$$

$$u_x^* = 0, \quad u_y^* = 0 \quad \text{at} \quad y^* = 0 \tag{3.4-31}$$

$$u_x^* = 1, \quad u_y^* = 0 \quad \text{at} \quad y^* = \infty \tag{3.4-32}$$

The system of equations above is difficult to solve, for two reasons. First, these are elliptic differential equations that require specifying some downstream boundary conditions that in practice are generally not known. Second, equations (3.4-26) and (3.4-27) are coupled, owing to the pressure appearing in both equations; note that coupling through the velocity components does not cause any problems since the

[3] R. B. Bird, W. E. Stewart, and E. N. Lightfoot, *Transport Phenomena*, 2nd ed., Wiley, Hoboken, NJ, 2002, p. 137.

continuity equation given by equation (3.4-28) permits defining a new dependent variable, the *stream function*, that can be used to eliminate the two velocity components. Hence, we seek to explore the conditions required to eliminate these two complications.

Note that in the limit of a very large Reynolds number, the system of equations above reduces to

$$u_x^* \frac{\partial u_x^*}{\partial x^*} + u_y^* \frac{\partial u_x^*}{\partial y^*} = \frac{\partial^2 u_x^*}{\partial y^{*2}} \tag{3.4-33}$$

$$\frac{\partial u_x^*}{\partial x^*} + \frac{\partial u_y^*}{\partial y^*} = 0 \tag{3.4-34}$$

$$u_x^* = 1 \qquad \text{at} \quad x^* = 0 \tag{3.4-35}$$

$$u_x^* = 0, \quad u_y^* = 0 \qquad \text{at} \quad y^* = 0 \tag{3.4-36}$$

$$u_x^* = 1 \qquad \text{at} \quad y^* = \infty \tag{3.4-37}$$

Equations (3.4-33) through (3.4-37) are the classical boundary-layer equations for flow over a flat plate. Note that by showing that the pressure term in equation (3.4-26) is negligible in the limit of a large Reynolds number, we have eliminated the coupling between this equation and equation (3.4-27). Moreover, we have shown that the axial viscous term in equation (3.4-26) is negligible in the limit of a large Reynolds number and thereby have converted the system of elliptic differential equations into a parabolic differential equation that requires only an upstream boundary condition. By introducing a stream function and similarity variable, equations (3.4-33) and (3.4-34) can be transformed into a nonlinear ordinary differential equation that can be solved via approximate techniques such as the Blasius series solution or numerically.[4]

Note that the criterion for applicability of the hydrodynamic boundary-layer approximation is

$$\text{Re}_L \equiv \frac{U_\infty \rho L}{\mu} \gg 1 \qquad \text{hydrodynamic boundary-layer flow} \tag{3.4-38}$$

Since L is merely some fixed value of the axial coordinate x, the criterion above always breaks down in the vicinity of the leading edge of the flat plate. Hence, if one is seeking to determine an integral quantity such as the total drag on the flat plate, the error will not be significant if equation (3.4-38) is satisfied over most of the plate. However, the error incurred by invoking the boundary-layer approximation can be quite large in the vicinity of the leading edge of the plate for point quantities such as the local velocity components or local shear stress. Note that for 90% of the flat plate to satisfy the condition that $\text{Re}_L \geqslant O(100)$, the Reynolds number at the end of the plate must be 1000. Our scaling analysis results for assessing the error incurred in making the boundary-layer approximation are

[4] H. Schlichting, *Boundary Layer Theory*, McGraw-Hill, New York, 1960, pp. 116–124.

consistent with the results obtained from numerical solutions. For example, Janssen found that the error in the drag coefficient was 40% at $\mathrm{Re}_L = 100$ and negligible at $\mathrm{Re}_L = 1000$.[5]

3.5 QUASI-STEADY-STATE-FLOW APPROXIMATION

Thus far, the three examples that we have considered have involved steady-state flows. Here we consider how to use o(1) scaling to analyze unsteady-state problems, in particular how to determine when the *quasi-steady-state approximation* is applicable. The latter implies that the unsteady-state term does not appear explicitly in the describing equations; however, the time dependence enters through the boundary conditions. In this example we see that there are several possible time scales. Choosing the proper time scale depends on the conditions being considered. In particular, we will see that for studying transient phenomena, the proper time scale is the instantaneous observation time.

Consider the unsteady-state two-dimensional flow of a viscous Newtonian fluid with constant physical properties between two infinitely wide parallel flat plates. The upper plate is stationary, whereas the lower plate is initially at rest and then set into oscillatory motion as shown in Figure 3.5-1. We use o(1) scaling to address three different approximations that we might make in modeling this flow: (1) when we can ignore the transient startup effect on the flow, (2) when we can assume

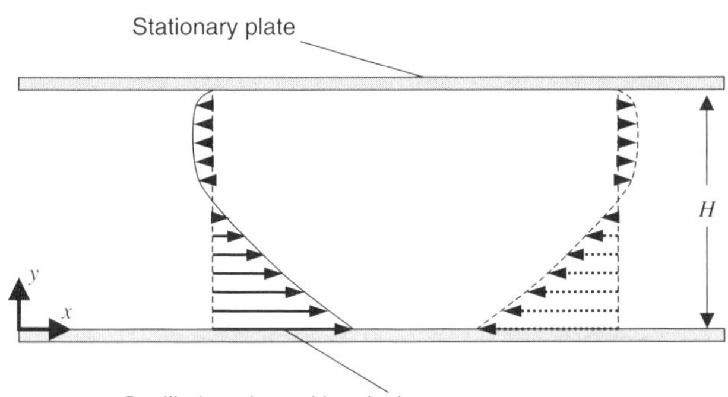

Figure 3.5-1 Unsteady-state two-dimensional flow of a viscous Newtonian fluid that has constant physical properties between two infinitely wide parallel flat plates; the upper plate is stationary, whereas the lower plate is initially at rest and then set into oscillatory motion with a velocity given by $u_x = U_0 \cos \omega t$, where U_0 is the amplitude and ω is the angular frequency; the velocity profiles shown by the solid and dashed lines correspond to two different times during the oscillatory motion.

[5] F. Janssen, *J. Fluid Mech.*, **3**, 329 (1958).

that this flow is quasi-steady-state, and (3) when we can consider the effect of the oscillating plate to be confined to a region of influence.

The describing equations are obtained by appropriately simplifying equations (C.1-1), (D.1-10), and (D.1-11) in the Appendices (step 1):

$$\rho \frac{\partial u_x}{\partial t} = \mu \frac{\partial^2 u_x}{\partial y^2} \tag{3.5-1}$$

$$0 = -\frac{dP}{dy} - \rho g \tag{3.5-2}$$

$$\frac{\partial u_x}{\partial x} = 0 \tag{3.5-3}$$

$$u_x = 0 \quad \text{at} \quad t = 0 \tag{3.5-4}$$

$$u_x = U_0 \cos \omega t \quad \text{at} \quad y = 0 \tag{3.5-5}$$

$$u_x = 0 \quad \text{at} \quad y = H \tag{3.5-6}$$

Equation (3.5-4) states that the fluid is initially stationary. Equations (3.5-5) and (3.5-6) are the no-slip conditions applied at the oscillated lower and stationary upper boundaries, respectively.

Introduce the following scale factors and dimensionless variables (steps 2, 3, and 4):

$$u_x^* \equiv \frac{u_x}{u_{xs}}; \qquad t^* \equiv \frac{t}{t_s}; \qquad y^* \equiv \frac{y}{y_s} \tag{3.5-7}$$

Note that we do not have to scale the pressure since equation (3.5-2) indicates that it is purely hydrostatic and not coupled with equation (3.5-1). Introduce these dimensionless variables into the describing equations and divide each equation through by the dimensional coefficient of a term that must be retained to ensure that the problem has physical significance (steps 5 and 6):

$$\frac{\rho y_s^2}{\mu t_s} \frac{\partial u_x^*}{\partial t^*} = \frac{\partial^2 u_x^*}{\partial y^{*2}} \tag{3.5-8}$$

$$u_x^* = 0 \quad \text{at} \quad t^* = 0 \tag{3.5-9}$$

$$u_x^* = \frac{U_0}{u_{xs}} \cos \omega t_s t^* \quad \text{at} \quad y^* = 0 \tag{3.5-10}$$

$$u_x^* = 0 \quad \text{at} \quad y^* = \frac{H}{y_s} \tag{3.5-11}$$

Now let us set appropriate dimensionless groups in equations (3.5-8) through (3.5-11) equal to 1 to ensure that our dimensionless variables are o(1) (step 7).

The dimensionless groups in equations (3.5-10) and (3.5-11) determine the following scale factors:

$$u_{xs} = U_0; \qquad y_s = H \tag{3.5-12}$$

One might be tempted to set the dimensionless group in equation (3.5-8) equal to 1 to determine the time scale. However, this choice implies that our time scale is equal to the time required for the motion of the lower oscillating plate to be felt at the upper stationary plate by the action of viscosity since this implies that $t_s = H^2/\nu$, where $\nu \equiv \mu/\rho$ is the kinematic viscosity; if this is the appropriate time scale, quasi-steady-state can never be achieved for this flow. Alternatively, one could obtain the time scale from the dimensionless group in equation (3.5-10). However, this time scale, $t_s = 2\pi/\omega$, would characterize the oscillatory motion, which again might not be proper for the conditions being considered.[6] The latter would certainly not be the correct time scale to characterize the transient period during which the fluid is accelerated from rest. The latter time scale is the instantaneous time at which we "observe" the flow; we call this the *observation time*, t_o. Hence, we have three possible time scales:

$$\begin{aligned} t_{st} &= t_o && \text{time scale corresponding to the observation time} \\ t_{sv} &= \frac{H^2}{\nu} && \text{time scale characterizing the viscous penetration} \\ t_{sp} &= \frac{2\pi}{\omega} && \text{time scale characterizing the periodic motion} \end{aligned} \tag{3.5-13}$$

Clearly, $0 \leqslant t_{st} < \infty$, since this scale is the actual time beginning at the inception of unsteady-state flow. In contrast, the time scales t_{sv} and t_{sp} have fixed values that depend on the values of the parameters in equation (3.5-13). If $t_{sv} < t_{sp}$, the effect of the oscillatory plate motion will penetrate across the entire fluid to the upper stationary plate. If $t_{sv} > t_{sp}$, the effect of the plate motion will be confined to a region of influence whose thickness is less than H. We can determine the thickness of this region of influence using o(1) scaling analysis, as will be shown.

The time scales defined in equation (3.5-13) permit us to determine the criterion for assuming that the transients associated with fluid motion induced during startup of the oscillatory motion have died out. This criterion is merely that t_{st} must be much greater than the characteristic time for the oscillatory motion; that is,

$$t_{st} = t_o \gg \frac{2\pi}{\omega} \Rightarrow \frac{\omega t_o}{2\pi} \gg 1 \quad \text{condition to ignore transient flow effects} \tag{3.5-14}$$

[6]Note that we choose $t_s = 2\pi/\omega$ rather than $1/\omega$ since the oscillatory motion is characterized by the time it takes to complete one cycle.

Now let us assume that the transient effects have died out [i.e., that the condition defined in equation (3.5-14) is satisfied]. We now seek to determine when quasi-steady-state can be assumed, that is, when the unsteady-state term in equation (3.5-8) can be neglected. When the velocity and length scales defined by equation (3.5-12) and the time scale t_{sp} defined in equation (3.5-13) are substituted into equations (3.5-8) through (3.5-11), we obtain

$$\frac{\omega\rho H^2}{2\pi\mu}\frac{\partial u_x^*}{\partial t^*} = \frac{\partial^2 u_x^*}{\partial y^{*2}} \tag{3.5-15}$$

$$u_x^* = 0 \quad \text{at} \quad t^* = 0 \tag{3.5-16}$$

$$u_x^* = \cos 2\pi t^* \quad \text{at} \quad y^* = 0 \tag{3.5-17}$$

$$u_x^* = 0 \quad \text{at} \quad y^* = 1 \tag{3.5-18}$$

We can now assess when the set of describing equations above can be simplified (step 8). In particular, we see from equation (3.5-15) that this flow can be considered to be quasi-steady-state when the following condition holds:

$$\frac{\omega\rho H^2}{2\pi\mu} = \frac{H^2/\nu}{2\pi/\omega} = \frac{t_{sv}}{t_{sp}} \ll 1 \Rightarrow \text{quasi-steady-state} \tag{3.5-19}$$

For quasi-steady-state, the system of equations above can be solved quite simply analytically to obtain the following solution:

$$u_x^* = (1 - y^*)\cos 2\pi t^* \tag{3.5-20}$$

The physical implication of the above is that the viscous time scale must be sufficiently short so that the motion of the lower plate can penetrate across the entire fluid to the upper stationary plate within a time that is much shorter than that characterizing the periodic motion of the plate. Another way to state this is that if the motion of the lower plate is sufficiently slow, its effect can penetrate all the way to the upper plate. Under such conditions, the acceleration of the fluid is relatively insignificant.

In the scaling analysis that led to the criterion for assuming quasi-steady-state given by equation (3.5-19), we have assumed that the effect of the lower oscillating plate penetrates the entire cross-section; that is, we have assumed that the velocity goes from its minimum to maximum value over a length scale on the order of the spacing between the two plates. This may not necessarily be true; that is, there may be a region of influence whose thickness we again denote by $y_s = \delta_m$. Using this length scale then recasts equation (3.5-15) into the form

$$\frac{\omega\rho\delta_m^2}{2\pi\mu}\frac{\partial u_x^*}{\partial t^*} = \frac{\partial^2 u_x^*}{\partial y^{*2}} \tag{3.5-21}$$

Since there is insufficient time for the effect of the oscillating plate to penetrate through the entire fluid layer, this is inherently an unsteady-state flow. Hence, the two terms in equation (3.5-21) must balance; this implies the following:

$$\frac{\omega\rho\delta_m^2}{2\pi\mu} = 1 \Rightarrow \delta_m^2 = \frac{2\pi\mu}{\omega\rho} \Rightarrow \frac{\delta_m}{H} = \sqrt{\frac{2\pi/\omega}{H^2/\nu}} = \sqrt{\frac{t_{sp}}{t_{sv}}} \tag{3.5-22}$$

We see that the thickness of the region of influence relative to the spacing between the two plates is proportional to the square root of the ratio of the characteristic time for the periodic plate motion to that for viscous penetration. Smaller values of t_{sp} correspond to higher-frequency oscillations and hence to a thinner region of influence or boundary-layer thickness. Note that when $t_{sp} = t_{sv}$, the region of influence penetrates across the entire fluid layer. For values of $t_{sp} > t_{sv}$, this is no longer a region of influence or boundary-layer problem but an unsteady-state-state flow for which the oscillating plate affects the entire fluid layer. For values of $t_{sp} \gg t_{sv}$, we achieve the quasi-steady-state condition defined by equation (3.5-19).

Note that when $t_{sp} \ll t_{sv}$, the describing equations simplify to

$$\frac{\partial u_x^*}{\partial t^*} = \frac{\partial^2 u_x^*}{\partial y^{*2}} \tag{3.5-23}$$

$$u_x^*\big|_{t^*} = u_x^*\big|_{t^*+1} \qquad \text{for} \quad t^* > 0 \tag{3.5-24}$$

$$u_x^* = \cos 2\pi t^* \qquad \text{at} \quad y^* = 0 \tag{3.5-25}$$

$$u_x^* = 0 \qquad \text{at} \quad y^* = \frac{H}{\delta_m} = \sqrt{\frac{t_{sv}}{t_{sp}}} \cong \infty \tag{3.5-26}$$

Note also that for this region of influence scaling, for which we are assuming that the transients have died out ($\omega t_o \gg 1$), we have replaced the initial condition with the periodic flow condition given by equation (3.5-24). The solution to this simplified set of describing equations is straightforward. We know that u_x^* must be periodic at all values of y^*; however, u_x^* will lag u_x^* at the lower plate by an increasing amount as y^* increases. We also note that u_x^* must damp out as $y^* \to \infty$. Hence, we assume a solution of the form

$$u_x^* = e^{-\alpha y^*} \cos[2\pi t^* - g(y^*)] \tag{3.5-27}$$

where α and $g(y^*)$ are an undetermined constant and function, respectively. Substituting equation (3.5-27) into equations (3.5-23) through (3.5-26) then gives the following values for these unknown quantities:

$$\alpha = \frac{1}{\sqrt{2}}, \qquad g(y^*) = \frac{y^*}{\sqrt{2}} \tag{3.5-28}$$

In summary, we see that the scaling of unsteady-state problems can be complicated by several time scales. It is important that the implications of each time scale be considered carefully when using scaling analysis to simplify such problems.

3.6 FLOWS WITH END AND SIDEWALL EFFECTS

The examples considered in Sections 3.2 through 3.5 involved flows that were assumed to be infinitely wide in the lateral direction. In this example we use o(1) scaling to determine a criterion for ignoring sidewall effects; that is, to permit one to assume that the flow is infinitely wide in the lateral direction. The same type of scaling arguments used to justify ignoring sidewall effects can also be used to justify ignoring end effects. We also seek to determine the thickness of the region of influence within which one cannot ignore the effect of the sidewalls on point or local quantities such as the velocity profile or drag at the sidewalls.

Consider the steady-state fully developed gravity-driven flow of a Newtonian liquid film having thickness H and constant physical properties down a channel inclined at an angle θ to the horizontal and having width W as shown in Figure 3.6-1. The describing equations are obtained by appropriately simplifying equations (C.1-1), (D.1-10), (D.1-11), (D.1-12) in the Appendices for a flow that is caused by a gravitational body force (step 1):

$$0 = \mu \frac{\partial^2 u_z}{\partial x^2} + \mu \frac{\partial^2 u_z}{\partial y^2} + \rho g \sin\theta \tag{3.6-1}$$

$$0 = -\frac{\partial P}{\partial y} + \rho g \cos\theta \tag{3.6-2}$$

$$0 = -\frac{\partial P}{\partial z} \tag{3.6-3}$$

$$\frac{\partial u_z}{\partial z} = 0 \tag{3.6-4}$$

$$u_z = 0 \quad \text{at} \quad x = \pm\frac{1}{2} W \tag{3.6-5}$$

$$\frac{\partial u_z}{\partial y} = 0 \quad \text{at} \quad y = 0 \tag{3.6-6}$$

$$u_z = 0 \quad \text{at} \quad y = H \tag{3.6-7}$$

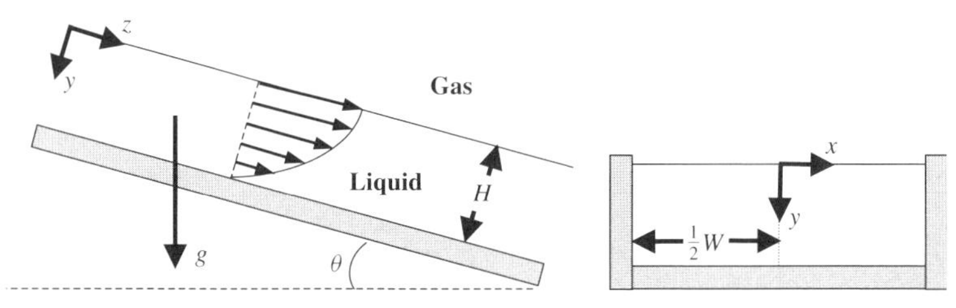

Figure 3.6-1 Steady-state fully developed gravity-driven flow of a Newtonian liquid film of thickness H and constant physical properties flowing down a flat plate inclined at an angle θ to the horizontal; this flow is bounded laterally by two parallel flat plates spaced a distance W apart.

Equations (3.6-5) and (3.6-7) are no-slip conditions at the solid boundaries. Equation (3.6-6) is the implication of assuming negligible drag by the gas on the liquid–gas interface. Equations (3.6-2) and (3.6-3) imply that the pressure is purely hydrostatic and need not be considered further in the scaling analysis.

Introduce the following scale factors and dimensionless variables (steps 2, 3, and 4):

$$u_z^* \equiv \frac{u_z}{u_{zs}}; \qquad x^* \equiv \frac{x}{x_s}; \qquad y^* \equiv \frac{y}{y_s} \tag{3.6-8}$$

Substitute these variables into the describing equations and divide through by the dimensional coefficient of a term that must be retained to obtain the following set of dimensionless describing equations (steps 5 and 6):

$$0 = \frac{y_s^2}{x_s^2}\frac{\partial^2 u_z^*}{\partial x^{*2}} + \frac{\partial^2 u_z^*}{\partial y^{*2}} + \frac{\rho g y_s^2}{\mu u_{zs}}\sin\theta \tag{3.6-9}$$

$$u_z^* = 0 \quad \text{at} \quad x^* = \pm\frac{1}{2}\frac{W}{x_s} \tag{3.6-10}$$

$$\frac{\partial u_z^*}{\partial y^*} = 0 \quad \text{at} \quad y^* = 0 \tag{3.6-11}$$

$$u_z^* = 0 \quad \text{at} \quad y^* = \frac{H}{y_s} \tag{3.6-12}$$

To bound our dimensionless variables to be o(1), we set the dimensionless group in equation (3.6-9) that is a measure of the ratio of the gravitational body force to the principal viscous drag force equal to 1 since gravity causes the flow. The lateral and vertical length scales over which the velocity goes from its minimum to its maximum value are obtained by setting the dimensionless groups in equations (3.6-10) and (3.6-12) equal to 1. This yields the following values for the length and velocity scales (step 7):

$$x_s = \frac{W}{2}, \qquad y_s = H, \qquad \frac{\rho g y_s^2 \sin\theta}{\mu u_{zs}} = \frac{\rho g H^2 \sin\theta}{\mu u_{zs}} = 1 \Rightarrow u_{zs} = \frac{\rho g H^2 \sin\theta}{\mu} \tag{3.6-13}$$

Note that the velocity scale in equation (3.6-13) is within a multiplicative constant of O(1) of the surface velocity for fully developed laminar film flow down an inclined plate.

When the velocity and length scales defined by equation (3.6-13) are substituted into equations (3.6-9) through (3.6-12), we obtain

$$0 = \frac{4H^2}{W^2}\frac{\partial^2 u_z^*}{\partial x^{*2}} + \frac{\partial^2 u_z^*}{\partial y^{*2}} + 1 \tag{3.6-14}$$

$$u_z^* = 0 \quad \text{at} \quad x^* = \pm 1 \tag{3.6-15}$$

$$\frac{\partial u_z^*}{\partial y^*} = 0 \qquad \text{at} \quad y^* = 0 \tag{3.6-16}$$

$$u_z^* = 0 \qquad \text{at} \quad y^* = 1 \tag{3.6-17}$$

Note in the set of describing equations above that the dimensionless dependent and independent variables are bounded of o(1). We see from the above that we can ignore the term in equation (3.6-14) that accounts for the effect of the sidewalls if the following dimensionless group is very small; that is (step 8),

$$\frac{4H^2}{W^2} \ll 1 \Rightarrow \text{sidewall effects can be neglected} \tag{3.6-18}$$

The condition in equation (3.6-18) will be satisfied if $4H^2/W^2 = \text{o}(0.01)$. If this condition is satisfied, the solution to the appropriately simplified form of equation (3.6-14) and the boundary conditions given by equations (3.6-16) and (3.6-17) can be obtained quite simply analytically. This solution will be accurate for predicting any quantities for which H and W are the appropriate length scales; that is, integral quantities that depend on the velocity profile across the entire cross section of the flow, such as the average velocity, volumetric flow rate, and overall drag force.

If one seeks to determine the drag force or velocity in the vicinity of the sidewalls, the simplified form of equation (3.6-14) cannot be used. Clearly, there is a region of influence within which the effect of the sidewalls on the flow cannot be ignored. Within this region of influence, the viscous stress term in equation (3.6-14) arising from the drag at the sidewalls is just as important as the principal viscous term; that is, τ_{zx} is approximately the same magnitude as τ_{yz}. This means that the dimensionless group multiplying the term arising from τ_{zx} in equation (3.6-14) must be set equal to 1. This provides a measure of the thickness of the region of influence within which one cannot ignore the effect of the sidewalls; that is,

$$\frac{4H^2}{x_s^2} = \frac{4H^2}{\delta_m^2} = 1 \Rightarrow \delta_m = 2H \tag{3.6-19}$$

In summary, the effect of the sidewalls on the flow can be ignored if the flow channel is much wider than the depth of the liquid film. Ignoring the sidewall effects is a reasonable approximation under such conditions provided that one is not interested in predicting some quantity in the immediate vicinity of the sidewalls. Scaling analysis provides both the criterion for ignoring the sidewall effects as well as an estimate of the region in which these effects will be important.

3.7 FREE SURFACE FLOW

The flow considered in Section 3.6 involved a free surface, the liquid–gas interface. However, the complications introduced were minimal in that example because this free surface was planar. In this example we consider a nonplanar two-dimensional free surface flow involving the unsteady-state draining of a viscous Newtonian

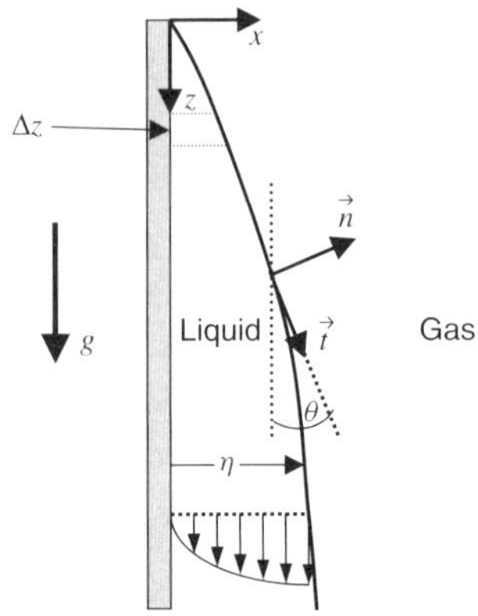

Figure 3.7-1 Draining of a two-dimensional viscous Newtonian liquid film that has constant physical properties due to a gravitational body force; the liquid film has a free surface at which the effects of both viscous drag and surface-tension forces are assumed to be negligible.

liquid film having constant physical properties due to a gravitational body force, as shown in Figure 3.7-1; we assume that the effects of viscous drag and surface-tension forces at the free surface can be neglected. We use o(1) scaling to determine when the describing equations can be simplified; in particular, when the effects of surface curvature can be ignored.

The describing equations are obtained by appropriately simplifying equations (C.1-1), (D.1-10), and (D.1-12) in the Appendices for a flow that is caused by a gravitational body force (step 1):

$$\rho \frac{\partial u_z}{\partial t} + \rho u_z \frac{\partial u_z}{\partial z} + \rho u_x \frac{\partial u_z}{\partial x} = -\frac{\partial P}{\partial z} + \mu \frac{\partial^2 u_z}{\partial z^2} + \mu \frac{\partial^2 u_z}{\partial x^2} + \rho g \tag{3.7-1}$$

$$\rho \frac{\partial u_x}{\partial t} + \rho u_z \frac{\partial u_x}{\partial z} + \rho u_x \frac{\partial u_x}{\partial x} = -\frac{\partial P}{\partial x} + \mu \frac{\partial^2 u_x}{\partial z^2} + \mu \frac{\partial^2 u_x}{\partial x^2} \tag{3.7-2}$$

$$\frac{\partial u_z}{\partial z} + \frac{\partial u_x}{\partial x} = 0 \tag{3.7-3}$$

$$u_z = 0, \quad u_x = 0, \quad \eta = \infty \quad \text{at} \quad t = 0 \tag{3.7-4}$$

$$u_z = 0, \quad u_x = 0 \quad \text{at} \quad x = 0 \tag{3.7-5}$$

$$\left.\begin{aligned} 4\frac{\partial u_z}{\partial z}\sin\theta\cos\theta - \left(\frac{\partial u_x}{\partial z} + \frac{\partial u_z}{\partial x}\right)\cos 2\theta = 0 \\ -P_{\text{atm}} + P + 2\mu\frac{\partial u_z}{\partial z}\cos 2\theta - 2\mu\left(\frac{\partial u_x}{\partial z} + \frac{\partial u_z}{\partial x}\right)\sin\theta\cos\theta = 0 \end{aligned}\right\} \quad \text{at} \quad x = \eta(z,t) \tag{3.7-6}$$

$$u_z = 0, \quad u_x = 0, \quad \eta = 0 \quad \text{at} \quad z = 0 \tag{3.7-7}$$

$$u_z = f_1(x,t), \quad u_x = f_2(x,t) \quad \text{at} \quad z = L \tag{3.7-8}$$

Note that pressure terms are included in both equations (3.7-1) and (3.7-2) even though we are ignoring surface-tension effects. These pressure terms must be included because this is a developing flow; that is, flow in the x-direction is caused by an induced pressure force. The first of equations (3.7-6) is a statement that the adjacent gas phase does not exert any viscous drag force on the liquid interface; that is, $\vec{t} \cdot \vec{n}\underline{\underline{\sigma}} = 0$ at $x = \eta(z, t)$, where $\vec{n}$ and $\vec{t}$ are the local normal and tangential unit vectors at the free surface, respectively, and $\underline{\underline{\sigma}} = P\underline{\underline{\delta}} + \underline{\underline{\tau}}$ is the total stress tensor in which $\underline{\underline{\delta}}$ is the identity tensor and $\underline{\underline{\tau}}$ is the viscous stress tensor defined for a Newtonian fluid by equations (D.1-4) through (D.1-9) in the Appendices. For the coordinate system shown in Figure 3.7-1, the normal and tangential unit vectors are expressed in terms of the unit vectors in the x- and z-coordinate directions, $\vec{\delta}_x$ and $\vec{\delta}_z$, respectively, as follows: $\vec{n} = \vec{\delta}_x \cos\theta - \vec{\delta}_z \sin\theta$ and $\vec{t} = \vec{\delta}_x \sin\theta + \vec{\delta}_z \cos\theta$. The functions $f_1(x, t)$ and $f_2(x, t)$ in equation (3.7-8) merely indicate that to solve this system of differential equations we would need to specify some downstream boundary conditions; often, these are unknown, which precludes solving these equations. This downstream boundary condition is applied at $z = L$, where L can be any specified value of z; hence, this is another example of *local scaling*.

Equations (3.7-1) through (3.7-8) constitute three differential equations and their associated initial and boundary conditions to determine four unknown dependent variables: u_z, u_x, P, and η. Hence, an auxiliary equation is needed to determine the location of the interface η. This is obtained via an integral mass balance over a differential length of the film Δz having local thickness $\eta(z, t)$, as shown in Figure 3.7-1. The following development of this auxiliary condition employs Leibnitz's rule for differentiating an integral given by equation (H.1-2) in the Appendices:

$$\left.\left(\int_0^{\eta} \rho u_z\, dx\right)\right|_z - \left.\left(\int_0^{\eta} \rho u_z\, dx\right)\right|_{z+\Delta z} = \frac{d}{dt}\int_0^{\eta} \rho\, dx\, \Delta z \Rightarrow -\frac{d}{dz}\int_0^{\eta} u_z\, dx = \frac{\partial \eta}{\partial t}$$

$$\Rightarrow -\int_0^{\eta} \frac{\partial u_z}{\partial z}\, dx - u_z\frac{\partial \eta}{\partial z} = \int_0^{\eta} \frac{\partial u_x}{\partial x}\, dx - u_z\frac{\partial \eta}{\partial z}$$

$$= u_x - u_z\frac{\partial \eta}{\partial z} = \frac{\partial \eta}{\partial t} \quad \text{at} \quad x = \eta(z, t) \tag{3.7-9}$$

Equation (3.7-9) is referred to as the *kinematic surface condition*. Note that the solution of the kinematic surface condition requires both an initial and a boundary condition for η; these are included in equations (3.7-4) and (3.7-7). The former states that the film is infinitely thick prior to the inception of draining; the latter states that the film thins to zero thickness at its leading edge as soon as draining begins.

Introduce the following scale factors and dimensionless variables (steps 2, 3, and 4):

$$u_z^* \equiv \frac{u_z}{u_{zs}}; \quad u_x^* \equiv \frac{u_x}{u_{xs}}; \quad P^* \equiv \frac{P - P_r}{P_s}; \quad \left(\frac{\partial \eta}{\partial t}\right)^* \equiv \frac{1}{\eta_{ts}}\frac{\partial \eta}{\partial t};$$

$$\left(\frac{\partial \eta}{\partial z}\right)^* \equiv \frac{1}{\eta_{zs}}\frac{\partial \eta}{\partial z}; \quad \eta^* \equiv \frac{\eta}{\eta_s}; \quad z^* \equiv \frac{z}{z_s}; \quad x^* \equiv \frac{x}{x_s}; \quad t^* \equiv \frac{t}{t_s} \tag{3.7-10}$$

Note that we allow for a reference pressure P_r in our definition of the dimensionless pressure since the dimensional pressure is not naturally referenced to zero. In addition, we have introduced separate scale factors denoted by η_{ts} and η_{zs} for the temporal and spatial derivatives, respectively, of the film thickness η since these do not necessarily scale with x_s/t_s and x_s/z_s. Substitute these variables into the describing equations and divide through by the dimensional coefficient of a term that must be retained to obtain the following set of dimensionless describing equations (steps 5 and 6):

$$\frac{\rho x_s^2}{\mu t_s}\frac{\partial u_z^*}{\partial t^*} + \frac{\rho u_{zs} x_s^2}{\mu z_s} u_z^* \frac{\partial u_z^*}{\partial z^*} + \frac{\rho u_{xs} x_s}{\mu} u_x^* \frac{\partial u_z^*}{\partial x^*} = -\frac{P_s x_s^2}{\mu u_{zs} z_s}\frac{\partial P^*}{\partial z^*} + \frac{x_s^2}{z_s^2}\frac{\partial^2 u_z^*}{\partial z^{*2}} + \frac{\partial^2 u_z^*}{\partial x^{*2}} + \frac{\rho g x_s^2}{\mu u_{zs}} \tag{3.7-11}$$

$$\frac{\rho x_s^2}{\mu t_s}\frac{\partial u_x^*}{\partial t^*} + \frac{\rho u_{zs} x_s^2}{\mu z_s} u_z^* \frac{\partial u_x^*}{\partial z^*} + \frac{\rho u_{xs} x_s}{\mu} u_x^* \frac{\partial u_x^*}{\partial x^*} = -\frac{P_s x_s}{\mu u_{xs}}\frac{\partial P^*}{\partial x^*} + \frac{x_s^2}{z_s^2}\frac{\partial^2 u_x^*}{\partial z^{*2}} + \frac{\partial^2 u_x^*}{\partial x^{*2}} \tag{3.7-12}$$

$$\frac{\partial u_z^*}{\partial z^*} + \frac{u_{xs} z_s}{u_{zs} x_s}\frac{\partial u_x^*}{\partial x^*} = 0 \tag{3.7-13}$$

$$u_z^* = 0, \quad u_x^* = 0, \quad \eta^* = \infty \quad \text{at} \quad t^* = 0 \tag{3.7-14}$$

$$u_z^* = 0, \quad u_x^* = 0 \quad \text{at} \quad x^* = 0 \tag{3.7-15}$$

$$\left.\begin{aligned} &\frac{4x_s}{z_s}\frac{\partial u_z^*}{\partial z^*}\sin\theta\cos\theta - \left(\frac{u_{xs}x_s}{u_{zs}z_s}\frac{\partial u_x^*}{\partial z^*} + \frac{\partial u_z^*}{\partial x^*}\right)\cos 2\theta = 0 \\ &-\frac{P_{\text{atm}}x_s}{\mu u_{zs}} + \frac{P_s x_s}{\mu u_{zs}}P^* + \frac{P_r x_s}{\mu u_{zs}} + 2\frac{x_s}{z_s}\frac{\partial u_z^*}{\partial z^*}\cos 2\theta \\ &\quad - 2\left(\frac{u_{xs}x_s}{u_{zs}z_s}\frac{\partial u_x^*}{\partial z^*} + \frac{\partial u_z^*}{\partial x^*}\right)\sin\theta\cos\theta = 0 \end{aligned}\right\} \quad \text{at} \quad x^* = \frac{\eta(z,t)}{x_s} \tag{3.7-16}$$

$$u_z^* = 0, \quad u_x^* = 0, \quad \eta^* = 0 \quad \text{at} \quad z^* = 0 \tag{3.7-17}$$

$$u_z^* = f_1^*(x^*, t^*), \quad u_x^* = f_2^*(x^*, t^*) \quad \text{at} \quad z^* = \frac{L}{z_s} \tag{3.7-18}$$

$$\frac{u_{xs}}{u_{zs}\eta_{zs}}u_x^* - u_z^*\left(\frac{\partial \eta}{\partial z}\right)^* = \frac{\eta_{ts}}{u_{zs}\eta_{zs}}\left(\frac{\partial \eta}{\partial t}\right)^* \quad \text{at} \quad x^* = \frac{\eta(z,t)}{x_s} \tag{3.7-19}$$

Since the gravitational body force causes the flow, we balance the latter term with the principal viscous term by setting the dimensionless group that constitutes the last term in equation (3.7-11) equal to 1; this determines the axial velocity scale u_{zs}. The axial and transverse length scales, z_s and x_s, are determined by setting the appropriate dimensionless groups in equations (3.7-16) and (3.7-18) equal to 1.

Since this is a developing flow, the transverse velocity scale u_{xs} is determined by setting the dimensionless group in equation (3.7-13) equal to 1. Since pressure causes the flow in the transverse direction, the pressure scale P_s is determined by setting the dimensionless group multiplying the pressure term in equation (3.7-12) equal to 1. The first and third terms in the surface normal stress boundary condition given by equation (3.7-16) indicate that the dimensionless pressure can be referenced to zero if we set $P_r = P_{\text{atm}}$. Since this is inherently an unsteady-state flow, the time scale t_s is equal to the observation time t_o, that is, the arbitrary time at which the flow is "observed"; setting the time scale to be the observation time is the time-scaling analog of local scaling for spatial variables. Finally, since all three terms in equation (3.7-19) should be of equal magnitude, we set the dimensionless groups multiplying the first and third terms equal to 1 in order to determine the η_{zs} and η_{ts} scales, respectively. These considerations then determine the following scale and reference factors (step 7):

$$u_{zs} = \frac{\rho g \eta^2}{\mu}; \qquad u_{xs} = \frac{\rho g \eta^3}{\mu L}; \qquad P_s = \frac{\rho g \eta^2}{L}; \qquad P_r = P_{\text{atm}};$$

$$\eta_{ts} = \frac{\rho g \eta^3}{\mu L}; \qquad \eta_{zs} = \frac{\eta}{L}; \qquad z_s = L; \qquad x_s = \eta; \qquad t_s = t_o \tag{3.7-20}$$

Note that the z-velocity scale in equation (3.7-20) is within a multiplicative constant of O(1) of the surface velocity for fully developed laminar film flow down a vertical plate. Note also that we found that $\partial\eta/\partial z$ scales with η/L, which is the ratio of the transverse and longitudinal length scales. That is, although we introduced η_{zs} as the scale for $\partial\eta/\partial z$ to allow for the possibility that the latter might not scale with the ratio of the length scales, scaling analysis justified what might appear to be an obvious choice for this scale. However, scaling analysis also indicated that the proper scale for $\partial\eta/\partial t$ is $\rho g \eta^3/\mu L$ rather than η/t_o, which would be the intuitive choice. In fact, one could introduce separate scales for all the derivatives in the describing equations and then use the systematic scaling method to determine these. However, this is cumbersome in practice for the more complex describing equations encountered in scientific research and engineering practice. Hence, if a more limited set of scales confined to the dependent and independent variables but not their derivatives is chosen, one can rely on the forgiving nature of scaling to discern that separate scales might need to be introduced on one or more derivatives in order to achieve o(1) scaling.

When the scales defined by equation (3.7-20) are substituted into equations (3.7-11) through 3.7-19), we obtain

$$\frac{\eta^2}{\nu t_o}\frac{\partial u_z^*}{\partial t^*} + \text{Re}\frac{\eta}{L}u_z^*\frac{\partial u_z^*}{\partial z^*} + \text{Re}\frac{\eta}{L}u_x^*\frac{\partial u_z^*}{\partial x^*} = -\frac{\eta^2}{L^2}\frac{\partial P^*}{\partial z^*} + \frac{\eta^2}{L^2}\frac{\partial^2 u_z^*}{\partial z^{*2}} + \frac{\partial^2 u_z^*}{\partial x^{*2}} + 1 \tag{3.7-21}$$

$$\frac{\eta^2}{\nu t_o}\frac{\partial u_x^*}{\partial t^*} + \text{Re}\frac{\eta}{L}u_z^*\frac{\partial u_x^*}{\partial z^*} + \text{Re}\frac{\eta}{L}u_x^*\frac{\partial u_x^*}{\partial x^*} = -\frac{\partial P^*}{\partial x^*} + \frac{\eta^2}{L^2}\frac{\partial^2 u_x^*}{\partial z^{*2}} + \frac{\partial^2 u_x^*}{\partial x^{*2}} \tag{3.7-22}$$

$$\frac{\partial u_z^*}{\partial z^*} + \frac{\partial u_x^*}{\partial x^*} = 0 \tag{3.7-23}$$

$$u_z^* = 0, \quad u_x^* = 0, \quad \eta^* = \infty \qquad \text{at} \quad t^* = 0 \tag{3.7-24}$$

$$u_z^* = 0, \quad u_x^* = 0 \qquad \text{at} \quad x^* = 0 \tag{3.7-25}$$

$$\left.\begin{aligned} &\frac{4\eta}{L}\frac{\partial u_z^*}{\partial z^*}\sin\theta\cos\theta - \left(\frac{\eta^2}{L^2}\frac{\partial u_x^*}{\partial z^*} + \frac{\partial u_z^*}{\partial x^*}\right)\cos 2\theta = 0 \\ \frac{\eta}{L}P^* + 2&\frac{\eta}{L}\frac{\partial u_z^*}{\partial z^*}\cos 2\theta - 2\left(\frac{\eta^2}{L^2}\frac{\partial u_x^*}{\partial z^*} + \frac{\partial u_z^*}{\partial x^*}\right)\sin\theta\cos\theta = 0 \end{aligned}\right\} \quad \text{at} \quad x^* = 1 \tag{3.7-26}$$

$$u_z^* = 0, \quad u_x^* = 0, \quad \eta^* = 0 \qquad \text{at} \quad x^* = 0 \tag{3.7-27}$$

$$u_z^* = f_1^*(x^*, t^*), \quad u_x^* = f_2^*(x^*, t^*) \qquad \text{at} \quad z^* = 1 \tag{3.7-28}$$

$$u_x^* - u_z^*\left(\frac{\partial \eta}{\partial z}\right)^* = \left(\frac{\partial \eta}{\partial t}\right)^* \qquad \text{at} \quad x^* = 1 \tag{3.7-29}$$

where $\nu \equiv \mu/\rho$ is the kinematic viscosity and $\text{Re} \equiv \rho u_{zs}\eta/\mu = \rho^2 g\eta^3/\mu^2$ is the Reynolds number. The dimensionless group $\eta^2/\nu t_o$ is a measure of the ratio of the characteristic time for the diffusion of vorticity[7] to the observation time. We see from the above that our describing equations can be simplified significantly if we can make the creeping-flow approximation; that is, if $\text{Re} = \text{o}(0.01)$, we can ignore the nonlinear inertia terms in equations (3.7-21) and (3.7-22). These equations can be further simplified if we can make the lubrication-flow approximation; that is, if $\eta^2/L^2 = \text{o}(0.01)$ as well as $\text{Re} = \text{o}(0.01)$, we can ignore the axial diffusion of vorticity terms in equations (3.7-21) and (3.7-22). If $\eta^2/L^2 = \text{o}(0.01)$, the tangential and normal stress boundary conditions given by equation (3.7-26) also simplify somewhat. If the aspect ratio satisfies the condition $\eta/L = \text{o}(0.01)$ (a more demanding condition than that required for lubrication flow), the *quasi-parallel-flow approximation* can be made; that is, the effect of surface curvature on the flow can be ignored. Finally, if $\eta^2/\nu t_o = \text{o}(0.01)$, which implies that the observation time is long in comparison to the characteristic time for the diffusion of vorticity, the quasi-steady-state assumption can be made. In summary, the following conditions justify simplifying the describing equations for this flow:

$$\text{Re} = \frac{\rho^2 g\eta^3}{\mu^2} \ll 1 \Rightarrow \text{creeping-flow approximation} \tag{3.7-30}$$

$$\text{Re} \ll 1 \quad \text{and} \quad \frac{\eta^2}{L^2} \ll 1 \Rightarrow \text{lubrication-flow approximation} \tag{3.7-31}$$

[7]The vorticity is defined to be the curl of the velocity field, $\nabla \times \vec{u}$; if the curl of the equations of motion is taken, the vorticity appears as a diffused quantity, the transport coefficient being the kinematic viscosity; for this reason we refer to the diffused quantity in the equations of motion as the vorticity.

$$\frac{\eta}{L} \ll 1 \Rightarrow \text{quasi-parallel-flow approximation} \qquad (3.7\text{-}32)$$

$$\frac{\eta^2}{\nu t_o} \ll 1 \Rightarrow \text{quasi-steady-state approximation} \qquad (3.7\text{-}33)$$

If the conditions in equations (3.7-30) through (3.7-33) apply, equations (3.7-21) through (3.7-29) simplify to (step 8)

$$0 = \frac{\partial^2 u_z^*}{\partial x^{*2}} + 1 \qquad (3.7\text{-}34)$$

$$\frac{\partial u_z^*}{\partial z^*} + \frac{\partial u_x^*}{\partial x^*} = 0 \qquad (3.7\text{-}35)$$

$$\eta^* = \infty \quad \text{at} \quad t^* = 0 \qquad (3.7\text{-}36)$$

$$u_z^* = 0, \quad u_x^* = 0 \quad \text{at} \quad x^* = 0 \qquad (3.7\text{-}37)$$

$$\frac{\partial u_z^*}{\partial x^*} = 0 \quad \text{at} \quad x^* = 1 \qquad (3.7\text{-}38)$$

$$u_x^* - u_z^* \left(\frac{\partial \eta}{\partial z}\right)^* = \left(\frac{\partial \eta}{\partial t}\right)^* \quad \text{at} \quad x^* = 1 \qquad (3.7\text{-}39)$$

An estimate of the instantaneous local film thickness η can be obtained from η_{ts}, the scale for $\partial\eta/\partial t$; that is,

$$\eta_{ts} \cong \frac{\partial \eta}{\partial t} \cong -\frac{\rho g \eta^3}{\mu L} \Rightarrow \eta^2 \cong \frac{\mu L}{2\rho g t} \qquad (3.7\text{-}40)$$

Note that the negative sign was inserted in equation (3.7-40) because η decreases with increasing t; that is, only magnitudes are involved in scaled variables. This estimate for η can be used in evaluating the criteria described by equations (3.7-30) through (3.7-33) to assess the applicability of the various assumptions that can be invoked to simplify the describing equations for this flow. Note that the estimate for η given by equation (3.7-40) is within a multiplicative constant of $O(1)$ of that obtained by actually solving the describing equations. This is a particular advantage of scaling analysis; that is, it can provide an estimate of the answer we seek by solving the describing equations.

Let us assume that the approximations indicated by equations (3.7-30) through (3.7-33) are justified, at least over a reasonable range of t_o and L. The solution to the resulting simplified describing equations given by equations (3.7-34) through (3.7-39) is straightforward. One can integrate equation (3.7-34) analytically subject to the boundary conditions given by equations (3.7-37) and (3.7-38). The resulting solution for axial velocity u_z^* is the same as that for fully developed film flow down a vertical plate, but with the local film thickness $\eta(z^*, t^*)$ replacing the constant film thickness; hence, the t^*- and z^*- dependence of u_z^* enters implicitly through the boundary conditions. The transverse velocity u_x^* can then be obtained by substituting this solution for u_z^* into the continuity equation given by equation (3.7-35) and integrating across the flow. The solutions for u_z^* and u_x^* then can be substituted into the kinematic surface condition given by equation (3.7-39) to obtain a partial differential equation in terms of only one dependent variable, $\eta(z^*, t^*)$. This can be solved for the instantaneous local film thickness $\eta(z^*, t^*)$ either by the method of separation of variables or via the method of combination of variables (i.e., a similarity solution).

An ad hoc solution to this flow problem was developed by van Rossum[8] and is given by Bird et al.[9] Van Rossum assumes quasi-parallel flow and hence uses the axial velocity profile for fully developed film flow down a vertical plate. He then carries out an integral mass balance on a control volume that consists of a differential axial length and the entire cross-section of the film in which he uses the average axial velocity. The solution he obtains for $\eta(z, t)$ is within a multiplicative constant of O(1) of that obtained via scaling analysis given by equation (3.7-40). The scaling analysis developed here provides a systematic method for justifying the assumptions used in the ad hoc solution of van Rossum. Moreover, scaling analysis leads to an improved solution since the exact form of the kinematic surface condition (integral mass balance) is solved that incorporates contributions from both the axial and transverse velocity components.

3.8 POROUS MEDIA FLOW

Thus far, all the flows that we have considered have involved homogeneous media, that is, media consisting of a single phase. Here we consider steady-state fully developed pressure-driven flow of a Newtonian fluid with constant physical properties through a heterogeneous medium consisting of a microporous solid contained within a cylindrical tube of radius R, as shown in Figure 3.8-1.

Flow through a porous medium is described by a modified form of the equations of motion in order to accommodate the heterogeneity introduced by the microporous

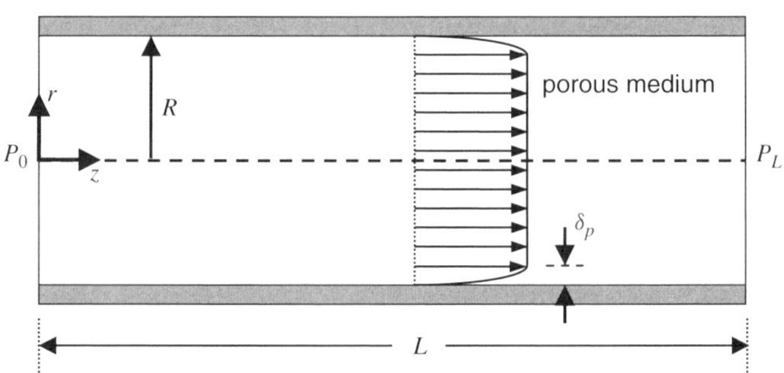

Figure 3.8-1 Steady-state fully developed pressure-driven flow of a viscous Newtonian fluid with constant physical properties through a heterogeneous medium consisting of a microporous solid contained within a cylindrical tube of radius R; the axial profile is shown that satisfies the no-slip condition at the tube wall, whose region of influence is δ_p.

[8] J. J. van Rossum, *Appl. Sci. Res.*, **A7**, 121–144 (1958).

[9] Bird et al., *Transport Phenomena*, 2nd ed., Wiley, Hoboken, NJ. 2002, Problem 2D.2, pp. 73–74.

solid. The local Reynolds number that characterizes flow through the pores is well into the creeping-flow regime, due to the small pore size; hence, one can safely ignore the inertia terms in the equations of motion. Moreover, the resistance to flow offered by the pore walls within the porous medium is generally much greater than that of the solid walls that bound the entire porous medium (the wall of the cylindrical tube in the present problem); the former can be described by Darcy's law, whereas the latter is usually neglected. However, the effect of the boundaries of the porous medium cannot be ignored within some region of influence adjacent to the boundaries. We seek to use scaling analysis to determine the criterion for when the effect of the lateral boundaries on the flow through the porous medium can be neglected and to determine the thickness of the region of influence within which the effect of drag on the lateral boundaries on the flow must be considered.

The appropriately simplified form of equations (E.2-1) and (E.2-3) in the Appendices for steady-state creeping flow through a porous medium are given by (step 1)

$$0 = -\frac{\partial P}{\partial z} - \frac{\mu}{k_p}\widehat{u}_z + \mu\frac{1}{r}\frac{d}{dr}\left(r\frac{d\widehat{u}_z}{dr}\right) \tag{3.8-1}$$

$$0 = -\frac{\partial P}{\partial r} \tag{3.8-2}$$

$$\widehat{u}_z = 0 \quad \text{at} \quad r = R \tag{3.8-3}$$

$$\frac{d\widehat{u}_z}{dr} = 0 \quad \text{at} \quad r = 0 \tag{3.8-4}$$

$$P = P_0 \quad \text{at} \quad z = 0 \tag{3.8-5}$$

where $\widehat{u}_z$ is the superficial flow velocity through the porous medium (i.e., the flow velocity averaged over a differential cross-sectional area of the heterogeneous medium in contrast to the flow velocity through a pore) and k_p is the Darcy permeability of the microporous solid. Note that we have ignored any effect of hydrostatic pressure in equation (3.8-2). Equations (3.8-1) and (3.8-2) imply that the axial pressure gradient is a constant. Hence, our describing equations simplify to

$$0 = \frac{\Delta P}{L} - \frac{\mu}{k_p}\widehat{u}_z + \mu\frac{1}{r}\frac{d}{dr}\left(r\frac{d\widehat{u}_z}{dr}\right) \tag{3.8-6}$$

$$\widehat{u}_z = 0 \quad \text{at} \quad r = R \tag{3.8-7}$$

$$\frac{d\widehat{u}_z}{dr} = 0 \quad \text{at} \quad r = 0 \tag{3.8-8}$$

where $\Delta P \equiv P_0 - P_L$, in which P_0 and P_L are the pressures at $z = 0$ and $z = L$, respectively. The equations above differ from the conventional equations of motion by the inclusion of the Darcy term, which accounts for the resistance to flow offered by the porous medium.

Introduce the following scale factors and dimensionless variables (steps 2, 3, and 4):

$$\widehat{u}_z^* \equiv \frac{\widehat{u}_z}{\widehat{u}_{zs}}; \qquad r^* \equiv \frac{r}{r_s} \tag{3.8-9}$$

Substitute these variables into the describing equations and divide through by the dimensional coefficient of a term that must be retained to obtain the following set of dimensionless describing equations (steps 5 and 6):

$$0 = 1 - \frac{\mu L \widehat{u}_{zs}}{k_p \Delta P} \widehat{u}_z^* + \frac{\mu L \widehat{u}_{zs}}{\Delta P r_s} \frac{1}{r^*} \frac{d}{dr^*} \left(r^* \frac{d\widehat{u}_z^*}{dr^*} \right) \tag{3.8-10}$$

$$\widehat{u}_z^* = 0 \qquad \text{at} \quad r^* = \frac{R}{r_s} \tag{3.8-11}$$

$$\frac{d\widehat{u}_z^*}{dr^*} = 0 \qquad \text{at} \quad r^* = 0 \tag{3.8-12}$$

The radial length scale r_s is determined by setting the dimensionless group in equation (3.8-11) equal to 1. Since pressure causes the flow and the porous medium is assumed to offer the primary resistance to flow, the velocity scale $\widehat{u}_{zs}$ is determined by setting the dimensionless group in the Darcy term in equation (3.8-10) equal to 1. These considerations then determine the following scale and reference factors (step 7):

$$r_s = R; \qquad \widehat{u}_{zs} = \frac{k_p \Delta P}{\mu L} \tag{3.8-13}$$

Note that the velocity scale in equation (3.8-13) is the axial velocity that would be predicted if just the pressure and Darcy flow terms were retained in equation (3.8-10).

When the scales defined by equation (3.8-13) are substituted into equations (3.8-10) through (3.8-12), we obtain

$$0 = 1 - \widehat{u}_z^* + \frac{k_p}{R^2} \frac{1}{r^*} \frac{d}{dr^*} \left(r^* \frac{d\widehat{u}_z^*}{dr^*} \right) \tag{3.8-14}$$

$$\widehat{u}_z^* = 0 \qquad \text{at} \quad r^* = 1 \tag{3.8-15}$$

$$\frac{d\widehat{u}_z^*}{dr^*} = 0 \qquad \text{at} \quad r^* = 0 \tag{3.8-16}$$

Hence, we see that the effects of viscous drag at the inner wall of the cylindrical tube can be neglected if the following condition is satisfied (step 8):

$$\frac{k_p}{R^2} \ll 1 \Rightarrow \text{viscous drag at boundary of porous medium can be neglected} \tag{3.8-17}$$

The velocity obtained by solving equation (3.8-14) while dropping the last term will be accurate provided that the condition indicated in equation (3.8-17) is satisfied; that is, if $k_p/R^2 = \text{o}(0.01)$ and provided that one is not interested in predicting the velocity very close to the tube wall or the viscous drag at the wall. It is of interest to determine the region of influence δ_p, wherein the effect of the tube wall cannot be ignored. To do this it is necessary to translate the coordinate system to the tube wall via the transformation $\tilde{r} \equiv R - r$. Since we are considering a region of influence, $\tilde{r}_s = \delta_p$; the velocity scale still is given by equation (3.8-13). The resulting transformed dimensionless describing equations are given by

$$0 = 1 - \widehat{u}_z^* + \frac{k_p}{\delta_p^2}\frac{1}{R/\delta_p - \tilde{r}^*}\frac{d}{dr^*}\left[\left(\frac{R}{\delta_p} - \tilde{r}^*\right)\frac{d\widehat{u}_z^*}{d\tilde{r}^*}\right] \tag{3.8-18}$$

$$\widehat{u}_z^* = 0 \qquad \text{at} \quad \tilde{r}^* = 0 \tag{3.8-19}$$

$$\frac{d\widehat{u}_z^*}{dr^*} = 0 \qquad \text{at} \quad \tilde{r}^* = \frac{R}{\delta_p} \tag{3.8-20}$$

Within the region of influence the last term in equation (3.8-18) cannot be ignored; hence, we set the dimensionless group multiplying this term equal to 1 to determine the thickness of the region of influence:

$$\frac{k_p}{\delta_p^2} = 1 \Rightarrow \delta_p = \sqrt{k_p} \tag{3.8-21}$$

For typical Darcy permeabilities, δ_p is on the order of 10 to 100 pore diameters. Hence, in most cases one can ignore the effect of the system boundaries on the relationship between the average superficial velocity through the porous medium and the pressure gradient used. The curvature effects can be ignored within the region of influence near the wall if the following criterion is satisfied:

$$\frac{R}{\delta_p} \gg 1 \Rightarrow \text{curvature effects can be ignored} \tag{3.8-22}$$

in which case the describing equations simplify to

$$0 = 1 - \widehat{u}_z^* + \frac{d^2\widehat{u}_z^*}{dr^{*2}} \tag{3.8-23}$$

$$\widehat{u}_z^* = 0 \qquad \text{at} \quad \tilde{r}^* = 0 \tag{3.8-24}$$

$$\frac{d\widehat{u}_z^*}{dr^*} = 0 \qquad \text{as} \quad \tilde{r}^* \to \infty \tag{3.8-25}$$

3.9 COMPRESSIBLE FLUID FLOW

In the examples we have considered thus far, we have assumed that the flowing fluid was incompressible. This is certainly a tenuous assumption for gas flows and can be questioned even for complex liquid flows.[10] In this example we use systematic scaling analysis to assess when a flow can be considered to be incompressible. This example will again involve introducing separate scale factors for a spatial derivative rather than assuming that it scales with the ratio of the characteristic value for the dependent variable divided by the characteristic length.

Consider the steady-state pressure-driven flow in a cylindrical tube of a compressible gas whose other physical properties will be assumed to be constant as shown in Figure 3.9-1. Due to the compressibility, the density of the gas will decrease due to the pressure drop in the axial direction; it also changes in the radial direction, however, this effect is usually quite small (although it will be included in this scaling analysis). Correspondingly, there will be an increase in the axial velocity, thereby implying that this is a developing flow with both nonzero axial and radial velocity components. Hence, we must use the appropriate form of the steady-state equations of motion that allow for a compressible flow. We use scaling analysis to determine the conditions for which this flow can be considered to be incompressible and fully developed. Although we allow only for a variable density in this problem, the manner in which we use scaling to assess when this

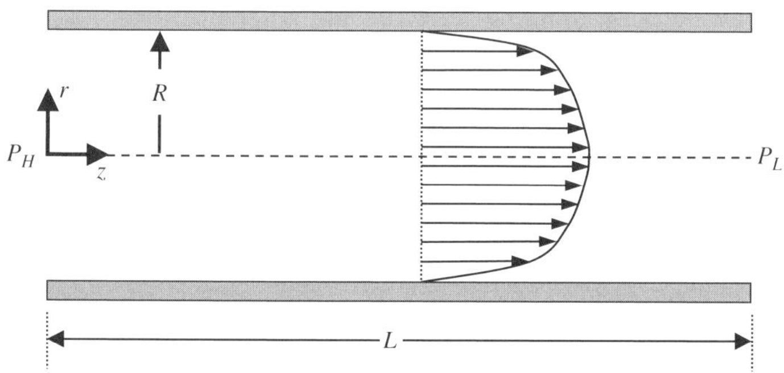

Figure 3.9-1 Steady-state pressure-driven flow in a cylindrical tube of radius R and length L of a compressible gas for which the other physical properties are assumed to be constant.

[10]Examples of complex fluids include the flow of microemulsions, proteins, micellar solutions, and suspensions, as well as other nonhomogeneous liquids.

property can be assumed to be constant can be used to handle other variable properties, such as the viscosity.

The describing equations in terms of the components of the viscous stress tensor are obtained by appropriately simplifying the continuity equations and equations of motion given by equations (C.2-1), (D.2-1), and (D.2-3) in the Appendices:

$$\rho u_r \frac{\partial u_z}{\partial r} + \rho u_z \frac{\partial u_z}{\partial z} = -\frac{\partial P}{\partial z} - \frac{1}{r}\frac{\partial}{\partial r}(r\tau_{rz}) - \frac{\partial \tau_{zz}}{\partial z} \tag{3.9-1}$$

$$\rho u_r \frac{\partial u_r}{\partial r} + \rho u_z \frac{\partial u_r}{\partial z} = -\frac{\partial P}{\partial r} - \frac{1}{r}\frac{\partial}{\partial r}(r\tau_{rr}) - \frac{\tau_{\theta\theta}}{r} - \frac{\partial \tau_{rz}}{\partial z} \tag{3.9-2}$$

$$\frac{1}{r}\frac{\partial}{\partial r}(\rho r u_r) + \frac{\partial}{\partial z}(\rho u_z) = 0 \tag{3.9-3}$$

where τ_{ij} is the viscous shear stress associated with the transfer of j-momentum in the i-direction. These equations have been given in terms of the components of the viscous stress tensor to emphasize that the term $\tau_{\theta\theta}$ is not zero even though this is an axisymmetric flow. To complete the specification of the describing equations, we need to provide an equation of state that relates the density to the pressure (the temperature is assumed to be constant) and need to specify appropriate boundary conditions; this will be done after we have rearranged the equations above into a more convenient form.

When the appropriate components of the viscous stress tensor given by equations (D.2-4) through (D.2-9) in the Appendices are substituted into equations (3.9-1) and (3.9-2) and the resulting equations simplified using the continuity equation given by equation (3.9-3), we obtain

$$\rho u_r \frac{\partial u_z}{\partial r} + \rho u_z \frac{\partial u_z}{\partial z} = -\frac{\partial P}{\partial z} + \frac{\mu}{r}\frac{\partial}{\partial r}\left(r\frac{\partial u_z}{\partial r}\right) - \frac{\mu}{3}\frac{\partial}{\partial z}\left(\frac{u_r}{\rho}\frac{\partial \rho}{\partial r} + \frac{u_z}{\rho}\frac{\partial \rho}{\partial z}\right) + \mu\frac{\partial^2 u_z}{\partial z^2} \tag{3.9-4}$$

$$\rho u_r \frac{\partial u_r}{\partial r} + \rho u_z \frac{\partial u_r}{\partial z} = -\frac{\partial P}{\partial r} + \mu\frac{\partial}{\partial r}\left[\frac{1}{r}\frac{\partial}{\partial r}(r u_r)\right] - \frac{\mu}{3}\frac{\partial}{\partial r}\left(\frac{u_r}{\rho}\frac{\partial \rho}{\partial r} + \frac{u_z}{\rho}\frac{\partial \rho}{\partial z}\right) + \mu\frac{\partial^2 u_r}{\partial z^2} \tag{3.9-5}$$

The derivatives of the density can be expressed in terms of derivatives of the pressure as follows:

$$\frac{\partial \rho}{\partial r} = \left.\frac{\partial \rho}{\partial P}\right|_T \frac{\partial P}{\partial r} = \frac{\gamma}{c^2}\frac{\partial P}{\partial r}; \qquad \frac{\partial \rho}{\partial z} = \left.\frac{\partial \rho}{\partial P}\right|_T \frac{\partial P}{\partial z} = \frac{\gamma}{c^2}\frac{\partial P}{\partial z} \tag{3.9-6}$$

in which c is the speed of sound in the gas and γ is the ratio of the heat capacity at constant pressure to that at constant volume. To assess the effect of pressure on the density, we represent the density in terms of a Taylor series expansion about some reference density ρ_0, which is chosen to be at the downstream low pressure P_L:

$$\rho = \rho_0 + \left.\frac{\partial \rho}{\partial P}\right|_T (P - P_L) + \mathrm{O}(P - P_L)^2 = \rho_0 + \frac{\gamma}{c^2}(P - P_L) + \mathrm{O}(P - P_L)^2 \tag{3.9-7}$$

Note that it is sufficient to represent the density as a Taylor series truncated at the second term since we need consider only small variations in the density. That is, because we seek to determine the conditions for which incompressible flow can be assumed, we need explore only small variations in the density since large variations in the density most certainly will be associated with compressible flow. Hence, our describing equations and associated boundary conditions are given by (step 1)

$$\rho u_r \frac{\partial u_z}{\partial r} + \rho u_z \frac{\partial u_z}{\partial z} = -\frac{\partial P}{\partial z} + \frac{\mu}{r}\frac{\partial}{\partial r}\left(r\frac{\partial u_z}{\partial r}\right) - \frac{\mu}{3}\frac{\partial}{\partial z}\left(\frac{\gamma u_r}{\rho c^2}\frac{\partial P}{\partial r} + \frac{\gamma u_z}{\rho c^2}\frac{\partial P}{\partial z}\right) + \mu\frac{\partial^2 u_z}{\partial z^2} \tag{3.9-8}$$

$$\rho u_r \frac{\partial u_r}{\partial r} + \rho u_z \frac{\partial u_r}{\partial z} = -\frac{\partial P}{\partial r} + \mu\frac{\partial}{\partial r}\left[\frac{1}{r}\frac{\partial}{\partial r}(r u_r)\right] - \frac{\mu}{3}\frac{\partial}{\partial r}\left(\frac{\gamma u_r}{\rho c^2}\frac{\partial P}{\partial r} + \frac{\gamma u_z}{\rho c^2}\frac{\partial P}{\partial z}\right) + \mu\frac{\partial^2 u_r}{\partial z^2} \tag{3.9-9}$$

$$\frac{1}{r}\frac{\partial}{\partial r}(\rho r u_r) + \frac{\partial}{\partial z}(\rho u_z) = 0 \tag{3.9-10}$$

$$\rho = \rho_0 + \frac{\gamma}{c^2}(P - P_L) \tag{3.9-11}$$

$$\frac{\partial u_z}{\partial r} = 0, \quad u_r = 0, \quad P = p(z) \quad \text{at} \quad r = 0 \tag{3.9-12}$$

$$u_z = 0, \quad u_r = 0 \quad \text{at} \quad r = R \tag{3.9-13}$$

$$u_z = f_1(r), \quad u_r = f_2(r), \quad P = P_H \quad \text{at} \quad z = 0 \tag{3.9-14}$$

$$u_z = g_1(r), \quad u_r = g_2(r) \quad \text{at} \quad z = L \tag{3.9-15}$$

in which $p(z)$, $f_1(r)$, $g_1(r)$, $f_2(r)$, and $g_2(r)$ are unspecified functions that in principle would need to be known in order to integrate the full set of describing equations. If the conditions that we seek to determine for assuming fully developed incompressible flow are satisfied, it will not be necessary to know these unspecified functions.

Introduce the following scale factors, reference factors, and dimensionless variables (steps 2, 3, and 4):

$$u_z^* \equiv \frac{u_z}{u_{zs}}; \qquad u_r^* \equiv \frac{u_r}{u_{rs}}; \qquad P^* \equiv \frac{P - P_r}{P_s}; \qquad \left(\frac{\partial P}{\partial r}\right)^* \equiv \frac{1}{P_{rs}}\frac{\partial P}{\partial r};$$
$$\rho^* \equiv \frac{\rho}{\rho_s}; \qquad r^* \equiv \frac{r}{r_s}; \qquad z^* \equiv \frac{z}{z_s} \tag{3.9-16}$$

We again allow for a reference pressure P_r in our definition of the dimensionless pressure since the latter is not naturally referenced to zero. Note that we must also scale the density in this problem since it is one of the dependent variables. However, we do not need to introduce a reference factor for the density even though it is not referenced to zero. The reason for this is that we are considering only small variations in density; that is, the density does not vary significantly in either coordinate direction. We have introduced a scale for the radial derivative of the pressure denoted by P_{rs} since we do not anticipate that this will scale in the same way as the axial pressure gradient. The question might arise as to how one knows whether to scale a derivative as the ratio of some dependent variable scale divided by some independent variable scale or to introduce a separate scale for the entire derivative. The answer is contained simply in the forgiving nature of scaling. That is, if we were to assume that the radial pressure derivative scales as the pressure scale P_s divided by the radial length scale r_s, we would find that the dimensionless group in front of the dimensionless radial pressure derivative was much larger than that of any other term in the r-component of the equations of motion. This clearly would indicate that we scaled incorrectly. Hence, determining whether a derivative needs its own scale is often a matter of trial and error. If any term is scaled incorrectly, the forgiving nature of scaling will indicate a contradiction in the dimensionless equations. One then rescales until a self-consistent set of dimensionless equations is obtained; that is, a system of equations for which balancing terms are of O(1) and all other terms, including those multiplied by dimensionless groups, are of o(1). The consequence of not introducing a separate scale for the radial pressure derivative is explored in Practice Problem 3.P.31.

Substitute the variables defined in equation (3.9-16) into the describing equations and divide through by the dimensional coefficient of a term that must be retained to obtain the following set of dimensionless describing equations (steps 5 and 6):

$$\frac{\rho_s u_{rs} r_s}{\mu}\rho^* u_r^* \frac{\partial u_z^*}{\partial r^*} + \frac{\rho_s u_{zs} r_s^2}{\mu z_s}\rho^* u_z^* \frac{\partial u_z^*}{\partial z^*} = -\frac{P_s r_s^2}{\mu u_{zs} z_s}\frac{\partial P^*}{\partial z^*} + \frac{1}{r^*}\frac{\partial}{\partial r^*}\left(r^* \frac{\partial u_z^*}{\partial r^*}\right)$$
$$-\frac{1}{3}\frac{\gamma u_{rs} P_{rs} r_s^2}{\rho_s c^2 u_{zs} z_s}\frac{\partial}{\partial z^*}\left[\frac{u_r^*}{\rho^*}\left(\frac{\partial P}{\partial r}\right)^*\right] - \frac{1}{3}\frac{\gamma P_s r_s^2}{\rho_s c^2 z_s^2}\frac{\partial}{\partial z^*}\left(\frac{u_z^*}{\rho^*}\frac{\partial P^*}{\partial z^*}\right) + \frac{r_s^2}{z_s^2}\frac{\partial^2 u_z^*}{\partial z^{*2}} \tag{3.9-17}$$

$$\frac{\rho_s u_{rs} r_s}{\mu}\rho^* u_r^* \frac{\partial u_r^*}{\partial r^*} + \frac{\rho_s u_{zs} r_s^2}{\mu z_s}\rho^* u_z^* \frac{\partial u_r^*}{\partial z^*} = -\frac{P_{rs} r_s^2}{\mu u_{rs}}\left(\frac{\partial P}{\partial r}\right)^* + \frac{\partial}{\partial r^*}\left[\frac{1}{r^*}\frac{\partial}{\partial r^*}(r^* u_r^*)\right]$$
$$-\frac{1}{3}\frac{\gamma P_{rs} r_s}{\rho_o c^2}\frac{\partial}{\partial r^*}\left[\frac{u_r^*}{\rho^*}\left(\frac{\partial P}{\partial r}\right)^*\right] - \frac{1}{3}\frac{\gamma u_{zs} P_s r_s}{\rho_o c^2 u_{rs} z_s}\frac{\partial}{\partial r^*}\left(\frac{u_z^*}{\rho^*}\frac{\partial P^*}{\partial z^*}\right) + \frac{r_s^2}{z_s^2}\frac{\partial^2 u_r^*}{\partial z^{*2}} \tag{3.9-18}$$

$$\frac{1}{r^*}\frac{\partial}{\partial r^*}(\rho^* r^* u_r^*) + \frac{u_{zs} r_s}{u_{rs} z_s}\frac{\partial}{\partial z^*}(\rho^* u_z^*) = 0 \tag{3.9-19}$$

$$\rho^* = \frac{\rho_0}{\rho_s} + \frac{\gamma P_s}{c^2 \rho_s}\left(P^* + \frac{P_r - P_L}{P_s}\right) \tag{3.9-20}$$

$$\frac{\partial u_z^*}{\partial r^*} = 0, \quad u_r^* = 0, \quad P^* = \frac{p(z^*) - P_r}{P_s} \quad \text{at} \quad r^* = 0 \tag{3.9-21}$$

$$u_z^* = 0, \quad u_r^* = 0 \quad \text{at} \quad r^* = \frac{R}{r_s} \tag{3.9-22}$$

$$u_z^* = f_1^*(r^*), \quad u_r^* = f_2^*(r^*), \quad P^* = \frac{P_H - P_r}{P_s} \quad \text{at} \quad z^* = 0 \tag{3.9-23}$$

$$u_z^* = g_1^*(r^*), \quad u_r^* = g_2^*(r^*) \quad \text{at} \quad z^* = \frac{L}{z_s} \tag{3.9-24}$$

The radial and axial length scales, r_s and z_s, can be bounded between zero and 1 by setting the dimensionless groups in equations (3.9-22) and (3.9-24) equal to 1. The dimensionless pressure can be bounded between zero and 1 by setting the dimensionless group containing the reference pressure in equation (3.9-20) equal to zero and the dimensionless group containing the pressure scale in equation (3.9-23) equal to 1. Since pressure causes the axial flow and the principal viscous term (i.e., involving the second-order radial derivative) must be retained, the velocity scale u_{zs} is determined by setting the dimensionless group multiplying the pressure term in equation (3.9-17) equal to 1. Since compressibility implies a developing flow, the dimensionless group in the continuity equation given by (3.9-19) must be equal to 1. The density scale is obtained by setting the dimensionless group in the principal term in equation (3.9-20) equal to 1. Since the radial pressure gradient causes the flow in the radial direction, the radial pressure gradient scale P_{rs} is determined by setting the dimensionless group multiplying the pressure term in equation (3.9-18) equal to 1. These considerations then determine the following scale and reference factors (step 7):

$$r_s = R; \quad z_s = L; \quad P_r = P_L; \quad P_s = P_H - P_L \equiv \Delta P; \quad u_{zs} = \frac{R^2 \Delta P}{\mu L};$$
$$u_{rs} = \frac{R^3 \Delta P}{\mu L^2}; \quad \rho_s = \rho_0; \quad P_{rs} = \frac{R \Delta P}{L^2} \tag{3.9-25}$$

Note that the axial velocity scale u_{zs} is within a multiplicative constant of O(1) of the value for the maximum velocity in fully developed incompressible flow in a cylindrical tube.

When the scales defined by equation (3.9-25) are substituted into equations (3.9-17) through (3.9-24), we obtain

$$\mathrm{Re}\frac{R}{L}u_r^*\frac{\partial u_z^*}{\partial r^*}+\mathrm{Re}\frac{R}{L}u_z^*\frac{\partial u_z^*}{\partial z^*}=-\frac{\partial P^*}{\partial z^*}+\frac{1}{r^*}\frac{\partial}{\partial r^*}\left(r^*\frac{\partial u_z^*}{\partial r^*}\right)$$
$$-\frac{1}{3}\gamma\frac{\mathrm{Ma}^2}{\mathrm{Re}}\frac{R^3}{L^3}\frac{\partial}{\partial z^*}\left[\frac{u_r^*}{\rho^*}\left(\frac{\partial P}{\partial r}\right)^*\right]-\frac{1}{3}\gamma\frac{\mathrm{Ma}^2}{\mathrm{Re}}\frac{R}{L}\frac{\partial}{\partial z^*}\left(\frac{u_z^*}{\rho^*}\frac{\partial P^*}{\partial z^*}\right)+\frac{R^2}{L^2}\frac{\partial^2 u_z^*}{\partial z^{*2}} \tag{3.9-26}$$

$$\mathrm{Re}\frac{R}{L}u_r^*\frac{\partial u_r^*}{\partial r^*}+\mathrm{Re}\frac{R}{L}u_z^*\frac{\partial u_r^*}{\partial z^*}=-\left(\frac{\partial P}{\partial r}\right)^*+\frac{\partial}{\partial r^*}\left[\frac{1}{r^*}\frac{\partial}{\partial r^*}(r^*u_r^*)\right]$$
$$-\frac{1}{3}\gamma\frac{\mathrm{Ma}^2}{\mathrm{Re}}\frac{R}{L}\frac{\partial}{\partial r^*}\left[\frac{u_r^*}{\rho^*}\left(\frac{\partial P}{\partial r}\right)^*\right]-\frac{1}{3}\gamma\frac{\mathrm{Ma}^2}{\mathrm{Re}}\frac{L}{R}\frac{\partial}{\partial r^*}\left(\frac{u_z^*}{\rho^*}\frac{\partial P^*}{\partial z^*}\right)+\frac{R^2}{L^2}\frac{\partial^2 u_r^*}{\partial z^*} \tag{3.9-27}$$

$$\frac{1}{r^*}\frac{\partial}{\partial r^*}(\rho^*r^*u_r^*)+\frac{\partial}{\partial z^*}(\rho^*u_z^*)=0 \tag{3.9-28}$$

$$\rho^*=1+\gamma\frac{\mathrm{Ma}^2}{\mathrm{Re}}\frac{L}{R}P^* \tag{3.9-29}$$

$$\frac{\partial u_z^*}{\partial r^*}=0,\quad u_r^*=0,\quad P^*=\frac{p^*(z^*)-P_L}{\Delta P}\qquad \text{at}\quad r^*=0 \tag{3.9-30}$$

$$u_z^*=0,\quad u_r^*=0\qquad \text{at}\quad r^*=1 \tag{3.9-31}$$

$$u_z^*=f_1^*(r^*),\quad u_r^*=f_2^*(r),\ P^*=1\qquad \text{at}\quad z^*=0 \tag{3.9-32}$$

$$u_z^*-g_1^*(r^*),\quad u_r^*=g_2^*(r^*)\qquad \text{at}\quad z^*=1 \tag{3.9-33}$$

where $\mathrm{Re}\equiv R\rho_s v_{zs}/\mu$ is the Reynolds number and $\mathrm{Ma}\equiv u_{zs}/c$ is the Mach number; the latter is the ratio of the characteristic velocity of the fluid divided by the speed of sound in the medium. We see that equations (3.9-26), (3.9-27), and (3.9-29) can be simplified significantly if we can make the *incompressible flow approximation*, which requires that the Mach number be very small; that is, $\mathrm{Ma}\ll 1$. Note that the size of the Mach number required to ensure incompressible flow depends on both the Reynolds number and the aspect ratio. These equations can be further simplified if we can assume fully developed flow, which requires that $\mathrm{Re}(R/L)\ll 1$; this condition is ensured if $\mathrm{Re}(R/L)=\mathrm{O}(0.01)$. In summary, the following conditions justify simplifying the describing equations for this flow:

$$\text{Ma} = \frac{u_{zs}}{c} = \frac{R^2 \, \Delta P}{\mu L c} \ll 1 \Rightarrow \text{incompressible flow approximation} \quad (3.9\text{-}34)$$

$$\text{Re}\frac{R}{L} \ll 1 \Rightarrow L = 100\frac{\rho u_{zs} R^2}{\mu} \Rightarrow \text{fully developed flow approximation} \quad (3.9\text{-}35)$$

If the conditions in equations (3.9-34) and (3.9-35) apply, equations (3.9-26) through (3.9-33) simplify to (step 8)

$$0 = 1 + \frac{1}{r^*}\frac{d}{dr^*}\left(r^*\frac{du_z^*}{dr^*}\right) \quad (3.9\text{-}36)$$

$$\frac{du_z^*}{dr^*} = 0 \qquad \text{at} \quad r^* = 0 \quad (3.9\text{-}37)$$

$$u_z^* = 0 \qquad \text{at} \quad r^* = 1 \quad (3.9\text{-}38)$$

Of course, these are just the describing equations for fully developed flow of a Newtonian fluid with constant physical properties in a cylindrical tube.

3.10 DIMENSIONAL ANALYSIS CORRELATION FOR THE TERMINAL VELOCITY

In dimensional analysis we seek to determine the dimensionless groups required to correlate data or to scale a process up or down. These dimensionless groups can always be determined by means of o(1) scaling analysis since this procedure leads to the minimum parametric representation for a set of describing equations. However, in the preceding sections we indicated that carrying out an o(1) scaling analysis can be somewhat complicated and time consuming. In contrast, the scaling analysis approach to dimensional analysis illustrated in this section is much easier and quicker to implement. Note, however, that it does not provide as much information as does o(1) scaling analysis for achieving the minimum parametric representation. In particular, it does not lead to groups whose magnitude can be used to assess the relative importance of particular terms in the describing equations. It also does not identify regions of influence or boundary layers, whose identification in some cases can reduce the number of dimensionless groups. This first example of the use of scaling for dimensional analysis in fluid-dynamics applications will provide more details on the steps involved. We also compare the results of scaling analysis to those obtained from using the Pi theorem, underscore the advantages of using the former to achieve the minimum parametric representation. The steps referred to here are those outlined in Section 2.4 for the scaling approach to dimensional analysis; these differ from those used in Sections 3.2 through 3.9 since no attempt is made to achieve o(1) scaling.

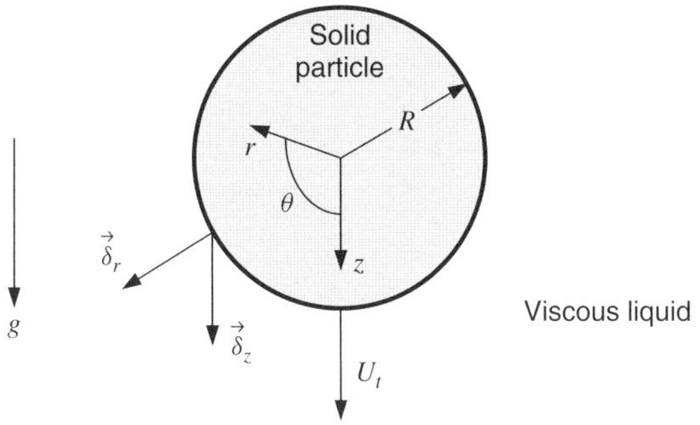

Figure 3.10-1 Solid sphere of radius R falling at its terminal velocity U_t through a viscous Newtonian fluid with constant physical properties.

In this first example of the use of scaling analysis for dimensional analysis in fluid dynamics, we consider developing a correlation for the terminal velocity U_t of a spherical particle having radius R and density ρ_p falling due to gravitational acceleration g through an incompressible Newtonian liquid having density ρ and viscosity μ as shown in Figure 3.10-1. We begin by writing the equation that in principle would have to be solved to obtain the terminal velocity. This constitutes a force balance on the sphere involving form drag due to the pressure, viscous drag on the sphere boundaries, and a net gravitational force that causes the flow:

$$-\iint_S \vec{\delta}_z \cdot \vec{\delta}_r \left[P\underline{\underline{\delta}} - \mu(\nabla\vec{u} + \nabla\vec{u}^\dagger)\right]\Big|_{r=R} dS + (\rho_p - \rho)g\frac{4}{3}\pi R^3 = 0 \qquad (3.10\text{-}1)$$

where $\vec{\delta}_i$ is the unit vector in the i-direction, S the surface area, $\underline{\underline{\delta}}$ the identity tensor, P the dynamic pressure, $\vec{u}$ the fluid velocity, and $\dagger$ denotes the transpose of a second-order tensor. The sign convention employed in arriving at equation (3.10-1) is consistent with defining the force on a fluid particle as described by equation (A.1-1) in the Appendices. To carry out the integration in equation (3.10-1), one would have to solve the axisymmetric equations of motion in spherical coordinates with boundary conditions consisting of no-slip at the sphere surface and a far-field velocity condition, which are given by

$$\rho\vec{u} \cdot \nabla\vec{u} = -\nabla P + \mu\nabla^2\vec{u} \qquad (3.10\text{-}2)$$

$$\vec{u} = 0 \qquad \text{at} \quad r = R \qquad (3.10\text{-}3)$$

$$\vec{u} \cdot \vec{\delta}_r = -U_t\cos\theta, \quad \vec{u} \cdot \vec{\delta}_\theta = -U_t\sin\theta \qquad \text{as} \quad r \to \infty \qquad (3.10\text{-}4)$$

where r and θ denote the radial and circumferential coordinates, respectively. Since there is no need to achieve o(1) scaling in dimensional analysis, it is sufficient to express the describing differential equations in the generalized vector–tensor form given by equation (3.10-2), which is the appropriately simplified form of equation (B.2-3) in the Appendices. Equations (3.10-1) through (3.10-4) constitute step 1 in the procedure for dimensional analysis outlined in Section 2.4.

Define the following dimensionless variables (steps 2, 3, and 4):

$$\vec{u}^* \equiv \frac{\vec{u}}{u_s}; \qquad P^* \equiv \frac{P}{P_s}; \qquad \nabla^* \equiv L_s \nabla; \qquad S^* \equiv \frac{S}{L_s^2} \tag{3.10-5}$$

where $*$ denotes a dimensionless variable and u_s, P_s, and L_s denote velocity, pressure, and length scales, respectively, that will be chosen to obtain the minimum parametric representation. Introducing these into equations (3.10-1) through (3.10-4) and dividing through by the dimensional coefficient of one term in each equation yields (steps 5 and 6)

$$-\iint_{S^*} \vec{\delta}_z \cdot \vec{\delta}_n \left[\frac{P_s L_s}{\mu_s u_s} P^* \underline{\underline{\delta}} - (\nabla^* \vec{u}^* + \nabla^* \vec{u}^{*\dagger}) \right] \Bigg|_{r^*=\frac{R}{L_s}} dS^* + \frac{4(\rho_p - \rho) g \pi R^3}{3\mu u_s L_s} = 0 \tag{3.10-6}$$

$$\frac{\rho u_s L_s}{\mu} \vec{u}^* \cdot \nabla^* \vec{u}^* = -\frac{P_s L_s}{\mu u_s} \nabla P^* + \nabla^{*2} \vec{u}^* \tag{3.10-7}$$

$$\vec{u}^* = 0 \qquad \text{at} \quad r^* = \frac{R}{L_s} \tag{3.10-8}$$

$$\vec{u}^* \cdot \vec{\delta}_r = -\frac{U_t}{u_s} \cos\theta, \quad \vec{u}^* \cdot \vec{\delta}_\theta = -\frac{U_t}{u_s} \sin\theta \qquad \text{as} \quad r^* \to \infty \tag{3.10-9}$$

One possible set of scale factors is obtained by setting the following dimensionless groups equal to 1; note that no attempt is made to ensure that any of the dimensionless variables are $o(1)$ since we are merely seeking to determine the minimum parametric representation rather than to assess what assumptions might be made to simplify the describing equations (step 7):

$$\frac{R}{L_s} = 1 \Rightarrow L_s = R; \qquad \frac{U_t}{u_s} = 1 \Rightarrow u_s = U_t; \qquad \frac{P_s L_s}{\mu u_s} = 1 \Rightarrow P_s = \frac{\mu U_t}{R} \tag{3.10-10}$$

When these scale factors are substituted into equations (3.10-6) through (3.10-9), the latter assume the form

$$-\iint_{S^*} \vec{\delta}_z \cdot \vec{\delta}_n [P^* \underline{\underline{\delta}} - (\nabla^* \vec{u}^* + \nabla^* \vec{u}^{*\dagger})] \Big|_{r^*=1} dS^* + \frac{4(\rho_p - \rho) g \pi R^2}{3\mu U_t} = 0 \tag{3.10-11}$$

$$\frac{\rho U_t R}{\mu} \vec{u}^* \cdot \nabla^* \vec{u}^* = -\nabla P^* + \nabla^{*2} \vec{u}^* \tag{3.10-12}$$

$$\vec{u}^* = 0 \qquad \text{at} \quad r^* = 1 \tag{3.10-13}$$

$$\vec{u}^* \cdot \vec{\delta}_r = -\cos\theta, \quad \vec{u}^* \cdot \vec{\delta}_\theta = -\sin\theta \qquad \text{as} \quad r^* \to \infty \tag{3.10-14}$$

Equations (3.10-11) through (3.10-14) represent the minimum parametric representation of the describing equations. Hence, the solution to equation (3.10-12) for the dimensionless velocity $\vec{u}^*$ will depend on r^*, θ, and the dimensionless group $\rho U_t R/\mu$, which is seen to be the Reynolds number. When this solution is substituted into equation (3.10-11), evaluated at $r^* = 1$ and integrated over the surface area S^*, the resulting solution for the dimensionless terminal velocity can be correlated in terms of the following two dimensionless groups:

$$\Pi_1 \equiv \frac{(\rho_p - \rho)gR^2}{\mu U_t} \qquad \text{and} \quad \Pi_2 \equiv \frac{\rho U_t R}{\mu} \text{ (a Reynolds number)} \tag{3.10-15}$$

Hence, either data or a numerical solution for U_t can be correlated in terms of Π_1 and Π_2; that is, the correlation for the terminal velocity involves only two dimensionless parameters and is of the general form

$$f_1(\Pi_1, \Pi_2) = 0 \Rightarrow \frac{\rho U_t R}{\mu} = f_2\left[\frac{(\rho_p - \rho)gR^2}{\mu U_t}\right] \tag{3.10-16}$$

The two dimensionless groups appearing in equation (3.10-16) are not optimal if one is seeking a correlation for U_t since it appears in both groups; that is, determining U_t from known values of the physical properties and sphere radius would require a trial-and-error solution. By invoking the transformation in step 8 with $a = 1$ and $b = 1$ in equation (2.4-2), a new dimensionless group Π_3 not containing U_t can be obtained:

$$\Pi_3 = \Pi_1 \times \Pi_2 = \frac{(\rho_p - \rho)gR^2}{\mu U_t}\frac{\rho U_t R}{\mu} = \frac{(\rho_p - \rho)\rho g R^3}{\mu^2} \tag{3.10-17}$$

Hence, data or numerical results for U_t can be correlated in terms of Π_3 and either Π_1 or Π_2; that is, our correlation for the terminal velocity can be expressed in the general form

$$f_3(\Pi_2, \Pi_3) = 0 \Rightarrow \frac{\rho U_t R}{\mu} = f_4\left[\frac{(\rho_p \quad \rho)\rho g R^3}{\mu^2}\right] \tag{3.10-18}$$

It is instructive to rework this problem in dimensional analysis using the Pi theorem approach. A naive application of the Pi theorem with $n = 6$ quantities and $m = 3$ units (mass, length, and time) to be considered in the dimensional analysis indicates that the correlation for U_t requires $n - m = 3$ rather than two dimensionless groups. Note that if force is also considered as a unit, $n = 7$, but then $m = 4$ because the dimensional constant g_c in Newton's law of motion must also be included since this law interrelates force, mass, length, and time units; hence, the Pi theorem would still predict three dimensionless groups rather than two. Hence, it would appear that the Pi theorem gives a less general result than does the scaling approach to dimensional analysis. To obtain the most general result from the Pi theorem, one must recognize that the gravitational acceleration g appears as the product $g(\rho_p - \rho)$, and hence n is really only 5; hence, with

$m = 3$ the Pi theorem will also predict two dimensionless groups. In contrast, the scaling approach naturally generates the grouping $g(\rho_p - \rho)$ and does not require addressing the subtleties associated with choosing the correct number of units and dimensional constants.

Standard references[11] suggest that U_t can be correlated in terms of Π_1 alone; that is,

$$U_t = \frac{2R^2 g(\rho_p - \rho)}{9\mu} \Rightarrow \Pi_1 = \frac{(\rho_p - \rho) g R^2}{\mu U_t} = \frac{9}{2} \tag{3.10-19}$$

However, this is for the special case of creeping or very low Reynolds number flow for which the inertia terms can be neglected.[12] Hence, Π_2 (or equivalently, Π_3) no longer appears in the correlation. The correlation given by equation (3.10-19) can be obtained (to within a multiplicative constant) by invoking the formalism in step 9 in the systematic scaling analysis method for dimensional analysis; that is, equation (3.10-16) can be expanded in a Taylor series in the parameter Π_2, the Reynolds number, which is a small parameter for creeping flow:

$$f_1(\Pi_1, \Pi_2) = f\Big|_{\Pi_2=0} + \frac{\partial f}{\partial \Pi_2}\Big|_{\Pi_2=0} \Pi_2 + \mathrm{O}(\Pi_2^2) \tag{3.10-20}$$

Truncating the expansion in equation (3.10-20) at the first term implies that the terminal velocity U_t can be correlated in terms of only the dimensionless group Π_1. Note that this same result could have been obtained by making the creeping-flow approximation in equation (3.10-2), in which case the Reynolds number would not have appeared as a dimensionless group.

A correlation in terms of just one group for the special case of creeping flow can also be obtained from the Pi theorem; however, it requires some subtle reasoning to achieve this result. In using the Pi theorem for creeping flow, as stated earlier, one must recognize that the quantities g, ρ_p, and ρ appear only as the product $g(\rho_p - \rho)$ and not individually. One must also be aware that creeping flow constitutes a problem in statics since there is no acceleration of the fluid. This implies that force must be introduced as a unit in addition to mass, length, and time. However, for problems in statics, one does not introduce the dimensional constant g_c since Newton's law of motion is not involved; that is, there is no fundamental relationship between force, mass, length, and time units. These considerations then infer that $n = 4$ and $m = 4$, thereby suggesting zero dimensionless groups. However, $n = m$ is a degenerate case of the Pi theorem that can be shown to imply one dimensionless group in this case. Again one sees that scaling analysis obviates the need to understand these subtle concepts in achieving the minimum number of dimensionless parameters via the Pi theorem approach.

[11] See Bird et al., *Transport Phenomena*, 2nd ed. p. 186.

[12] See Section 3.3 for determining the conditions required for creeping flow.

3.11 SUMMARY

The example in Section 3.2 provided an introduction to several features of scaling analysis. In this problem, scaling was used to assess under what conditions the flow could be assumed to be caused either by the applied pressure or by the upper moving boundary. It also introduced region of influence scaling by determining the thickness of the zone wherein the effects of the upper moving plate could never be ignored when predicting certain quantities; that is, thin boundary-layer regions might be safely ignored when predicting integral quantities such as the volumetric flow rate or average velocity but must be considered when predicting local quantities such as the drag at the upper moving plate. This example provided a means for estimating the error incurred when the assumptions suggested by scaling analysis are invoked since an analytical solution was available for this flow. Demanding that a quantity be one order of magnitude smaller in order to ignore some term in the describing equations typically results in an error of 40 to 50%; demanding two orders of magnitude reduces the error to less than 10%. However, the error that is encountered also depends on the quantity being considered; for example, point quantities within a region of influence might incur very large errors even when the relevant dimensionless group is several orders of magnitude less than 1. Finally, this example illustrated the forgiving nature of scaling; that is, if an incorrect assumption is made in determining one or more of the scales, a contradiction will emerge in the final dimensionless describing equations. This usually takes the form of having one term much larger than any of the others, thus implying that there is no term to balance it.

In Section 3.3 we introduced the creeping- and lubrication-flow approximations. The former requires that the Reynolds number be small, whereas the latter requires that in addition, some aspect ratio be small. This example illustrated that the dimensionless groups that emerge from scaling analysis have a physical interpretation in that they provide a measure of relative effects. For example, the Reynolds number is seen to be a measure of the ratio of the convection of momentum to the principal viscous stress. In this example it was necessary to prescribe unspecified downstream boundary conditions, due to the elliptic nature of the describing equations. Scaling provided a means for assessing when these troublesome terms can be ignored; that is, in practice these downstream conditions might not be known, which would preclude obtaining a solution to the describing equations. Finally, this example introduced the concept of local scaling, whereby one considers the flow at some arbitrary distance in the principal direction of flow that is considered to be constant during the scaling analysis.

Hydrodynamic boundary-layer theory, which is a special case of region of influence scaling, was considered in Section 3.4. The need for considering a region of influence arose naturally in scaling this problem. Indeed, when the problem was scaled without introducing any small transverse length scale over which the dependent variables experienced a characteristic change, a contradiction resulted in that the viscous terms dropped out of the dimensionless describing equations. A proper scaling analysis that introduces a region of influence or boundary layer provides a straightforward way to resolve this classical problem known as d'Alembert's paradox. This example also involved local scaling in which one of the scales is the local axial distance in the direction of flow, which is considered to be a constant in the change of variables involved in scaling analysis. We found that the boundary-layer approximation is reasonable when the Reynolds number based on the local axial length scale becomes large (i.e., $\text{Re} \equiv L\rho U_\infty/\mu \gg 1$). However, it is clear that the boundary-layer approximation must break down in the vicinity of the leading edge where L becomes small. This limitation of boundary-layer theory, which emerges from scaling analysis, is not mentioned in some transport and fluid mechanics textbooks.

In Section 3.5, scaling analysis was applied to an unsteady-state flow problem in order to ascertain when the quasi-steady-state approximation can be invoked. In this problem we found that there were several possible time scales, depending on the conditions being considered. If the transient effects associated with initiating this flow were important, the appropriate time scale was the observation time, the particular time from the start of the process at which the flow was being observed. This particular flow was still time-dependent even after the transient flow effects died out, owing to the periodic oscillation of the lower plate. Scaling analysis led to the condition required to assume quasi-steady-state, whereby the unsteady-state term could be ignored in the describing equations. However, the flow was still unsteady-state since the time dependence entered implicitly through the boundary conditions. If the time scale for the viscous penetration of vorticity was much longer than the time scale for the oscillating plate motion, the effects of the latter on the flow were confined to a region of influence near the oscillating plate.

Scaling analysis was used in Section 3.6 to determine when end and sidewall effects could be ignored. When the appropriate aspect ratio is sufficiently small, the corresponding sidewall or end effects can be ignored when determining the maximum velocity or integral quantities, such as the average flow rate. However, there is always some region of influence that can be assessed by scaling wherein one cannot ignore the effect of the lateral boundaries on quantities such as the local velocity or drag at the boundary.

In Section 3.7 we considered free surface flows, flows for which the location of some interface between adjacent phases is unknown initially and must be determined by solving the describing equations. The latter require an additional equation, referred to as the kinematic surface condition, to relate the location of the free surface to the local velocity components; this is obtained from an integral mass balance. This was the first problem we considered that required introducing a scale factor for a derivative: the time derivative of the film thickness. This was necessary because this derivative did not scale with the characteristic length scale divided by the characteristic time scale. If one did not recognize this, the forgiving nature of scaling would have led to a contradiction in the dimensionless describing equations. Scaling analysis indicated that the curvature effects could be ignored if the quasi-parallel-flow approximation was applicable; this is the spatial analog to the quasi-steady-state approximation considered in Section 3.5, whereby the effects of the nonconstant film thickness enter only through the boundary conditions. However, scaling analysis indicated that the quasi-parallel-flow approximation always breaks down sufficiently close to the leading edge of the flow. This was the first problem in which it was necessary to introduce a reference factor, since in this case the pressure was not naturally referenced to zero.

Porous media flows were considered in Section 3.8 for which the flow was described by the Darcy flow equation incorporating the Brinkman term to allow satisfying the no-slip condition at the solid boundaries. Scaling analysis was used to determine the conditions for which the effect of the bounding walls on the flow through the porous media could be ignored. Scaling again indicated that there was a region of influence wherein the effect of the bounding walls could not be ignored on quantities such as the local velocity or drag on these boundaries.

In Section 3.9, scaling was applied to a compressible flow for which an equation of state is required to relate the density to the state variables, in this case the nonconstant pressure. This problem also required introducing a separate scale for a derivative, the radial pressure gradient. However, if one did not recognize this, the forgiving nature of scaling would have led to a contradiction in the dimensionless describing equations. This is explored further in Practice Problem 3.P.31. Scaling indicated that this flow could be assumed to be incompressible if the dimensionless group known as the Mach number was much less than 1; the Mach number is the ratio of the velocity of the fluid divided by the speed of sound in the medium. The effects of compressibility were explored in this problem by expanding the density in a Taylor series about some mean value characteristic of the flow. This same technique can be used in conjunction with scaling to explore the effects of other nonconstant physical and transport properties, such as surface tension and viscosity. This is considered in more depth in Chapter 4 when we consider the application of scaling to heat transfer and allow for the variation of the viscosity with temperature.

Scaling was applied to dimensional analysis in Section 3.10. The critical difference between o(1) scaling and dimensional analysis is that in the latter there is no attempt to define dimensional variables that are bounded of o(1). The goal in dimensional analysis is to arrive at the minimum parametric representation of the problem; that is, to obtain a set of dimensionless describing equations in terms of the minimum number of dimensionless parameters. Whereas o(1) scaling leads to a unique set of dimensionless groups for a prescribed set of physical variables and system parameters, dimensional analysis does not. However, any one set of dimensionless groups can be converted into any other set of dimensionless groups by defining new dimensionless groups that are obtained by multiplying products of the original groups raised to appropriate positive or negative powers. In some cases it is advantageous to use this property of dimensional analysis to isolate one or more dimensional quantities into a particular dimensionless group; for example, by isolating all the quantities that are varied in an experiment into one dimensionless group, the dependence of the phenomenon being studied on this particular group can be determined since all the other dimensionless groups will be constant. Isolating certain dimensional quantities into one group is also advantageous when developing correlations for experimental data since it allows determining that particular dimensional quantity directly rather than by means of a trial-and-error procedure. The systematic scaling procedure offers several advantages relative to the Pi theorem approach for dimensional analysis. The problem considered in Section 3.10 demonstrated that scaling analysis led to the minimum parametric representation directly, whereas obtaining it using the Pi theorem required considerable intuitive knowledge. The scaling procedure naturally groups variables that always appear in some combination, whereas the Pi theorem requires that one somehow recognize this intuitively. Scaling also obviates the need to know which dimensional variables need to be considered as primary and secondary quantities and when dimensional constants need to be introduced into the dimensional analysis.

3.E EXAMPLE PROBLEMS

3.E.1 Gravity-Driven Laminar Film Flow down a Vertical Wall

A film of an incompressible viscous Newtonian liquid that has constant physical properties is in fully developed laminar flow down a vertical wall, due to a gravitational body force in the presence of an insoluble gas phase as shown in Figure 3.E.1-1. Scaling analysis will be used to determine when the effect of the drag exerted by the gas phase on the velocity profile in the liquid film can be neglected.

The describing equations obtained by appropriately simplifying equation (D.1-10) in the Appendices (step 1)

$$\mu_1 \frac{d^2 u_x}{dy^2} + \rho_1 g = 0 \qquad \text{for} \quad 0 \le y \le h_1 \tag{3.E.1-1}$$

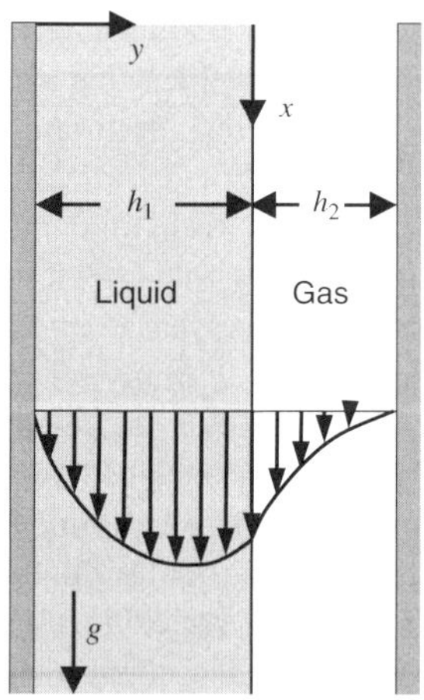

Figure 3.E.1-1 Fully developed laminar flow of an incompressible viscous Newtonian liquid film with constant physical properties down a vertical wall under the influence of gravity in the presence of a viscous gas phase.

$$\frac{d^2u_x}{dy^2} = 0 \qquad \text{for} \quad h_1 \le y \le h_1 + h_2 \tag{3.E.1-2}$$

$$u_x = 0 \qquad \text{at} \quad y = 0 \tag{3.E.1-3}$$

$$u_x|_{h_1^-} = u_x|_{h_1^+} \qquad \text{at} \quad y = h_1 \tag{3.E.1-4}$$

$$\mu_1 \left.\frac{du_x}{dy}\right|_{h_1^-} = \mu_2 \left.\frac{du_x}{dy}\right|_{h_1^+} \qquad \text{at} \quad y = h_1 \tag{3.E.1-5}$$

$$u_x = 0 \qquad \text{at} \quad y = h_1 + h_2 \tag{3.E.1-6}$$

Introduce the following scale factors, reference factor, and dimensionless variables (steps 2, 3, and 4):

$$u_{x1}^* \equiv \frac{u_{x1}}{u_{s1}}; \qquad u_{x2}^* \equiv \frac{u_{x2}}{u_{s2}}; \qquad y_1^* \equiv \frac{y}{y_{s1}}; \qquad y_2^* \equiv \frac{y - y_{r2}}{y_{s2}} \tag{3.E.1-7}$$

where y_{s1} is the length scale in the liquid film and y_{r2} and y_{s2} are the reference and length scales in the gas phase; these are introduced to ensure that we can achieve o(1) scaling for the spatial coordinates in both phases. A reference length scale is needed in the gas phase because y is not naturally referenced to zero in this phase. Substitute these dimensionless variables into the describing equations and divide through by the dimensional coefficient of a term that must be retained in each equation (steps 5 and 6):

$$\frac{d^2u_{x1}^*}{dy_1^{*2}} + \frac{\rho_1 g y_{s1}^2}{\mu_1 u_{s1}} = 0 \qquad \text{for} \quad 0 \le y_1^* \le \frac{h_1}{y_{s1}} \tag{3.E.1-8}$$

$$\frac{d^2u_{x2}^*}{dy_2^{*2}} = 0 \qquad \text{for} \quad \frac{h_1 - y_{r2}}{y_{s2}} \le y_2^* \le \frac{h_1 + h_2 - y_{r2}}{y_{s2}} \tag{3.E.1-9}$$

$$u_{x1}^* = 0 \qquad \text{at} \quad y_1^* = 0 \tag{3.E.1-10}$$

$$u_{x1}^* = \frac{u_{s2}}{u_{s1}} u_{x2}^* \qquad \text{at} \quad y_1^* = \frac{h_1}{y_{s1}} \tag{3.E.1-11}$$

$$\frac{du_{x1}^*}{dy_1^*} = \frac{\mu_2 u_{s2} y_{s1}}{\mu_1 u_{s1} y_{s2}} \frac{du_{x2}^*}{dy_2^*} \qquad \text{at} \quad y_1^* = \frac{h_1}{y_{s1}} \tag{3.E.1-12}$$

$$u_{x2}^* = 0 \qquad \text{at} \quad y_2^* = \frac{h_1 + h_2 - y_{r2}}{y_{s2}} \tag{3.E.1-13}$$

Since gravity causes the flow in the liquid film, the dimensionless group in equation (3.E.1-8) provides u_{s1}. Since the drag at the interface causes the flow in the gas phase, the dimensionless group in equation (3.E.1-11) gives u_{s2}. We bound the dimensionless spatial variables in the liquid and gas phases to be o(1) by setting the dimensionless geometric ratio groups in equations (3.E.1-8) and (3.E.1-9) equal to zero (for determining y_{r2}) or 1 (for determining y_{s1} and y_{s2}). This results in the following scale and reference factors (step 7):

$$u_{s1} = u_{s2} = \frac{\rho_1 g h_1^2}{\mu_1}; \qquad y_{s1} = h_1; \qquad y_{s2} = h_2; \qquad y_{r2} = h_1 \tag{3.E.1-14}$$

The resulting scaled dimensionless describing equations then are given by

$$\frac{d^2 u_{x1}^*}{dy_1^{*2}} + 1 = 0 \quad \text{for} \quad 0 \leq y_1^* \leq 1 \tag{3.E.1-15}$$

$$\frac{d^2 u_{x2}^*}{dy_2^{*2}} = 0 \quad \text{for} \quad 0 \leq y_2^* \leq 1 \tag{3.E.1-16}$$

$$u_{x1}^* = 0 \quad \text{at} \quad y_1^* = 0 \tag{3.E.1-17}$$

$$u_{x1}^* = u_{x2}^* \quad \text{at} \quad y_1^* = 1 \tag{3.E.1-18}$$

$$\frac{du_{x1}^*}{dy_1^*} = \frac{\mu_2 h_1}{\mu_1 h_2} \frac{du_{x2}^*}{dy_2^*} \quad \text{at} \quad y_1^* = 1 \tag{3.E.1-19}$$

$$u_{x_2}^* = 0 \quad \text{at} \quad y_2^* = 1 \tag{3.E.1-20}$$

We see that the effect of the gas phase drag on the liquid film will be negligible if the following condition is satisfied (step 8):

$$\frac{\mu_2 h_1}{\mu_1 h_2} \ll 1 \Rightarrow \frac{\mu_2 h_1}{\mu_1 h_2} = \mathrm{o}(0.01) \tag{3.E.1-21}$$

3.E.2 Flow Between Two Approaching Parallel Circular Flat Plates

Consider the unsteady-state laminar flow of an incompressible Newtonian liquid having constant physical properties between two parallel circular flat plates that slowly approach each other with an axial velocity given by

$$U = U_0 e^{-\beta t} \tag{3.E.2-1}$$

where U_0 is the initial axial velocity and β is a time constant; a schematic of this flow is shown in Figure 3.E.2-1. We use scaling to simplify the describing equations for this relatively complex flow.

The describing equations are obtained by appropriately simplifying equations (C.2-1), (D.2-10), and (D.2-12) in the Appendices (step 1):

$$\frac{1}{r}\frac{\partial}{\partial r}(r u_r) + \frac{\partial u_z}{\partial z} = 0 \tag{3.E.2-2}$$

$$\rho \frac{\partial u_r}{\partial t} + \rho u_r \frac{\partial u_r}{\partial r} + \rho u_z \frac{\partial u_r}{\partial z} = -\frac{\partial P}{\partial r} + \mu \frac{\partial}{\partial r}\left[\frac{1}{r}\frac{\partial}{\partial r}(r u_r)\right] + \mu \frac{\partial^2 u_r}{\partial z^2} \tag{3.E.2-3}$$

$$\rho \frac{\partial u_z}{\partial t} + \rho u_r \frac{\partial u_z}{\partial r} + \rho u_z \frac{\partial u_z}{\partial z} = -\frac{\partial P}{\partial z} + \mu \frac{\partial}{\partial r}\left[\frac{1}{r}\frac{\partial}{\partial r}\left(r \frac{\partial u_z}{\partial r}\right)\right] + \mu \frac{\partial^2 u_z}{\partial z^2} \tag{3.E.2-4}$$

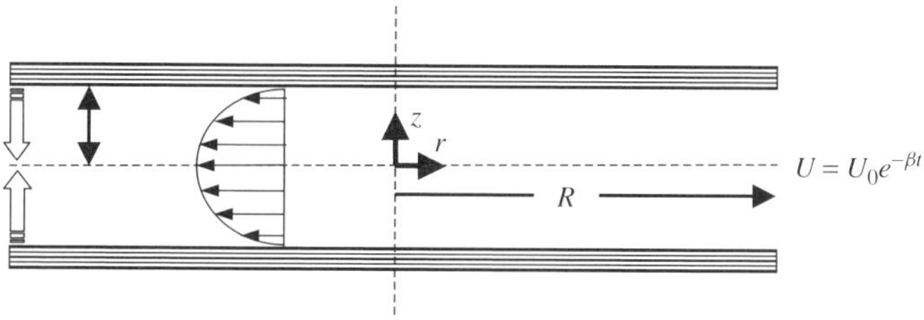

Figure 3.E.2-1 Laminar flow of an incompressible Newtonian liquid that has constant physical properties between two parallel circular plates that are approaching each other slowly.

$$u_r = 0, \quad u_z = -\frac{1}{2}U(t) = -\frac{1}{2}U_0 e^{-\beta t} \quad \text{at} \quad z = H(t) \tag{3.E.2-5}$$

$$u_r = 0, \quad u_z = \frac{1}{2}U(t) = \frac{1}{2}U_0 e^{-\beta t} \quad \text{at} \quad z = -H(t) \tag{3.E.2-6}$$

$$\int_{-H}^{H} u_r \cdot 2\pi r\, dz = -\frac{d}{dt}(\pi r^2 H) = -2\pi r^2 \frac{dH}{dt} = \pi r^2 U_0 e^{-\beta t}, \quad u_z = f_1(z, t) \quad \text{at} \quad r = R \tag{3.E.2-7}$$

$$u_r = 0, \quad \frac{\partial u_z}{\partial r} = 0 \quad \text{at} \quad r = 0 \tag{3.E.2-8}$$

$$P = P_{\text{atm}} \quad \text{at} \quad r = R \tag{3.E.2-9}$$

The boundary condition given by equation (3.E.2-7) is a statement that the mass squeezed out by the plates moving together must flow out the periphery of the circular plates. The function f_1 merely denotes that to solve these differential equations, one would need to specify a boundary condition for u_z at the periphery of the circular plates. The simplifications justified by scaling analysis will obviate the need to know f_1.

Introduce the following scale factors, reference factor, and dimensionless variables (steps 2, 3, and 4):

$$u_r^* \equiv \frac{u_r}{u_{rs}}; \quad u_z^* \equiv \frac{u_z}{u_{zs}}; \quad P^* \equiv \frac{P - P_r}{P_s}; \quad z^* \equiv \frac{z}{z_s}; \quad r^* \equiv \frac{r}{r_s}; \quad t^* \equiv \frac{t}{t_s} \tag{3.E.2-10}$$

Substitute these dimensionless variables into the describing equations and divide through by the dimensional coefficient of a term that must be retained in each equation to obtain (steps 5 and 6)

$$\frac{1}{r^*}\frac{\partial}{\partial r^*}(r^* u_r^*) + \frac{u_{zs} r_s}{u_{rs} z_s}\frac{\partial u_z^*}{\partial z^*} = 0 \tag{3.E.2-11}$$

$$\frac{\rho z_s^2}{\mu t_s}\frac{\partial u_r^*}{\partial t^*} + \frac{\rho u_{rs} z_s^2}{\mu r_s} u_r^* \frac{\partial u_r^*}{\partial r^*} + \frac{\rho u_{zs} z_s}{\mu} u_z^* \frac{\partial u_r^*}{\partial z^*}$$

$$= -\frac{P_s z_s^2}{\mu u_{rs} r_s}\frac{\partial P^*}{\partial r^*} + \frac{z_s^2}{r_s^2}\frac{\partial}{\partial r^*}\left[\frac{1}{r^*}\frac{\partial}{\partial r^*}(r^* u_r^*)\right] + \frac{\partial^2 u_r^*}{\partial z^{*2}} \tag{3.E.2-12}$$

$$\frac{\rho z_s^2}{\mu t_s}\frac{\partial u_z^*}{\partial t^*} + \frac{\rho u_{rs} z_s^2}{\mu r_s} u_r^* \frac{\partial u_z^*}{\partial r^*} + \frac{\rho u_{zs} z_s}{\mu} u_z^* \frac{\partial u_z^*}{\partial z^*}$$

$$= -\frac{P_s z_s}{\mu u_{zs}}\frac{\partial P^*}{\partial z^*} + \frac{z_s^2}{r_s^2}\frac{\partial}{\partial r}\left[\frac{1}{r^*}\frac{\partial}{\partial r^*}\left(r^* \frac{\partial u_z^*}{\partial r^*}\right)\right] + \frac{\partial^2 u_z^*}{\partial z^{*2}} \tag{3.E.2-13}$$

$$u_r^* = 0, \quad u_z^* = -\frac{1}{2}\frac{U_0}{u_{zs}} e^{-\beta t_s t^*} \quad \text{at} \quad z^* = \frac{H(t)}{z_s} \tag{3.E.2-14}$$

$$u_r^* = 0, \quad u_z^* = \frac{1}{2}\frac{U_0}{u_{zs}} e^{-\beta t_s t^*} \quad \text{at} \quad z^* = -\frac{H(t)}{z_s} \tag{3.E.2-15}$$

$$\int_{-H/z_s}^{H/z_s} u_r^* dz^* = \frac{U_0 r_s}{2 u_{rs} z_s} r^* e^{-\beta t_s t^*}, \quad u_z^* = f_1^*(z^*, t^*) \quad \text{at} \quad r^* = \frac{R}{r_s} \tag{3.E.2-16}$$

$$u_r^* = 0, \quad \frac{\partial u_z^*}{\partial r^*} = 0 \quad \text{at} \quad r^* = 0 \tag{3.E.2-17}$$

$$P^* = \frac{P_{\text{atm}} - P_r}{P_s} \quad \text{at} \quad r^* = \frac{R}{r_s} \tag{3.E.2-18}$$

The dimensionless group in equation (3.E.2-14) is set equal to 1 to obtain the axial velocity scale u_{zs}. The geometric ratio groups in equations (3.E.2-14) and (3.E.2-16), when set equal to 1, provide the axial and radial length scales z_s and r_s. Equation (3.E.2-11) then provides the radial velocity scale u_{rs}. Since pressure causes the radial flow, P_s is obtained by setting the dimensionless group multiplying the pressure term in equation (3.E.2-12) equal to 1. The reference pressure P_r is obtained from equation (3.E.2-18). Finally, the time scale t_s is dictated by the approach velocity of the two parallel plates and is obtained by setting the dimensionless group in the argument of the exponential in equation (3.E.2-14) equal to 1. This results in the following scale and reference factors (step 7):

$$u_{rs} = \frac{RU_0}{2H}; \quad u_{zs} = \frac{U_0}{2}; \quad P_s = \frac{\mu U_0 R^2}{2H^3}; \quad P_r = P_{\text{atm}}; \quad z_s = H;$$
$$r_s = R; \quad t^* = \frac{1}{\beta} \tag{3.E.2-19}$$

Note in this case that several of our scales are time-dependent, owing to the presence of H in their definition.

The resulting scaled dimensionless describing equations then are given by

$$\frac{1}{r^*}\frac{\partial}{\partial r^*}(r^* u_r^*) + \frac{\partial u_z^*}{\partial z^*} = 0 \tag{3.E.2-20}$$

$$\frac{\rho\beta H^2}{\mu}\frac{\partial u_r^*}{\partial t^*} + \mathrm{Re}u_r^*\frac{\partial u_r^*}{\partial r^*} + \mathrm{Re}u_z^*\frac{\partial u_r^*}{\partial z^*} = -\frac{\partial P^*}{\partial r^*} + \frac{H^2}{R^2}\frac{\partial}{\partial r^*}\left[\frac{1}{r^*}\frac{\partial}{\partial r^*}(r^*u_r^*)\right] + \frac{\partial^2 u_r^*}{\partial z^{*2}} \tag{3.E.2-21}$$

$$\frac{\rho\beta H^2}{\mu}\frac{\partial u_z^*}{\partial t^*} + \mathrm{Re}u_r^*\frac{\partial u_z^*}{\partial r^*} + \mathrm{Re}u_z^*\frac{\partial u_z^*}{\partial z^*} = -\frac{R^2}{H^2}\frac{\partial P^*}{\partial z^*} + \frac{H^2}{R^2}\frac{\partial}{\partial r}\left[\frac{1}{r^*}\frac{\partial}{\partial r^*}\left(r^*\frac{\partial u_z^*}{\partial r^*}\right)\right] + \frac{\partial^2 u_z^*}{\partial z^{*2}} \tag{3.E.2-22}$$

$$u_r^* = 0, \quad u_z^* = -\,e^{-t^*} \quad \text{at} \quad z^* = 1 \tag{3.E.2-23}$$

$$u_r^* = 0, \quad u_z^* = e^{-t^*} \quad \text{at} \quad z^* = -1 \tag{3.E.2-24}$$

$$\int_{-1}^{1} u_r^*\, dz^* = r^* e^{-t^*}, \quad u_z^* = f_1^*(z^*, t^*) \quad \text{at} \quad r^* = 1 \tag{3.E.2-25}$$

$$u_r^* = 0, \quad \frac{\partial u_z^*}{\partial r^*} = 0 \quad \text{at} \quad r^* = 0 \tag{3.E.2-26}$$

$$P^* = 0 \quad \text{at} \quad r^* = 1 \tag{3.E.2-27}$$

where $\mathrm{Re} \equiv \rho U_0 H/2\mu$ is the Reynolds number. We see that equations (3.E.2-20) through (3.E.2-27) can be greatly simplified if the following conditions hold (step 8):

$$\mathrm{Re} = \frac{\rho U_0 H}{2\mu} \ll 1 \Rightarrow \text{creeping flow} \tag{3.E.2-28}$$

$$\frac{H^2}{R^2} \ll 1 \Rightarrow \text{lubrication flow if Re} \ll 1 \text{ as well} \tag{3.E.2-29}$$

$$\frac{\rho\beta H^2}{\mu} \ll 1 \Rightarrow \text{quasi-steady-state flow} \tag{3.E.2-30}$$

All of these conditions are satisfied for sufficiently close approach distances H between the two plates. Note that our scaling ensured that the dimensionless radial pressure gradient was o(1); this scaling does not necessarily ensure that the axial pressure derivative is also o(1). In fact, the condition given by equation (3.E.2-29) implies that the dimensionless axial pressure gradient in equation (3.E.2-22) is of the following order:

$$\frac{\partial P^*}{\partial z^*} \cong \frac{H^2}{R^2}\frac{\partial^2 u_z^*}{\partial z^{*2}} = \mathrm{O}\left(\frac{H^2}{R^2}\right) \ll 1 \tag{3.E.2-31}$$

This implies that the axial pressure gradient is essentially zero for the case of lubrication flow; this in turn implies that the radial pressure gradient is not a function of z.

If the conditions given by equations (3.E.2-28) through (3.E.2-30) are satisfied, equations (3.E.2-20) through (3.E.2-27) simplify to

$$0 = -\frac{\partial P^*}{\partial r^*} + \frac{\partial^2 u_r^*}{\partial z^{*2}} \tag{3.E.2-32}$$

$$\frac{\partial P^*}{\partial z^*} \cong 0 \tag{3.E.2-33}$$

$$u_r^* = 0, \quad u_z^* = -e^{-t^*} \quad \text{at} \quad z^* = 1 \tag{3.E.2-34}$$

$$u_r^* = 0, \quad u_z^* = e^{-t^*} \quad \text{at} \quad z^* = -1 \tag{3.E.2-35}$$

$$\int_{-1}^{1} u_r^* \, dz^* = r^* e^{-t^*}, \quad u_z^* = f_1^*(z^*, t^*) \quad \text{at} \quad r^* = 1 \tag{3.E.2-36}$$

$$P^* = 0 \quad \text{at} \quad r^* = 1 \tag{3.E.2-37}$$

Equations (3.E.2-36) and (3.E.2-37) are required to obtain the radial pressure distribution. This simplified set of describing equations can be solved analytically in closed form.

3.E.3 Design of a Hydraulic Ram

Consider the operation of a hydraulic ram shown in Figure 3.E.3-1. This device consists of a piston of radius R_1 that is free to slide within a cylinder of inner radius R_2 containing a viscous oil that can be assumed to be an incompressible Newtonian liquid with constant physical properties. An applied force causes the piston to be pushed into the cylinder, which in turn causes a high pressure P_0 to be generated within the oil confined between the closed end of the cylinder and the piston head. The difference between this high pressure and the ambient pressure P_{atm} then causes oil to flow in the gap between the piston and the cylinder. The force that must be applied to push the piston into the cylinder is equal to the sum of the force exerted by P_0 on the piston head and the drag force caused by the

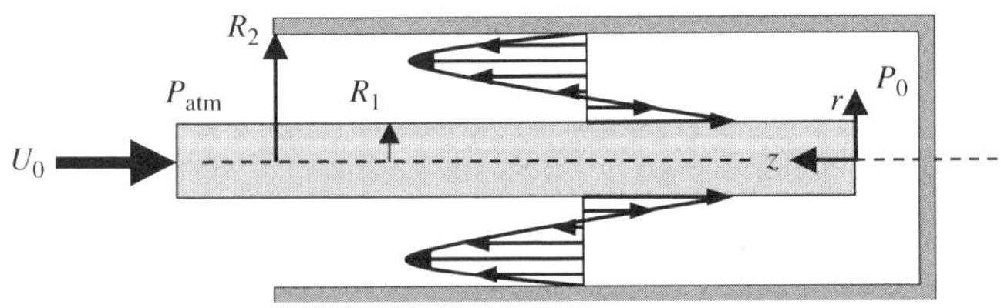

Figure 3.E.3-1 Hydraulic ram consisting of a cylindrical piston of radius R_1 that slides within a cylindrical housing of radius R_2; the cylindrical housing and piston assembly contains a viscous oil that can be considered to be an incompressible Newtonian liquid with constant physical properties; a time-dependent force is applied to the piston, causing it to move into the cylindrical housing at a constant velocity U_0; the instantaneous axial velocity profile in the annular gap is sketched in the figure.

oil being pushed through the annular gap. Let us assume that the piston is being pushed in at a constant velocity U_0. We ignore any end effects and assume that the lubrication-flow approximation can be made for the flow in the thin annular gap.[13] We employ scaling analysis to determine when curvature effects in the describing equations for the flow in the annular gap can be neglected; that is, the conditions for which the describing equations in cylindrical coordinates reduce to those in rectangular coordinates.

The describing equations are obtained by simplifying appropriately the equations of motion in cylindrical coordinates given by equations (C.2-1), (D.2-10), and (D.2-12) in the Appendices (step 1). In writing these describing equations we place our coordinate system on the moving piston. In doing so, it appears that the cylinder is moving in the direction opposite to that of the piston.

$$0 = -\frac{\partial P}{\partial z} + \mu \frac{1}{r}\frac{d}{dr}\left(r\frac{du_z}{dr}\right) \tag{3.E.3-1}$$

$$u_z = 0 \quad \text{at} \quad r = R_1 \tag{3.E.3-2}$$

$$u_z = U_0 \quad \text{at} \quad r = R_2 \tag{3.E.3-3}$$

$$P = P_{\text{atm}} \quad \text{at} \quad z = L(t) \tag{3.E.3-4}$$

where P_{atm} is the prevailing atmospheric pressure and $L(t)$ is the wetted length of the piston; that is, the instantaneous length of the piston that is in contact with the flowing oil. Note that we retain the partial derivative $\partial P/\partial z$ in equation (3.E.3-1) because the pressure gradient is also a function of time, since the wetted length increases as the piston is pushed into the cylinder.

Introduce the following scale factors and dimensionless variables (steps 2, 3, and 4):

$$u_z^* \equiv \frac{u_z}{u_{zs}}; \qquad P^* \equiv \frac{P - P_r}{P_s}; \qquad r^* \equiv \frac{r - r_r}{r_s}; \qquad z^* \equiv \frac{z}{z_s} \tag{3.E.3-5}$$

Note that we have introduced a reference factor for the radial coordinate since it is not naturally referenced to zero within the region wherein the flow is occurring; that is, within the annular gap. We have also introduced a reference pressure since the pressure is also not naturally referenced to zero at either end of the annular gap. Substitute these dimensionless variables into the describing equations and divide through by the dimensional coefficient of a term that must be retained in each equation to obtain (steps 5 and 6)

$$0 = -\frac{P_s r_s^2}{\mu u_{zs} z_s}\frac{\partial P^*}{\partial z^*} + \frac{1}{\left(\frac{r_s}{r_r}r^* + 1\right)}\frac{d}{dr^*}\left[\left(\frac{r_s}{r_r}r^* + 1\right)\frac{du_z^*}{dr^*}\right] \tag{3.E.3-6}$$

[13]The lubrication-flow approximation for this flow can also be justified using scaling analysis; this is considered in practice Problem 3.P.11.

$$u_z^* = 0 \quad \text{at} \quad r^* = \frac{R_1 - r_r}{r_s} \tag{3.E.3-7}$$

$$u_z^* = \frac{U_0}{u_{zs}} \quad \text{at} \quad r^* = \frac{R_2 - r_r}{r_s} \tag{3.E.3-8}$$

$$P^* = \frac{P_{\text{atm}} - P_r}{P_s} \quad \text{at} \quad z^* = \frac{L}{z_s} \tag{3.E.3-9}$$

The dimensionless radial coordinate can be bounded between zero and 1 if we set the dimensionless group in equation (3.E.3-7) equal to zero and that in equation (3.E.3-8) equal to 1, thereby obtaining $r_r = R_1$ and $r_s = R_2 - R_1$. The velocity scale is obtained by setting the dimensionless group in equation (3.E.3-8) equal to 1 to obtain $u_{zs} = U_0$. The axial length scale can be obtained by setting the dimensionless group in equation (3.E.3-9) equal to 1 to obtain $z_s = L(t)$. The dimensionless pressure can be referenced to zero by setting the dimensionless group in equation (3.E.3-9) equal to zero to obtain $P_r = P_{\text{atm}}$. Finally, since pressure causes the flow in the axial direction, P_s can be obtained by setting the dimensionless group in equation (3.E.3-6) equal to 1 to obtain $P_s = \mu U_0 L/(R_2 - R_1)^2$ (step 7).

The resulting scaled dimensionless describing equations then are given by

$$0 = -\frac{\partial P^*}{\partial z^*} + \frac{1}{[(R_2 - R_1)/R_1]r^* + 1}\frac{d}{dr^*}\left[\left(\frac{R_2 - R_1}{R_1}r^* + 1\right)\frac{du_z^*}{dr^*}\right] \tag{3.E.3-10}$$

$$u_z^* = 0 \quad \text{at} \quad r^* = 0 \tag{3.E.3-11}$$

$$u_z^* = 1 \quad \text{at} \quad r^* = 1 \tag{3.E.3-12}$$

$$P^* = 0 \quad \text{at} \quad z^* = 1 \tag{3.E.3-13}$$

We see that if the dimensionless group $(R_2 - R_1)/R_1 = \text{o}(0.01)$, equation (3.E.3-10) reduces to (step 8)

$$0 = -\frac{\partial P^*}{\partial z^*} + \frac{d^2 u_z^*}{dr^{*2}} \tag{3.E.3-14}$$

This is the same equation that would apply to steady-state pressure-driven lubrication flow between two parallel flat plates. Hence, we conclude that curvature effects can be ignored in problems such as that considered here if the following condition is satisfied:

$$\frac{R_2 - R_1}{R_1} \ll 1 \Rightarrow \text{curvature effects can be neglected} \tag{3.E.3-15}$$

A closed-form analytical solution for the axial velocity profile can easily be obtained for equation (3.E.3-14) subject to the boundary conditions given by equations (3.E.3-11) through (3.E.3-13). The unknown axial pressure gradient in the solution for the velocity profile can be obtained from the known volumetric flow rate corresponding to the volume swept out by the piston moving in at a constant velocity U_0. The force required to push the piston in at a constant velocity can be obtained by a force balance on the piston that includes the pressure force at each end of the piston and the drag force exerted by the flowing oil.

3.E.4 Rotating Disk Flow

Consider a disk rotating at a constant angular frequency ω (radians per second) in an infinite fluid having constant physical properties, as shown in Figure 3.E.4-1. The rotation of the disk causes both an angular fluid velocity and flow in the radial direction. This in turn causes flow toward the disk in the axial direction. Infinitely far into the fluid the velocity must be purely axial toward the disk. This flow geometry is of considerable practical value since the rotating disk is referred to as a *uniformly accessible surface*. By this we mean that the heat- or mass-transfer flux to the surface of a rotating disk is invariant with position along its surface. For example, this makes the rotating disk an ideal geometry for determining reaction kinetics in electrochemical systems for which the rotating disk can serve as one of the electrodes.[14] The rotating disk was first analyzed by von Kármán[15]; a more

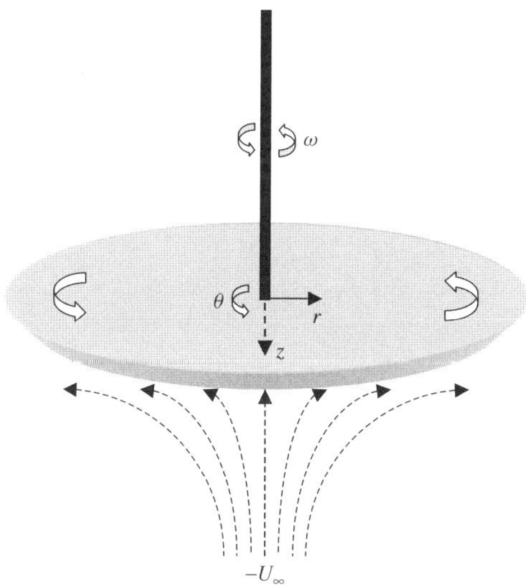

Figure 3.E.4-1 Flow created by a flat disk rotating at an angular velocity ω (rad/s) in an unbounded fluid; the rotational motion of the disk draws fluid toward the disk; the axial velocity infinitely far removed from the disk is U_∞.

[14]The rotating disk electrode apparatus was first exhibited at the Brussels World's Fair in 1958. The application of scaling analysis for using the rotating disk to study mass transfer is considered in Chapter 5.

[15]T. von Kármán, *Z. Angew. Math. Mech.*, **1**, 244–247 (1921).

complete analysis was done by Cochran[16]; an overview of the analysis and use of this instrument is given by Levich.[17] This problem can be solved analytically for laminar flow conditions in the absence of free convection effects. However, under proper operating conditions, the hydrodynamics take on a boundary-layer character such that the change in velocity components occurs within a thin region of influence near the rotating disk. Establishing a thin boundary layer is important for heat- or mass-transfer characterization using this apparatus since it ensures that container boundary effects are minimized. We use scaling analysis to ascertain the conditions for which the boundary layer will be thin.

A surprising aspect of laminar rotating disk flow is that there is no radial pressure gradient and the radial and angular velocities are directly proportional to the radial position, whereas the axial velocity depends only on the axial position. Classical treatments of this flow begin by recognizing these considerations intuitively and then proceeding to develop the rigorous analytical solution for this flow. However, these considerations can be ascertained via simple integral mass and momentum balances. Consider a mass balance on a control volume of arbitrary radius r extending from the surface of the disk far into the fluid, where there is only an axial velocity component given by $u_z = -U_\infty$. Note that in practice U_∞ is unknown; however, it can be determined from the solution for the hydrodynamics and a specified disk rotation rate. A mass balance on this control volume then yields

$$U_\infty \pi r^2 = 2\pi r \int_0^\infty u_r\, dz \Rightarrow U_\infty = \frac{2}{r}\int_0^\infty u_r\, dz \Rightarrow u_r = r f_1(z) \qquad \text{(3.E.4-1)}$$

If the general form of the radial velocity profile given by equation (3.E.4-1) is substituted into the continuity equation given by equation (C.2-1), we can conclude that the axial velocity is independent of r:

$$\frac{1}{r}\frac{\partial}{\partial r}(r u_r) + \frac{\partial u_z}{\partial z} = 0 \Rightarrow 2 f_1(z) + \frac{\partial u_z}{\partial z} = 0 \Rightarrow u_z = -2\int_0^z f_1(\tilde{z})\, d\tilde{z} = f_2(z) \qquad \text{(3.E.4-2)}$$

where $\tilde{z}$ denotes a dummy integration variable. Hence, we conclude that u_z is a function only of z. We can now prove that there is no radial pressure gradient by considering an integral z-momentum balance on a control volume of arbitrary radius r extending from an arbitrary height z into the fluid far from the rotating disk where $u_z = -U_\infty$:

$$\rho U_\infty^2 \pi r^2 - \left(\int_0^r \rho u_z^2 \cdot 2\pi \tilde{r}\, d\tilde{r}\right)\Bigg|_z - P_\infty \pi r^2 + \left(\int_0^r P \cdot 2\pi \tilde{r}\, d\tilde{r}\right)\Bigg|_z$$
$$+ \left(\int_z^\infty \tau_{rz} \cdot 2\pi r\, d\tilde{z}\right)\Bigg|_z - \left(\int_0^r \tau_{zz} \cdot 2\pi \tilde{r}\, d\tilde{r}\right)\Bigg|_z = 0 \qquad \text{(3.E.4-3)}$$

[16] W. G. Cochran, *Proc. Cambridge Philos. Soc.*, **30**, 365–375 (1934).

[17] V. G. Levich, *Physicochemical Hydrodynamics*, Prentice-Hall, Englewood Cliffs, NJ 1962, pp. 60–78.

where $\tilde{r}$ and $\tilde{z}$ again denote dummy integration variables. When the components of Newton's constitutive equation given by equations (D.2-6) and (D.2-9) in the Appendices are substituted, we obtain

$$\begin{aligned}\rho U_\infty^2 r^2 - \left(\int_0^r \rho u_z^2 \cdot 2\tilde{r}\, d\tilde{r}\right)\bigg|_z - P_\infty r^2 + \left(\int_0^r P \cdot 2\tilde{r}\, d\tilde{r}\right)\bigg|_z \\ - \left(\int_z^\infty \mu \frac{\partial u_r}{\partial \tilde{z}} \cdot 2r d\tilde{z}\right)\bigg|_z + \left(\int_0^r \mu \frac{\partial u_z}{\partial z} \cdot 4\tilde{r}\, d\tilde{r}\right)\bigg|_z = 0\end{aligned} \tag{3.E.4-4}$$

Substitute the functional forms for u_r and u_z into equation (3.E.4-4) to obtain

$$\rho U_\infty^2 r^2 - \rho f_2^2 r^2 - P_\infty r^2 + 2\left(\int_0^r P \cdot \tilde{r}\, d\tilde{r}\right)\bigg|_z + 2\mu r^2 f_1 + 2\mu \frac{df_2}{dz} r^2 = 0 \tag{3.E.4-5}$$

Hence, we conclude that

$$\frac{2}{r^2}\int_0^r P \cdot \tilde{r}\, d\tilde{r} = -\rho U_\infty^2 + \rho f_2^2 + P_\infty - 2\mu f_1 - 2\mu \frac{df_2}{dz} \Rightarrow P = f_3(z) \tag{3.E.4-6}$$

When the functional forms for u_r, u_z, and P given by equations (3.E.4-1), (3.E.4-2), and (3.E.4-6) are substituted into the radial component of the equations of motion given by equation (D.2-10) in the Appendices, we can conclude that $u_\theta = rf_4(z)$; that is,

$$\rho r f_1^2 - \frac{u_\theta^2}{r} + f_2 r \frac{df_1}{dz} = -\underbrace{\frac{df_3}{dr}}_{=0} + \mu \underbrace{\frac{\partial}{\partial r}\left[\frac{1}{r}\frac{\partial}{\partial r}(r^2 f_1)\right]}_{=0} + r\frac{d^2 f_1}{dz^2} \Rightarrow u_\theta = rf_4(z) \tag{3.E.4-7}$$

Hence, we conclude that for rotating disk laminar flow, u_z, P, ru_r, and u_θ/r are functions only of the axial coordinate z.

The prior considerations are rigorous for a rotating disk, causing laminar flow in an infinite fluid. However, in practice the rotating disk is placed in a finite container, which implies that the velocity might not be purely axial far above the rotating disk, due to recirculation. When one is using the rotating disk to characterize some heat- or mass-transfer process, one would like to minimize this finite container effect. This will be minimized when there is a thin boundary layer adjacent to the rotating disk across which the radial and angular velocities components decay. Hence, we use scaling to ascertain the criteria required to assure that this boundary layer is thin.

In view of the considerations for a rotating disk in an infinite fluid, the equations of motion in cylindrical coordinates given by equations (C.2-1), (D.2-10), (D.2-11), and (D.2-12) in the Appendices simplify to (step 1)

$$\rho u_r \frac{\partial u_r}{\partial r} - \rho \frac{u_\theta^2}{r} + \rho u_z \frac{\partial u_r}{\partial z} = \mu \frac{\partial^2 u_r}{\partial z^2} \tag{3.E.4-8}$$

$$\rho u_r \frac{\partial u_\theta}{\partial r} + \rho \frac{u_r u_\theta}{r} + \rho u_z \frac{\partial u_\theta}{\partial z} = \mu \frac{\partial^2 u_\theta}{\partial z^2} \tag{3.E.4-9}$$

$$\rho u_z \frac{du_z}{dz} = -\frac{dP}{dz} + \mu \frac{d^2 u_z}{dz^2} \tag{3.E.4-10}$$

$$\frac{1}{r}\frac{\partial}{\partial r}(r u_r) + \frac{\partial u_z}{\partial z} = 0 \tag{3.E.4-11}$$

The corresponding boundary conditions are given by

$$u_r = 0, \quad u_z = 0, \quad u_\theta = \omega r \qquad \text{at} \quad z = 0 \tag{3.E.4-12}$$

$$u_z = -U_\infty, \quad u_r = 0, \quad u_\theta = 0, \quad P = P_\infty \qquad \text{as} \quad z \to \infty \tag{3.E.4-13}$$

$$u_r = 0, \quad u_\theta = 0 \qquad \text{at} \quad r = 0 \tag{3.E.4-14}$$

Note that equations (3.E.4-8) through (3.E.4-14) are rigorous for a rotating disk in an unbounded fluid.

Introduce the following dimensionless variables containing appropriate reference and scale factors (steps 2, 3, and 4):

$$u_r^* \equiv \frac{u_r}{u_{rs}}; \qquad u_z^* \equiv \frac{u_z}{u_{zs}}; \qquad u_\theta^* \equiv \frac{u_\theta}{u_{\theta s}}; \qquad P^* \equiv \frac{P - P_r}{P_s}; \qquad r^* \equiv \frac{r}{r_s};$$
$$z^* \equiv \frac{z}{z_s} \tag{3.E.4-15}$$

Substituting these dimensionless variables into equations (3.E.4-8) through (3.E.4-14) and dividing through by the dimensional coefficient of one of the principal terms in each equation then yields the following dimensionless describing equations (steps 5 and 6):

$$\frac{\rho u_{rs} z_s^2}{\mu r_s} u_r^* \frac{\partial u_r^*}{\partial r^*} - \frac{\rho u_{\theta s}^2 z_s^2}{\mu u_{rs} r_s} \frac{u_\theta^{*2}}{r^*} + \frac{\rho u_{zs} z_s}{\mu} u_z^* \frac{\partial u_r^*}{\partial z^*} = \frac{\partial^2 u_r^*}{\partial z^{*2}} \tag{3.E.4-16}$$

$$\frac{\rho u_{rs} z_s^2}{\mu r_s} u_r^* \frac{\partial u_\theta^*}{\partial r^*} + \frac{\rho u_{rs} z_s^2}{\mu r_s} \frac{u_r^* u_\theta^*}{r^*} + \frac{\rho u_{zs} z_s}{\mu} u_z^* \frac{\partial u_\theta^*}{\partial z^*} = \frac{\partial^2 u_\theta^*}{\partial z^{*2}} \tag{3.E.4-17}$$

$$\frac{\rho u_{zs} z_s}{\mu} u_z^* \frac{du_z^*}{dz^*} = -\frac{P_s z_s}{\mu u_{zs}} \frac{dP^*}{dz^*} + \frac{d^2 u_z^*}{dz^{*2}} \tag{3.E.4-18}$$

$$\frac{u_{rs}z_s}{u_{zs}r_s}\frac{1}{r^*}\frac{\partial}{\partial r^*}(r^*u_r^*) + \frac{\partial u_z^*}{\partial z^*} = 0 \qquad (3.E.4\text{-}19)$$

$$u_r^* = 0, \quad u_z^* = 0, \quad u_\theta^* = \frac{\omega r_s}{u_{\theta s}}r^* \qquad \text{at} \quad z^* = 0 \qquad (3.E.4\text{-}20)$$

$$u_z^* = -\frac{U_\infty}{u_{zs}}, \quad u_r^* = 0, \quad u_\theta^* = 0, \quad P^* = \frac{P_\infty - P_r}{P_s} \qquad \text{as} \quad z^* \to \infty \qquad (3.E.4\text{-}21)$$

$$u_r^* = 0, \quad u_\theta^* = 0 \qquad \text{at} \quad r^* = 0 \qquad (3.E.4\text{-}22)$$

We now apply step 7 to achieve o(1) scaling. Since we seek to describe the flow at any arbitrary position along the disk, we set $r_s = R$, the local radial coordinate; that is, we are considering a local scaling analysis. The appropriate dimensionless groups in equations (3.E.4-20) and (3.E.4-21) suggest the following scale and reference factors:

$$\frac{u_{\theta s}}{\omega r_s} = \frac{u_{\theta s}}{\omega R} = 1 \Rightarrow u_{\theta s} = \omega R; \qquad \frac{P_\infty - P_r}{P_s} = 0 \Rightarrow P_r = P_\infty \qquad (3.E.4\text{-}23)$$

Since this is a developing flow, the continuity equation given by (3.E.4-19) implies that

$$\frac{u_{rs}z_s}{u_{zs}r_s} = \frac{u_{rs}z_s}{u_{zs}R} = 1 \Rightarrow u_{zs} = \frac{u_{rs}z_s}{R} \qquad (3.E.4\text{-}24)$$

The scale for u_r is obtained from equation (3.E.4-16) since the inertia and viscous terms must balance:

$$\frac{\rho u_{\theta s}^2 z_s^2}{\mu u_{rs} r_s} = \frac{\rho R\omega^2 z_s^2}{\mu u_{rs}} = 1 \Rightarrow u_{rs} = \frac{\rho R\omega^2 z_s^2}{\mu} \Rightarrow u_{zs} = \frac{\rho\omega^2 z_s^3}{\mu} \qquad (3.E.4\text{-}25)$$

Similarly, the inertia and viscous terms in equation (3.E.4-18) must balance, which provides the axial length scale:

$$\frac{\rho u_{zs} z_s}{\mu} = \frac{\rho^2\,\omega^2 z_s^4}{\mu^2} = 1 \Rightarrow z_s \equiv \delta_m = \sqrt{\frac{\mu}{\rho\omega}} = \sqrt{\frac{\nu}{\omega}} \qquad (3.E.4\text{-}26)$$

where ν is the kinematic viscosity. The axial length scale z_s has been identified with the momentum boundary-layer thickness δ_m, the region of influence within which the fluid is affected by the rotating disk. Note that in contrast to most boundary-layer problems in fluid dynamics, δ_m is constant over the entire surface of the rotating disk. Finally, the pressure scale is also obtained from equation (3.E.4-18):

$$\frac{P_s z_s}{\mu u_{zs}} = \frac{P_s R}{\mu u_{rs}} = \frac{P_s R}{\rho\omega^2\delta_m^3} = \frac{P_s R}{\rho\omega^{1/2}\nu^{3/2}} = 1 \Rightarrow P_s = \frac{\rho\omega^{1/2}\nu^{3/2}}{R} \qquad (3.E.4\text{-}27)$$

Substitution of the scale and reference factors then yields the following set of dimensionless describing equations:

$$u_r^* \frac{\partial u_r^*}{\partial r^*} - \frac{u_\theta^{*2}}{r^*} + u_z^* \frac{\partial u_r^*}{\partial z^*} = \frac{\partial^2 u_r^*}{\partial z^{*2}} \tag{3.E.4-28}$$

$$u_r^* \frac{\partial u_\theta^*}{\partial r^*} + \frac{u_r^* u_\theta^*}{r^*} + u_z^* \frac{\partial u_\theta^*}{\partial z^*} = \frac{\partial^2 u_\theta^*}{\partial z^{*2}} \tag{3.E.4-29}$$

$$u_z^* \frac{du_z^*}{dz^*} = -\frac{dP^*}{dz^*} + \frac{d^2 u_z^*}{dz^{*2}} \tag{3.E.4-30}$$

$$\frac{1}{r^*}\frac{\partial}{\partial r^*}(r^* u_r^*) + \frac{\partial u_z^*}{\partial z^*} = 0 \tag{3.E.4-31}$$

$$u_r^* = 0, \quad u_z^* = 0, \quad u_\theta^* = r^* \quad \text{at} \quad z^* = 0 \tag{3.E.4-32}$$

$$u_z^* = -\frac{U_\infty}{\sqrt{\omega\nu}}, \quad u_r^* =, 0, \quad u_\theta^* = 0, \quad P^* = 0 \quad \text{as} \quad z^* \to \infty \tag{3.E.4-33}$$

$$u_r^* = 0, \quad u_\theta^* = 0 \quad \text{at} \quad r^* = 0 \tag{3.E.4-34}$$

Equations (3.E.4-28) through (3.E.4-34) have been solved via a series-expansion method (see footnote 17). The resulting series can be truncated at the first term if $\delta_m \ll 1$ corresponding to a very thin momentum boundary layer. The resulting solution indicates that $\delta_m = 3.6\sqrt{\nu/\omega}$ and $U_\infty = 0.88447\sqrt{\nu\omega}$. These agree to within a constant of O(1) with the estimates obtained for δ_m given by equation (3.E.4-26) and for U_∞ obtained by setting the dimensionless group in equation (3.E.4-33) equal to 1. Note that scaling has provided reliable estimates of both the momentum boundary-layer thickness δ_m and the far-field axial velocity U_∞ without the need to actually solve the describing equations.

To assume that the rotating disk is effectively in an unbounded fluid, it is necessary for the boundary layer to be very thin; that is, the following criterion must be satisfied (step 8):

$$\frac{\delta_m}{H} = \sqrt{\frac{\nu}{\omega H^2}} \ll 1 \quad \text{or} \quad \sqrt{\frac{\nu}{\omega H^2}} = \text{o}(0.01) \tag{3.E.4-35}$$

where H denotes the distance of the rotating disk from the nearest parallel boundary. This condition will be satisfied when the kinematic viscosity is low, the angular rotation rate is high, or the boundary is far removed from the rotating disk.

3.E.5 Entry Region Flow Between Parallel Plates

Figure 3.E.5-1 shows a schematic of pressure-driven steady-state laminar entry-region flow of an incompressible Newtonian fluid with constant physical properties between two parallel flat plates spaced a distance $2H$ apart. The constant flow velocity at the entrance is U_0. This is assumed to be a high Reynolds number laminar flow for which the inertia or convection terms cannot be ignored in the entry

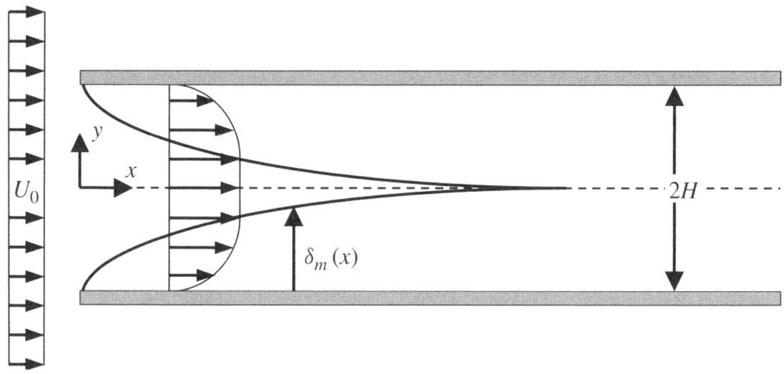

Figure 3.E.5-1 High Reynolds number steady-state pressure-driven entry region flow of an incompressible Newtonian fluid with constant physical properties between two parallel flat plates spaced a distance $2H$ apart.

region. We will use scaling to estimate the entry length required to achieve fully developed laminar flow. Note that this flow differs somewhat from the boundary-layer flow considered in Section 3.4 in that this is a confined boundary-layer flow. Hence, owing to the deceleration of the flow within the boundary layer at the walls, the uniform or plug flow in the center must accelerate. Moreover, in contrast to the boundary-layer flow considered in Section 3.4 that was caused by the velocity of the flow external to the boundary layer, the flow in the present example is caused by an applied pressure gradient.

The describing equations are obtained by simplifying equations (C.1-1), (D.1-10), and (D.1-11) in the Appendices appropriately (step 1):

$$\rho u_x \frac{\partial u_x}{\partial x} + \rho u_y \frac{\partial u_x}{\partial y} = -\frac{\partial P}{\partial x} + \mu \frac{\partial^2 u_x}{\partial x^2} + \mu \frac{\partial^2 u_x}{\partial y^2} \tag{3.E.5-1}$$

$$\rho u_x \frac{\partial u_y}{\partial x} + \rho u_y \frac{\partial u_y}{\partial y} = -\frac{\partial P}{\partial y} + \mu \frac{\partial^2 u_y}{\partial x^2} + \mu \frac{\partial^2 u_y}{\partial y^2} \tag{3.E.5-2}$$

$$\frac{\partial u_x}{\partial x} + \frac{\partial u_y}{\partial y} = 0 \tag{3.E.5-3}$$

$$u_x = U_0, \quad u_y = 0, \quad P = P_0 \quad \text{at} \quad x = 0 \tag{3.E.5-4}$$

$$u_x = f_1(y), \quad u_y = f_2(y) \quad \text{at} \quad x = L \tag{3.E.5-5}$$

$$u_x = 0, \quad u_y = 0 \quad \text{at} \quad y = \pm H \tag{3.E.5-6}$$

$$u_x = f_3(x), \quad u_y = f_4(x) \quad \text{at} \quad y = \pm(H - \delta_m) \tag{3.E.5-7}$$

where $f_1(y)$ and $f_2(y)$ are unspecified functions of y and $f_3(x)$ and $f_4(x)$ are unspecified functions of x that in principle would have to be known in order to integrate the set of differential equations above. Note that we have introduced a region of influence $\delta_m(x)$ within which the effect of the viscous shear induced by the presence of each wall is confined. The rationale for introducing this region of influence or hydrodynamic boundary-layer thickness was discussed in Section 3.4. The set of equations above also introduces the pressure P_0 that would have to be specified at the entry region in order to integrate the system of equations above.

Introduce the following scale factors, reference factors, and dimensionless variables (steps 2, 3, and 4):

$$u_x^* \equiv \frac{u_x}{u_{xs}}; \qquad u_y^* \equiv \frac{u_y}{u_{ys}}; \qquad P^* \equiv \frac{P}{P_s}; \qquad x^* \equiv \frac{x}{x_s}; \qquad y^* \equiv \frac{y - y_r}{y_s} \tag{3.E.5-8}$$

We have introduced a reference factor y_r in the definition of y^* to force this dimensionless variable to zero at the wall. Note that the symmetry of this problem permits considering only the region $-H \leq y \leq 0$. Substituting these dimensionless variables into equations (3.E.5-1) through (3.E.5-7) and dividing through by the dimensional coefficient of one of the principal terms in each equation then yields the following equations that describe the flow in the region $-H \leq y \leq 0$ (steps 5 and 6):

$$u_x^* \frac{\partial u_x^*}{\partial x^*} + \frac{u_{ys} x_s}{u_{xs} y_s} u_y^* \frac{\partial u_x^*}{\partial y^*} = -\frac{P_s}{\rho u_{xs}^2} \frac{\partial P^*}{\partial x^*} + \frac{\mu}{\rho u_{xs} x_s} \frac{\partial^2 u_x^*}{\partial x^{*2}} + \frac{\mu x_s}{\rho u_{xs} y_s^2} \frac{\partial^2 u_x^*}{\partial y^{*2}} \tag{3.E.5-9}$$

$$u_x^* \frac{\partial u_y^*}{\partial x^*} + \frac{u_{ys} x_s}{u_{xs} y_s} u_y^* \frac{\partial u_y^*}{\partial y^*} = -\frac{P_s x_s}{\rho u_{xs} u_{ys} y_s} \frac{\partial P^*}{\partial y^*} + \frac{\mu}{\rho u_{xs} x_s} \frac{\partial^2 u_y^*}{\partial x^{*2}} + \frac{\mu x_s}{\rho u_{xs} y_s^2} \frac{\partial^2 u_y^*}{\partial y^{*2}} \tag{3.E.5-10}$$

$$\frac{\partial u_x^*}{\partial x^*} + \frac{u_{ys} x_s}{u_{xs} y_s} \frac{\partial u_y^*}{\partial y^*} = 0 \tag{3.E.5-11}$$

$$u_x^* = \frac{U_0}{u_{xs}}, \quad u_y^* = 0, \quad P^* = \frac{P_0}{P_s} \quad \text{at} \quad x^* = 0 \tag{3.E.5-12}$$

$$u_x^* = f_1^*(y^*), \quad u_y^* = f_2^*(y^*) \quad \text{at} \quad x^* = \frac{L}{x_s} \tag{3.E.5-13}$$

$$u_x^* = 0, \quad u_y^* = 0 \quad \text{at} \quad y^* = -\frac{H - y_r}{y_s} \tag{3.E.5-14}$$

$$u_x^* = f_3^*(x^*), \quad u_y^* = f_4^*(x^*) \quad \text{at} \quad y^* = -\frac{H - \delta_m - y_r}{y_s} \tag{3.E.5-15}$$

We now apply step 7 to bound the variables to be $\circ(1)$. We can bound y^* to be between 0 and 1 by setting the dimensionless groups in equations (3.E.5-14) and (3.E.5-15) equal to zero and 1, respectively, to obtain $y_r = H$ and $y_s = \delta_m$. The axial length scale can be bounded to be between 0 and 1 by setting the dimensionless group in equation (3.E.5-13) equal to 1, thereby obtaining $x_s = L$. The axial velocity can be bounded to be between 0 and 1 by setting the dimensionless group in equation (3.E.5-12) equal to 1, thereby obtaining $u_{xs} = U_0$. Since this is a developing flow, both terms in the dimensionless continuity equation should be of the same order; hence, we require that the dimensionless group in equation (3.E.5-11) be equal to 1, thereby obtaining $u_{ys} = U_0\delta_m/L$. Since this is a pressure-driven high Reynolds number flow, the dimensionless pressure term should be of the same order as the inertia terms; hence, we require that dimensionless group multiplying the pressure term in equation (3.E.5-9) be equal to 1, thereby obtaining $P_s = \rho U_0^2$.

The resulting scaled dimensionless describing equations are given by

$$u_x^* \frac{\partial u_x^*}{\partial x^*} + u_y^* \frac{\partial u_x^*}{\partial y^*} = -\frac{\partial P^*}{\partial x^*} + \frac{1}{\text{Re}} \frac{\delta_m}{L} \frac{\partial^2 u_x^*}{\partial x^{*2}} + \frac{1}{\text{Re}} \frac{L}{\delta_m} \frac{\partial^2 u_x^*}{\partial y^{*2}} \tag{3.E.5-16}$$

$$u_x^* \frac{\partial u_y^*}{\partial x^*} + u_y^* \frac{\partial u_y^*}{\partial y^*} = -\frac{L^2}{\delta_m^2} \frac{\partial P^*}{\partial y^*} + \frac{1}{\text{Re}} \frac{\delta_m}{L} \frac{\partial^2 u_y^*}{\partial x^{*2}} + \frac{1}{\text{Re}} \frac{L}{\delta_m} \frac{\partial^2 u_y^*}{\partial y^{*2}} \tag{3.E.5-17}$$

$$\frac{\partial u_x^*}{\partial x^*} + \frac{\partial u_y^*}{\partial y^*} = 0 \tag{3.E.5-18}$$

$$u_x^* = 1, \quad u_y^* = 0, \quad P^* = \frac{P_0}{\rho U_0^2} \quad \text{at} \quad x^* = 0 \tag{3.E.5-19}$$

$$u_x^* = f_1^*(y^*), \quad u_y^* = f_2^*(y^*) \quad \text{at} \quad x^* = 1 \tag{3.E.5-20}$$

$$u_x^* = 0, \quad u_y^* = 0 \quad \text{at} \quad y^* = 0 \tag{3.E.5-21}$$

$$u_x^* = f_3^*(x^*), \quad u_y^* = f_4^*(x^*) \quad \text{at} \quad y^* = 1 \tag{3.E.5-22}$$

where $\text{Re} \equiv \delta_m \rho U_0/\mu$ is the Reynolds number. The principal viscous term in equation (3.E.5-16) must be of the same size as the pressure and inertia terms within the region of influence (hydrodynamic boundary layer) if we are to satisfy the boundary conditions given by equations (3.E.5-21) and (3.E.5-22). Hence, we require that the dimensionless group multiplying the principal viscous term in equation (3.E.5-16) be equal to 1, which provides an estimate of the boundary-layer thickness $\delta_m(x^*)$:

$$\frac{1}{\text{Re}} \frac{L}{\delta_m} = 1 \Rightarrow \delta_m = \sqrt{\frac{\mu L}{\rho U_0}} \tag{3.E.5-23}$$

When the boundary layer thickness $\delta_m = H$, the flow is fully developed. Hence, we can use equation (3.E.5-23) evaluated at $\delta_m = H$ to obtain an estimate of the entrance length L_e:

$$L_e \cong \frac{\rho U_0 H^2}{\mu} \tag{3.E.5-24}$$

This agrees to within a multiplicative constant of order 1 with the entrance length required to achieve fully developed flow obtained from solving the boundary-layer equations that yields[18]

$$L_e = 0.16 \frac{\rho U_0 H^2}{\mu} \tag{3.E.5-25}$$

[18] See H. Schlichting, *Boundary Layer Theory*, 4th ed., McGraw-Hill, New York, 1980, p. 168.

Whereas for this well-studied flow, an equation is available to predict the entrance length that obviates the need to use scaling analysis, the latter provides an invaluable tool for estimating the entry region for flows for which no such results are available.

Equations (3.E.5-16) through (3.E.5-23) can be greatly simplified if $\delta_m^2/L^2 = \text{o}(0.01)$ (step 8). This permits ignoring the axial diffusion of vorticity term in equation (3.E.5-16), thereby obviating the need to satisfy any downstream boundary condition. Moreover, in view of equation (3.E.5-17), this condition implies that the dimensionless derivative $\partial P^*/\partial y^*$ is very small. Note that we have not scaled the dimensionless derivative $\partial P^*/\partial y^*$ in equation (3.E.5-17) to be $\text{o}(1)$ since there is no reason for this derivative to scale as P_s/y_s. However, since we have scaled $u_x^* \partial u_y^*/\partial x^*$ to be of order $\text{o}(1)$, equation (3.E.5-17) implies that $\partial P^*/\partial y^*$ is of $\text{O}(\delta_m^2/L^2)$ and hence that it is very small. This decouples the solution of equation (3.E.5-16) from equation (3.E.5-17) and implies that the dimensionless describing equations for the entry-region flow problem can be reduced to

$$u_x^* \frac{\partial u_x^*}{\partial x^*} + u_y^* \frac{\partial u_x^*}{\partial y^*} = -\frac{dP^*}{dx^*} + \frac{\partial^2 u_x^*}{\partial y^{*2}} \tag{3.E.5-26}$$

$$\frac{\partial u_x^*}{\partial x^*} + \frac{\partial u_y^*}{\partial y^*} = 0 \tag{3.E.5-27}$$

$$u_x^* = 1, \quad u_y^* = 0, \quad P^* = \frac{P_0}{\rho U_0^2} \quad \text{at} \quad x^* = 0 \tag{3.E.5-28}$$

$$u_x^* = 0, \quad u_y^* = 0 \quad \text{at} \quad y^* = 0 \tag{3.E.5-29}$$

$$u_x^* = f_3^*(x^*) \quad \text{at} \quad y^* = 1 \tag{3.E.5-30}$$

To solve equations (3.E.5-26) through (3.E.5-30), it is necessary to know the unspecified function $f_3^*(x^*)$ in equation (3.E.5-30). This is obtained by solving the ideal flow (inviscid) flow equations[19] outside the boundary-layer region for which viscous effects can be ignored, due to the high Reynolds number. In doing this, one carries out integral mass and momentum balances that account for the acceleration of the flow due to the thinning of the inviscid core region that is caused by the growing boundary layer at each wall. These equations can then be solved analytically to determine the unspecified function $f_3^*(x^*)$ in equation (3.E.5-30). Equations (3.E.5-26) and (3.E.5-27) can then be solved numerically. The resulting solution will yield the entrance length given by equation (3.E.5-25).

3.E.6 Rotating Flow in an Annulus with End Effects

Consider the steady-state flow of an incompressible Newtonian liquid with constant physical properties in the annular region between two concentric cylinders of length L, shown in Figure 3.E.6-1. The inner cylinder has radius R_1 and is

[19]The ideal or inviscid flow equations correspond to an infinite Reynolds number, which implies no viscous effects whatsoever; in the case of hydrodynamic boundary-layer flows, the flow region outside the boundary layer is described by the ideal flow equations.

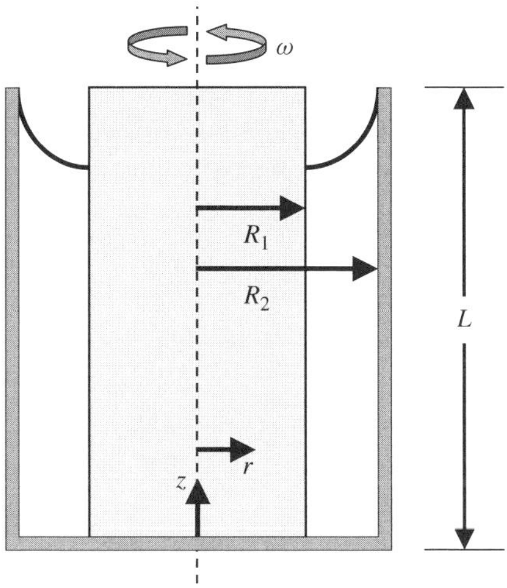

Figure 3.E.6-1 Steady-state incompressible laminar flow of a Newtonian liquid with constant physical properties in the annular region between a stationary inner and a rotating outer cylinder.

stationary. The outer cylinder has radius R_2 and rotates at a constant angular velocity ω (radians per second). The flow is caused primarily by the rotation of the outer cylinder. However, the bottom of the outer cylinder is also rotating and dragging the adjacent liquid with it, causing an end effect. We neglect any effect of the small gap between the bottom of the stationary inner cylinder and the rotating outer cylinder. We use scaling analysis to derive criteria for ignoring the end effect on the primary rotational flow in the annulus.

The describing equations for this flow are obtained by appropriately simplifying the equations of motion in cylindrical coordinates given by equations (D.2-10) through (D.2-12) in the Appendices (step 1):

$$\frac{\rho u_\theta^2}{r} = \frac{\partial P}{\partial r} \tag{3.E.6-1}$$

$$0 = \frac{\partial}{\partial r}\left[\frac{1}{r}\frac{\partial}{\partial r}(r u_\theta)\right] + \frac{\partial^2 u_\theta}{\partial z^2} \tag{3.E.6-2}$$

$$0 = \frac{\partial P}{\partial z} + \rho g \tag{3.E.6-3}$$

$$u_\theta = 0 \quad \text{at} \quad r = R_1 \tag{3.E.6-4}$$

$$u_\theta = \omega R_2 \quad \text{at} \quad r = R_2 \tag{3.E.6-5}$$

$$u_\theta = \omega r \quad \text{at} \quad z = 0 \tag{3.E.6-6}$$

$$\frac{\partial u_\theta}{\partial z} = 0 \quad \text{at} \quad z = f_1(r) \tag{3.E.6-7}$$

The boundary condition given by equation (3.E.6-7) allows for the fact that the rotation causes a centrifugal pressure force that increases with the radius. At any point in the liquid the centrifugal pressure has to be balanced by the hydrostatic pressure. Hence, the liquid depth will increase with increasing radius. The location of this interface can be determined from the solution to the pressure profile. The function $f_1(r)$ must satisfy the integral conservation of mass for an incompressible liquid given by

$$\int_{R_1}^{R_2} 2\pi r f_1(r)\,dr = \pi(R_2^2 - R_1^2)L_0 \tag{3.E.6-8}$$

where L_0 is the liquid depth in the absence of any rotation.

Introduce the following scale factors, reference factor, and dimensionless variables (steps 2, 3, and 4):

$$u_\theta^* \equiv \frac{u_\theta}{u_s}; \qquad P^* \equiv \frac{P}{P_s}; \qquad r^* \equiv \frac{r - r_r}{r_s}; \qquad z^* \equiv \frac{z}{z_s} \tag{3.E.6-9}$$

We have introduced a reference factor for the dimensionless radial coordinate since r is not naturally referenced to zero within the region where flow is occurring. Substitute these dimensionless variables into the describing equations and divide through by the dimensional coefficient of a term that must be retained (steps 5 and 6):

$$\frac{u_\theta^{*2}}{r^* + r_r/r_s} = \frac{P_s}{\rho u_s^2}\frac{\partial P^*}{\partial r^*} \tag{3.E.6-10}$$

$$0 = \frac{\partial}{\partial r^*}\left\{\frac{1}{r^* + r_r/r_s}\frac{\partial}{\partial r^*}\left[\left(r^* + \frac{r_r}{r_s}\right)u_\theta^*\right]\right\} + \frac{r_s^2}{z_s^2}\frac{\partial^2 u_\theta^*}{\partial z^{*2}} \tag{3.E.6-11}$$

$$0 = \frac{P_s}{\rho g z_s}\frac{\partial P^*}{\partial z^*} + 1 \tag{3.E.6-12}$$

$$u_\theta^* = 0 \quad \text{at} \quad r^* = \frac{R_1 - r_r}{r_s} \tag{3.E.6-13}$$

$$u_\theta^* = \frac{\omega R_2}{u_s} \quad \text{at} \quad r^* = \frac{R_2 - r_r}{r_s} \tag{3.E.6-14}$$

$$u_\theta^* = \frac{\omega r_s r^* + r_r/r_s}{u_s} \quad \text{at} \quad z^* = 0 \tag{3.E.6-15}$$

$$\frac{\partial u_\theta^*}{\partial z^*} = 0 \quad \text{at} \quad z^* = f_1^*(r^*) \tag{3.E.6-16}$$

$$\int_{\frac{R_1 - r_r}{r_s}}^{\frac{R_2 - r_r}{r_s}} 2\left(r^* + \frac{r_r}{r_s}\right) f_1^*\left(r^*\right) dr^* = \frac{\left(R_2^2 - R_1^2\right) L_0}{r_s^2 z_s} \tag{3.E.6-17}$$

We now apply step 7 to bound the variables to be $\text{o}(1)$. We can bound r^* to be between zero and 1 by setting the dimensionless group in equation (3.E.6-13) equal to zero and that in equation (3.E.6-14) equal to 1, thereby obtaining $r_r = R_1$

and $r_s = R_2 - R_1$. Our dimensionless axial coordinate can be bounded between zero and 1 by setting the dimensionless group L_0/z_s that appears in equation (3.E.6-17) equal to 1, thereby obtaining $z_s = L_0$. Our velocity scale is obtained from the dimensionless group in equation (3.E.6-14) to obtain $u_s = \omega R_2$. Finally, our pressure scale is obtained from the dimensionless group in equation (3.E.6-10) to obtain $P_s = \rho\omega^2 R_2^2$. When these scale and reference factors are substituted in equations (3.E.6-10) through (3.E.6-17), we obtain the following set of dimensionless describing equations:

$$\frac{u_\theta^{*2}}{r^* + R_1/(R_2 - R_1)} = \frac{\partial P^*}{\partial r^*} \tag{3.E.6-18}$$

$$0 = \frac{\partial}{\partial r^*}\left\{\frac{1}{r^* + R_1/(R_2 - R_1)}\frac{\partial}{\partial r^*}\left[\left(r^* + \frac{R_1}{R_2 - R_1}\right)u_\theta^*\right]\right\} + \frac{(R_2 - R_1)^2}{L_0^2}\frac{\partial^2 u_\theta^*}{\partial z^{*2}} \tag{3.E.6-19}$$

$$0 = \frac{\omega^2 R_2^2}{gL}\frac{\partial P^*}{\partial z^*} + 1 \tag{3.E.6-20}$$

$$u_\theta^* = 0 \quad \text{at} \quad r^* = 0 \tag{3.E.6-21}$$

$$u_\theta^* = 1 \quad \text{at} \quad r^* = 1 \tag{3.E.6-22}$$

$$u_\theta^* = \frac{R_2 - R_1}{R_2}\left(r^* + \frac{R_1}{R_2 - R_1}\right) \quad \text{at} \quad z^* = 0 \tag{3.E.6-23}$$

$$\frac{\partial u_\theta^*}{\partial z^*} = 0 \quad \text{at} \quad z^* = f_1^*(r^*) \tag{3.E.6-24}$$

$$\int_0^1 2\left(r^* + \frac{R_1}{R_2 - R_1}\right) f_1^*(r^*)\, dr^* = \frac{R_2 + R_1}{R_2 - R_1} \tag{3.E.6-25}$$

We see that if the dimensionless group $(R_2 - R_1)^2/L_0^2 = \mathrm{o}(0.01)$ the end effect can be ignored in equation (3.E.6-19) (step 8). The resulting simplified set of describing equations can be solved analytically.[20] A further simplification is possible if the group $(R_2 - R_1)/R_1 = \mathrm{o}(0.01)$, in which case the describing equations reduce to

$$0 = \frac{\partial P^*}{\partial r^*} \tag{3.E.6-26}$$

$$0 = \frac{d^2 u_\theta^*}{dr^{*2}} \tag{3.E.6-27}$$

$$0 = \frac{\omega^2 R_2^2}{gL_0}\frac{\partial P^*}{\partial z^*} + 1 \tag{3.E.6-28}$$

[20]This simplified set of describing equations has been solved in Bird et al., *Transport Phenomena*, 2nd ed., pp. 93–95; however, no attempt is made to justify when these simplified equations are applicable.

$$u_\theta^* = 0 \quad \text{at} \quad r^* = 0 \tag{3.E.6-29}$$

$$u_\theta^* = 1 \quad \text{at} \quad r^* = 1 \tag{3.E.6-30}$$

$$\int_0^1 f_1^*(r^*)\, dr^* = 1 \tag{3.E.6-31}$$

This simplified set of describing equations also admits an analytical solution. The integral mass-balance condition is retained in the form given by equation (3.E.6-31), which permits obtaining the liquid depth profile in the annular region.

3.E.7 Impulsively Initiated Pressure-Driven Laminar Tube Flow

Consider an incompressible Newtonian liquid with constant physical properties contained in a semi-infinitely long cylindrical tube having radius R. Initially, the liquid in the tube is at rest. At time $t = 0$, a constant pressure drop $\Delta P \equiv P_0 - P_L$ is applied across some length L of the tube to initiate continuous flow, as shown in Figure 3.E.7-1. We will ignore any entrance effects, in which case this is an interesting example of an unsteady-state fully developed flow. Figure 3.E.7-1 shows the axial velocity profiles at times t_1 and t_2, where $t_2 > t_1$. If the entrance effects are neglected, the velocity profile at any time applies along the entire length of the tube. The unsteady-state acceleration of the flow is suggested by the increased area under the velocity profile at t_2 relative to t_1. We use scaling analysis to determine the criterion for assuming that this impulsively initiated flow has achieved steady-state conditions.

The describing equations for this flow are obtained by appropriately simplifying the unsteady-state equations of motion in cylindrical coordinates given by equation (D.2-12) in the Appendices to obtain (step 1)

$$\rho \frac{\partial u_z}{\partial t} = \frac{\Delta P}{L} + \mu \frac{1}{r}\frac{\partial}{\partial r}\left(r \frac{\partial u_z}{\partial r}\right) \tag{3.E.7-1}$$

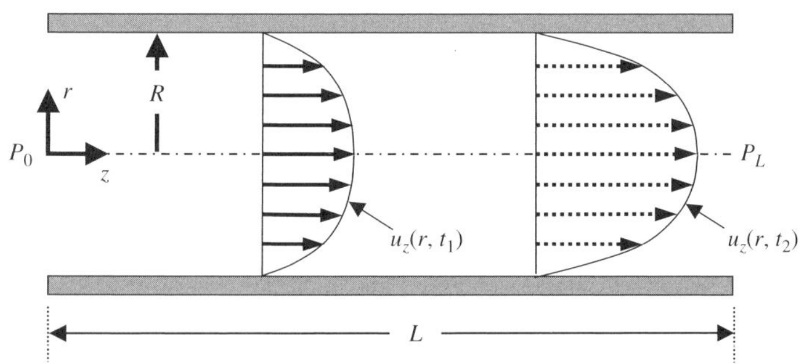

Figure 3.E.7-1 Laminar flow of an incompressible Newtonian liquid with constant physical properties in a circular tube of radius R due to an impulsively applied pressure difference $\Delta P \equiv P_0 - P_L$, showing the axial velocity profiles at times t_1 and t_2, where $t_2 > t_1$.

$$u_z = 0 \quad \text{at} \quad t = 0 \tag{3.E.7-2}$$

$$u_z = 0 \quad \text{at} \quad r = R \tag{3.E.7-3}$$

$$\frac{\partial u_z}{\partial r} = 0 \quad \text{at} \quad r = 0 \tag{3.E.7-4}$$

In writing equation (3.E.7-1) we have used the fact that $u_z = f_1(r, t)$ and that the radial component of the equations of motion establishes that $\partial P/\partial r = 0$, which in turn implies that $P = f_2(z, t)$. In view of these considerations, the axial component of the equations of motion then implies that $\partial P/\partial z = -\Delta P/L$.

Introduce the following dimensionless variables (steps 2, 3, and 4):

$$u_z^* \equiv \frac{u_z}{u_s}; \qquad r^* \equiv \frac{r}{r_s}; \qquad t^* \equiv \frac{t}{t_s} \tag{3.E.7-5}$$

Substitute these dimensionless variables into equations (3.E.7-1) through (3.E.7-4) and divide through by the dimensional coefficient of a term that must be retained to obtain (steps 5 and 6)

$$\frac{\rho r_s^2}{\mu t_s}\frac{\partial u_z^*}{\partial t^*} = \frac{\Delta P r_s^2}{L\mu u_s} + \frac{1}{r^*}\frac{\partial}{\partial r^*}\left(r^*\frac{\partial u_z^*}{\partial r^*}\right) \tag{3.E.7-6}$$

$$u_z^* = 0 \quad \text{at} \quad t^* = 0 \tag{3.E.7-7}$$

$$u_z^* = 0 \quad \text{at} \quad r^* = \frac{R}{r_s} \tag{3.E.7-8}$$

$$\frac{\partial u_z^*}{\partial r^*} = 0 \quad \text{at} \quad r^* = 0 \tag{3.E.7-9}$$

We can bound our radial coordinate between zero and 1 by setting the dimensionless group in equation (3.E.7-8) equal to 1 thereby obtaining $r_s = R$ (step 7). Since pressure causes this flow, we obtain our velocity scale by setting the dimensionless group in equation (3.E.7-6) equal to 1 thereby obtaining $u_s = \Delta P R^2/L\mu$. Our time scale in this case is the observation time t_o; that is, the arbitrary time at which we chose to observe this flow after it is impulsively initiated. When these scale factors are substituted into equations (3.E.7-6) through (3.E.7-9), we obtain

$$\frac{\rho R^2}{\mu t_o}\frac{\partial u_z^*}{\partial t^*} = 1 + \frac{1}{r^*}\frac{\partial}{\partial r^*}\left(r^*\frac{\partial u_z^*}{\partial r^*}\right) \tag{3.E.7-10}$$

$$u_z^* = 0 \quad \text{at} \quad t^* = 0 \tag{3.E.7-11}$$

$$u_z^* = 0 \quad \text{at} \quad r^* = 1 \tag{3.E.7-12}$$

$$\frac{\partial u_z^*}{\partial r^*} = 0 \quad \text{at} \quad r^* = 0 \tag{3.E.7-13}$$

We see from equation (3.E.7-10) that the unsteady-state term becomes insignificant when the condition $\rho R^2/\mu t_o = \text{o}(0.01)$ applies (step 8). This in turn implies that steady-state will be achieved for observation times that satisfy the condition

$$t_o \gg \frac{\rho R^2}{\mu} \Rightarrow \text{steady-state is achieved} \tag{3.E.7-14}$$

The unsteady-state flow problem described by equations (3.E.7-1) through (3.E.7-4) has been solved analytically[21]; the solution indicates that the centerline (maximum) velocity will be within 10% of its steady-state value when

$$t_o = 0.45\frac{\rho R^2}{\mu} \Rightarrow \left\{ \begin{array}{l} \text{velocity is within 10\%} \\ \text{of its steady-state value} \end{array} \right. \tag{3.E.7-15}$$

It is surprising that the criterion that we derived from scaling analysis for achieving steady-state conditions is far more demanding than that obtained from an exact solution to the describing equations. However, the criterion given by equation (3.E.7-14) is based on the condition required for the pressure force to exactly balance the viscous force in equation (3.E.7-6). The latter is proportional to the derivative of the velocity profile. When the centerline (maximum) velocity is within 10% of its steady-state value, the slope of the velocity profile at the wall, which is proportional to the pressure applied, is nowhere near 10% of its steady-state value. Our more demanding criterion ensures that we predict not only the maximum velocity accurately via a steady-state solution, but also the drag at the wall.

3.E.8 Laminar Cylindrical Jet Flow

Consider the steady-state laminar flow of an incompressible Newtonian liquid jet with constant physical properties issuing from a circular orifice of initial velocity U_0 falling vertically under the influence of gravity in an inviscid gas as shown in Figure 3.E.8-1. We assume that curvature and surface-tension effects can be ignored in the tangential and normal stress boundary conditions at the interface between the liquid jet and ambient gas phase.[22] We use scaling analysis to explore the conditions for which quasi-parallel flow can be assumed; that is, when the axial velocity profile can be assumed to depend only on the axial coordinate.

The appropriately simplified equations of motion in cylindrical coordinates given by equations (C.2-1), (D.2-10), and (D.2-12) in the Appendices along with the boundary and kinematic conditions are given by (step 1)

$$\rho u_r \frac{\partial u_r}{\partial r} + \rho u_z \frac{\partial u_r}{\partial z} = -\frac{\partial P}{\partial r} + \mu \frac{\partial}{\partial r}\left[\frac{1}{r}\frac{\partial}{\partial r}(r u_r)\right] + \mu \frac{\partial^2 u_r}{\partial z^2} \tag{3.E.8-1}$$

[21]R. B. Bird, W. E. Stewart, and E. N. Lightfoot, *Transport Phenomena*, Wiley, New York, 1960 pp. 126–130.

[22]Note that scaling analysis could be used to determine when surface-tension and curvature effects can be neglected; the latter were considered in Section 3.7.

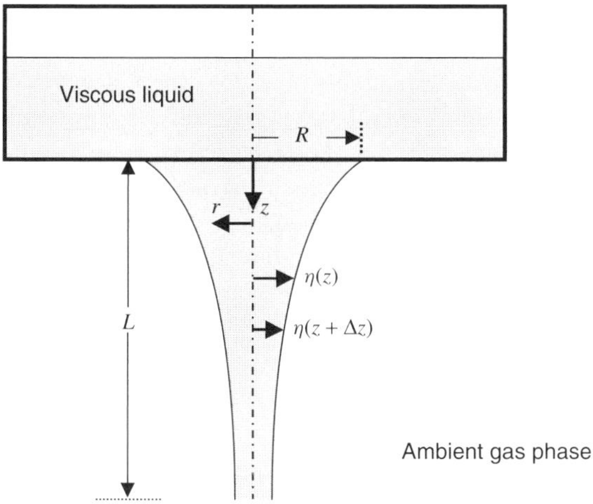

Figure 3.E.8-2 Steady-state flow of an incompressible Newtonian liquid that has constant physical properties issuing as a jet from a circular orifice of radius R with an initial velocity U_0 into an inviscid ambient gas phase.

$$\rho u_r \frac{\partial u_z}{\partial r} + \rho u_z \frac{\partial u_z}{\partial z} = -\frac{\partial P}{\partial z} + \frac{\mu}{r}\frac{\partial}{\partial r}\left(r \frac{\partial u_z}{\partial r}\right) + \mu \frac{\partial^2 u_r}{\partial z^2} + \rho g \tag{3.E.8-2}$$

$$\frac{1}{r}\frac{\partial}{\partial r}(r u_r) + \frac{\partial u_z}{\partial z} = 0 \tag{3.E.8-3}$$

$$u_z = U_0, \quad u_r = 0, \quad P = P_{\text{atm}}, \quad \eta = R \qquad \text{at} \quad z = 0 \tag{3.E.8-4}$$

$$u_z = f_1(r), \quad u_r = f_2(r) \qquad \text{at} \quad z = L \tag{3.E.8-5}$$

$$\frac{\partial u_z}{\partial r} = 0, \quad u_r = 0 \qquad \text{at} \quad r = 0 \tag{3.E.8-6}$$

$$\left.\begin{aligned} &\tau_{rz} = -\mu\left(\frac{\partial u_z}{\partial r} + \frac{\partial u_r}{\partial z}\right) = 0 \\ &\sigma_{rr} = P - 2\mu \frac{\partial u_r}{\partial r} = P_{\text{atm}} \\ &\frac{\partial P}{\partial z} - 2\mu \frac{\partial^2 u_r}{\partial z\, \partial r} = 0 \end{aligned}\right\} \quad \text{at} \quad r = \eta(z) \tag{3.E.8-7}$$

$$\frac{d\eta}{dz} = \frac{u_r}{u_z} \qquad \text{at} \quad r = \eta(z) \tag{3.E.8-8}$$

where f_1 and f_2 are unknown functions of r that are included for completeness since in principle downstream boundary conditions are required for the velocity components. Equation (3.E.8-7) encompasses three boundary conditions at the interface between the liquid jet and the ambient gas phase required for the three dependent variables: pressure and the two velocity components. The first two of these equations are the no-drag and continuity of the normal stress, respectively.

The third of these equations is obtained by differentiating the normal stress balance with respect to z; this provides an independent condition interrelating the pressure and velocity at the interface. Equation (3.E.8-8) is the kinematic surface condition that is obtained by an integral mass balance on a differential volume element over an arbitrary cross section of the jet. This is needed as an auxiliary condition to obtain the location of the surface at which the no-slip and continuity of normal stress boundary conditions must be applied.

Introduce the following scale factors, reference factors, and dimensionless variables (steps 2, 3, and 4):

$$u_z^* \equiv \frac{u_z - u_{zr}}{u_{zs}}; \qquad u_r^* \equiv \frac{u_r}{u_{rs}}; \qquad P^* \equiv \frac{P - P_r}{P_s}; \qquad \eta^* \equiv \frac{\eta}{\eta_s};$$
$$\left(\frac{\partial u_z}{\partial r}\right)^* \equiv \frac{1}{\beta_s}\frac{\partial u_z}{\partial r}; \qquad z^* \equiv \frac{z}{z_s}; \qquad r^* \equiv \frac{r}{r_s} \tag{3.E.8-9}$$

We have introduced reference factors for both the axial velocity and pressure since neither of these variables is naturally referenced to zero. Note that we have also introduced a scale β_s for $\partial u_z/\partial r$ since this derivative does not scale as u_{zs}/r_s. If we had used the latter to scale this derivative, the forgiving nature of scaling would have indicated a contradiction. However, we anticipate the need to scale this derivative with its own scale since u_z does not change significantly across the jet. Introduce these dimensionless variables into the describing equations and divide through by the dimensional coefficient of a term that must be retained to obtain (steps 5 and 6)

$$u_r^* \frac{\partial u_r^*}{\partial r^*} + \frac{u_{zs} r_s}{u_{rs} z_s}\left(u_z^* + \frac{u_{zr}}{u_{zs}}\right)\frac{\partial u_r^*}{\partial z^*}$$
$$= -\frac{P_s}{\rho u_{rs}^2}\frac{\partial P^*}{\partial r^*} + \frac{\mu}{\rho u_{rs} r_s}\frac{\partial}{\partial r^*}\left[\frac{1}{r^*}\frac{\partial}{\partial r^*}(r^* u_r^*)\right] + \frac{\mu r_s}{\rho u_{rs} z_s^2}\frac{\partial^2 u_r^*}{\partial z^{*2}} \tag{3.E.8-10}$$

$$\frac{u_{rs}\beta_s z_s}{u_{zs}^2} u_r^* \frac{\partial u_z^*}{\partial r^*} + \left(u_z^* + \frac{u_{zr}}{u_{zs}}\right)\frac{\partial u_z^*}{\partial z^*}$$
$$= -\frac{P_s}{\rho u_{zs}^2}\frac{\partial P^*}{\partial z^*} + \frac{\mu \beta_s z_s}{\rho u_{zs}^2 r_s}\frac{1}{r^*}\frac{\partial}{\partial r^*}\left[r^*\left(\frac{\partial u_z}{\partial r}\right)^*\right] + \frac{\mu}{\rho u_{zs} z_s}\frac{\partial^2 u_z^*}{\partial z^{*2}} + \frac{g z_s}{u_{zs}^2} \tag{3.E.8-11}$$

$$\frac{1}{r^*}\frac{\partial}{\partial r^*}(r^* u_r^*) + \frac{u_{zs} r_s}{u_{rs} z_s}\frac{\partial u_z^*}{\partial z^*} = 0 \tag{3.E.8-12}$$

$$u_z^* = \frac{U_0 - u_{zr}}{u_{zs}}, \quad u_r^* = 0, \quad P^* = \frac{P_{\text{atm}} - P_r}{P_s}, \quad \eta^* = \frac{R}{\eta_s} \qquad \text{at} \quad z^* = 0 \tag{3.E.8-13}$$

$$u_z^* = f_1^*(r^*), \quad u_r^* = f_2^*(r^*) \qquad \text{at} \quad z^* = \frac{L}{z_s} \tag{3.E.8-14}$$

$$\frac{\partial u_z^*}{\partial r^*} = 0, \quad u_r^* = 0 \qquad \text{at} \quad r^* = 0 \tag{3.E.8-15}$$

$$\left.\begin{aligned}
&\left(\frac{\partial u_z}{\partial r}\right)^* + \frac{u_{rs}}{\beta_s z_s}\frac{\partial u_r^*}{\partial z^*} = 0\\
&P^* = 2\frac{\mu u_{rs}}{P_s r_s}\frac{\partial u_r^*}{\partial r^*} + \frac{P_{\text{atm}} - P_r}{P_s}\\
&\frac{\partial P^*}{\partial z^*} - 2\frac{\mu u_{rs}}{P_s r_s}\frac{\partial^2 u_r^*}{\partial z^* \partial r^*} = 0
\end{aligned}\right\} \quad \text{at} \quad r^* = \frac{\eta_s}{r_s}\eta^* \tag{3.E.8-16}$$

$$\frac{d\eta^*}{dz^*} = \frac{u_{rs} z_s}{u_{zs}\eta_s}\frac{u_r^*}{u_z^*} \quad \text{at} \quad r^* = \frac{\eta_s}{r_s}\eta^* \tag{3.E.8-17}$$

Note $\partial^2 u_z/\partial r^2$ scales as β_s/r_s since $\partial u_z/\partial r$ goes from its minimum value of zero at $r = 0$ to its maximum value of β_s at $r = \eta(z)$.

We now apply step 7 to bound the variables to be $\circ(1)$. The dimensionless groups consisting of geometric ratios in equations (3.E.8-13), (3.E.8-14), and (3.E.8-16), when set equal to 1, imply that $\eta_s = r_s = R$ and $z_s = L$. The dimensionless groups in equation (3.E.8-13) containing the reference velocity and pressure, when set equal to zero, imply that $u_{zr} = U_0$ and $P_r = P_{\text{atm}}$. Since gravity causes flow in the axial direction, the dimensionless group that is a measure of the ratio of the gravity force to the axial acceleration must be set equal to 1, thereby obtaining the axial velocity scale $u_{zs} = \sqrt{gL}$. Since this is a developing flow, the dimensionless group in the continuity equation must be set equal to 1, thereby obtaining the radial velocity scale $u_{rs} = R\sqrt{gL}/L$. Since the two remaining terms in the normal stress balance in equation (3.E.8-16) must balance, the dimensionless group in this equation must be set equal to 1, thereby obtaining the pressure scale $P_s = \mu\sqrt{gL}/L$. Finally, since the two terms in the zero-drag condition in equation (3.E.8-16) must balance, the dimensionless group in this equation must be set equal to 1, thereby obtaining the derivative scale $\beta_s = R\sqrt{gL}/L^2$. When these values for the scale and reference factors are substituted into equations (3.E.8-10) through (3.E.8-17), we obtain the following minimum parametric representation of the describing equations:

$$u_r^*\frac{\partial u_r^*}{\partial r^*} + \left(u_z^* + \frac{U_0}{\sqrt{gL}}\right)\frac{\partial u_r^*}{\partial z^*}$$

$$= -\frac{1}{\text{Re}}\frac{L}{R}\frac{\partial P^*}{\partial r^*} + \frac{1}{\text{Re}}\frac{L}{R}\frac{\partial}{\partial r^*}\left[\frac{1}{r^*}\frac{\partial}{\partial r^*}(r^* u_r^*)\right] + \frac{1}{\text{Re}}\frac{R}{L}\frac{\partial^2 u_r^*}{\partial z^{*2}} \tag{3.E.8-18}$$

$$\frac{R^2}{L^2}u_r^*\frac{\partial u_z^*}{\partial r^*} + \left(u_z^* + \frac{U_0}{\sqrt{gL}}\right)\frac{\partial u_z^*}{\partial z^*}$$

$$= -\frac{1}{\text{Re}}\frac{R}{L}\frac{\partial P^*}{\partial z^*} + \frac{1}{\text{Re}}\frac{R}{L}\frac{1}{r^*}\frac{\partial}{\partial r^*}\left[r^*\left(\frac{\partial u_z}{\partial r}\right)^*\right] + \frac{1}{\text{Re}}\frac{R}{L}\frac{\partial^2 u_z^*}{\partial z^{*2}} + 1 \tag{3.E.8-19}$$

$$\frac{1}{r^*}\frac{\partial}{\partial r^*}(r^* u_r^*) + \frac{\partial u_z^*}{\partial z^*} = 0 \tag{3.E.8-20}$$

$$u_z^* = 0, \quad u_r^* = 0, \quad P^* = 0, \quad \eta^* = 1 \quad \text{at} \quad z^* = 0 \tag{3.E.8-21}$$

$$u_z^* = f_1^*(r^*), \quad u_r^* = f_2^*(r^*) \qquad \text{at} \quad z^* = 1 \tag{3.E.8-22}$$

$$\frac{\partial u_z^*}{\partial r^*} = 0, \quad u_r^* = 0 \qquad \text{at} \quad r^* = 0 \tag{3.E.8-23}$$

$$\left.\begin{aligned} &\left(\frac{\partial u_z}{\partial r}\right)^* + \frac{\partial u_r^*}{\partial z^*} = 0 \\ &P^* = 2\frac{\partial u_r}{\partial r} \\ &\frac{\partial P^*}{\partial z^*} - 2\frac{\partial^2 u_r^*}{\partial z^* \partial r^*} = 0 \end{aligned}\right\} \quad \text{at} \quad r^* = \eta^* \tag{3.E.8-24}$$

$$\frac{d\eta^*}{dz^*} = \frac{u_r^*}{u_z^*} \qquad \text{at} \quad r^* = \eta^* \tag{3.E.8-25}$$

where $\text{Re} \equiv \rho R u_{zs}/\mu = \rho R\sqrt{gL}/\mu$ is the Reynolds number.

Inspection of equations (3.E.8-18) through (3.E.8-25) indicates that the criteria for assuming quasi-parallel flow are the following (step 8):

$$\left.\begin{aligned} &\text{Re} = \frac{\rho R\sqrt{gL}}{\mu} \gg 1 \\ &\frac{R}{L} \ll 1 \end{aligned}\right\} \Rightarrow \text{quasi-parallel flow} \tag{3.E.8-26}$$

When the conditions above apply, equations (3.E.8-18) through (3.E.8-25) simplify to:

$$\left(u_z^* + \frac{U_0}{\sqrt{gL}}\right)\frac{\partial u_z^*}{\partial z^*} = 1 \tag{3.E.8-27}$$

$$\frac{1}{r^*}\frac{\partial}{\partial r^*}(r^* u_r^*) + \frac{\partial u_z^*}{\partial z^*} = 0 \tag{3.E.8-28}$$

$$u_z^* = 0, \quad \eta^* = 1 \qquad \text{at} \quad z^* = 0 \tag{3.E.8-29}$$

$$u_r^* = 0 \qquad \text{at} \quad r^* = 0 \tag{3.E.8-30}$$

$$\frac{d\eta^*}{dz^*} = \frac{u_r^*}{u_z^*} \qquad \text{at} \quad r^* = \eta^* \tag{3.E.8-31}$$

The solution to the system of equations above is straightforward and yields the following solution for the axial velocity:

$$u_z^* = -\frac{U_0}{\sqrt{gL}} + \sqrt{\frac{U_0^2}{gL} + 2z^*} \Rightarrow u_z = \sqrt{U_0^2 + 2gz} \tag{3.E.8-32}$$

This solution for the axial velocity profile corresponds to acceleration in free fall, which of course is a reasonable solution under the assumption that the viscous effects are negligible. Equations (3.E.8-32) can be substituted into equations (3.E.8-28) and (3.E.8-31) to obtain the corresponding radial velocity and jet diameter as a function of axial position.

3.E.9 Gravity-Driven Film Flow over a Saturated Porous Medium

Consider the steady-state fully developed flow of an incompressible Newtonian liquid film over an inclined liquid-saturated porous medium due to a gravitationally induced body force, as shown in Figure 3.E.9-1. This flow could correspond to runoff down water-saturated sloped ground. Because of the slope, gravity will cause flow of both the liquid in the film and that within the porous medium. We use scaling to determine when the flow through the porous medium has a negligible effect on the flow of the liquid film.

The describing equations for this flow are obtained by simplifying equations (D.1-10) and (E.1-1) in the Appendices appropriately (step 1):

$$0 = \mu \frac{d^2 u_x}{dy^2} + \rho g \sin\theta \qquad 0 \le y \le H \tag{3.E.9-1}$$

$$0 = -\frac{\mu}{k_p} \widehat{u}_x + \mu \frac{d^2 \widehat{u}_x}{dy^2} + \rho g \sin\theta \qquad -\infty < y \le 0 \tag{3.E.9-2}$$

$$\frac{du_x}{dy} = 0 \quad \text{at} \quad y = H \tag{3.E.9-3}$$

$$u_x = \widehat{u}_x \quad \text{at} \quad y = 0 \tag{3.E.9-4}$$

$$\frac{d\widehat{u}_x}{dy} = \frac{du_x}{dy} \quad \text{at} \quad y = 0 \tag{3.E.9-5}$$

$$\widehat{u}_x = 0 \quad \text{as} \quad y \to -\infty \tag{3.E.9-6}$$

where μ is the shear viscosity and k_p is the Darcy permeability. Note that $\widehat{u}_x$ in the porous medium is the superficial velocity; that is, the velocity through the porous medium treated as if it were homogeneous. Equations (3.E.9-4) and (3.E.9-5) are the continuity of velocity and shear at the interface between the porous medium and the liquid film; the latter equation assumes that the effective viscosity of the liquid flowing through the porous medium is the same as that of the liquid in the film.

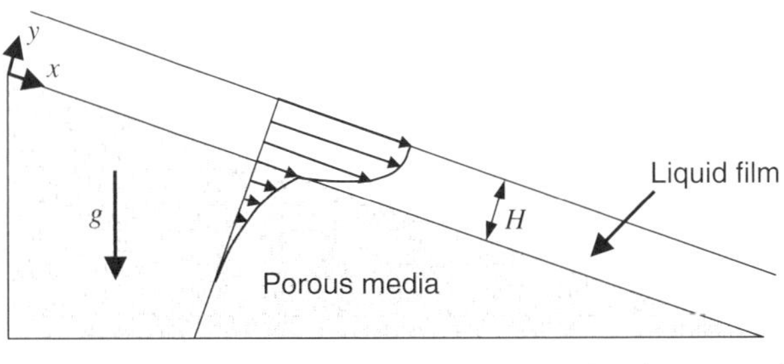

Figure 3.E.9-1 Steady-state fully developed laminar flow of an incompressible Newtonian liquid film of thickness H with constant physical properties over an inclined liquid-saturated porous medium due to a gravitationally induced body force.

Define the following dimensionless variables (steps 2, 3, and 4):

$$u_x^* \equiv \frac{u_x}{u_{xs}}; \qquad \widehat{u}_x{}^* \equiv \frac{\widehat{u}_x}{\widehat{u}_{xs}}; \qquad y^* \equiv \frac{y}{y_s} \quad \text{for } 0 \le y \le H;$$

$$\widehat{y}^* \equiv \frac{y}{\widehat{y}_s} \quad \text{for } -\infty < y \le 0 \tag{3.E.9-7}$$

We have introduced separate scales for the velocity as well as for the spatial coordinate in the two regions. The need for this can be seen by considering the fact that the maximum velocity in the porous medium is the minimum velocity in the liquid film; hence, these two velocities must be scaled differently to achieve o(1) scaling. The different spatial coordinate scales are necessary because the velocity goes between its maximum and minimum values over vastly different length scales in the two regions. Note again that had we not done this, we would have arrived at a contradiction in our scaled equations; the forgiving nature of scaling would then indicate that we had not scaled some quantity so that it was bounded of o(1). Introduce these dimensionless variables and divide through by the dimensional coefficient of one term in each of equations (3.E.9-1) through (3.E.9-6) to obtain (steps 5 and 6)

$$0 = \frac{d^2u_x^*}{dy^{*2}} + \frac{\rho g y_s^2 \sin\theta}{\mu u_{xs}}, \qquad 0 \le y^* \le \frac{H}{y_s} \tag{3.E.9-8}$$

$$0 = -\widehat{u}_x^* + \frac{k_p}{\widehat{y}_s^2}\frac{d^2\widehat{u}_x^*}{d\widehat{y}_2^*} + \frac{\rho g k_p \sin\theta}{\mu \widehat{u}_{xs}}, \qquad -\infty < \widehat{y}^* \le 0 \tag{3.E.9-9}$$

$$\frac{du_x^*}{dy^*} = 0 \quad \text{at} \quad y^* = \frac{H}{y_s} \tag{3.E.9-10}$$

$$u_x^* = \frac{\widehat{u}_{xs}}{u_{xs}}\widehat{u}_x{}^* \quad \text{at} \quad y^* = \widehat{y}^* = 0 \tag{3.E.9-11}$$

$$\frac{\widehat{u}_{xs}\, y_s}{u_{xs}\, \widehat{y}_s}\frac{d\widehat{u}_x{}^*}{d\widehat{y}^*} = \frac{du_x^*}{dy^*} \quad \text{at} \quad y^* = \widehat{y}^* = 0 \tag{3.E.9-12}$$

$$\widehat{u}_x{}^* = 0 \quad \text{as} \quad \widehat{y}^* \to -\infty \tag{3.E.9-13}$$

The dimensionless group in equation (3.E.9-10) is set equal to 1, thereby obtaining our length scale in the liquid film, $y_s = H$ (step 7). Since gravity causes the flow in the liquid film, we set the dimensionless group that is a measure of the ratio of the gravity force to the viscous drag in equation (3.E.9-8) equal to 1, thereby obtaining our velocity scale $u_{xs} = \rho g H^2 \sin\theta/\mu$. Since the principal viscous term in equation (3.E.9-9) must be retained, we set its dimensionless coefficient equal to 1, thereby obtaining the length scale in the porous medium, $\widehat{y}_s = \sqrt{k_p}$. The maximum velocity in the porous medium occurs at its interface with the liquid film. Hence, we set the dimensionless group in equation (3.E.9-12) equal to 1, thereby obtaining our velocity scale in the porous medium, $\widehat{u}_{xs} = \rho g H\sqrt{k_p}\sin\theta/\mu$. When these values for the scale factors are substituted into equations (3.E.9-8) through (3.E.9-13), we obtain the following minimum parametric representation of the describing equations:

$$0 = \frac{d^2 u_x^*}{dy^{*2}} + 1, \qquad 0 \le y^* \le 1 \tag{3.E.9-14}$$

$$0 = -\widehat{u}_x^* + \frac{d^2 \widehat{u}_x{}^*}{d\widehat{y}^{*2}} + \frac{\sqrt{k_p}}{H}, \qquad -\infty < \widehat{y}^* \le 0 \tag{3.E.9-15}$$

$$\frac{du_x^*}{dy^*} = 0 \quad \text{at} \quad y^* = 1 \tag{3.E.9-16}$$

$$u_x^* = \frac{\sqrt{k_p}}{H}\widehat{u}_x{}^* \quad \text{at} \quad y^* = \widehat{y}^* = 0 \tag{3.E.9-17}$$

$$\frac{d\widehat{u}_x{}^*}{d\widehat{y}^*} = \frac{du_x^*}{dy^*} \quad \text{at} \quad y^* = \widehat{y}^* = 0 \tag{3.E.9-18}$$

$$\widehat{u}_x{}^* = 0 \quad \text{as} \quad \widehat{y}^* \to -\infty \tag{3.E.9-19}$$

If the following condition holds, the form of the no-slip boundary condition given by equation (3.E.9-17), (i.e., continuity of velocity across the interface between the porous medium and the liquid film), reduces to the familiar zero velocity condition at a stationary solid boundary (step 8):

$$\frac{\sqrt{k_p}}{H} \ll 1 \Rightarrow \text{liquid film velocity} \cong 0 \text{ at the interface with porous medium} \tag{3.E.9-20}$$

Hence, if $\sqrt{k_p}/H = \text{o}(0.01)$, the liquid film flow is described by the following set of simplified describing equations:

$$0 = \frac{d^2 u_x^*}{dy^{*2}} + 1, \qquad 0 \le y^* \le 1 \tag{3.E.9-21}$$

$$\frac{du_x^*}{dy^*} = 0 \quad \text{at} \quad y^* = 1 \tag{3.E.9-22}$$

$$u_x^* = 0 \quad \text{at} \quad y^* = 0 \tag{3.E.9-23}$$

These are the equations describing film flow down an impermeable stationary solid surface.

3.E.10 Flow in a Hollow-Fiber Membrane with Permeation

A membrane is a semipermeable medium that permits the passage of some molecules, colloidal aggregates, or particles relative to others. A hollow-fiber module is one form of a membrane contactor that consists of hundreds to thousands of small hollow fibers encased in a cylindrical shell. In one configuration of this module the parallel bundle of hollow fibers is sealed off at one end so that flow is possible in only one direction. The feed is introduced on the shell side of the module. The permeable components pass through the walls into the core of the hollow-fiber membrane. They then proceed to flow in parallel through all the fibers, after which they are collected as the product stream in the case of purification or concentration of solutes or as a waste stream in the case of removing contaminants. Since the permeable components flow in parallel through the hollow fibers, one can model the hydrodynamics in a hollow-fiber module of this type by considering the flow in a single fiber, as shown in Figure 3.E.10-1. Consider the case of a permeate stream that consists of an incompressible Newtonian liquid with constant physical properties. The flow through the core (called the *lumen*) of the hollow fiber is complex since the mass flow increases as the permeate stream flows toward the open end. We use scaling analysis to explore the conditions under which the describing equations for this flow can be simplified.

The appropriately simplified equations of motion in cylindrical coordinates given by equations (C.2-1), (D.2-10), and (D.2-12) in the Appendices along with the boundary and auxiliary conditions are given by (step 1)

$$\rho u_r \frac{\partial u_r}{\partial r} + \rho u_z \frac{\partial u_r}{\partial z} = -\frac{\partial P}{\partial r} + \mu \frac{\partial}{\partial r}\left[\frac{1}{r}\frac{\partial}{\partial r}(r u_r)\right] + \mu \frac{\partial^2 u_r}{\partial z^2} \tag{3.E.10-1}$$

$$\rho u_r \frac{\partial u_z}{\partial r} + \rho u_z \frac{\partial u_z}{\partial z} = -\frac{\partial P}{\partial z} + \mu \frac{1}{r}\frac{\partial}{\partial r}\left(r \frac{\partial u_z}{\partial r}\right) + \mu \frac{\partial^2 u_z}{\partial z^2} \tag{3.E.10-2}$$

$$\frac{1}{r}\frac{\partial}{\partial r}(r u_r) + \frac{\partial u_z}{\partial z} = 0 \tag{3.E.10-3}$$

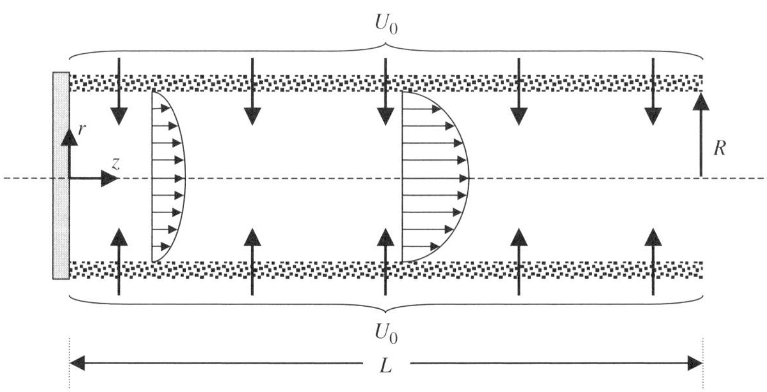

Figure 3.E.10-1 Flow of an incompressible Newtonian liquid in a cylindrical hollow-fiber membrane of radius R, one end of which is closed and the other open, due to permeation through the wall at a velocity U_0; the axial velocity profiles shown at two axial positions illustrate the acceleration caused by the mass addition due to the radial permeation.

$$\frac{\partial u_z}{\partial r} = 0, \quad u_r = 0 \qquad \text{at} \quad r = 0 \tag{3.E.10-4}$$

$$u_z = 0, \quad u_r = -U_0 \qquad \text{at} \quad r = R \tag{3.E.10-5}$$

$$u_z = 0, \quad u_r = 0 \qquad \text{at} \quad z = 0 \tag{3.E.10-6}$$

$$u_z = f_1(r), \quad u_r = f_2(r), \quad P = P_{\text{atm}} \qquad \text{at} \quad z = L \tag{3.E.10-7}$$

$$2\pi R z U_0 = \int_0^R u_z 2\pi r \, dr \tag{3.E.10-8}$$

where $f_1(r)$ and $f_2(r)$ are undetermined functions that are required for completeness in specifying the boundary conditions. The fact that these functions are generally unknown in practice significantly complicates solving the full set of describing equations. Equation (3.E.10-5) accounts for the fact that permeation through the walls of the hollow-fiber membrane results in a nonzero radial velocity component. This permeation also causes this to be a developing flow. Equation (3.E.10-8) is a statement that for an incompressible liquid the total permeation over a length z of the fiber wall is equal to the volumetric flow rate at that axial position. This serves as an auxiliary condition to determine the unspecified axial pressure gradient.

Define the following scale factors, reference factor, and dimensionless variables (steps 2, 3, and 4):

$$u_z^* \equiv \frac{u_z}{u_{zs}}; \qquad u_r^* \equiv \frac{u_r}{u_{rs}}; \qquad P^* \equiv \frac{P - P_r}{P_s}; \qquad r^* \equiv \frac{r}{r_s}; \qquad z^* \equiv \frac{z}{z_s} \tag{3.E.10-9}$$

Introduce these dimensionless variables into equations (3.E.10-1) through (3.E.10-8) and divide each equation through by the dimensional coefficient of a term that must be retained to obtain (steps 5 and 6)

$$\frac{\rho u_{rs}^2}{P_s} u_r^* \frac{\partial u_r^*}{\partial r^*} + \frac{\rho u_{zs} u_{rs} r_s}{P_s z_s} u_z^* \frac{\partial u_r^*}{\partial z^*} = -\frac{\partial P^*}{\partial r^*} + \frac{\mu u_{rs}}{P_s r_s} \frac{\partial}{\partial r^*}\left[\frac{1}{r^*}\frac{\partial}{\partial r^*}(r^* u_r^*)\right] + \frac{\mu u_{rs} r_s}{P_s z_s^2} \frac{\partial^2 u_r^*}{\partial z^{*2}} \tag{3.E.10-10}$$

$$\frac{\rho u_{zs} r_s}{\mu} u_r^* \frac{\partial u_z^*}{\partial r^*} + \frac{\rho u_{zs} r_s^2}{\mu z_s} u_z^* \frac{\partial u_z^*}{\partial z^*} = -\frac{P_s r_s^2}{\mu u_{zs} z_s} \frac{\partial P^*}{\partial z^*} + \frac{1}{r^*}\frac{\partial}{\partial r^*}\left(r^* \frac{\partial u_z^*}{\partial r^*}\right) + \frac{r_s^2}{z_s^2} \frac{\partial^2 u_z^*}{\partial z^{*2}} \tag{3.E.10-11}$$

$$\frac{u_{rs} z_s}{u_{zs} r_s} \frac{1}{r^*} \frac{\partial}{\partial r^*}(r^* u_r^*) + \frac{\partial u_z^*}{\partial z^*} = 0 \tag{3.E.10-12}$$

$$\frac{\partial u_z^*}{\partial r^*} = 0, \quad u_r^* = 0 \qquad \text{at} \quad r^* = 0 \tag{3.E.10-13}$$

$$u_z^* = 0, \quad u_r^* = -\frac{U_0}{u_{rs}} \qquad \text{at} \quad r^* = \frac{R}{r_s} \tag{3.E.10-14}$$

$$u_z^* = 0, \quad u_r^* = 0 \quad \text{at} \quad z^* = 0 \tag{3.E.10-15}$$

$$u_z^* = f_1^*(r^*), \quad u_r^* = f_2^*(r^*), \quad P^* = \frac{P_{\text{atm}} - P_r}{P_s} \quad \text{at} \quad z^* = \frac{L}{z_s} \tag{3.E.10-16}$$

$$\frac{U_0 R z_s}{u_{zs} r_s^2} z^* = \int_0^{R/r_s} u_z^* r^* dr^* \tag{3.E.10-17}$$

The dimensionless geometric ratios in equations (3.E.10-14) and (3.E.10-16) can be set equal to 1, thereby determining the length scales $r_s = R$ and $z_s = L$ (step 7). Since the permeation through the hollow-fiber membrane wall causes the radial flow, we set the dimensionless group containing U_0 in equation (3.E.10-14) equal to 1, thereby obtaining $u_{rs} = U_0$. Since this is a developing flow, setting the dimensionless group equal to 1 in equation (3.E.10-12) gives $u_{zs} = LU_0/R$. Since the axial flow is caused by the axial pressure gradient, we set the dimensionless group containing P_s in equation (3.E.10-11) equal to 1, thereby obtaining $P_s = \mu U_0 L^2/R^3$. Finally, our dimensionless pressure will be bounded between zero and 1 if we set the dimensionless group in equation (3.E.10-16) equal to zero, which gives $P_r = P_{\text{atm}}$. When these values for the scale and reference factors are substituted into equations (3.E.10-10) through (3.E.10-17), we obtain the following minimum parametric representation of the describing equations:

$$\text{Re}\frac{R^2}{L^2} u_r^* \frac{\partial u_r^*}{\partial r^*} + \text{Re}\frac{R^2}{L^2} u_z^* \frac{\partial u_r^*}{\partial z^*} = -\frac{\partial P^*}{\partial r^*} + \frac{R^2}{L^2}\frac{\partial}{\partial r^*}\left[\frac{1}{r^*}\frac{\partial}{\partial r^*}(r^* u_r^*)\right] + \frac{R^4}{L^4}\frac{\partial^2 u_r^*}{\partial z^{*2}} \tag{3.E.10-18}$$

$$\text{Re} u_r^* \frac{\partial u_z^*}{\partial r^*} + \text{Re} u_z^* \frac{\partial u_z^*}{\partial z^*} = -\frac{\partial P^*}{\partial z^*} + \frac{1}{r^*}\frac{\partial}{\partial r^*}\left(r^* \frac{\partial u_z^*}{\partial r^*}\right) + \frac{R^2}{L^2}\frac{\partial^2 u_z^*}{\partial z^{*2}} \tag{3.E.10-19}$$

$$\frac{1}{r^*}\frac{\partial}{\partial r^*}(r^* u_r^*) + \frac{\partial u_z^*}{\partial z^*} = 0 \tag{3.E.10-20}$$

$$\frac{\partial u_z^*}{\partial r^*} = 0, \quad u_r^* = 0 \quad \text{at} \quad r^* = 0 \tag{3.E.10-21}$$

$$u_z^* = 0, \quad u_r^* = -1 \quad \text{at} \quad r^* = 1 \tag{3.E.10-22}$$

$$u_z^* = 0, \quad u_r^* = 0 \quad \text{at} \quad z^* = 0 \tag{3.E.10-23}$$

$$u_z^* = f_1^*(r^*), \quad u_r^* = f_2^*(r^*), \quad P^* = 0 \quad \text{at} \quad z^* = 1 \tag{3.E.10-24}$$

$$z^* = \int_0^1 u_z^* r^* dr^* \tag{3.E.10-25}$$

The fact that equation (3.E.10-25) does not contain any dimensionless groups is an indication that we have scaled the describing equations properly. That is, the scaling has normalized the integral mass balance in this equation to be $\circ(1)$, as it should be, since the two terms in this equation must balance each other.

Inspection of equations (3.E.9-18) through (3.E.10-25) indicates that considerable simplification is possible if the following conditions apply (step 8):

$$\mathrm{Re} \equiv \frac{\rho U_0 R}{\mu} \ll 1 \quad \text{and} \quad \frac{R^2}{L^2} \ll 1 \tag{3.E.10-26}$$

We recognize these conditions to be those for assuming lubrication flow that was considered in Section 3.3. However, for this flow the Reynolds number characterizes the ratio of the radial convection to viscous force. If the conditions above apply, our describing equations simplify to

$$0 = \frac{\partial P^*}{\partial r^*} \tag{3.E.10-27}$$

$$0 = -\frac{dP^*}{dz^*} + \frac{1}{r^*}\frac{\partial}{\partial r^*}\left(r^*\frac{\partial u_z^*}{\partial r^*}\right) \tag{3.E.10-28}$$

$$\frac{1}{r^*}\frac{\partial}{\partial r^*}(r^* u_r^*) + \frac{\partial u_z^*}{\partial z^*} = 0 \tag{3.E.10-29}$$

$$\frac{\partial u_z^*}{\partial r^*} = 0 \quad \text{at} \quad r^* = 0 \tag{3.E.10-30}$$

$$u_z^* = 0, \quad u_r^* = -1 \quad \text{at} \quad r^* = 1 \tag{3.E.10-31}$$

$$u_z^* = 0 \quad \text{at} \quad z^* = 0 \tag{3.E.10-32}$$

$$P^* = 0 \quad \text{at} \quad z^* = 1 \tag{3.E.10-33}$$

$$z^* = \int_0^1 u_z^* r^* dr^* \tag{3.E.10-34}$$

Equation (3.E.10-27) implies that $P^* = P^*(z^*)$, which permits integrating equation (3.E.10-28) directly to obtain the axial velocity profile in terms of the unspecified axial pressure gradient. The latter can be obtained by substituting the axial velocity profile into equation (3.E.10-34). The corresponding radial velocity profile can be obtained from equation (3.E.10-29). The resulting solutions for the dimensionless velocity and pressure profiles are given by

$$u_z^* = 4z^*(1 - r^{*2}); \qquad u_r^* = r^*(r^{*2} - 2); \qquad P^* = 8(1 - z^{*2}) \tag{3.E.10-35}$$

3.E.11 Falling Head Method for Determining Soil Permeability

The falling head method is used to determine the permeability of soils. This test, shown in Figure 3.E.11-1, involves driving a pipe of radius R into the soil until it penetrates the water table, assumed here to be at the end of the tube. The pipe is filled with water to a height L_0 and the time t_d required to drain it to a height L_d is measured. We use the scaling method for dimensional analysis to develop a correlation for the draining time.

Step 1, the scaling procedure for dimensional analysis, involves writing the appropriate describing equations to determine the quantity of interest. The draining

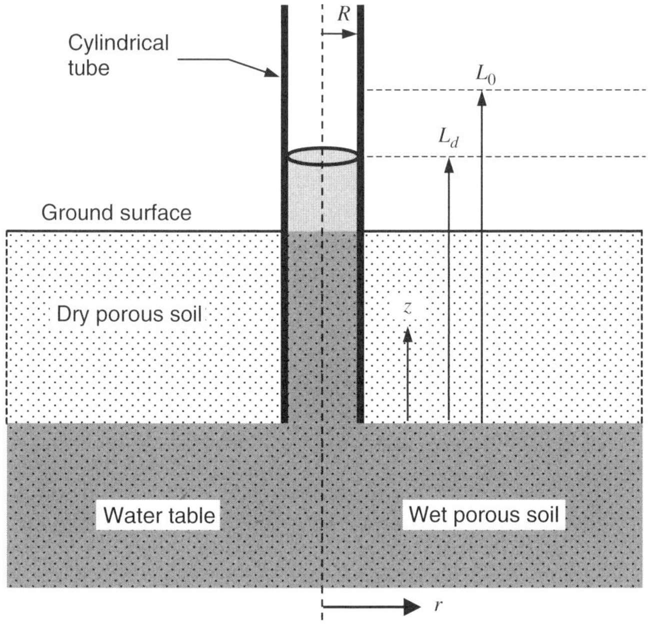

Figure 3.E.11-1 Falling head apparatus for measuring the permeability of soils; a cylindrical tube of radius R is pushed to the depth of the water table, which is defined to be at $z = 0$, and filled to an initial depth of L_0 with water; the instantaneous water depth in the tube $L(t)$ decreases due to permeation into the soil.

time t_d is related to the axial velocity u_z and Darcy's law via an instantaneous mass balance on the water in the tube:

$$L_0 - L_d = -\int_0^{t_d} \widehat{u}_z\Big|_{z=0} dt = \int_0^{t_d} \frac{k_p}{\mu}\left(\frac{dP}{dz} + \rho g\right)\Bigg|_{z=0} dt, \qquad 0 \le r \le R \tag{3.E.11-1}$$

where k_p is the permeability, μ the viscosity, ρ the density, and g the gravitational acceleration. Note that we have ignored the effect of the Brinkman term for flow through the porous medium based on the assumption that $k_p/R^2 \ll 1$, as discussed in Section 3.8. To evaluate the integral above it would be necessary to solve for the pressure distribution in the porous medium. This is obtained in turn by solving the Darcy flow equations in the porous medium subject to the appropriate boundary conditions. The incompressible continuity equation given by equation (C.2-1) in the Appendices in combination with Darcy's law for flow through porous media given by equation (E.2-3) in the Appendices implies that the pressure P is obtained from a solution to the axisymmetric form of Laplace's equation in cylindrical coordinates:

$$\frac{1}{r}\frac{\partial}{\partial r}\left(r\frac{\partial P}{\partial r}\right) + \frac{\partial^2 P}{\partial z^2} = 0 \tag{3.E.11-2}$$

The pressure is subject to the following boundary conditions:

$$P = P_{\text{atm}} + \rho g L(t) \quad \text{at} \quad z = 0, \quad 0 \le r \le R, \quad L_d \le L(t) \le L_0 \tag{3.E.11-3}$$

$$P = P_{\text{atm}} \quad \text{at} \quad z = 0, \quad R \le r < \infty \tag{3.E.11-4}$$

$$\widehat{u}_z = -\frac{k_p}{\mu}\frac{\partial P}{\partial z} = 0 \quad \text{as} \quad z \to -\infty, \quad 0 \le r < \infty \tag{3.E.11-5}$$

$$P = P_{\text{atm}} - \rho g z \quad \text{as} \quad r \to \infty, \quad -\infty < z \le 0 \tag{3.E.11-6}$$

Note that in specifying the boundary conditions for this partial differential equation, one must also specify the domain of any other variables that are not specified in the particular boundary condition; in some cases, such as for the domain of the boundary condition given by equation (3.E.11-3), this introduces additional parameters into the dimensional analysis. The boundary conditions given by equation (3.E.11-3) involve the instantaneous liquid depth $L(t)$ in the tube. This can be obtained from a mass balance for the amount of liquid flowing from the tube up to the instantaneous time t and is given by

$$L_0 - L = -\int_0^t \widehat{u}_z\Big|_{z=0} dt = \int_0^t \frac{k_p}{\mu}\left(\frac{dP}{dz} + \rho g\right)\Bigg|_{z=0} dt \qquad 0 \le r \le R \tag{3.E.11-7}$$

Define the following scale and reference factors and corresponding dimensionless variables (steps 2, 3, and 4):

$$P^* \equiv \frac{P - P_r}{P_s}; \qquad L^* \equiv \frac{L - L_r}{L_s}; \qquad r^* \equiv \frac{r}{r_s}; \qquad z^* \equiv \frac{z}{z_s}; \qquad t^* \equiv \frac{t}{t_s} \tag{3.E.11-8}$$

Substitute these dimensionless variables into equations (3.E.11-1) through (3.E.11-7) and divide each equation by the dimensional coefficient of one of its terms (steps 5 and 6). Since this is dimensional analysis rather than o(1) scaling, we can divide through by the dimensional coefficient of any arbitrary term in each equation:

$$\frac{\mu z_s (L_0 - L_d)}{k_p P_s t_s} = \int_0^{t_d/t_s} \left(\frac{dP^*}{dz^*} + \frac{\rho g z_s}{P_s}\right)\Bigg|_{z^*=0} dt^*, \qquad 0 \le r^* \le \frac{R}{r_s} \tag{3.E.11-9}$$

$$\frac{1}{r^*}\frac{\partial}{\partial r^*}\left(r^* \frac{\partial P^*}{\partial r^*}\right) + \frac{r_s^2}{z_s^2}\frac{\partial^2 P^*}{\partial z^{*2}} = 0 \tag{3.E.11-10}$$

$$P^* = \frac{P_{\text{atm}} + \rho g L_s\left(L^* + \dfrac{L_r}{L_s}\right) - P_r}{P_s} \quad \text{at } z = 0 \begin{cases} 0 \le r^* \le \dfrac{R}{r_s} \\ \dfrac{L_d}{L_s} \le L^*\left(1 + \dfrac{L_r}{L_s}\right) \le \dfrac{L_0}{L_s} \end{cases} \tag{3.E.11-11}$$

$$P^* = \frac{P_{\text{atm}} - P_r}{P_s} \quad \text{at} \quad z^* = 0, \quad \frac{R}{r_s} \le r^* < \infty \tag{3.E.11-12}$$

$$\frac{\partial P^*}{\partial z^*} = 0 \quad \text{as} \quad z^* \to -\infty, \quad 0 \le r^* < \infty \tag{3.E.11-13}$$

$$P^* = \frac{P_{\text{atm}} - \rho g z_s z^* - P_r}{P_s} \quad \text{as} \quad r^* \to \infty, \quad -\infty < z^* \le 0 \tag{3.E.11-14}$$

$$\frac{\mu z_s L_0 \left(1 - \frac{L_s}{L_0} L^*\right)}{k_p P_s t_s} = \int_0^{t^*} \left(\frac{dP^*}{dz^*} + \frac{\rho g z_s}{P_s}\right)\bigg|_{z=0} dt^*, \qquad 0 \le r^* \le \frac{R}{r_s} \tag{3.E.11-15}$$

We are free to choose which groups we set equal to zero or 1 in order to determine our scale and reference factors since we are not concerned about scaling our variables to be of order one. Hence, let us arbitrarily make the following choices (step 7):

$$\frac{R}{r_s} = 1 \Rightarrow r_s = R; \qquad \frac{L_0}{L_s} = 1 \Rightarrow L_s = L_0; \qquad \frac{L_r}{L_0} = \frac{L_d}{L_0} \Rightarrow L_r = L_d,$$

$$\frac{\rho g L_s}{P_s} = \frac{\rho g L_0}{P_s} = 1 \Rightarrow P_s = \rho g L_0; \qquad \frac{P_{\text{atm}} - P_r}{P_s} = 0 \Rightarrow P_r = P_{\text{atm}};$$

$$\frac{\rho g z_s}{P_s} = \frac{z_s}{L_0} = 1 \Rightarrow z_s = L_0; \qquad \frac{k_p \rho g t_s}{\mu(L_0 - L_d)} = 1 \Rightarrow t_s = \frac{\mu(L_0 - L_d)}{k_p \rho g} \tag{3.E.11-16}$$

When these scale and reference factors are substituted into the equations (3.E.11-9) through (3.E.10.15), we obtain the following minimum parametric representation of the dimensionless describing equations:

$$1 = \int_0^{\Pi_1} \left(\frac{dP^*}{dz^*} + 1\right)\bigg|_{z^*=0} dt^*, \qquad 0 \le r^* \le 1 \tag{3.E.11-17}$$

$$\frac{1}{r^*}\frac{\partial}{\partial r^*}\left(r^* \frac{\partial P^*}{\partial r^*}\right) + \Pi_2 \frac{\partial^2 P^*}{\partial z^{*2}} = 0 \tag{3.E.11-18}$$

$$P^* = 1 \quad \text{at} \quad z^* = 0, \quad 0 \le r^* \le 1, \quad 0 \le L^* \le 1 - \Pi_3 \tag{3.E.11-19}$$

$$P^* = 0 \quad \text{at} \quad z^* = 0, \quad 1 \le r^* < \infty \tag{3.E.11-20}$$

$$\frac{\partial P^*}{\partial z^*} = 0 \quad \text{as} \quad z^* \to -\infty, \quad 0 \le r^* < \infty \tag{3.E.11-21}$$

$$P^* = 1 \quad \text{as} \quad r^* \to \infty, \quad -\infty < z^* \le 0 \tag{3.E.11-22}$$

$$1 - L^* = \int_0^{t^*} \left(\frac{dP^*}{dz^*} + 1\right)\bigg|_{z=0} dt^*, \qquad 0 \le r^* \le 1 \tag{3.E.11-23}$$

where the following dimensionless groups have been defined:

$$\Pi_1 \equiv \frac{k_p \rho g t_d}{\mu(L_0 - L_d)}; \qquad \Pi_2 \equiv \frac{R}{L_0}; \qquad \Pi_3 \equiv \frac{L_d}{L_0} \tag{3.E.11-24}$$

A solution to equation (3.E.11-18) subject to the boundary conditions given by equations (3.E.11-19) through (3.E.11-23) will yield the dimensionless pressure P^* as a function of r^*, z^*, and t^* and the dimensionless groups Π_2 and Π_3. When the axial pressure gradient is evaluated at exit of the tube, where it is not a function of r^*, and substituted into equation (3.E.11-17), the resulting solution for the dimensionless draining time Π_1 will be a function only of the dimensionless groups Π_2 and Π_3; that is, the minimum parametric representation is given by a general correlation of the form

$$\Pi_1 \equiv \frac{k_p \rho g t_s}{\mu(L_0 - L_d)} = f_1(\Pi_2, \Pi_3) = f\left(\frac{R}{L_0}, \frac{L_d}{L_0}\right) \tag{3.E.11-25}$$

where $f_1(\Pi_2, \Pi_3)$ denotes some unspecified function of the dimensionless groups Π_2 and Π_3 that would have to be determined empirically. Note that a naive application of the Pi theorem for eight quantities in three units (i.e., $n = 8$ and $m = 3$) would have suggested that five ($m - n = 5$) rather than three dimensionless groups were necessary to correlate the draining time. For the Pi theorem to yield the minimum parametric representation, it is necessary to recognize that the grouping $k_p \rho g/\mu$ can be considered as a single quantity, in which case only two units need to be considered.

It would be necessary to take a considerable number of data to establish the correlation indicated in equation (3.E.11-25). We show here how a general correlation for the draining time can be established from very limited data for a specific tube, using water in a particular soil. The following empirical correlation has been obtained using water and a 5-cm-radius pipe for a soil having a permeability $k_p = 5.9\times 10^{-6}$ cm^2 [23]:

$$t_d = 4.94 \ln \frac{L_0}{L_d} \tag{3.E.11-26}$$

This empirical result is a special case of equation (3.E.11-25) and can be used to determine the functional form of a more general correlation for the draining time that can be used for different tubes, fluids, and soils. This can be seen more easily by using step 8 of the scaling procedure for dimensional analysis outlined in Section 2.4 in order to eliminate both L_0 and L_d from the dimensionless group that contains the draining time t_d. This involves generating a new dimensionless group formed from the original set of groups by means of the operation

$$\Pi_4 \equiv \frac{\Pi_1}{\Pi_2} = \frac{t_d k_p \rho g}{\mu R\,(1 - L_d/L_0)} = f_2(\Pi_2, \Pi_3) \tag{3.E.11-27}$$

where $f_2(\Pi_2, \Pi_3)$ denotes an unspecified function of the dimensionless groups Π_2 and Π_3 that would have to be determined empirically. Equation (3.E.11-27) can

[23] R. J. Ray, A Rayleigh free convection compliant ice front model for sorted patterned ground, M.S. thesis, University of Colorado, Boulder, Co, 1981.

be recast into the following form using step 11:

$$\frac{t_d k_p \rho g}{\mu R} \equiv \Pi_5 = f_3(\Pi_2, \Pi_3) \tag{3.E.11-28}$$

where $f_3(\Pi_2, \Pi_3)$ denotes an unspecified function of the dimensionless groups Π_2 and Π_3 that would have to be determined empirically. In general, the radius R of the tubes used to determine soil permeability is much less than the initial fill height L_0. If $R \ll L_0$, then $\Pi_2 \ll 1$ and we can use the procedure in step 9 of the scaling analysis procedure for dimensional analysis to eliminate group Π_2 from the correlation; hence,

$$\frac{t_d k_p \rho g}{\mu R} \equiv \Pi_5 = f_4(\Pi_3) \tag{3.E.11-29}$$

where $f_4(\Pi_3)$ is an unspecified function of Π_3 only. A comparison of equation (3.E.11-29) with equation (3.E.11-26) implies the following:

$$t_d = \Pi_5 \frac{\mu R}{k_p \rho g} = -4.94 \ln \Pi_3 \tag{3.E.11-30}$$

A generalized correlation relating soil permeability k_p and draining time t_d can be obtained by substituting values for the quantities in equation (3.E.11-30) to obtain the following correlation:

$$\begin{aligned} \Pi_5 &= -\frac{4.94(5.9\times10^{-6}\ \text{cm}^2)\left(1\ \text{g/cm}^3\right)\left(980\ \text{cm/s}^2\right)}{(0.01\ \text{g/cm}\cdot\text{s})\ (5\ \text{cm})} \ln \Pi_3 \\ &\Rightarrow \Pi_5 = -0.572 \ln \Pi_3 \Rightarrow \frac{t_d k_p \rho g}{\mu R} = -0.572 \ln \frac{L_0}{L_d} \end{aligned} \tag{3.E.11-31}$$

In this case, using the scaling analysis approach for dimensional analysis in combination with data for a specific falling head test gives the functional form of a generalized correlation that relates the measured draining time t_d to the soil permeability k_p and relevant physical properties and process parameters. The generalized correlation given by equation (3.E.11-31) applies for any falling head test, irrespective of the fluid, pipe size, and soil, provided that the dimensionless group $\Pi_2 \ll 1$.

3.P PRACTICE PROBLEMS

3.P.1 Alternative Scales for Laminar Flow Between Stationary and Moving Parallel Plates

Consider the steady-state fully developed laminar flow of an incompressible viscous Newtonian fluid with constant physical properties between two infinitely wide parallel flat plates due to both an applied axial pressure gradient and to the upper plate moving at a constant velocity U_0 as shown in Figure 3.2-1. In Section 3.2 we introduce scales for the velocity and y-coordinate to determine the criterion necessary to ignore the effect of the motion of the upper plate.

(a) Rescale this problem by introducing an additional scale for the second derivative as well as the velocity and y-coordinate; you will find that there is no dimensionless group to determine the velocity scale; however, this can be determined by integrating the scale for the second derivative.

(b) Rescale this problem by introducing an additional scale for the first derivative as well as the velocity and y-coordinate. Again you will find that there is no dimensionless group to determine the velocity scale; however, this can be determined by integrating the scale for the first derivative.

3.P.2 Laminar Flow Between Stationary and Moving Parallel Plates

Consider the steady-state fully developed laminar flow of an incompressible viscous Newtonian fluid with constant physical properties between two infinitely wide parallel flat plates due to both an applied axial pressure gradient and to the upper plate moving at a constant velocity U_0, as shown in Figure 3.2-1. In Section 3.2 we scaled this flow to determine the criterion necessary to ignore the effect of the motion of the upper plate. We found that the motion of the upper plate would not affect quantities such as the average velocity, volumetric flow rate, or drag at the stationary plate if equation (3.2-19) were satisfied. However, there was a region of influence next to the upper plate within which the motion of the plate could never be ignored. In this problem we explore complementary flow conditions for which the flow is caused primarily by the motion of the upper plate.

(a) Determine the criterion necessary to neglect the effect of the applied pressure on quantities such as the average velocity or volumetric flow rate.

(b) Determine if there is a region of influence within which the effect of the pressure on the flow can never be ignored in determining point quantities such as the local velocity or drag at the wall.

(c) Solve the simplified describing equations for the velocity profile for conditions such that the criterion you derived in part (a) is satisfied.

3.P.3 Gravity and Pressure-Driven Laminar Flow in a Vertical Tube

Consider the steady-state fully developed laminar flow of a Newtonian liquid with constant physical properties in a vertical tube of radius R that is subject to both gravity and a constant axial pressure gradient, as shown in Figure 3.P.3-1.

(a) Write the appropriate form of the simplified equations of motion for this flow.

(b) Write the boundary conditions required for the differential equations above.

(c) Scale the describing equations to determine the criterion for ignoring the effect of the applied pressure gradient on the velocity profile.

(d) Solve the resulting simplified describing equations to obtain the velocity profile.

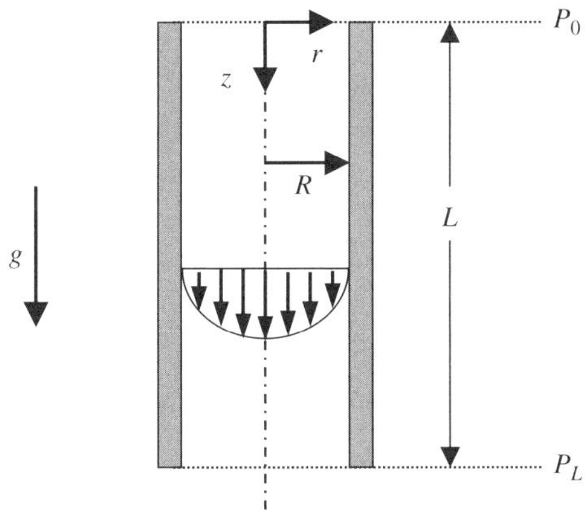

Figure 3.P.3-1 Steady-state fully developed laminar flow of a Newtonian liquid with constant physical properties in a vertical tube of radius R and length L subject to both gravity and a constant axial pressure gradient $(P_0 - P_L)/L$.

3.P.4 Axial Flow in a Rotating Tube

Consider the steady-state fully developed laminar flow of an incompressible Newtonian liquid with constant physical properties in a vertical cylindrical tube of radius R and length L subject to a constant axial pressure gradient $(P_0 - P_L)/L$ and a constant angular rotation about its axis of symmetry at ω radians per second, as shown in Figure 3.P.4-1.

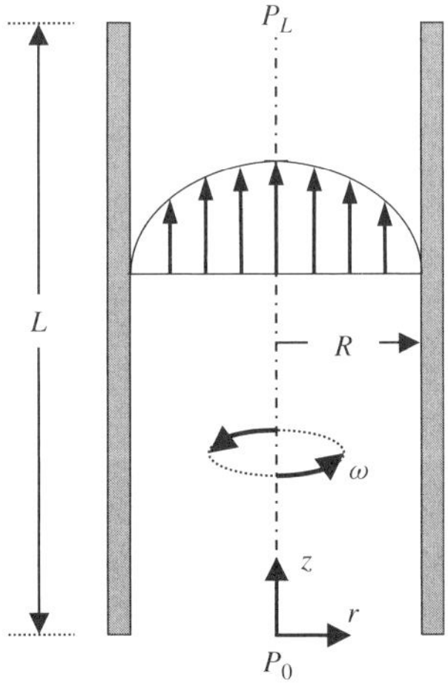

Figure 3.P.4-1 Steady-state fully developed laminar flow of an incompressible Newtonian liquid with constant physical properties in a vertical cylindrical tube of radius R and length L subject to a constant axial pressure gradient $(P_0 - P_L)/L$ and a constant angular rotation at ω radians per second.

(a) Write the appropriate form of the simplified equations of motion for this flow.
(b) Write the boundary conditions required for the above differential equations.
(c) Scale the describing equations to determine the dimensionless criteria for ignoring the effect of the gravitational body force on the flow.
(d) Develop a dimensionless criterion for assuming that the radial pressure gradient is much less than the axial pressure gradient.
(e) Develop a dimensionless criterion for assuming that the circumferential velocity is much less than the axial velocity.

3.P.5 Laminar Flow Between Converging Flat Plates

In Section 3.3 we developed the criteria for invoking the lubrication-flow approximation for laminar flow between two converging flat plates. However, these criteria break down near the upstream region of this flow. Use scaling analysis to determine the thickness of the region of influence in which the lubrication-flow approximation breaks down.

3.P.6 Laminar Flow Between Diverging Flat Plates

Consider the pressure-driven steady-state one-dimensional laminar plug flow of an incompressible Newtonian fluid with constant physical properties and constant velocity U_0 impinging on two nonparallel infinitely wide diverging flat plates as shown in Figure 3.P.6-1.

(a) Write the appropriate form of the simplified equations of motion for this flow.
(b) Write the boundary conditions required for the differential equations above.
(c) Scale the describing equations to determine the dimensionless criteria for making the lubrication-flow approximation.
(d) Solve the resulting describing equations appropriate to lubrication flow for the x- and y-component velocity profiles.

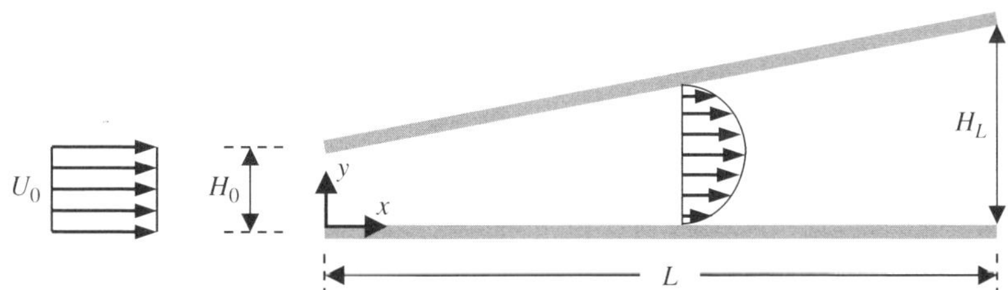

Figure 3.P.6-1 Steady-state developing laminar flow of an incompressible Newtonian fluid with constant physical properties between two infinitely wide diverging flat plates; only the local axial velocity profile is shown.

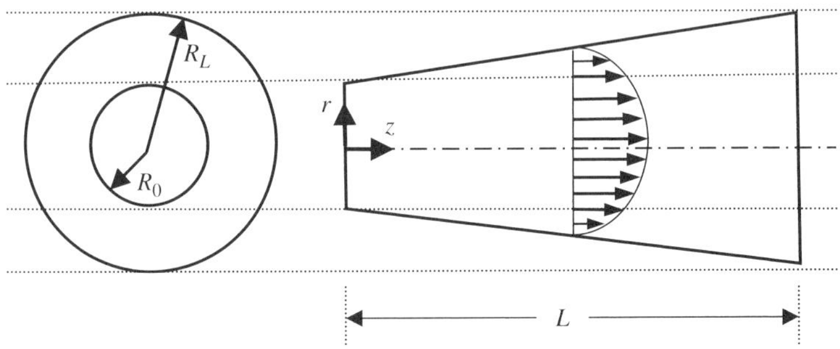

Figure 3.P.7-1 Pressure-driven steady-state laminar flow of an incompressible Newtonian fluid with constant physical properties through a diverging nozzle of length L and a circular cross-section with radii R_0 and R_L.

3.P.7 Laminar Flow in a Diverging Nozzle

Consider the pressure-driven steady-state laminar flow of an incompressible Newtonian fluid with constant physical properties through a diverging nozzle of length L and a circular cross section as shown in Figure 3.P.7-1. Assume plug flow at $z = 0$ with a constant velocity U_0.

(a) Write the appropriate form of the equations of motion.
(b) Write the boundary conditions that are necessary to solve the describing equations.
(c) Scale the describing equations to determine the criteria necessary to assume lubrication flow.
(d) Solve the resulting simplified lubrication-flow equations for the z- and r-velocity and axial pressure profiles. Note that the unspecified axial pressure gradient can be shown to be a constant and can be obtained from an integral mass balance and the known inlet velocity.

3.P.8 Steady-State Flow Between Parallel Circular Disks

Two parallel disks of outer radius R_2 are separated by a distance H as shown in Figure 3.P.8-1. An incompressible Newtonian liquid with constant physical properties is injected through a porous tube of radius R_1 located concentric with the axis of symmetry of the two disks.

(a) If the liquid is injected at a constant volumetric flow rate Q, write the appropriate form of the equations of motion.
(b) Write the boundary conditions required to solve the equations of motion.
(c) Scale the describing equations to determine the criteria necessary to assume lubrication flow.

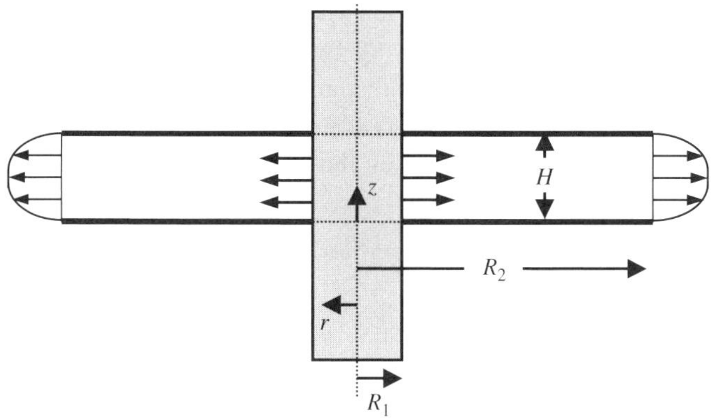

Figure 3.P.8-1 Steady-state radial flow of an incompressible Newtonian liquid with constant physical properties between two parallel disks of radius R_2 separated by a distance H due to injection of liquid at a volumetric flow rate Q through a porous tube of radius R_1.

(d) Solve the lubrication-flow equations to obtain the velocity profile as a function of r and z.

(e) Determine the pressure drop $P_1 - P_2$ necessary to inject this liquid at the constant volumetric flow rate Q.

3.P.9 Unsteady-State Flow Between Parallel Circular Disks

Consider the radial flow of an incompressible Newtonian liquid with constant physical properties between two parallel disks as shown in Figure 3.P.8-1. Assume that the liquid injection through the porous cylindrical tube is time-dependent and described by

$$Q = Q_0 e^{-\alpha t}, \qquad \text{where } Q_0 \text{ and } \alpha \text{ are constants} \tag{3.P.9-1}$$

(a) Write the boundary conditions required to solve the equations of motion.

(b) Scale the describing equations to determine the criterion necessary to assume that the flow is quasi-steady-state.

(c) Solve the quasi-steady-state lubrication-flow equations to obtain the velocity profile as a function of r, z, and t.

3.P.10 Steady-State Flow Between Spinning Parallel Circular Disks

Consider the steady-state radial flow of an incompressible Newtonian liquid with constant physical properties between two parallel disks, as shown in Figure 3.P.8-1. The radial flow is caused by liquid injection through the porous cylindrical tube at a constant volumetric flow rate Q. Assume now that both disks are spun at a constant angular velocity ω (radians per second).

(a) Write the appropriate form of the unsteady-state equations of motion assuming that the lubrication-flow approximation is applicable.
(b) Write the boundary conditions required to solve the equations of motion.
(c) Scale the describing equations to determine the criterion necessary to ignore the effect of the rotating disks on the pressure profile.

3.P.11 Lubrication-Flow Approximation for a Hydraulic Ram

In Example Problem 3.E.3 we considered lubrication flow in a hydraulic ram as shown in Figure 3.E.3-1 and used scaling to determine the criterion that must be satisfied in order to ignore curvature effects on the flow. However, we did not justify the lubrication-flow approximation that was made.

(a) Write the appropriate form of the equations of motion; however, do not make the lubrication-flow simplifications.
(b) Write the boundary conditions required to solve the equations of motion.
(c) Scale the describing equations to determine the criteria necessary to justify the lubrication-flow approximation.

3.P.12 Flow in a Rotating Disk Viscometer

Consider an incompressible Newtonian liquid with constant physical properties that fills a cylindrical container of radius R to a depth H. A circular plate contacts the liquid at its upper surface but does not contact the sidewalls of the container, as shown in Figure 3.P.12-1. By rotating the upper circular plate it is possible to obtain the viscosity of the liquid by measuring the *torque*, which is the force times the radial distance from the axis of rotation, on the upper plate. Operation of this instrument involves accelerating the upper plate from rest to a constant angular rotation rate of ω radians per second.

(a) Write the appropriate simplified form of the equations of motion for this unsteady-state fully developed flow; do not ignore the edge effects due to the presence of the sidewall of the container.
(b) Write the initial and boundary conditions required to solve the equations of motion.
(c) Scale the describing equations to determine a criterion for when steady-state flow conditions can be assumed.
(d) Scale the describing equations to determine a criterion for when the effects of the sidewall of the cylindrical container on the flow can be neglected.
(e) Solve the steady-state describing equations in the absence of sidewall effects for the angular velocity u_θ as a function of z and r.
(f) Use the velocity profile that you obtained in part (e) to obtain an equation for the torque exerted on the upper rotating plate by the liquid.

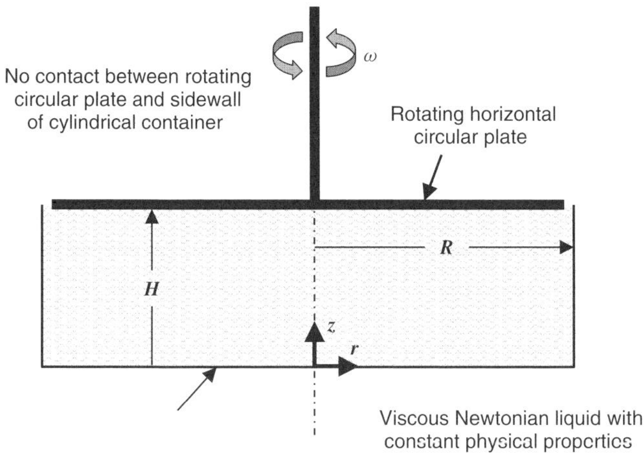

Figure 3.P.12-1 Flow of a viscous Newtonian liquid with constant physical properties in a rotating disk viscometer.

3.P.13 Flow in an Oscillating Disk Viscometer

Consider the viscometer shown in Figure 3.P.12-1, which is filled entirely with an incompressible Newtonian liquid with constant physical properties. Another way to operate this viscometer is to oscillate rather than rotate the upper circular plate. Assume that the upper plate is oscillated continuously at an angular rate of $\omega_0 \sin 2\pi f t$ radians per second, where ω_0 is the amplitude of the oscillation in radians per second and f is the frequency of the oscillation in cycles per second.

(a) Write the appropriate simplified form of the equations of motion for this unsteady-state, fully developed flow for which edge effects due to the sidewall can be ignored.
(b) Write the initial and boundary conditions required to solve the equations of motion.
(c) Scale the describing equations to determine a criterion for when quasi-steady-state flow conditions can be assumed.
(d) Consider the describing equations for the special case of very high frequency oscillations for which the effects of the oscillating upper circular plate are confined to a thin boundary layer near the upper plate. Use scaling to obtain an equation that can be used to estimate the thickness of the viscous boundary layer at the upper oscillating circular plate.
(e) Based on your scaling in part (d), develop a criterion for ignoring the effect of the bottom surface of the container on the oscillating flow.

3.P.14 Falling Needle Viscometer

The falling needle viscometer is useful for obtaining viscosity measurements when only small quantities of the liquid are available for measurement purposes. The viscosity is obtained using this viscometer by measuring the time it takes for a long cylindrical needle that is falling at its terminal velocity to pass between two known reference planes. This viscometer can be modeled quite well by considering the hydrodynamics to be steady-state fully developed laminar flow in the annular region between two cylinders, the inner of which is moving downward at a constant velocity U_0 and the outer of which is stationary; that is, end effects are ignored. Unfortunately, it is difficult to drop the needle exactly along the centerline of the outer cylinder. Hence, in general, the cylinders are not concentric, as shown in Figure 3.P.14-1. We seek to simplify the describing equations to permit a tractable solution.[24] It is convenient to use a moving cylindrical coordinate system located with its axis concentric with that of the falling cylinder.

(a) Consider the steady-state fully developed flow of an incompressible Newtonian liquid with constant physical properties in the annular region between two cylindrical tubes whose centers are displaced by a distance ε, as shown in Figure 3.P.14-1. Recall that this flow is driven by the moving boundary of

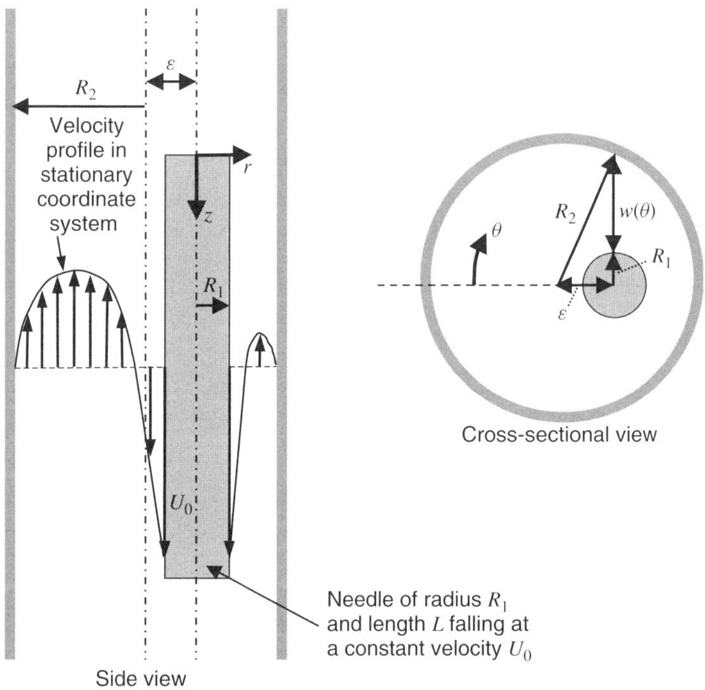

Figure 3.P.14-1 Falling needle viscometer showing a small inner cylinder having radius R_1 being dropped at a constant velocity U_0 a distance ε off-center in a cylindrical tube of radius R_2.

[24]This effect has been analyzed by D. B. Thiessen and W. B. Krantz, *Rev. Sci. Instrum.*, **63**(9), 4200–4204 (1992).

the inner cylinder (the needle), which is falling at its terminal velocity dictated by a balance between the gravity force on its volume and the drag force on its surface area. Show that the appropriate form of the equations of motion in cylindrical coordinates is given by

$$0 = \frac{\partial P}{\partial r} \tag{3.P.14-1}$$

$$0 = \frac{\partial P}{\partial \theta} \tag{3.P.14-2}$$

$$0 = -\frac{\partial P}{\partial z} + \mu \frac{1}{r}\frac{\partial}{\partial r}\left(r\frac{\partial u_z}{\partial r}\right) + \mu \frac{1}{r^2}\frac{\partial^2 u_z}{\partial \theta^2} + \rho g \tag{3.P.14-3}$$

(b) Show that equation (3.P.14-3) simplifies to

$$0 = -\frac{\Delta P}{L} + \mu \frac{1}{r}\frac{\partial}{\partial r}\left(r\frac{\partial u_z}{\partial r}\right) + \mu \frac{1}{r^2}\frac{\partial^2 u_z}{\partial \theta^2} + \rho g \tag{3.P.14-4}$$

where $\Delta P \equiv P_L - P_0$ is the pressure drop across the length of the falling needle for which P_0 is the pressure at $z = 0$ and P_L is the pressure at $z = L$.

(c) Write the appropriate boundary conditions required to solve the simplified equations of motion. Since the axial pressure drop across the length of the needle ΔP is unknown, an auxiliary condition is needed. This is determined from a force balance across the falling needle that involves the gravitational, viscous drag, and pressure forces. It will be helpful in specifying the no-slip condition at the outer cylinder to recall the law of cosines, which permits relating the local gap thickness $w(\theta)$ to the radii of the inner and outer cylinders, needle displacement ε, and local angular coordinate θ in the form

$$w(\theta) = \varepsilon \cos\theta - R_1 + (R_2^2 - \varepsilon^2 \sin^2\theta)^{1/2} \cong \varepsilon \cos\theta + (R_2 - R_1) \quad \text{if} \quad \varepsilon \ll R_2 \tag{3.P.14-5}$$

(d) Scale the describing equations to determine the criterion for assuming that the derivatives of the axial velocity in the circumferential direction can be ignored. Discuss the physical implications of this criterion.

(e) Use scaling to obtain an estimate of ΔP, the pressure driving force that causes flow in the annular gap between the inner cylindrical needle and the outer stationary tube wall.

(f) Use scaling and the result you obtained in part (e) for ΔP to obtain an estimate of the velocity U_0 of the falling needle.

3.P.15 Leading-Edge Considerations for Laminar Boundary-Layer Flow

In Section 3.4 we consider laminar boundary-layer flow over a semi-infinite flat plate and determined that the hydrodynamic boundary-layer approximation can be made when the criterion given by equation (3.4-38) is satisfied. However, as discussed in Section 3.4, the hydrodynamic boundary-layer approximation always breaks down near the leading edge of the plate. Use scaling analysis to estimate the region of influence wherein the criterion given by equation (3.4-38) is no longer satisfied.

3.P.16 Laminar Boundary-Layer Flow with Blowing

Consider a uniform plug flow of an incompressible viscous Newtonian liquid with constant physical properties and velocity U_∞ intercepting a stationary semi-infinite infinitely wide horizontal flat plate such as that considered in Section 3.4. Assume that the horizontal flat plate is porous such that there is a constant blowing velocity V_0 along its length, as shown in Figure 3.P.16-1. Blowing is used to increase heat and mass transfer in boundary-layer flows since it causes the local Reynolds number to increase, which in turn can cause a transition to turbulent boundary-layer flow.

(a) Use scaling to determine the condition for which the blowing effect on the boundary-layer flow can be neglected.

(b) Provide a physical interpretation of the condition on the dimensionless group that you obtained in part (a).

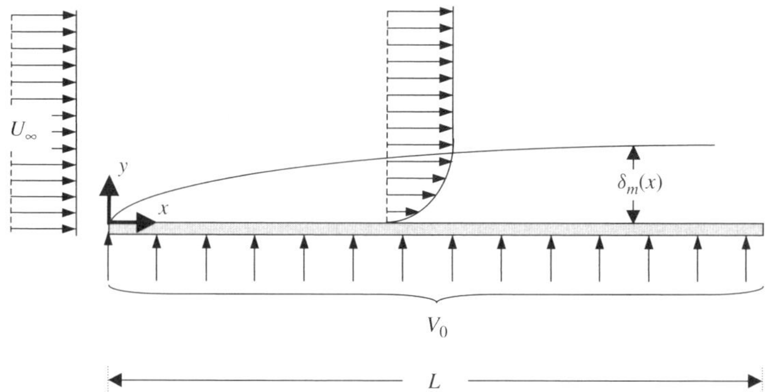

Figure 3.P.16-1 Uniform plug flow with velocity U_∞ of an incompressible viscous Newtonian fluid with constant physical properties intercepting a stationary semi-infinite infinitely wide horizontal porous flat plate along which there is a blowing velocity V_0; the blowing causes the boundary-layer thickness, δ_m, to increase.

3.P.17 Laminar Boundary-Layer Flow with Suction

Consider a uniform plug flow of an incompressible viscous Newtonian liquid with constant physical properties and velocity U_∞ intercepting a stationary semi-infinitely long infinitely wide horizontal flat plate such as that considered in Section 3.4. Assume that the horizontal flat plate is porous so that a constant suction

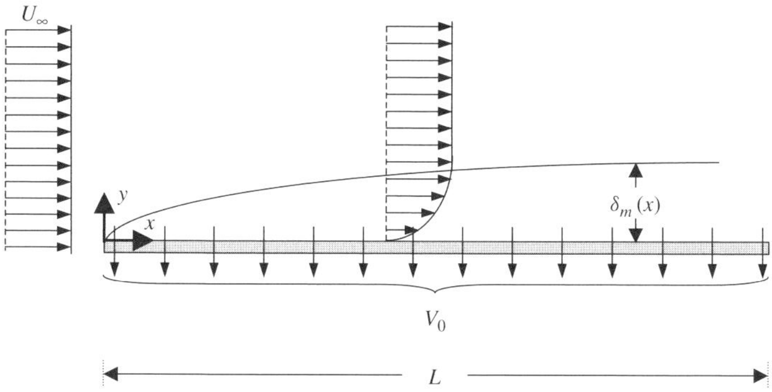

Figure 3.P.17-1 Uniform plug flow with velocity U_∞ of an incompressible viscous Newtonian fluid with constant physical properties intercepting a stationary semi-infinitely long infinitely wide horizontal porous flat plate along which there is a suction velocity V_0; the suction causes the boundary-layer thickness, δ_m, to decrease.

velocity V_0 can be applied along its length, as shown in Figure 3.P.17-1. Suction is used to decrease the boundary-layer thickness to decrease the local Reynolds number and thereby delay the transition to turbulence. It is also used to increase heat and mass transfer in the laminar boundary-layer flow regime by decreasing the thickness of the boundary-layer region, which provides the controlling resistance to conduction or diffusion.

(a) Write the appropriate forms of the equations of motion applicable to this boundary-layer flow; it is not necessary here to justify the form of these equations by scaling.
(b) Write the boundary conditions required to solve the equations of motion.
(c) We might anticipate that with boundary-layer suction such as we have in this problem, the boundary-layer thickness might ultimately become constant rather than grow without bound as it does for a boundary layer on a semi-infinitely long flat plate without suction or with blowing. Use scaling analysis to determine the criterion for obtaining a constant boundary-layer thickness; express your answer in terms of a dimensionless group that must be very small.
(d) For the constant boundary-layer condition obtained in part (c), determine the x- and y-velocity component profiles.

3.P.18 Entry-Region Laminar Flow in a Cylindrical Tube

Figure 3.P.18-1 shows a schematic of pressure-driven steady-state laminar entry-region flow of a viscous Newtonian fluid with constant physical properties in a cylindrical tube of radius R. The flow velocity at the entrance is assumed to be a constant U_0. This is assumed to be a high Reynolds number flow for which the inertia terms cannot be ignored in the entry region. Hence, in the entry region the

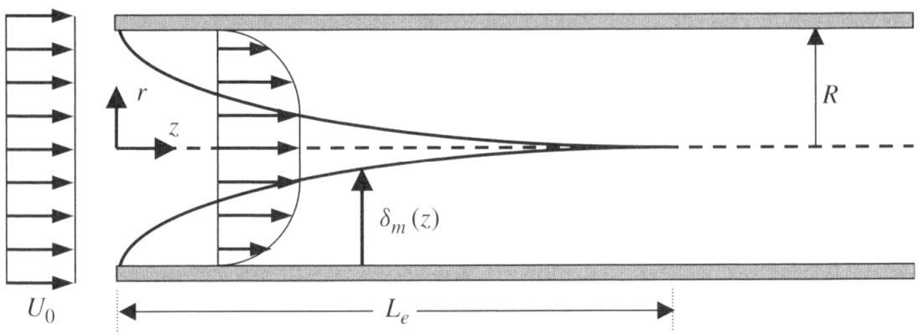

Figure 3.P.18-1 High Reynolds number steady-state pressure-driven entry-region laminar flow of a viscous Newtonian fluid with constant physical properties in a cylindrical tube of radius R, showing the developing boundary layer $\delta_m(z)$ and entrance length L_e required for the initial uniform flow U_0 to rearrange so as to become fully developed

action of viscosity will be confined to a region of influence near the tube wall denoted by the boundary-layer thickness $\delta_m(z)$. The latter will increase axially; when it reaches the center of the tube, the flow is fully developed. Note that since this is a confined flow, the fluid in the center of the tube must accelerate as the boundary layer grows.

(a) Write the appropriate simplified form of the equations of motion for this steady-state, developing flow.

(b) Write the boundary conditions required to solve the equations of motion.

(c) Scale the describing equations to determine the conditions required to invoke the hydrodynamic boundary-layer approximation; that is, for which the coupling between the axial and radial components of the equations of motion can be ignored and for which the axial diffusion of vorticity can be ignored.

(d) Use your scaling analysis to estimate the axial distance required to attain fully developed flow; reconcile your result with that obtained from the approximate analytical solution of Langhaar given by[25]

$$L_e = 0.227\frac{R^2 \rho U_0}{\mu} \tag{3.P.18-1}$$

3.P.19 Pressure-Driven Flow in an Oscillating Tube

Consider the unsteady-state laminar flow of an incompressible Newtonian fluid with constant physical properties through a horizontal cylindrical tube of radius R whose length L is sufficiently long to ensure that entrance and exit effects can be ignored; that is, the tube can be assumed to be essentially infinitely long. Initially, there is no flow. At time $t = 0$ a constant pressure gradient $(P_0 - P_L)/L = \Delta P/L$ is impressed across this tube. Simultaneously, the wall of this tube is oscillated at a velocity $u_z = U_0 \cos \omega t$, where ω is the angular frequency of the oscillation in radians per second. A schematic of this flow problem is shown in Figure 3.P.19-1.

[25]H. L. Langhaar, *Trans. ASME*, **64**, A55 (1942).

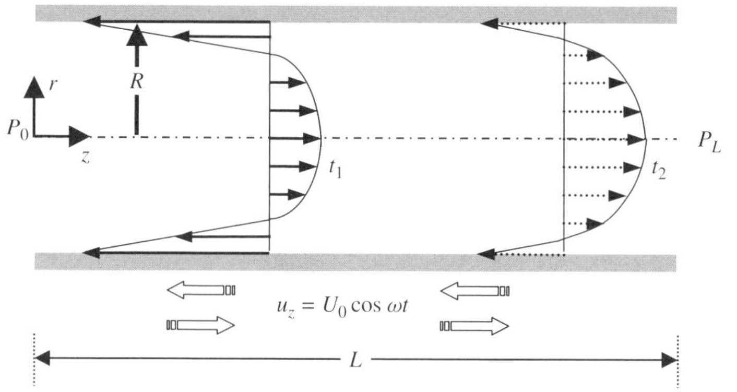

Figure 3.P.19-1 Unsteady-state laminar flow of an incompressible Newtonian liquid with constant physical properties in a circular tube of radius R due to an impulsively applied pressure difference $\Delta P \equiv P_0 - P_L$ and oscillation of the tube wall described by $u_z = U_0 \cos \omega t$; the axial velocity profiles are shown at times t_1 and t_2, where $t_2 > t_1$.

(a) Write the appropriate form of the equations of motion for this unsteady-state flow.

(b) Write the initial and boundary conditions required to solve the equations of motion.

(c) Determine the appropriate velocity scale for conditions such that the flow is mainly caused primarily by the applied pressure gradient.

(d) Determine the appropriate velocity scale for conditions such that the flow is caused primarily by the oscillating pipe wall.

(e) Scale the describing equations to determine when the effect of the wall oscillation can be neglected.

(f) Scale the describing equations to determine when this can be considered to be a quasi-steady-state creeping flow. Be careful to consider both time effects; i.e., the transients following startup and the periodic oscillations.

(g) Solve for the velocity profile $u_z(r, t)$ for the special case of quasi-steady-state creeping flow.

3.P.20 Countercurrent Liquid–Gas Flow in a Cylindrical Tube

Consider the steady-state fully developed laminar flow of a nonvolatile Newtonian liquid film of thickness H with constant physical properties at the inner wall of a vertical cylindrical tube of radius R and length L due to gravity, pressure, and interfacial drag arising from the fully developed upward pressure-induced flow of a gas having constant physical properties in the center of the tube, as shown in Figure 3.P.20-1.

(a) Write the appropriately simplified continuity and equations of motion for both the liquid and gas phases for this flow.

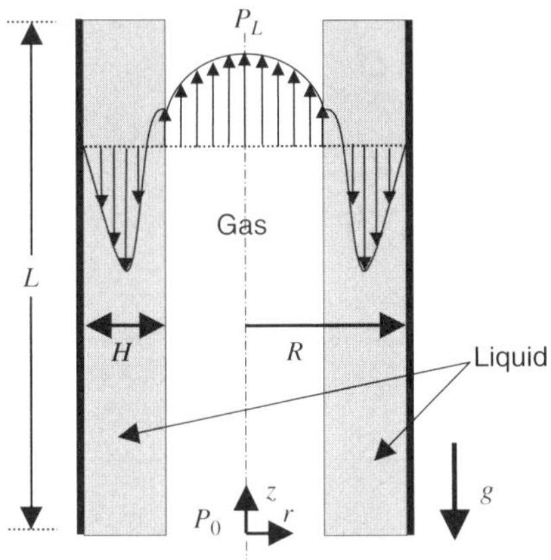

Figure 3.P.20-1 Steady-state fully developed laminar flow of a nonvolatile Newtonian liquid film of thickness H with constant physical properties at the inner wall of a vertical cylindrical tube of radius R and length L due to gravity, pressure, and interfacial drag arising from the upward fully developed pressure-induced flow of a gas with constant physical properties in the center of the tube.

(b) Write the boundary conditions required to solve the differential equations you obtained in part (a).

(c) Determine the appropriate velocity scale in the liquid film when this flow is caused primarily by the gravitational body force.

(d) Determine the appropriate velocity scale in the gas phase when this flow is caused primarily by the applied pressure.

(e) Use scaling to determine the criterion for when the effect of the gas flow on the liquid film flow can be ignored.

(f) Use scaling to determine the criterion for when the effect of the gravitational body force on the gas flow can be ignored.

(g) For the conditions considered in part (d), determine the criterion for ignoring curvature effects in the describing equations; that is, for when the describing equations in cylindrical coordinates reduce to the corresponding equations in rectangular coordinates.

(h) Determine the appropriate velocity scale in the liquid film when this flow is caused primarily by the drag exerted on it by the upward gas flow.

(i) Use scaling to determine the criterion for when the effect of gravity on the liquid film flow can be ignored.

(j) This flow is an idealization of that within a single packing element such as a Raschig ring in countercurrent liquid–gas contacting in a packed column. Flooding in a packed column begins when the upward gas flow causes a net amount of liquid to be carried upward to the top of the column. Use the results from your scaling analysis to determine the pressure gradient required to initiate flooding.

3.P.21 Stratified Flow of Two Immiscible Liquid Layers

Consider the steady-state fully developed flow of two immiscible Newtonian liquids with constant physical properties between two stationary parallel impermeable flat plates, as shown in Figure 3.P.21-1. These fluids have densities ρ_1 and ρ_2, viscosities μ_1 and μ_2, and thicknesses H_1 and H_2, respectively. A high pressure P_0 is applied at $x = 0$ and a low pressure P_L is applied at $x = L$. Liquid 1 flows symmetrically about the center plane between the two flat plates, whereas liquid 2 is confined to a layer having thickness $H_2 - H_1$ adjacent to each of the two flat plates. The velocities in liquids 1 and 2 are denoted by u_1 and u_2, respectively. For the scaling analysis in this problem, introduce the following dimensionless variables in terms of undefined scale and reference factors:

$$u_1^* \equiv \frac{u_1 - u_{1r}}{u_{1s}}; \qquad u_2^* \equiv \frac{u_2}{u_{2s}}; \qquad y_1^* \equiv \frac{y}{y_{1s}}; \qquad y_2^* \equiv \frac{y - y_{2r}}{y_{2s}} \qquad \text{(3.P.21-1)}$$

(a) Write the appropriate form of the simplified continuity and equations of motion along with the necessary boundary conditions for this flow; gravitational body forces may be neglected in your analysis.

(b) Explain why a reference factor is needed for the velocity in liquid 1 and why a reference factor is needed for the spatial coordinate in liquid 2.

(c) Determine the scale and reference factors in the dimensionless variables defined above for the case where the flow in both liquids is caused by the applied pressure $\Delta P \equiv P_0 - P_L$. Indicate why you set various dimensionless groups equal to zero or 1.

(d) What is the criterion for ignoring the effect of liquid 1 on the flow of liquid 2 for the conditions in part (c)?

(e) Determine the scale and reference factors in the dimensionless variables defined above for the case where the flow in liquid 1 is caused by the applied pressure $\Delta P \equiv P_0 - P_L$, whereas the flow in liquid 2 is caused primarily by the drag force exerted by liquid 1 at the interface. Indicate why you set various dimensionless groups equal to zero or 1.

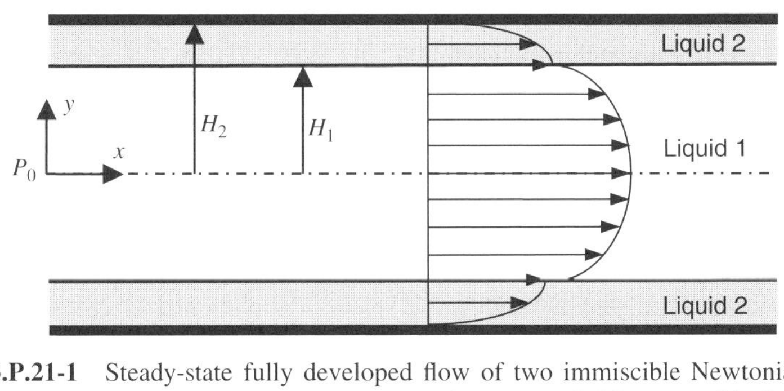

Figure 3.P.21-1 Steady-state fully developed flow of two immiscible Newtonian liquids with constant physical properties between two stationary parallel impermeable flat plates.

(f) What is the criterion for determining whether the scaling in part (c) or in part (e) is appropriate for the describing equations?

(g) Use the results of your scaling analysis in part (e) to estimate the velocity gradients in both liquids 1 and 2; that is, develop appropriate scales for the velocity gradients in each liquid.

(h) For the scaling analysis done in part (e), determine the dimensionless groups that are necessary to correlate the total rate of entropy production $\dot{S}$, which is given by

$$\dot{S} = -\int_0^{H_1} \frac{\mu_1}{T}\left(\frac{du_x}{d_y}\right)^2 WLdy - \int_{H_1}^{H_2} \frac{\mu_2}{T}\left(\frac{du_x}{d_y}\right)^2 WLdy \qquad (3.P.21\text{-}2)$$

where T is the absolute temperature and W and L denote the width and length, respectively, of the two flat plates.

(i) One means for reducing the viscous drag at the walls experienced in pumping viscous liquids such as petroleum is to inject a less viscous immiscible liquid such as water, which will form a layer at the wall. However, for this idea to work, we have to establish that the less viscous liquid (e.g., water) rather than the more viscous liquid (e.g., petroleum) will go to the wall region. This question can be answered by invoking the principle that continuous steady-state processes seek a state of minimum entropy production. Use this principle along with the results of your scaling analysis in part (e) to determine whether the viscous or the less viscous liquid will go to the wall region.

3.P.22 Laminar Cylindrical Jet Flow

In Example Problem 3.E.8 we considered a jet of an incompressible Newtonian liquid with constant physical properties issuing from a circular orifice with an initial velocity U_0 and falling vertically under the influence of gravity in an inviscid gas as shown in Figure 3.E.8-1. We scaled the describing equations to explore the conditions for which quasi-parallel flow can be assumed; that is, when the axial velocity profile can be assumed to depend only on the axial coordinate. In scaling the describing equations we introduced a scale for the radial derivative of the axial velocity since we did not anticipate that this derivative would scale as the ratio of the characteristic axial velocity scale divided by the characteristic radial length scale. We anticipated the need to scale this derivative with its own scale since the axial velocity does not change significantly across the jet. It was stated that if we had scaled this radial derivative with the ratio of the characteristic axial velocity scale to the characteristic radial length scale, the forgiving nature of scaling would have indicated a contradiction. To better understand what is meant by this, let us assume (incorrectly!) that the radial derivative of the axial velocity scales as the axial velocity scale u_{zs} divided by the radial length scale r_s. Show that this leads to an inconsistency in the resulting dimensionless equations in that the dimensionless group multiplying one term in the describing equations becomes very large, whereas the others terms are of o(1); this implies that there is no term to balance this very large term.

3.P.23 Free Surface Flow Down a Plate with Condensation

Consider the steady-state incompressible laminar film flow of a Newtonian liquid having constant physical properties down an infinitely wide plane surface inclined at an angle θ to the horizontal as shown in Figure 3.P.23-1. At the interface between the liquid film and the ambient gas phase, a constant amount of liquid $\tilde{W}_m$ is added to the film flow per unit area of free surface. For example, this might occur due to condensation occurring at the liquid–gas interface.

(a) Write the appropriate forms of the steady-state equations of motion assuming that the ambient gas phase exerts negligible drag on the liquid film.[26]

(b) Write the boundary conditions required to solve the equations of motion; note that the continuity of surface normal stresses and tangential stress balance must account for the surface-curvature effects.[27]

(c) Derive the kinematic surface condition for this flow; note that the mass addition at the interface must be taken into consideration.

(d) Scale the describing equations to determine the conditions required to make the lubrication-flow approximation and to ignore the surface-curvature effects.

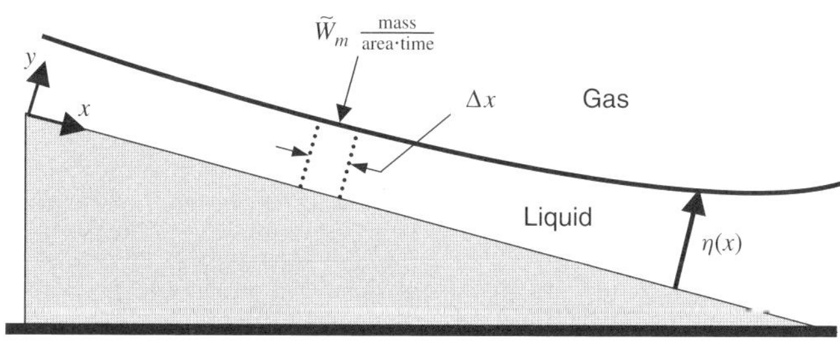

Figure 3.P.23-1 Steady-state incompressible laminar film flow of a Newtonian liquid with constant physical properties down an infinitely wide plane surface inclined at an angle θ to the horizontal. Mass addition at the free surface of $\tilde{W}_m$ units of mass per unit area per unit time causes the local film thickness η to increase as a function of axial distance x.

3.P.24 Free Surface Flow Over a Horizontal Filter

A filter in a large industrial plant operates by letting a solution flow across a large sheet of filter paper as shown in Figure 3.P.24-1. Assume steady-state developing incompressible laminar flow of a Newtonian liquid with constant physical properties

[26]Scaling was used to determine the criterion for making this assumption in Example Problem 3.E.1.

[27]See the example in Section 3.7 as a guide to scaling this problem.

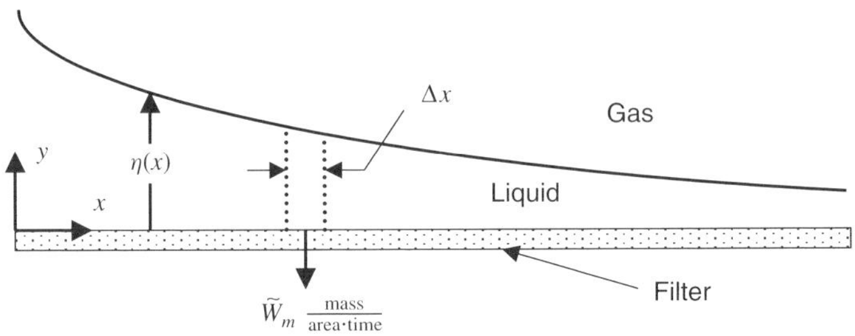

Figure 3.P.24-1 Gravitationally induced steady-state laminar flow of an incompressible Newtonian liquid with constant physical properties across a filter through which the mass flow rate per unit area is $\tilde{W}_m$.

that is driven by a constant hydrostatic pressure gradient due to a variable film thickness in the x-direction. Assume also that mass is removed from this flow at the filter boundary at a constant rate of $\tilde{W}_m$ units of mass per unit area per unit time.

(a) Write the appropriate forms of the steady-state equations of motion assuming that the ambient gas phase exerts negligible drag on the liquid film (see footnote 26).
(b) Write the boundary conditions required to solve the equations of motion; note that the continuity of surface normal stresses and tangential stress balance must account for the surface-curvature effects (see footnote 27).
(c) Derive the kinematic surface condition for this flow; note that the mass loss through the filter must be taken into consideration.
(d) Scale the describing equations to determine the condition(s) required to make the lubrication-flow approximation and to ignore the surface-curvature effects.
(e) Solve the simplified equations of motion that you obtained in part (d) to obtain the axial velocity profile in terms of the unknown hydrostatic pressure gradient.
(f) Determine the unknown hydrostatic pressure gradient given that the volumetric flow rate Q is specified.
(g) Determine the component of the velocity normal to the filter.

3.P.25 Curtain-Coating Flow

The process of curtain coating is shown in Figure 3.P.25-1. This process is used, for example, to apply protective polymer coatings at high speed to continuous steel or tin-plate strip. The curtain flow emanates from a slot at $x = 0$, at which point it has a velocity U_0. The solid guides at $z = \pm W/2$ maintain the film at a constant thickness in the z-direction. However, since the film accelerates due to the gravitational body force, it thins in the y-direction, as shown in the side view in

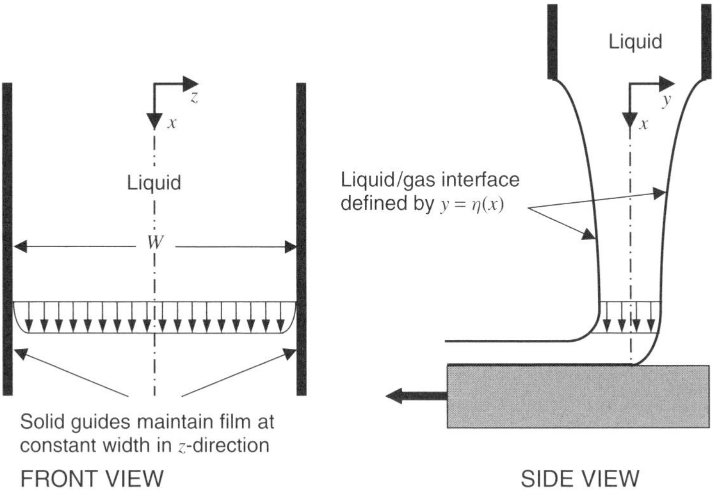

Figure 3.P.25-1 Gravitationally induced steady-state laminar curtain-coating flow of an incompressible Newtonian liquid with constant physical properties emanating from a rectangular slot with an initial velocity of U_0; front view shows solid guides that maintain the film at constant width; the axial velocity profile is shown in both views.

Figure 3.P.25-1. The liquid–gas interface is defined by $y = \eta(x)$, where η decreases with increasing x due to the increase in the x-velocity component. The object to be coated can be ignored entirely in this analysis since we are concerned only with the falling curtain flow. We assume steady-state laminar flow of an incompressible Newtonian liquid with constant physical properties and ignore edge effects at the sidewalls. Note that the applicability of all these assumptions could be determined using scaling analysis. We use scaling analysis to explore the conditions for which quasi-parallel flow can be assumed. We then test the applicability of our scaling analysis results using data from the curtain-coating process.

(a) Write the appropriate form of the three-dimensional equations of motion for this flow assuming that the ambient gas phase exerts a negligible drag on the liquid film (see footnote 26).

(b) Write the appropriate boundary conditions for this flow assuming that the surface-tension effects associated with the curvature can be ignored.[28] However, do not ignore the effects of curvature in specifying the tangential and normal stress boundary conditions at the free surface. In deriving the tangential and normal stress boundary conditions at the liquid–gas interface, use the convection for θ and the normal and tangential unit vectors, $\vec{n}$ and $\vec{t}$, respectively, shown in Figure 3.P.25-2.

(c) Write the appropriate form of the kinematic surface condition for this flow.

[28]Note that scaling analysis could be used to determine when surface-tension and curvature effects can be neglected; the latter were considered in Section 3.7.

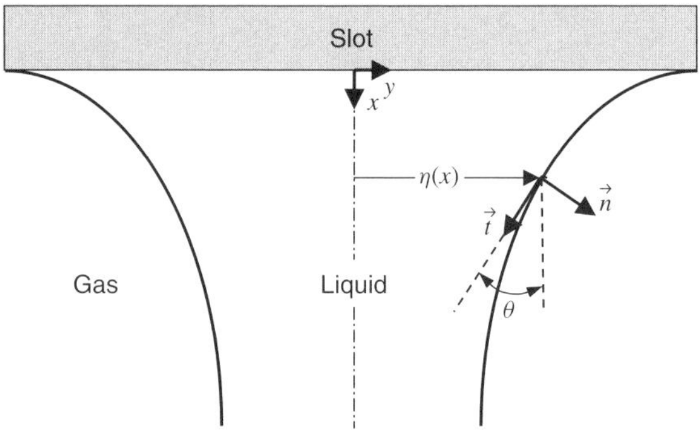

Figure 3.P.25-2 Enlarged view of the x–y plane for curtain flow, showing normal and tangential unit vectors, $\vec{n}$ and $\vec{t}$, respectively, at the liquid–gas interface.

(d) Introduce the following dimensionless variables involving unspecified scale and reference factors into the describing equations:

$$u_x^* \equiv \frac{u_x - u_{xr}}{u_{xs}}; \quad u_y^* \equiv \frac{u_y}{u_{ys}}; \quad u_z^* \equiv \frac{u_z}{u_{zs}}; \quad P^* \equiv \frac{P - P_r}{P_s};$$
$$\eta^* \equiv \frac{\eta}{\eta_s}; \quad \left(\frac{\partial u_x}{\partial y}\right)^* \equiv \frac{1}{\beta_s}\frac{\partial u_x}{\partial y}; \quad x^* \equiv \frac{x}{x_s}; \quad y^* \equiv \frac{y}{y_s}; \quad z^* \equiv \frac{z}{z_s} \tag{3.P.25-1}$$

Note that $\partial u_x/\partial y$ does not scale with u_{xs}/y_s since u_x does not undergo a characteristic change of u_{xs} over the distance y_s. The proper scale for β_s is obtained from the tangential stress boundary condition. Note that this scaling for $\partial u_x/\partial y$ implies that $\partial^2 u_x/\partial y^2$ scales with β_s/y_s. Use $x_s = L$, where L is some arbitrary but constant downstream distance. Your pressure scale P_s will come from balancing the pressure term with the convection terms in the y-component of the equations of motion. This follows from the fact that the pressure gradient arises from the acceleration of the flow.

(e) Simplify the tangential and normal stress boundary conditions appropriate to the assumption of small curvature; that is, for $d\eta/dx \ll 1$.

(f) Assume small curvature and use your scaling analysis in part (d) to determine the criteria for assuming that this is a quasi-parallel flow; that is, a flow that can be described by considering only the changes of the x-velocity component in the x-direction.

(g) Show that the solution to the simplified set of equations appropriate to the quasi-parallel-flow approximation is given by

$$u_x^2 = U_0^2 + 2xg \tag{3.P.25-2}$$

(h) Table 3.P.25-1 summarizes several data sets for a series of curtain flows for which the fall velocity u_x in cm/s at a position below the slot of $x = 5$ cm is

TABLE 3.P.25-1 Data Sets for Curtain Flow

Data Set	Viscosity (poise)	Slot Width (mm)	Mean Velocity at Slot Exit (cm/s)	Measured Velocity at $x = 5$ cm (cm/s)
1	1.2	0.6	18	96
2	1.5	0.6	56	110
3	1.95	0.2	38	100
4	2.0	0.2	47	109
5	2.6	0.6	18	97
6	2.7	0.3	33	93
7	2.8	0.3	13	87
8	3.45	0.6	6	86
9	3.7	1.5	5	86
10	5.3	1.5	12	85
11	9.9	0.6	8	75

reported as a function of the liquid viscosity in poise (g/cm ·s), the slot width in mm, and the mean velocity U_0 in cm/s at the exit of the slot. Compare the fall velocities predicted by the quasi-parallel-flow approximation in part (g) with the measured fall velocities and also assess the validity of the quasi-parallel-flow approximation. Discuss any significant deviations between the experimental and model results.

3.P.26 Flow in a Semi-infinite Porous Medium Bounded by a Flat Plate

Consider the steady-state fully developed pressure-driven flow of an incompressible Newtonian liquid through a liquid-saturated semi-infinite porous medium that is bounded by a horizontal flat plate as shown in Figure 3.P.26-1.

(a) Write the appropriate form of the equations of motion applicable to porous media for this flow.
(b) Write the boundary conditions required to solve the equations of motion.

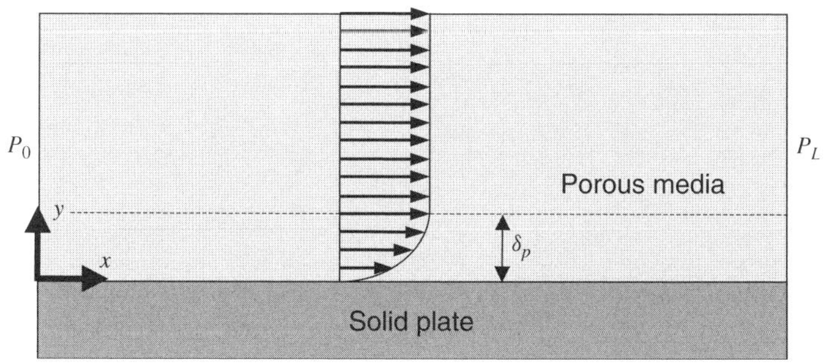

Figure 3.P.26-1 Steady-state fully developed flow of an incompressible Newtonian fluid with constant physical properties through a semi-infinite porous medium bounded by a horizontal flat plate due to an applied pressure gradient $(P_0 - P_L)/L$.

(c) Solve the equations you derived in parts (a) and (b) for the velocity profile.

(d) Scale the describing equations for this flow to determine the criterion for ignoring the effect of the flat plate on the velocity profile.

(e) Determine the thickness of the region of influence δ_p within which the effect of the flat plate on the velocity profile cannot be ignored.

(f) Solve the simplified form of the equations of motion for the velocity profile assuming that the effect of the flat plate can be ignored.

(g) Show how the solution for the velocity profile that you derived in part (c) reduces to the result that you obtained in part (f).

3.P.27 Porous Media Flow Between Parallel Flat Plates

Consider the steady-state fully developed pressure-driven flow of an incompressible Newtonian liquid through a liquid-saturated semi-infinite porous medium that is bounded by horizontal parallel flat plates as shown in Figure 3.P.27-1.

(a) Write the appropriate form of the equations of motion applicable to porous media for this flow.

(b) Write the boundary conditions required to solve the equations of motion.

(c) Solve the equations you derived in parts (a) and (b) for the velocity profile.

(d) Scale the describing equations for this flow to determine the criterion for ignoring the effect of the solid boundaries on the velocity profile.

(e) Determine the thickness of the region of influence δ_p within which the effect of the solid boundaries on the velocity profile cannot be ignored.

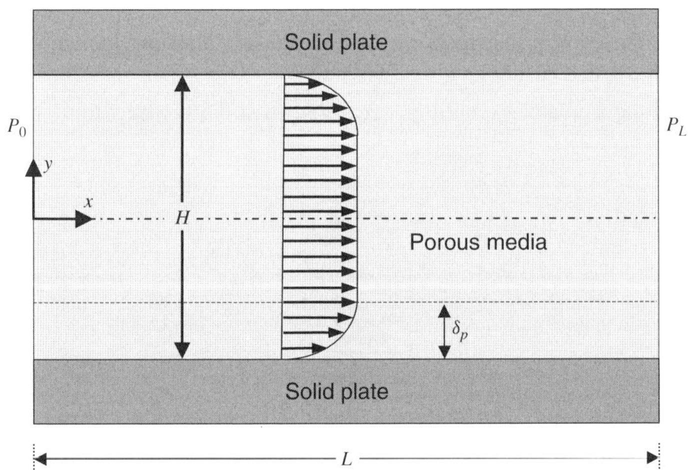

Figure 3.P.27-1 Steady-state fully developed flow of an incompressible Newtonian fluid with constant physical properties through a porous medium bounded by horizontal parallel flat plates due to an applied pressure gradient $(P_0 - P_L)/L$.

(f) Solve the simplified form of the equations of motion for the velocity profile assuming that the effect of the solid boundaries can be ignored.

(g) Show how the solution for the velocity profile that you derived in part (c) reduces to the result that you obtained in part (f).

3.P.28 Gravity-Driven Film Flow over a Saturated Porous Medium

In Example Problem 3.E.9 we considered the steady-state fully developed flow of an incompressible Newtonian liquid film over an inclined liquid-saturated porous medium due to a gravitationally induced body force as shown in Figure 3.E.9-1. We used scaling analysis to determine when the flow through the porous media has a negligible effect on the flow of the liquid film. We introduced one set of scales for the velocity and the length variables in the liquid film and another set of scales for velocity and length variables in the porous medium. We stated that if we had not done this, we would have arrived at a contradiction in our scaled equations. However, the forgiving nature of scaling would then indicate that we had not scaled some quantity so that it was $o(1)$. To understand better what is meant by the forgiving nature of scaling, rework this problem while assuming (incorrectly!) that the velocity and the length scales are the same in both the liquid film and the porous medium.

3.P.29 Radial Flow from a Porous Cylindrical Tube

Consider the steady-state laminar flow of a Newtonian liquid with constant physical properties that is caused by fluid emanating radially at a uniform volumetric flow rate Q from a cylindrical tube having length L, outer radius R, and porous walls that is immersed in an infinite pool of the same liquid as shown in Figure 3.P.29-1.

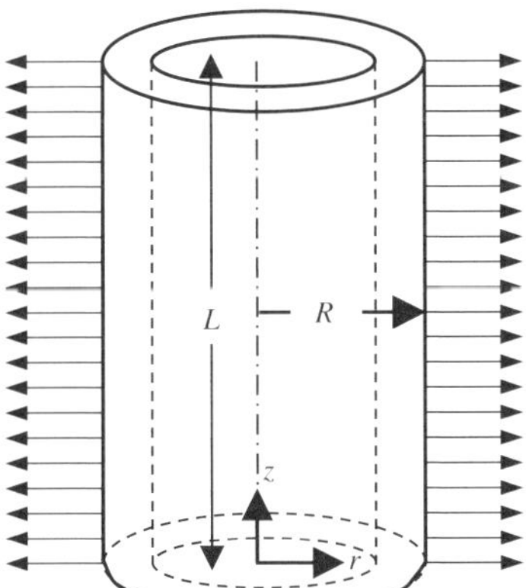

Figure 3.P.29-1 Steady-state laminar flow of a Newtonian liquid with constant physical properties that is caused by fluid emanating radially at a uniform volumetric flow rate Q from a cylindrical tube of length L, outer radius R, and porous walls that is immersed in an infinite pool of the same liquid; the radial velocity profile at the surface of the cylinder is shown at which the pressure is P_R; the pressure far removed from the cylinder is P_∞.

(a) Write the appropriate form of the simplified equations of motion for the flow exterior to the cylinder, allowing for the fact that the finite length of the cylinder implies that there will be end effects; note that this flow is caused by the pressure gradient $P_R - P_\infty$ generated between the outer surface of the porous cylindrical tube and infinity.

(b) Write the boundary conditions required for the differential equations above.

(c) Scale the describing equations to determine the criteria for ignoring the end effects on the radial velocity profile; note that the radial velocity scale is determined from the known volumetric flow rate.

(d) Solve the resulting simplified describing equations to obtain the radial velocity profile in the liquid exterior to the cylinder.

3.P.30 Entry-Region Flow in a Tube with a Porous Annulus

In this problem we explore the idea that one might be able to decrease the entry region length for steady-state laminar pipe flow by lining the wall of the pipe with an annular region of porous medium, as shown in Figure 3.P.30-1. The outer impermeable wall of the pipe is at R_2. The porous medium is confined in the annular region defined by radii R_1 and R_2, where $R_1 < R_2$. Assume that fluid enters the pipe in plug flow; that is, $u_z = U_0$ at $z = 0$ for $0 \leqslant r \leqslant R_2$. The flow within the pipe, including the porous annular region, is caused by an applied pressure difference over the length of the pipe L; the downstream pressure is known and denoted by P_L; however, the upstream pressure is not specified. This fluid may be assumed to be Newtonian and to have constant physical properties. The pipe is assumed to be horizontal such that gravitational body forces can be ignored.

(a) Write the appropriately simplified continuity and equations of motion for the flow within the pipe; in writing these equations, denote the velocity components as u_z, u_r, and u_θ.

(b) Write the appropriately simplified continuity and equations of motion for the flow within the porous pipe wall; in writing these equations, denote the velocity components as $\widehat{u}_z$, $\widehat{u}_r$, and $\widehat{u}_\theta$.

(c) Write the boundary conditions required to solve the differential equations you obtained in part (a).

(d) Write the boundary conditions required to solve the differential equations you obtained in part (b). Assume that the radial velocity profile within the porous medium becomes pluglike within a region of influence that has an unspecified thickness δ_p whose value will be estimated in the scaling process; that is, assume that the flow within the porous medium departs from being pluglike only very near the boundaries at $r = R_1$ and $r = R_2$.

(e) Determine the scale and reference factors for the flow in the pipe; estimate the thickness of the boundary layer or region of influence δ_m.

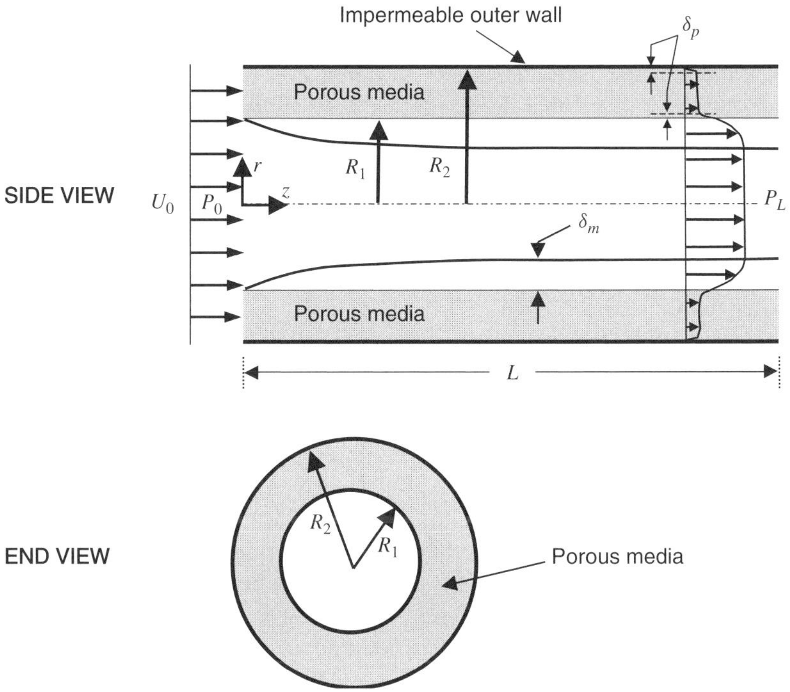

Figure 3.P.30-1 Side and end views of entry-region flow of an incompressible Newtonian fluid with constant physical properties in a pipe consisting of an open region of radius R_1 and an annular region of porous medium between R_1 and R_2.

(f) Determine the scale and reference factors for the flow within the porous medium; in particular, estimate the thickness of the region of influence δ_p.

(g) What is the criterion for ignoring the effect of the porous medium on flow in the pipe?

(h) What are the criteria for ignoring the effect of the curvature on the equations of motion for flow in both the nonporous and porous regions of the pipe?

(i) Assess the merits of using this outer annular region of the porous medium to reduce the entrance length required to achieve fully developed laminar flow in the pipe.

3.P.31 Steady-State Laminar Flow of a Compressible Gas

In Section 3.9 we considered the steady-state laminar flow of a compressible gas in a cylindrical tube as shown in Figure 3.9-1. We scaled the describing equations to determine the criterion for assuming that this flow was incompressible. This criterion was that the Mach number for the flow must be much less than 1. In scaling the describing equations we introduced a scale for the radial derivative of the pressure since we did not anticipate that this derivative would scale in the same way as the axial pressure gradient. We raised the question as to how one knows whether to scale a derivative as the ratio of some dependent variable scale divided by some independent variable scale or to introduce a separate scale for the entire derivative. We indicated that the answer to this rhetorical question was contained in the forgiving nature of scaling. To better understand what is meant by this, let us assume (incorrectly!) that the radial pressure derivative scales as the pressure scale P_s divided by the radial length scale r_s. Show that this leads to inconsistency in the resulting dimensionless equations in that one term becomes much greater than 1, with no other terms balancing it.

3.P.32 Velocity Profile Distortion Effects Due to Fluid Injection and Withdrawal

Flow-field-flow fractionation is a technique for separating small particles such as proteins and viruses from a carrier fluid such as water by combining a longitudinal laminar flow with a transverse flow. The latter can be imposed by making the closely spaced parallel lateral walls of the horizontal flow channel consist of two permeable membranes. Inflow and outflow at a constant velocity V_0 of the same carrier fluid (without any particles) occurs through the upper and lower membranes, respectively, as shown in Figure 3.P.32-1. This drives the particles, which are injected as a pulse in the axially flowing fluid, toward the lower membrane and thereby provides a means of separating them. Field-flow fractionation is considered in more detail in Chapter 5 when scaling is applied to mass transfer. In this problem we are concerned with developing a criterion for determining when the longitudinal velocity profile can be assumed to be unaffected by the transverse flow.

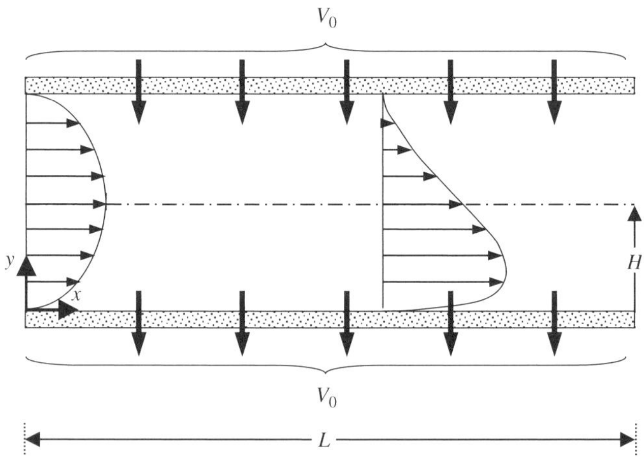

Figure 3.P.32-1 Flow-field-flow fractionation showing transverse injection of a fluid with a velocity V_0 into a longitudinal flow of the same fluid that is assumed to be initially in fully developed laminar flow; the distortion of the velocity profile that can occur due to transverse injection is shown schematically.

The entering carrier fluid can be assumed to be a Newtonian fluid with constant physical properties and to have a velocity profile given by

$$u_x = 2\overline{U}\left(\frac{2y}{H} - \frac{y^2}{H^2}\right) \tag{3.P.32-1}$$

where $\overline{U}$ is the average axial velocity.

(a) Write the appropriate form of the equations of motion that describe this flow.

(b) Write the boundary conditions that would be necessary to solve the equations of motion.

(c) Determine the criterion necessary to ignore the effect of the permeation through the upper and lower membrane boundaries on the solution for the x-component of the velocity if the latter is to be used to determine quantities such as the volumetric flow rate or average velocity.

(d) If one is interested in determining quantities in the vicinity of the membrane, such as the drag on its surface, one cannot ignore the effect of the permeation on the flow in this wall region (i.e., since the nonzero permeation velocity can have a significant influence on the small axial velocity near the membrane surface). Use scaling to determine the thickness of the region of influence near the lower membrane boundary wherein the effects of the permeation can never be ignored.

3.P.33 Flow Between Parallel Impermeable and Permeable Flat Plates

Consider the steady-state laminar flow of a Newtonian fluid with constant physical properties through a horizontal channel due to a pressure driving force $P_0 - P_L$ applied over the length L as shown in Figure 3.P.33-1. The upper surface of this channel at $y = H$ consists of an impermeable solid plate. The lower surface of this channel at $y = 0$ consists of a permeable membrane; the permeation velocity V_0 through this membrane is given by

$$V_0 = k_m[P(x) - P_{\text{atm}}] \tag{3.P.33-1}$$

where k_m is the permeability of the membrane, $P(x) - P_{\text{atm}}$ is the pressure drop across the membrane in which $P(x)$ is the local pressure on the high-pressure side of the membrane at $y = 0$, and P_{atm} is the constant pressure on the low-pressure side of the membrane, which is assumed to be atmospheric pressure. Note that, in general, $P_L - P_{\text{atm}} > 0$, to ensure that permeation occurs over the entire length L of the membrane. Ignore body forces and lateral edge effects (i.e., those in the z-direction perpendicular to the plane of Figure 3.P.33-1). Also assume that the flow is fully developed when it enters this membrane module at $x = 0$; note, however, that we are not given any information on the average or maximum velocity of the velocity profile at $x = 0$.

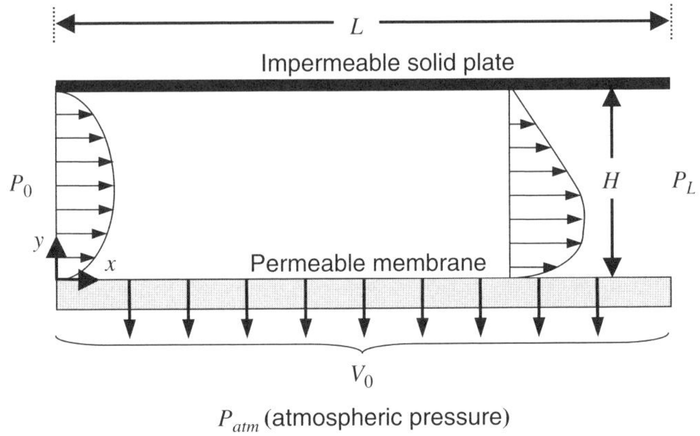

Figure 3.P.33-1 Steady-state laminar flow of a Newtonian fluid with constant physical properties through a horizontal channel due to a pressure driving force $P_0 - P_L$ applied over the length L; the upper surface at $y = H$ consists of an impermeable solid plate; the lower surface at $y = 0$ consists of a permeable membrane through which permeation occurs at a constant velocity V_0.

(a) Write the appropriate form of the equations of motion that describe this flow.

(b) Write the boundary conditions that would be necessary to solve the equations of motion.

(c) Determine the criterion for ignoring the axial diffusion of vorticity (momentum).

(d) Determine the criteria necessary to assume that this flow is essentially fully developed within the region $0 \leqslant x \leqslant L$.

(e) For the simplifying assumptions appropriate to parts (c) and (d), determine the solution for the dimensionless pressure and axial velocity profiles.

(f) For the simplifying assumptions appropriate to parts (c) and (d), derive an equation for determining the y-component of the velocity; it is not necessary to solve this equation.

(g) In part (e) you determined the solution for the dimensionless pressure profile $P(x)$. In fact, the pressure will have a slight dependence on y as well. For the simplifying assumptions appropriate to parts (c) and (d), determine the equations necessary to obtain the complete dimensionless pressure profile $P(x, y)$; that is, including the y-dependence as well. It is sufficient to express your result for the pressure profile in terms of the two velocity components u_x and u_y.

(h) Determine the criterion necessary to ignore the effect of the permeation through the membrane on the solution for the x-component of the velocity if the latter is to be used to determine quantities such as the volumetric flow rate or average velocity.

(i) If one is interested in determining quantities in the vicinity of the membrane, such as the drag on its surface, one cannot ignore the effect of the permeation on the flow in this wall region (i.e., since the nonzero permeation velocity can have a significant influence on the small axial velocity near the membrane surface). Use scaling to determine the thickness of the region of influence wherein the effects of the permeation can never be ignored.

3.P.34 Flow in an Annulus with Fluid Injection and Withdrawal

Consider the steady-state fully developed flow of an incompressible Newtonian fluid with constant physical properties in the annular region between radii R_1 and R_2, as shown in Figure 3.P.34-1. The flow in the axial direction is caused by a constant-pressure driving force given by $\Delta P \equiv P_0 - P_L$ applied across the length L. Both the inner wall at R_1 and the outer wall at R_2 are permeable membranes. Fluid is injected at a constant radial velocity V_0 into the annular region through the inner wall at R_1 and withdrawn at some constant unspecified velocity (not necessarily equal to V_0) from the annular region through the outer wall at R_2.

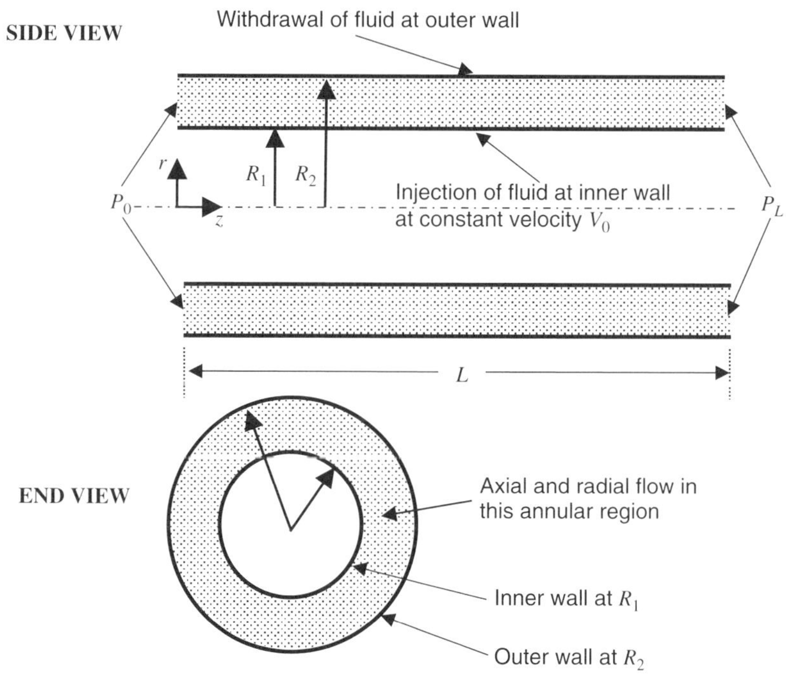

Figure 3.P.34-1 Steady-state fully developed laminar flow of a Newtonian fluid with constant physical properties through an annulus due to a pressure driving force $P_0 - P_L$ applied over the length L; fluid injection occurs at constant velocity V_0 through the inner permeable wall at $r = R_1$ and fluid withdrawal occurs at a constant unspecified velocity at the outer permeable wall at $r = R_2$.

This injection and withdrawal of fluid is done under conditions that maintain fully developed flow throughout the annulus.

(a) Simplify the continuity equation in cylindrical coordinates.

(b) What are the implications of the continuity equation for the case of fully developed flow for the axial component of velocity? Explore these mathematically by integrating the appropriate term in the continuity equation that contains the axial velocity. Be careful in considering the integration constant(s) you obtain in your partial integration.

(c) What are the implications of the continuity equation for the case of fully developed flow for the radial component of velocity? Explore these mathematically by integrating the appropriate term in the continuity equation that contains the radial velocity. Apply an appropriate boundary condition to this first-order differential equation for the axial velocity and obtain a equation for the radial velocity profile. Note in curvilinear coordinates that fully developed flow does not necessarily mean that the radial velocity component does not change in the radial direction, however, it does imply something about the radial variation of the radial velocity.

(d) Simplify the axial component of the equations of motion in cylindrical coordinates for this flow. Note that the axial pressure gradient is constant, as indicated in the problem statement.

(e) Write the appropriate boundary conditions needed to solve the differential equation that you derived in part (d).

(f) Simplify the radial component of the equations of motion in cylindrical coordinates for this flow; you may ignore the small effect of the gravitational body force.

(g) Scale the differential equations and boundary conditions that you derived for this flow in parts (d) and (e). Your answer should include specifying the scale and reference factors needed to ensure that your dependent and independent variables are bounded of order one.

(h) Use your scaling analysis in part (g) to determine the criterion for ignoring the effect of the fluid injection and withdrawal at the walls on the solution for the axial velocity.

(i) Use your scaling analysis in part (g) to determine the criterion for assuming that the inner wall is essentially at $r = 0$.

(j) Use your scaling analysis in part (g) to determine the criterion for ignoring the curvature effects. That is, what is the criterion for assuming that this is essentially the same as flow between two parallel flat plates at which there is injection at the lower plate and withdrawal at the upper plate?

(k) Solve the differential equation that you derived in part (f) to obtain an analytical solution for the pressure profile as a function of r and z; in carrying out this integration, do not forget that $\partial P/\partial z = -\Delta P/L$ since it will help you determine the integration "constant" that you obtain in your radial integration.

(l) Solve the differential equation and appropriate boundary conditions that you derived in parts (d) and (e) to obtain an analytical solution for the axial velocity profile as a function of r.

3.P.35 Flow Between Parallel Permeable Membranes

Consider the parallel permeable membranes shown in Figure 3.P.35-1, which are open at the downstream end at $x = L$ to atmospheric pressure P_{atm}, but closed at the upstream end at $x = 0$. All along the semipermeable walls of the parallel membranes, an incompressible Newtonian liquid with constant physical properties flows in at a constant velocity V_0. The fluid that flows in through the semipermeable walls ultimately exits to the atmosphere from the open end of the parallel membranes.

(a) Write the appropriate form of the equations of motion for this flow.

(b) Write the boundary conditions that are required to solve the equations of motion.

(c) Scale the describing equations to determine the criteria for making the lubrication-flow approximation.

(d) Solve for the axial velocity profile as a function of y and x; be certain that you express your answer entirely in terms of known quantities; that is, you must eliminate the pressure gradient from the equation that you obtain for the axial velocity profile.

(e) Solve for the pressure profile at any axial position along the flow.

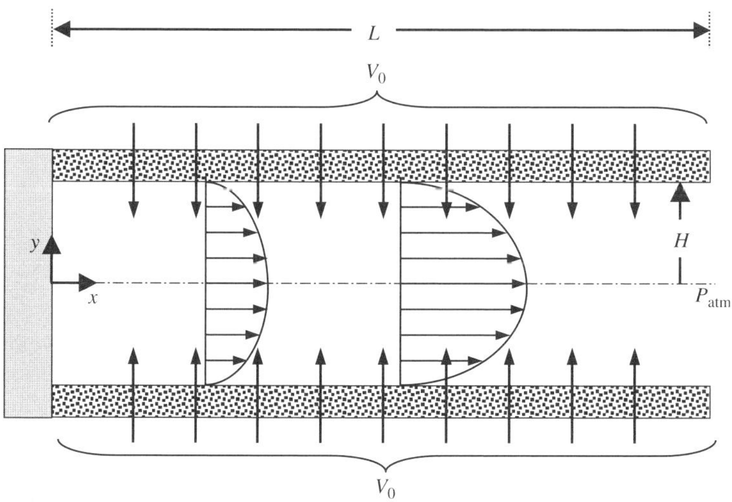

Figure 3.P.35-1 Steady-state laminar flow of an incompressible Newtonian liquid with constant physical properties between parallel permeable membranes owing to a constant radial velocity V_0 through the permeable wall at $y = \pm H$; the end between the parallel membranes at $x = 0$ is closed and impermeable; the end at $x = L$ is open to atmospheric pressure P_{atm}.

(f) Solve the simplified describing equations that you obtained in part (c) for the y-component of velocity as a function of y and x.

3.P.36 Dimensional Analysis for Flow Around a Falling Sphere

In Section 3.10 we used the scaling approach for dimensional analysis to develop a correlation for the terminal velocity U_t of a spherical particle having radius R and density ρ_p falling, owing to gravitational acceleration g through an incompressible Newtonian liquid having density ρ and viscosity μ, as shown in Figure 3.10-1.

(a) Use the scaling analysis approach to dimensional analysis to develop a correlation for the viscous drag force on the spherical particle.

(b) Consider how the correlation that you developed in part (a) simplifies for the special case of creeping flow.

3.P.37 Dimensional Analysis for Impulsively Initiated Laminar Tube Flow

In Example Problem 3.E.7 we considered the impulsively initiated laminar flow of an incompressible Newtonian fluid with constant physical properties in a cylindrical tube, as shown in Figure 3.E.7-1.

(a) Use the scaling approach to dimensional analysis to determine the dimensionless groups needed to correlate the instantaneous local velocity $u_z(r, t)$. Isolate u_z into just one dimensionless group.

(b) Use the scaling approach to dimensional analysis to determine the dimensionless groups needed to correlate the instantaneous viscous drag force on the wall. Be certain to write all the equations you would solve in order to obtain the viscous drag force. Isolate u_z into just one dimensionless group.

(c) How would the dimensional analysis in parts (a) and (b) simplify for the special case of fully developed flow?

3.P.38 Dimensional Analysis for Flow in an Oscillating Tube

In Practice Problem 3.P.19 we considered the unsteady-state flow of an incompressible Newtonian fluid with constant physical properties through a horizontal cylindrical tube of radius R and length L due to both an axial pressure gradient and an oscillating wall, as shown in Figure 3.P.19-1. Consider the special case of unsteady-state flow in this tube that is caused only by tube wall that is oscillated at a velocity $u_z = U_0 \cos \omega t$, where ω is the angular frequency of the oscillation in radians per second.

(a) Use the scaling approach to dimensional analysis to determine the dimensionless groups needed to correlate the instantaneous local velocity $u_z(z, t)$; isolate u_z into one dimensionless group, t into another, and the viscosity into a third dimensionless group; note that there may be more than three dimensionless groups.

(b) How would the dimensional analysis in part (a) simplify for quasi-steady-state creeping flow?

3.P.39 Dimensional Analysis for Curtain-Coating Flow

In Practice Problem 3.P.25 we consider the steady-state laminar curtain-coating flow of an incompressible Newtonian liquid with constant physical properties as shown in Figure 3.P.25-1. This problem involved the use of scaling analysis to simplify the describing equations appropriate to a quasi-parallel flow. In part (h) of this problem the analytical solution for the axial velocity u_x at 5 cm down from the entrance slot was compared with experimental measurements. We suspect that the result for the axial velocity that we obtained in part (g) of this problem will display considerably more error as the viscous terms that we ignored in our scaling analysis become more important. Assume that the pressure terms in the equations of motion can be ignored and use dimensional analysis to arrive at the dimensionless groups needed to correlate the axial velocity u_x at any point $x = L$ below the slot. Assume that the axial velocity that you are correlating represents an average value of u_x across the cross-section at any point $x = L$.

3.P.40 Dimensional Analysis for Flow Between Parallel Membranes

Consider the steady-state fully developed laminar flow of an incompressible Newtonian fluid with constant physical properties between two unbounded permeable membranes separated by a distance $2H$ due to a pressure driving force $\Delta P \equiv P_0 - P_L$ applied over a distance L, as shown in Figure 3.P.40-1. The same

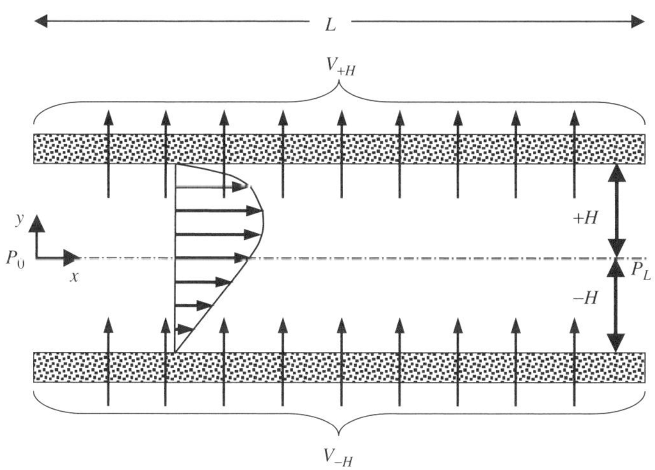

Figure 3.P.40-1 Steady-state fully developed laminar flow of an incompressible Newtonian fluid with constant physical properties through a horizontal channel due to a pressure driving force $P_0 - P_L$ applied over the length L; injection of this same fluid occurs through the membrane boundary at $y = -H$ at a constant permeation rate of V_{-H}; withdrawal of fluid occurs at the upper membrane boundary at $y = +H$ at a constant permeation rate of V_{+H}.

fluid is simultaneously injected at the lower plate at a velocity V_{-H} while uniform suction occurs at the upper plate to remove fluid at the velocity V_{+H}.

(a) Use the continuity equation to show that $V_{-H} = V_{+H} = V$, a constant.
(b) Use dimensional analysis to develop a correlation for the total drag force exerted in the $+z$-direction by the flowing fluid on the upper and lower plates.
(c) Use your dimensional analysis result to determine how much the total drag force will increase if the applied pressure force ΔP is doubled.
(d) Use dimensional analysis to develop a correlation for the volumetric flow rate Q.
(e) Use your dimensional analysis results in parts (b) and (d) to determine how much the total drag force will increase if the volumetric flow rate is doubled.

3.P.41 Dimensional Analysis for Flow in a Hollow-Fiber Membrane

In Example Problem 3.E.10, we consider the steady-state flow of an incompressible Newtonian fluid with constant physical properties in a hollow-fiber membrane, one end of which is closed and the other open to the atmosphere, which is caused by permeation through the wall at a constant velocity V_0, as shown in Figure 3.E.10-1. This problem involved the use of scaling analysis to assess the criteria for assuming that this is a lubrication flow. Assume now that we wish to correlate the drag force on the hollow-fiber wall for the general case when the lubrication-flow approximation cannot be made. Use dimensional analysis to infer the appropriate dimensionless groups to correlate the dimensional drag force. Express your answer in terms of the standard dimensionless groups used to correlate drag phenomena of this type; that is, in terms of an appropriately defined friction factor, Reynolds number, and aspect ratio.

4 Applications in Heat Transfer

If we assume that the ice is thin enough so that the temperature gradient can be considered as uniform from the upper to the lower surface, we can derive at once a very simple solution. . . .[1]

4.1 INTRODUCTION

The quotation cited above appeared in the classic text *Heat Conduction with Engineering and Geological Applications*, which still serves as a basic reference book in this field of research. In particular, this statement was made in connection with justifying when the unsteady-state freezing of water-saturated soil could be assumed to be quasi-steady-state. However, an appropriate rejoinder to the quote above would be: "How thin is thin?" The solution to the quasi-state-state problem indeed is "very simple". However, leaping to the conclusion that "the ice is thin enough" is intuitive. Alternatively, scaling analysis can be used to develop a quantitative criterion for assessing the applicability of the quasi-steady-state approximation. This is considered in Section 4.7 for this freezing problem involving heat transfer with phase change.

In this chapter we consider the application of scaling analysis to heat transfer. The organization of this chapter is the same as that used in Chapter 3. To understand fully the material in this chapter, it is necessary first to read Chapters 1 and 2. However, since some readers might be interested primarily in heat transfer rather than fluid dynamics, the first few examples are developed in the same detail as was done in Chapter 3; that is, it will be possible to understand how to apply scaling analysis to heat transfer without necessarily thoroughly understanding the material in Chapter 3. However, it will clearly be necessary to understand some aspects of fluid dynamics when convective heat transfer is considered. Note that in this chapter we again use the ordering symbols o(1) and O(1) introduced in Chapter 2.

[1]L. R. Ingersoll, O. J. Zobel, and A. C. Ingersoll, *Heat Conduction with Engineering and Geological Applications*, McGraw-Hill, New York, 1948, p. 197.

Scaling Analysis in Modeling Transport and Reaction Processes: A Systematic Approach to Model Building and the Art of Approximation, By William B. Krantz

Recall that the symbol o(1) implies that the magnitude of the quantity can range between 0 and more-or-less 1, whereas the symbol O(1) implies that the magnitude of the quantity is more-or-less 1 but not much less than 1.

Some of the scaling considerations in this chapter are similar to those encountered in scaling fluid dynamics problems: for example, quasi-steady-state and boundary-layer phenomena. However, in this chapter we apply scaling analysis to determine when other simplified models can be used, such as film theory and penetration theory for conductive heat transfer; there are no analogs to these approximations in fluid dynamics. The same disclaimer applies to this chapter as was stated for Chapter 3: namely, that no attempt will be made here to provide a detailed derivation of the describing equations that are used in the scaling analysis. Hence, the material in this chapter provides a useful supplement for a foundation course in heat transfer. The reader is referred to the appendices that summarize the energy equation in generalized vector–tensor notation as well as in rectangular, cylindrical, and spherical coordinates. These equations serve as the starting point for each example problem.

We begin by considering the use of o(1) scaling to simplify pure heat-conduction problems. Scaling analysis is then used to justify simplifications made in heat transfer, such as the penetration-theory and film-theory approximations, low Biot number heat transfer, conduction- and heat-generation-dominated convective heat transfer, low Peclet number convective heat transfer (the analog to the creeping-flow approximation in fluid dynamics) and high Peclet number convective heat transfer (the analog to high Reynolds number or boundary-layer flows). We then apply o(1) scaling to heat transfer with phase change, which introduces scaling of moving boundary problems. Applying scaling analysis to heat transfer now permits us to determine when the variation of physical and/or transport properties with temperature needs to be considered in developing models. Finally, the scaling analysis approach is applied to dimensional analysis for heat-transfer problems. Additional worked example and practice problems are included at the end of the chapter.

4.2 STEADY-STATE HEAT TRANSFER WITH END EFFECTS

This first example illustrates the application of the o(1) scaling analysis procedure to a steady-state conductive heat-transfer problem for which an exact analytical solution is available. If the describing equations can be solved analytically, there is no need to apply scaling analysis to explore how the problem can be simplified. However, this problem is instructive in that the solution to the simplified equations obtained via scaling can be compared with the analytical solution to the unsimplified describing equations to assess the error incurred as a function of the magnitude of the dimensionless group, which needs to be small to justify the approximation. It will also illustrate region-of-influence scaling whereby we seek to determine the thickness of a region wherein some important effect is concentrated. Region-of-influence scaling is particularly important since it forms the basis of thermal boundary-layer theory and penetration theory.

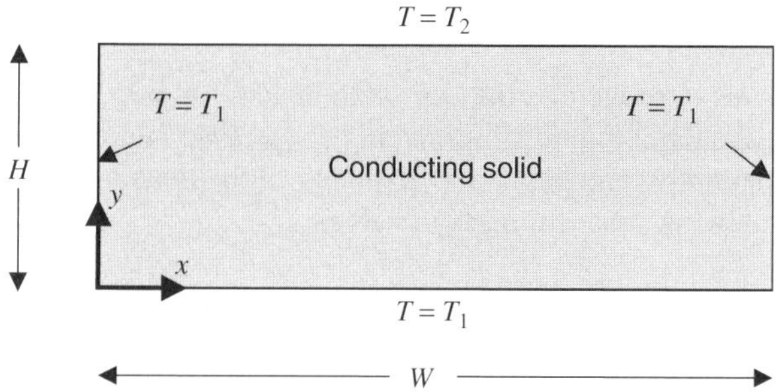

Figure 4.2-1 Steady-state two-dimensional heat conduction in a homogeneous solid with constant physical properties, width W, and height H such that $W > H$; the faces at $x = 0$, $x = W$, and $y = 0$ are held at a constant temperature T_1, whereas the face at $y = H$ is held at a constant temperature T_2.

Consider steady-state heat conduction in a homogeneous solid having constant physical properties that has width W in the x-direction, height H in the y-direction, and is infinitely thick in the z-direction, as shown in Figure 4.2-1; the geometry is such that $W > H$. The planar faces at $x = 0$, $x = W$, and $y = 0$ are held at a constant temperature T_1, whereas the planar face at $y = H$ is maintained at a constant temperature T_2. We anticipate that if $W \gg H$, we might be able to ignore the heat conduction in the x-direction. However, the question arises as to how much larger W has to be relative to H to ignore the lateral heat conduction. Another question is: How much error do we encounter if we ignore the lateral heat conduction? We employ scaling analysis to determine the criterion for when we can assume that this can be approximated as one-dimensional heat conduction in the y-direction; that is, we seek to determine when the lateral heat transfer or end effects can be neglected. We invoke the stepwise o(1) scaling analysis procedure outlined in Chapter 2. In this first example of o(1) scaling analysis applied to heat transfer, we show all the steps in detail and provide a discussion of the rationale for each step.

Step 1 involves writing the describing equations, in this case the thermal energy equation appropriately simplified for this problem statement and its boundary conditions. Equation F.1-2 in the Appendices simplifies to the following for the conditions specified for this heat-transfer problem:

$$\frac{\partial^2 T}{\partial x^2} + \frac{\partial^2 T}{\partial y^2} = 0 \tag{4.2-1}$$

$$T = T_1 \quad \text{at} \quad x = 0 \tag{4.2-2}$$

$$T = T_1 \quad \text{at} \quad x = W \tag{4.2-3}$$

$$T = T_1 \quad \text{at} \quad y = 0 \tag{4.2-4}$$

$$T = T_2 \quad \text{at} \quad y = H \tag{4.2-5}$$

These boundary conditions constitute the temperatures prescribed at the four boundaries.

Step 2 involves introducing arbitrary scale factors for each dependent and independent variable. Step 3 requires introducing a reference factor for the temperature since it is not naturally referenced to zero at any of the boundaries. Step 4 involves defining the following dimensionless variables:

$$T^* \equiv \frac{T - T_r}{T_s}; \qquad x^* \equiv \frac{x}{x_s}; \qquad y^* \equiv \frac{y}{y_s} \tag{4.2-6}$$

In step 5 these dimensionless variables are substituted into the describing equations (4.2-1) through (4.2-5):

$$\frac{1}{x_s^2}\frac{\partial^2 T^*}{\partial x^{*2}} + \frac{1}{y_s^2}\frac{\partial^2 T^*}{\partial y^{*2}} = 0 \tag{4.2-7}$$

$$T_s T^* + T_r = T_1 \quad \text{at} \quad x_s x^* = 0 \tag{4.2-8}$$

$$T_s T^* + T_r = T_1 \quad \text{at} \quad x_s x^* = W \tag{4.2-9}$$

$$T_s T^* + T_r = T_1 \quad \text{at} \quad y_s y^* = 0 \tag{4.2-10}$$

$$T_s T^* + T_r = T_2 \quad \text{at} \quad y_s y^* = H \tag{4.2-11}$$

Step 6 involves dividing through by the dimensional coefficient of the conduction term in the y-direction in equation (4.2-7) since this term must be retained to account for the principal direction for heat transfer. In the four boundary conditions we divide through by the dimensional coefficient of the dimensionless dependent variable, which yields

$$\frac{y_s^2}{x_s^2}\frac{\partial^2 T^*}{\partial x^{*2}} + \frac{\partial^2 T^*}{\partial y^{*2}} = 0 \tag{4.2-12}$$

$$T^* = \frac{T_1 - T_r}{T_s} \quad \text{at} \quad x^* = 0 \tag{4.2-13}$$

$$T^* = \frac{T_1 - T_r}{T_s} \quad \text{at} \quad x^* = \frac{W}{x_s} \tag{4.2-14}$$

$$T^* = \frac{T_1 - T_r}{T_s} \quad \text{at} \quad y^* = 0 \tag{4.2-15}$$

$$T^* = \frac{T_2 - T_r}{T_s} \quad \text{at} \quad y^* = \frac{H}{y_s} \tag{4.2-16}$$

Step 7 involves determining the scale and reference factors to ensure that the dimensionless variables are o(1); that is, that they are bounded between zero and more-or-less 1. This can be achieved by setting the dimensionless groups containing the reference temperature and temperature scale in equations (4.2-13), (4.2-14), or (4.2-15) equal to zero and in equation (4.2-16) equal to 1; that is,

$$\frac{T_1 - T_r}{T_s} = 0 \Rightarrow T_r = T_1 \tag{4.2-17}$$

$$\frac{T_2 - T_r}{T_s} = \frac{T_2 - T_1}{T_s} = 1 \Rightarrow T_s = T_2 - T_1 \tag{4.2-18}$$

Since our region of interest spans the solid between the four planar faces, the dimensionless spatial variables can be bounded between zero and 1 by setting the dimensionless groups containing x_s and y_s in equations (4.2-14) and (4.2-16) equal to 1; that is,

$$\frac{W}{x_s} = 1 \Rightarrow x_s = W \tag{4.2-19}$$

$$\frac{H}{y_s} = 1 \Rightarrow y_s = H \tag{4.2-20}$$

Our dimensionless equations now become

$$\frac{H^2}{W^2}\frac{\partial^2 T^*}{\partial x^{*2}} + \frac{\partial^2 T^*}{\partial y^{*2}} = 0 \tag{4.2-21}$$

$$T^* = 0 \quad \text{at} \quad x^* = 0 \tag{4.2-22}$$

$$T^* = 0 \quad \text{at} \quad x^* = 1 \tag{4.2-23}$$

$$T^* = 0 \quad \text{at} \quad y^* = 0 \tag{4.2-24}$$

$$T^* = 1 \quad \text{at} \quad y^* = 1 \tag{4.2-25}$$

Step 8 then involves using our scaled dimensionless describing equations to assess the conditions for which we can ignore lateral (x-direction) heat conduction. Equation (4.2-21) indicates that the lateral heat-conduction term will drop out of the describing equations if the following condition holds:

$$\frac{H^2}{W^2} \ll 1 \tag{4.2-26}$$

If the condition above is satisfied, equation (4.2-21) reduces to

$$\frac{\partial^2 T^*}{\partial y^{*2}} = 0 \tag{4.2-27}$$

for which the solution is given by

$$T^* = y^* \tag{4.2-28}$$

Note that the criterion that has emerged from our scaling analysis for ignoring the effect of lateral heat conduction is in terms of a dimensionless group. The physical significance of this dimensionless group is that it is the ratio of the magnitude of the lateral to vertical (x- to y-direction) heat conduction. The dimensionless groups that emerge from scaling analysis will always have a physical significance that can be determined by examining how a particular group was formed: in this case, by dividing the coefficient of the lateral heat-conduction term by that for the vertical heat-conduction term.

To gain a better feeling for the error incurred when scaling approximations are made, it is instructive to compare the approximate solution for small values of H^2/W^2 given by equation (4.2-28) to the exact analytical solution to equation (4.2-21) that is given by[2]

$$T^* = \frac{2}{\pi} \sum_{n=1}^{\infty} \frac{(-1)^{n+1}}{n} \sin n\pi x^* \frac{\sinh(n\pi y^* H/W)}{\sinh(n\pi H/W)} \tag{4.2-29}$$

Note that if the dimensionless group $H/W \ll 1$, equation (4.2-29) can be expanded in a Taylor series; the first nonzero term in this expansion is the approximate solution given by equation (4.2-28). Equation (4.2-28) predicts that the dimensionless temperature is constant along any horizontal plane corresponding to some specified value of y^*. The exact solution given by equation (4.2-29) predicts that the temperature along any horizontal plane varies with the lateral location. Figure 4.2-2 plots the error that is incurred when equation (4.2-28) is used to predict the dimensionless temperature at $y^* = 0.5$ (i.e., $T^* = 0.5$) as a function of the lateral location x^* for $H^2/W^2 = 0.01$ ($H/W = 0.1$) and $H^2/W^2 = 0.1$ ($H/W = 0.316$); that is,

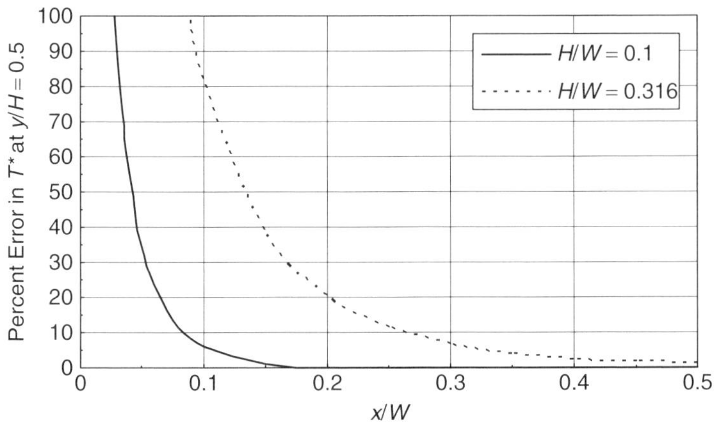

Figure 4.2-2 Percentage error in the dimensionless temperature at $y^* = 0.5$ that is incurred when the effect of lateral conduction is ignored as a function of the dimensionless lateral position x^* for $H/W = 0.1$ ($H^2/W^2 = 0.01$) and $H/W = 0.316$ ($H^2/W^2 = 0.1$).

[2]F. P. Incropera and D. P. DeWitt, *Fundamentals of Heat and Mass Transfer*, Wiley, New York, 1996, pp. 163–167.

the error incurred when lateral conduction is ignored.[3] Figure 4.2-2 is plotted only for $0 < x^* \leq 0.5$ since the temperature profile is symmetrical about the plane defined by $x^* = 0.5$. This figure indicates that for $H^2/W^2 = 0.01$ the error is less than 10% across 80% of the total width of the solid. In contrast, for $H^2/W^2 = 0.1$, the error is less than 10% across only 50% of the total width of the solid. This is to be expected since ignoring lateral conduction becomes a progressively better approximation as the dimensionless group H^2/W^2 decreases. These errors are typical of what one can anticipate for the approximations that emanate from scaling analysis. Since the dimensionless variables are scaled to be of order one, neglecting a term that is multiplied by a dimensionless group that is O(0.1) or O(0.01) should result in errors of approximately 10% and 1%, respectively.

The preceding analysis indicates that the error encountered in making some assumption depends not only on the magnitude of the dimensionless group that emanates from the scaling analysis, but also on the particular quantity that is being predicted. For example, ignoring the lateral conduction in the present problem yields accurate predictions for the temperature at the center of the solid even when H^2/W^2 is as large as 0.1. However, the prediction for the dimensionless temperature is considerably in error at $x^* = 0.05$ even when H^2/W^2 is as small as 0.01. Indeed, the error in the temperature predicted that is incurred when lateral heat conduction is neglected increases without bound at points progressively closer to the lateral boundaries. Moreover, the thickness of this wall region wherein lateral conduction is important is directly proportional to the value of H/W. Clearly, it would be of value to be able to estimate the thickness of this wall region wherein two-dimensional conductive heat transfer must be considered. This can be done using *region-of-influence scaling*, whereby we seek to determine the thickness of a region within which some effect is important; in this case, the thickness of the region near the lateral boundaries wherein lateral heat-conduction effects are significant.

To carry out region-of-influence scaling, the unspecified length scale factor in fact becomes the thickness of the region of influence, which we denote by the symbol δ_t to emphasize its particular physical significance. By this we mean that the relevant dependent variable, in this case $\partial T^*/\partial x^*$, is O(1) within this region. Let us rescale this problem to determine the magnitude of δ_t. It is sufficient here to consider only one of the lateral boundaries, due to the symmetry of the heat-transfer geometry. Equations (4.2-1), (4.2-2), (4.2-4), and (4.2-5) remain the same; however, equation (4.2-3) needs to be replaced by a boundary condition appropriate to the wall region, which is given by

$$\frac{\partial T}{\partial x} = 0 \qquad \text{at} \quad x = \delta_t \tag{4.2-30}$$

This boundary condition merely states that the lateral variation in temperature is confined to the wall region or region of influence whose thickness δ_t will be determined via scaling analysis.

[3]The infinite series in equation (4.2-29) converges rather slowly; hence, 50 terms in this series were retained in determining the dimensionless temperature predicted by this exact solution.

If the dimensionless variables defined by equations (4.2-6) are substituted into equations (4.2-1), (4.2-2), (4.2-4), (4.2-5), and (4.2-30) and step 6 in the scaling analysis procedure is applied, the following set of dimensionless describing equations is obtained:

$$\frac{y_s^2}{x_s^2}\frac{\partial^2 T^*}{\partial x^{*2}} + \frac{\partial^2 T^*}{\partial y^{*2}} = 0 \tag{4.2-31}$$

$$T^* = \frac{T_1 - T_r}{T_s} \quad \text{at} \quad x^* = 0 \tag{4.2-32}$$

$$\frac{\partial T^*}{\partial x^*} = 0 \quad \text{at} \quad x^* = \frac{\delta_t}{x_s} \tag{4.2-33}$$

$$T^* = \frac{T_1 - T_r}{T_s} \quad \text{at} \quad y^* = 0 \tag{4.2-34}$$

$$T^* = \frac{T_2 - T_r}{T_s} \quad \text{at} \quad y^* = \frac{H}{y_s} \tag{4.2-35}$$

Applying o(1) scaling then results in the following for the scale and reference factors:

$$T_r = T_1; \qquad T_s = T_2 - T_1; \qquad x_s = \delta_t; \qquad y_s = H \tag{4.2-36}$$

These scale factors differ from those obtained initially in that the lateral length scale is now the thickness of the region of influence near the lateral boundary rather than being the entire width of the solid. This is a reasonable result since the temperature goes through a characteristic change $T_2 - T_1$ over the distance δ_t near the lateral boundary rather than over the entire half-width of the solid.

If the scale and reference factors indicated in equations (4.2-36) are introduced into equations (4.2-31) through (4.2-35), we obtain the following set of dimensionless describing equations:

$$\frac{H^2}{\delta_t^2}\frac{\partial^2 T^*}{\partial x^{*2}} + \frac{\partial^2 T^*}{\partial y^{*2}} = 0 \tag{4.2-37}$$

$$T^* = 0 \quad \text{at} \quad x^* = 0 \tag{4.2-38}$$

$$\frac{\partial T^*}{\partial x^*} = 0 \quad \text{at} \quad x^* = 1 \tag{4.2-39}$$

$$T^* = 0 \quad \text{at} \quad y^* = 0 \tag{4.2-40}$$

$$T^* = 1 \quad \text{at} \quad y^* = 1 \tag{4.2-41}$$

If lateral heat conduction is important within the region of influence, both terms in equation (4.2-37) must be o(1). This implies that

$$\frac{H^2}{\delta_t^2} = 1 \Rightarrow \delta_t = H \tag{4.2-42}$$

That is, the thickness of the region of influence near the lateral boundary is equal to the vertical length scale. This implies that an approximate analysis that ignores lateral heat conduction will begin to incur significant error when $x^* \equiv x/W = \delta_t/W = H/W$. Note in Figure 4.2-2 that the percentage error in the dimensionless temperature begins to increase markedly when $x^* < 0.1$ for $H/W = 0.1$ $(H^2/W^2 = 0.01)$ and when $x^* < 0.316$ for $H/W = 0.316$ $(H^2/W^2 = 0.1)$. Hence, we see that scaling analysis not only determines the criterion for when lateral conduction can be ignored, but also provides a measure of the region near the lateral boundaries wherein this assumption breaks down.

A similar scaling analysis can be done to determine when three-dimensional heat conduction in a rectangular block can be simplified to a two- or one-dimensional problem. The criteria for ignoring axial conduction in a long thin solid cylinder or radial conduction in a short wide cylinder can also be determined using an analogous scaling analysis. These are considered in the practice problems at the end of the chapter.

4.3 FILM AND PENETRATION THEORY APPROXIMATIONS

Now that the procedure for o(1) scaling analysis has been illustrated in detail, we use this method to explore the various approximations made in classical heat-transfer modeling. The first problem that we consider is unsteady-state one-dimensional heat conduction in a solid that has constant physical properties and a thickness H as shown in Figure 4.3-1. This solid is initially at a constant

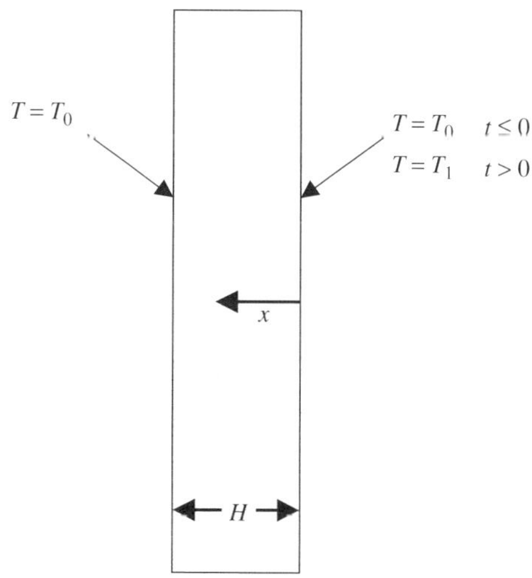

Figure 4.3-1 Unsteady-state one-dimensional heat conduction in a solid with constant physical properties and thickness H.

temperature T_0; however, one surface of this solid is then raised to a temperature T_1 while the other surface is maintained at T_0. We use scaling analysis to explore what approximations might be made to simplify this heat-transfer problem.

We begin by writing the thermal energy equation given by equation (F.1-2) in the Appendices appropriately simplified for the conditions defined in the problem statement along with the initial and boundary conditions (step 1):

$$\frac{\partial T}{\partial t} = \alpha \frac{\partial^2 T}{\partial x^2} \tag{4.3-1}$$

$$T = T_0 \quad \text{at} \quad t \leq 0, \quad 0 \leq x \leq H \tag{4.3-2}$$

$$T = T_1 \quad \text{at} \quad x = 0, \quad t > 0 \tag{4.3-3}$$

$$T = T_0 \quad \text{at} \quad x = H \tag{4.3-4}$$

where $\alpha \equiv k/\rho C_p$ is the thermal diffusivity in which k the thermal conductivity, ρ the mass density, and C_p the heat capacity at constant pressure. Equation (4.3-2) is the given initial temperature condition. Equations (4.3-3) and (4.3-4) are the prescribed constant temperatures at the two boundaries. This is a nontrivial problem to solve, due to the unsteady-state heat transfer and the finite thickness of the solid. We use o(1) scaling to explore when these describing equations might be simplified to permit a tractable solution.

We begin by defining dimensionless variables involving unspecified scale factors (steps 2, 3, and 4):

$$T^* \equiv \frac{T - T_r}{T_s}; \qquad x^* \equiv \frac{x}{x_s}; \qquad t^* \equiv \frac{t}{t_s} \tag{4.3-5}$$

Note that we have introduced a reference factor for the temperature since it is not naturally referenced to zero. We then introduce these dimensionless variables into the describing equations and divide through by the coefficient of one term in each equation that we believe should be retained (steps 5 and 6):

$$\frac{x_s^2}{\alpha t_s} \frac{\partial T^*}{\partial t^*} = \frac{\partial^2 T^*}{\partial x^{*2}} \tag{4.3-6}$$

$$T^* = \frac{T_0 - T_r}{T_s} \quad \text{at} \quad t^* \leq 0, \quad 0 \leq x^* \leq \frac{H}{x_s} \tag{4.3-7}$$

$$T^* = \frac{T_1 - T_r}{T_s} \quad \text{at} \quad x^* = 0, \quad t^* > 0 \tag{4.3-8}$$

$$T^* = \frac{T_0 - T_r}{T_s} \quad \text{at} \quad x^* = \frac{H}{x_s} \tag{4.3-9}$$

Now let us proceed to determine the scale factors (step 7). The dimensionless temperature can be bounded to be o(1) by setting the groups containing the temperature scale and reference factors in equations (4.3-7) or (4.3-9) and (4.3-8) equal to 1 and zero, respectively, to obtain

$$T^* = \frac{T_0 - T_r}{T_s} = 0 \Rightarrow T_r = T_0 \tag{4.3-10}$$

$$T^* = \frac{T_1 - T_r}{T_s} = 1 \Rightarrow T_s = T_1 - T_0 \tag{4.3-11}$$

The scale factor for the dimensionless time variable for unsteady-state problems is often the observation or contact time; that is, $t_s = t_o$, the particular time at which the process is being considered. The manner in which the length scale factor is determined depends on the observation time. Let us assume that the dimensionless groups containing the length scale in equations (4.3-7) or (4.3-9) determine x_s. Although this bounds the dimensionless spatial coordinate to be o(1), it does not necessarily bound the dimensionless temperature derivative to be o(1). Indeed, the temperature derivative could involve a much shorter length scale during the early stages of heat transfer when the conduction has not penetrated very far into the solid. However, let us assume that

$$\frac{H}{x_s} = 1 \Rightarrow x_s = H \tag{4.3-12}$$

Substitution of the scale and reference factors defined by equations (4.3-10) through (4.3-12) into the describing equations yields

$$\frac{H^2}{\alpha t_o}\frac{\partial T^*}{\partial t^*} = \frac{1}{\mathrm{Fo}_t}\frac{\partial T^*}{\partial t^*} = \frac{\partial^2 T^*}{\partial x^{*2}} \tag{4.3-13}$$

$$T^* = 0 \quad \text{at} \quad t^* \le 0, \quad 0 \le x^* \le 1 \tag{4.3-14}$$

$$T^* = 1 \quad \text{at} \quad x^* = 0, \quad t^* > 0 \tag{4.3-15}$$

$$T^* = 0 \quad \text{at} \quad x^* = 1 \tag{4.3-16}$$

The nature of this unsteady-state heat-transfer process is characterized by the dimensionless group in equation (4.3-13), which is referred to as the *Fourier number* for heat transfer, Fo_t. The physical significance of the Fourier number for heat transfer is that it is the ratio of the contact time available for heat transfer t_o divided by the characteristic time H^2/α required for heat conduction through the thickness H; that is,

$$\mathrm{Fo}_t \equiv \frac{\alpha t_o}{H^2} = \frac{\text{observation or contact time}}{\text{characteristic time for conduction}} \tag{4.3-17}$$

Now let us explore possible simplifications of the describing equations (step 8). Note that if $\mathrm{Fo}_t \gg 1$, the unsteady-state term becomes insignificant; hence, the heat transfer can be assumed to be steady-state; that is,

$$\mathrm{Fo}_t \equiv \frac{\alpha t_o}{H^2} \gg 1 \Rightarrow \text{steady-state heat transfer} \tag{4.3-18}$$

Note that a large Fourier number in this problem ensures that the heat transfer is truly steady-state, in contrast to *quasi-steady-state*. The latter implies that the unsteady-state term in the energy equation is negligible but that the problem is still unsteady-state, due to time dependence that enters through the boundary conditions. Quasi-steady-state heat transfer is considered in a subsequent section. If $\mathrm{Fo}_t \cong 10$, the error incurred in assuming steady state when determining quantities such as the heat flux into the solid at $x = 0$ will be on the order of 10%; if $\mathrm{Fo}_t \cong 100$, the error will be reduced to approximately 1%. However, unless the Fourier number is very large, the error incurred in predicting point quantities such as the temperature at $x = H$ could be quite large. Note that since heat is penetrating from the face at $x = 0$ toward the face at $x = H$, the solid in the region closer to $x = 0$ receives more heat than the region nearer to $x = H$ and therefore heats up more quickly. For this reason the error incurred in predicting the local temperature for moderate values of the Fourier number will be greater for planes nearer to $x = H$.

Let us now consider the special case of where $\mathrm{Fo}_t = \mathrm{O}(1)$, that is, when its value is essentially equal to 1. In this case the heat transfer is inherently unsteady state; however, the thermal penetration is through the entire thickness H of the solid; hence, H is the appropriate length scale to ensure that the dimensionless temperature derivative is $\mathrm{o}(1)$. Scaling permits estimating the time required for the heat penetration to reach the face at $x = H$; this will be denoted by t_H and is determined as

$$\mathrm{Fo}_t \equiv \frac{\alpha t_H}{H^2} = 1 \Rightarrow t_H = \frac{H^2}{\alpha} \tag{4.3-19}$$

Now let us explore another possible approximation that can be made in the describing equations. If $\mathrm{Fo}_t \ll \mathrm{O}(1)$, the contact time is so short that the thermal penetration will be confined to a region of influence or boundary layer whose thickness is less than H. Scaling can be used to determine the thickness of this region of influence. However, to do so, it is necessary to rescale the problem since the length scale over which the temperature experiences a characteristic change is no longer H. Let us denote the thickness of this region of influence by δ_t; that is, our length scale is now $x_s = \delta_t$. The dimensionless describing equations now are given by

$$\frac{\delta_t^2}{\alpha t_s}\frac{\partial T^*}{\partial t^*} = \frac{\partial^2 T^*}{\partial x^{*2}} \tag{4.3-20}$$

$$T^* = \frac{T_0 - T_r}{T_s} \quad \text{at} \quad t^* \le 0, \quad 0 \le x^* \le \frac{H}{\delta_t} \tag{4.3-21}$$

$$T^* = \frac{T_1 - T_r}{T_s} \quad \text{at} \quad x^* = 0, \quad t^* > 0 \tag{4.3-22}$$

$$T^* = \frac{T_0 - T_r}{T_s} \quad \text{at} \quad x^* = \frac{H}{\delta_t} \tag{4.3-23}$$

Equations (4.3-21) through (4.3-23) indicate that our reference and scale temperatures are still determined by equations (4.3-10) and (4.3-11). Moreover, due to the unsteady-state, the characteristic time again will be the observation time t_o. However, since this is inherently unsteady-state heat transfer, the unsteady-state term and the conduction term in equation (4.3-20) must balance each other and be of o(1). To ensure this, we set the dimensionless group multiplying the unsteady-state term in equation (4.3-20) equal to 1; this then provides an estimate for δ_t, the thickness of the region of influence:

$$\frac{\delta_t^2}{\alpha t_o} = 1 \Rightarrow \delta_t = \sqrt{\alpha t_o} \tag{4.3-24}$$

Note that equation (4.3-24) implies that the thickness of the region of influence or boundary-layer thickness increases with time. Note also that when $t = t_H = H^2/\alpha$, we obtain $\delta_t = H$; that is, the limiting value of δ_t is H, as expected.

Equation (4.3-24) implies that our describing equations can be written as

$$\frac{\partial T^*}{\partial t^*} = \frac{\partial^2 T^*}{\partial x^{*2}} \tag{4.3-25}$$

$$T^* = 0 \quad \text{at} \quad t^* \leq 0, \quad 0 \leq x^* \leq \frac{H}{\sqrt{\alpha t_o}} = \frac{1}{\sqrt{\text{Fo}_t}} \tag{4.3-26}$$

$$T^* = 1 \quad \text{at} \quad x^* = 0, \quad t^* > 0 \tag{4.3-27}$$

$$T^* = 0 \quad \text{at} \quad x^* = \frac{H}{\sqrt{\alpha t_o}} = \frac{1}{\sqrt{\text{Fo}_t}} \tag{4.3-28}$$

Note that if the Fourier number is very small, the thermal penetration thickness δ_t will be much less than the thickness of the solid H; therefore, the heat transfer will be confined to a thin boundary layer or region of influence near $x = 0$; that is,

$$\text{Fo}_t \equiv \frac{\alpha t_o}{H^2} \ll 1 \Rightarrow \text{heat transfer is contact-time limited} \tag{4.3-29}$$

If $\text{Fo}_t \ll 1$, it is reasonable to write the describing equations as

$$\frac{\partial T^*}{\partial t^*} = \frac{\partial^2 T^*}{\partial x^{*2}} \tag{4.3-30}$$

$$T^* = 0 \quad \text{at} \quad t^* \leq 0, \quad 0 \leq x^* < \infty \tag{4.3-31}$$

$$T^* = 1 \quad \text{at} \quad x^* = 0, \quad t^* > 0 \tag{4.3-32}$$

$$T^* = 0 \quad \text{as} \quad x^* \to \infty, 0 \leq t^* < \infty \tag{4.3-33}$$

This set of simplified equations admits an exact analytical solution via the method of combination of variables (this is also referred to as a *similarity solution*) in the form

$$T^* = \frac{T - T_0}{T_1 - T_0} = 1 - \text{erf}\left(\frac{1}{2}x^*\right) = 1 - \text{erf}\left(\frac{x}{\sqrt{4\alpha t_o}}\right) \tag{4.3-34}$$

where erf is the error function that is tabulated in standard references.[4] Note that $T^* = 0.01$ when $x^* = 3.64$, which implies that $x = 3.64\sqrt{\alpha t_o} = 3.64\delta_t$. Hence, we see that the exact solution for the unsteady-state thermal penetration confirms our scaling analysis result; that is, it predicts that the dimensionless temperature becomes essentially zero at a distance that is essentially (within a multiplicative constant of order one) equal to the thickness of the region of influence that we identified via scaling analysis.

Scaling analysis for this problem revealed the full spectrum of contact time behavior for a heat-transfer problem that can be characterized in terms of the magnitude of the Fourier number. In summary, if

$$\text{Fo}_t \gg 1 \Rightarrow \begin{cases} \text{the contact time is long relative to the conduction time} \\ \text{steady-state heat transfer} \end{cases} \tag{4.3-35}$$

$$\text{Fo}_t = 1 \Rightarrow \begin{cases} \text{the contact and conduction times are equal} \\ \text{unsteady-state heat-transfer between boundaries} \end{cases} \tag{4.3-36}$$

$$\text{Fo}_t \ll 1 \Rightarrow \begin{cases} \text{the contact time is short relative to the conduction time} \\ \text{unsteady-state heat transfer in a thin boundary layer} \end{cases} \tag{4.3-37}$$

In general, a model for a heat- (or mass-) transfer process based on assuming that the contact time for a conductive (or diffusive) heat- (or mass-) transfer process is long in comparison to the characteristic time for heat conduction (or species diffusion) is referred to as a *film theory model*. The latter terminology is commonly used in mass-transfer modeling but less commonly in heat transfer. A model for a heat- (or mass-) transport process based on assuming that the contact time for a conductive (or diffusive) heat- (or mass-) transfer process is very short in comparison to

[4]M. Abramowitz and I. A. Stegun, eds., *Handbook of Mathematical Functions with Formulas, Graphs, and Mathematical Tables*, National Bureau of Standards Applied Mathematics Series 55, U.S. Government Printing Office, Washington, DC, 1964.

the characteristic time for heat conduction (or species diffusion) is referred to as a *penetration theory model*. Film theory and penetration theory are used to model complex heat- (and mass-) transfer processes that preclude obtaining tractable analytical or numerical solutions. These models are particularly useful in determining heat- and mass-transfer coefficients for high-mass-transfer flux conditions. That is, the heat- and mass-transfer coefficients determined from correlations in the literature in general are valid only in the limit of very low-mass-transfer fluxes; indeed, correlating these coefficients for high-mass-transfer flux conditions would involve taking vastly more data and a far more complex correlation involving additional dimensionless groups. However, these heat- and mass-transfer coefficients for low-mass-transfer fluxes can be corrected for high-flux conditions using film theory and penetration theory; the former is used for long contact times, whereas the latter is used for very short contact times. The procedure for doing this is discussed by Bird et al.[5]

4.4 SMALL BIOT NUMBER APPROXIMATION

The two problems considered in Sections 4.2 and 4.3 involved only conductive heat transfer in a single phase. In this example we consider convective heat transfer involving two phases. *Convection* implies heat transfer by bulk flow coupled with heat conduction. Consider a solid sphere initially at temperature T_0, having constant physical properties, radius R, and falling at its constant terminal velocity U_t through a viscous liquid whose constant temperature is $T_\infty > T_0$, as shown in Figure 4.4-1. As a result of contact with the hot liquid, the temperature of the sphere gradually will increase. We characterize the heat transfer in the liquid via a *lumped-parameter approach*; that is, we assume that the heat transfer in the liquid can be described by a heat-transfer coefficient h. The latter can be obtained from correlations for the Nusselt number, a dimensionless heat-transfer coefficient, as a function of the Reynolds number for flow over a sphere that are available in standard references.[6]

Since we are representing the heat transfer in the liquid phase via a heat-transfer coefficient, describing equations need to be written only in the conducting solid sphere. The thermal energy equation in spherical coordinates given by equation (F.3-2) in the Appendices appropriately simplified for the conditions defined in the problem statement and the associated initial and boundary conditions are given by (step 1)

$$\frac{\partial T}{\partial t} = \alpha \frac{1}{r^2} \frac{\partial}{\partial r}\left(r^2 \frac{\partial T}{\partial r}\right) \tag{4.4-1}$$

$$T = T_0 \quad \text{at} \quad t \leq 0, \quad 0 \leq r \leq R \tag{4.4-2}$$

[5]R. B. Bird, W. E. Stewart, and E. L. Lightfoot, *Transport Phenomena*, 2nd ed., Wiley, Hoboken, NJ, 2002, pp. 703–708.

[6]See, for example, Bird et al., *Transport Phenomena*, 2nd ed., p. 439.

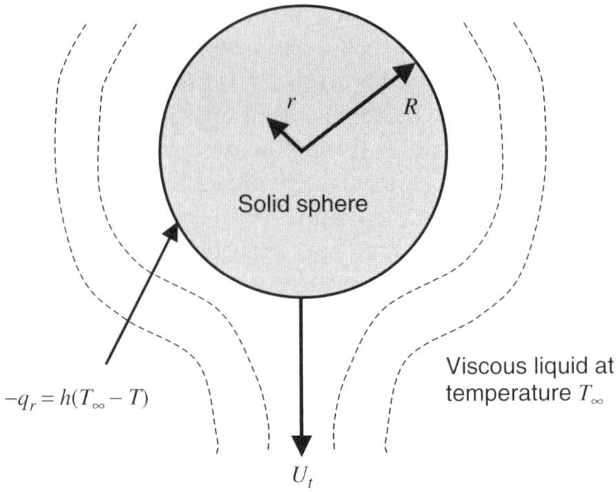

Figure 4.4-1 Solid sphere with constant physical properties, radius R, and initial temperature T_0 falling at its terminal velocity U_t in a viscous liquid whose temperature is T_∞ such that $T_\infty > T_0$.

$$\frac{\partial T}{\partial r} = 0 \quad \text{at} \quad r = 0, \quad t > 0 \tag{4.4-3}$$

$$k\frac{\partial T}{\partial r} = h(T_\infty - T) \quad \text{at} \quad r = R, \quad t > 0 \tag{4.4-4}$$

where $\alpha \equiv k/\rho C_p$ is the thermal diffusivity. Equation (4.4-2) is the given initial temperature condition. Equation (4.4-3) is a boundary condition frequently used when there is a point or axis of symmetry; it states that the temperature is at an extremum (in this case a minimum) at the center of the sphere. Equation (4.4-4) states that the heat-transfer flux from the surrounding liquid must be equal to the conductive heat flux into the solid sphere at its surface.

Define the following dimensionless variables involving unspecified scale and reference factors (steps 2, 3, and 4):

$$T^* \equiv \frac{T - T_r}{T_s}; \qquad r^* \equiv \frac{r}{r_s}; \qquad t^* \equiv \frac{t}{t_s} \tag{4.4-5}$$

Note that we have introduced a reference factor for the temperature since it is not naturally referenced to zero. Introduce these dimensionless variables into the describing equations and divide through by the coefficient of one term in each equation that should be retained (steps 5 and 6):

$$\frac{r_s^2}{\alpha t_s}\frac{\partial T^*}{\partial t^*} = \frac{1}{r^{*2}}\frac{\partial}{\partial r^*}\left(r^{*2}\frac{\partial T^*}{\partial r^*}\right) \tag{4.4-6}$$

$$T^* = \frac{T_0 - T_r}{T_s} \quad \text{at} \quad t^* \leq 0, \quad 0 \leq r^* \leq \frac{R}{r_s} \tag{4.4-7}$$

$$\frac{\partial T^*}{\partial r^*} = 0 \quad \text{at} \quad r^* = 0, \quad t^* > 0 \tag{4.4-8}$$

$$\frac{\partial T^*}{\partial r^*} = \frac{hr_s}{k}\left(\frac{T_\infty - T_r}{T_s} - T^*\right) \quad \text{at} \quad r^* = \frac{R}{r_s}, \quad t^* > 0 \tag{4.4-9}$$

We can bound the dimensionless temperature to be o(1) by setting the dimensionless group in equation (4.4-7) equal to zero to determine the reference temperature and by setting the dimensionless temperature ratio in equation (4.4-9) equal to 1 to determine the temperature scale (step 7); that is,

$$\frac{T_0 - T_r}{T_s} = 0 \Rightarrow T_r = T_0; \qquad \frac{T_\infty - T_0}{T_s} = 1 \Rightarrow T_s = T_\infty - T_0 \tag{4.4-10}$$

We can bound the dimensionless radial coordinate to be o(1) by setting the dimensionless group containing r_s in equation (4.4-7) or (4.4-9) equal to 1; that is,

$$\frac{R}{r_s} = 1 \Rightarrow r_s = R \tag{4.4-11}$$

There are two possible time scales, the observation time t_o and the characteristic time dictated by the dimensionless group in equation (4.4-6) given by

$$\frac{r_s^2}{\alpha t_s} = 1 \Rightarrow t_s = \frac{r_s^2}{\alpha} = \frac{R^2}{\alpha} \tag{4.4-12}$$

The time scale given by equation (4.4-12) is appropriate when this is inherently an unsteady-state heat-transfer problem for which both terms in equation (4.4-6) must be retained.

If we choose the observation time as our characteristic time, we obtain the following dimensionless describing equations:

$$\frac{1}{\text{Fo}_t}\frac{\partial T^*}{\partial t^*} = \frac{1}{r^{*2}}\frac{\partial}{\partial r^*}\left(r^{*2}\frac{\partial T^*}{\partial r^*}\right) \tag{4.4-13}$$

$$T^* = 0 \quad \text{at} \quad t^* \leq 0, \quad 0 \leq r^* \leq 1 \tag{4.4-14}$$

$$\frac{\partial T^*}{\partial r^*} = 0 \quad \text{at} \quad r^* = 0, \quad t^* > 0 \tag{4.4-15}$$

$$\frac{\partial T^*}{\partial r^*} = \text{Bi}_t(1 - T^*) \quad \text{at} \quad r^* = 1, \quad t^* > 0 \tag{4.4-16}$$

where $\mathrm{Fo}_t \equiv \alpha t_o/R^2$ is the thermal Fourier number, which is the ratio of the observation time to the characteristic time for heat conduction, and where $\mathrm{Bi}_t \equiv hR/k$ is the *thermal Biot number*, which is the ratio of the total heat transfer external to the sphere to the conductive heat transfer within the sphere.[7] For steady-state to be achieved, we must have

$$\mathrm{Fo}_t \gg 1 \Rightarrow t_o \gg \frac{R^2}{\alpha} \tag{4.4-17}$$

If the condition given by equation (4.4-17) is satisfied, the unsteady-state term in equation (4.4-13) can be ignored. Integration of the steady-state form of equation (4.4-13) subject to the boundary condition given by equation (4.4-15) implies that the temperature gradient is zero throughout the sphere; this is turn implies that the temperature is constant throughout the sphere. To satisfy the boundary condition given by equation (4.4-16), we must have $T^* = 1$ throughout the sphere; this implies that $T = T_\infty$ throughout the sphere. This result should not be surprising since steady-state implies that the sphere has come to thermal equilibrium with the surrounding liquid. Note that satisfying the steady-state condition occurs at earlier times as the Biot number increases, corresponding to improved heat transfer in the liquid phase.

If we choose the time scale given by equation (4.4-12), our dimensionless thermal energy equation assumes the form

$$\frac{\partial T^*}{\partial t^*} = \frac{1}{r^{*2}}\frac{\partial}{\partial r^*}\left(r^{*2}\frac{\partial T^*}{\partial r^*}\right) \tag{4.4-18}$$

The solution to equation (4.4-18) can be simplified for the special case of very small Biot numbers, that is, for $\mathrm{Bi}_t \ll 1$. For this case equation (4.4-16) implies that the dimensionless temperature gradient within the sphere is negligibly small, thereby implying that the temperature within the sphere is uniform. Hence, equation (4.4-18) can be integrated as follows:

$$\int_0^1 \frac{\partial T^*}{\partial t^*} 4\pi r^{*2}\, dr^* = \int_0^1 \frac{1}{r^{*2}}\frac{\partial}{\partial r^*}\left(r^{*2}\frac{\partial T^*}{\partial r^*}\right) 4\pi r^{*2}\, dr^* \tag{4.4-19}$$

$$\int_0^1 \frac{\partial T^*}{\partial t^*} r^{*2}\, dr^* = \int_0^{\mathrm{Bi}_t(1-T^*)} \partial\left(r^{*2}\frac{\partial T^*}{\partial r^*}\right) \tag{4.4-20}$$

Use of Liebnitz's rule given in Appendix H in the Appendices to integrate the first term and the fact that the temperature is essentially uniform within the sphere for very small Biot numbers then yields

$$\frac{d}{dt^*}\int_0^1 T^* r^{*2}\, dr^* = \frac{dT^*}{dt^*}\int_0^1 r^{*2}\, dr^* = \frac{1}{3}\frac{dT^*}{dt^*} = \mathrm{Bi}_t(1 - T^*) \tag{4.4-21}$$

[7]Do not confuse the Biot number with the Nusselt number; they are defined similarly and represent the ratio of convective to conductive heat transfer; however, in contrast to the Biot number, the Nusselt number involves the ratio of convective to conductive heat transfer in the fluid phase.

This equation can be integrated to give the following solution for the temperature of the sphere as a function of time:

$$T^* = 1 - e^{-3\mathrm{Bi}_t \cdot t^*} \tag{4.4-22}$$

For very small Biot numbers, equation (4.4-22) simplifies to

$$T^* \cong 3\mathrm{Bi}_t \cdot t^* \tag{4.4-23}$$

The error in the solution given by equation (4.4-23) will be in the range 10 to 100% if $\mathrm{Bi}_t = \mathrm{O}(0.1)$ and 1 to 10% if $\mathrm{Bi}_t = \mathrm{O}(0.01)$.

Scaling this problem has illustrated an important simplification: the small Biot number approximation that can be made when considering convective heat transfer to or from a solid object having finite dimensions.[8] The Biot number is a measure of the resistance to heat conduction in the solid object relative to the resistance to convective heat transfer in the surrounding fluid. For sufficiently small Biot numbers the heat transfer will be controlled totally by convection into the surrounding fluid. Under these conditions there will be a uniform temperature in the solid object, which permits a straightforward analytical solution. Note that most heat-transfer textbooks do not provide any rigorous justification for the equations appropriate to the small Biot number approximation.

4.5 SMALL PECLET NUMBER APPROXIMATION

The problem considered in Section 4.4 involved convective heat transfer in the fluid phase adjacent to a solid sphere. However, a lumped-parameter approach was used to account for this convection. In the present example problem we consider the convective heat transfer explicitly. Consider the steady-state fully developed laminar flow of a viscous Newtonian fluid with constant physical properties between two infinitely wide parallel plates separated by a distance $2H$ and of length L, as shown in Figure 4.5-1. The upstream (entering) temperature of the fluid is T_0. The temperature of the upper and lower plates is also maintained at T_0. Since the flow is assumed to be laminar and fully developed, the velocity profile is given by

$$u_x = U_m\left[1 - \left(\frac{y}{H}\right)^2\right] \tag{4.5-1}$$

where U_m is the maximum velocity. As a result of this shear flow, there will be heat generation via viscous dissipation. The latter will cause both radial and axial conduction as well as axial convection of heat.

[8]Note that the small Biot number approximation is sometimes referred to as the *lumped capacitance approximation*.

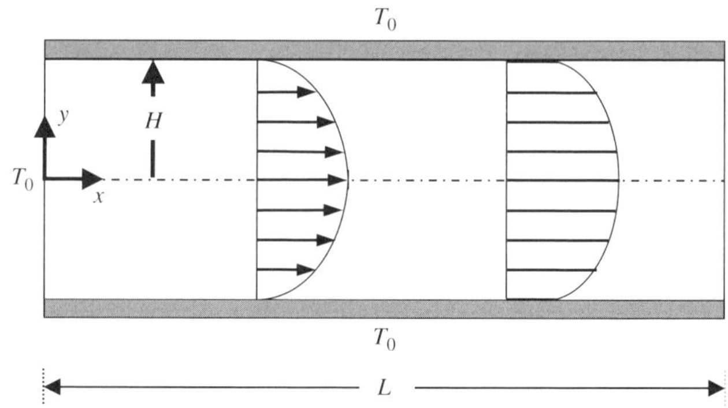

Figure 4.5-1 Steady-state fully developed laminar flow of a viscous Newtonian fluid with constant physical properties between two infinitely wide parallel plates separated by a distance $2H$ and of length L; the temperature of the entering fluid as well as that of the upper and lower plates is T_0; the shear flow causes viscous heat generation; the fully developed velocity profile is shown as well as a representative developing temperature profile.

The thermal energy equation given by equation (F.1-2) in the Appendices, appropriately simplified using equation (4.5-1) and for the conditions defined in the problem statement and the associated boundary conditions are given by (step 1)

$$\rho C_p U_m \left[1 - \left(\frac{y}{H}\right)^2\right] \frac{\partial T}{\partial x} = k\frac{\partial^2 T}{\partial x^2} + k\frac{\partial^2 T}{\partial y^2} + \frac{4\mu U_m^2}{H^4} y^2 \qquad (4.5\text{-}2)$$

$$T = T_0 \quad \text{at} \quad x = 0 \qquad (4.5\text{-}3)$$

$$T = f(y) \quad \text{at} \quad x = L \qquad (4.5\text{-}4)$$

$$T = T_0 \quad \text{at} \quad y = \pm H \qquad (4.5\text{-}5)$$

$$\frac{\partial T}{\partial y} = 0 \quad \text{at} \quad y = 0 \qquad (4.5\text{-}6)$$

where $f(y)$ is some function of y, which might be unknown. This is a nontrivial problem to solve, due to the elliptic nature of the describing equations. The presence of the second-order axial derivative requires that a downstream boundary condition be specified. In many problems such as this, these downstream conditions are not known, which precludes solving the describing equations numerically. Clearly, one would like to know when and how these describing equations might be simplified to permit a tractable solution. In particular, one would like to know when the axial conduction and convection terms might be neglected. We use o(1) scaling to determine the criteria for neglecting these terms.

Define the following dimensionless variables involving unspecified scale factors (steps 2, 3, and 4):

$$T^* \equiv \frac{T - T_r}{T_s}; \qquad x^* \equiv \frac{x}{x_s}; \qquad y^* \equiv \frac{y}{y_s} \tag{4.5-7}$$

We have introduced a reference factor for the temperature since it is not naturally referenced to zero. Introduce these dimensionless variables into the describing equations and divide through by the coefficient of one term in each equation that should be retained (steps 5 and 6):

$$\frac{\rho C_p U_m y_s^2}{k x_s}\left[1 - \left(\frac{y_s}{H} y^*\right)^2\right]\frac{\partial T^*}{\partial x^*} = \frac{y_s^2}{x_s^2}\frac{\partial^2 T^*}{\partial x^{*2}} + \frac{\partial^2 T^*}{\partial y^{*2}} + \frac{4\mu U_m^2 y_s^4}{k H^4 T_s} y^{*2} \tag{4.5-8}$$

$$T^* = \frac{T_0 - T_r}{T_s} \quad \text{at} \quad x^* = 0 \tag{4.5-9}$$

$$T^* = f(y^*) \quad \text{at} \quad x^* = \frac{L}{x_s} \tag{4.5-10}$$

$$T^* = \frac{T_0 - T_r}{T_s} \quad \text{at} \quad y^* = \pm\frac{H}{y_s} \tag{4.5-11}$$

$$\frac{\partial T^*}{\partial y^*} = 0 \quad \text{at} \quad y^* = 0 \tag{4.5-12}$$

Step 7 involves bounding the independent and dependent dimensionless variables to be ○(1). This can be done for the dimensionless spatial coordinates by setting the dimensionless groups containing x_s and y_s in equations (4.5-10) and (4.5-11) equal to 1; that is,

$$\frac{L}{x_s} = 1 \Rightarrow x_s = L; \qquad \frac{H}{y_s} = 1 \Rightarrow y_s = H \tag{4.5-13}$$

To determine the reference and scale factors for the temperature, we need to consider the conditions for which we are scaling. We are seeking to determine when axial conduction and convection can be ignored relative to heat generation by viscous dissipation and transverse conduction. If the former are negligible, the heat generation by viscous dissipation must be balanced by the transverse heat conduction. Hence, we can bound our dimensionless temperature to be ○(1) by setting the dimensionless group in equation (4.5-9) or (4.5-10) equal to zero, thereby determining the reference temperature, and by setting the dimensionless group multiplying the heat generation term in equation (4.4-8) equal to 1, to determine the temperature scale; that is,

$$\frac{T_0 - T_r}{T_s} = 0 \Rightarrow T_r = T_0; \qquad \frac{\mu U_m^2}{k T_s} = 1 \Rightarrow T_s = \frac{\mu U_m^2}{k} \tag{4.5-14}$$

Substitution of the scale and reference factors defined by equations (4.5-13) and (4.5-14) into the dimensionless describing equations given by equations (4.5-8) through (4.5-12) yields

$$\frac{\rho C_p U_m H^2}{kL}(1-y^{*2})\frac{\partial T^*}{\partial x^*} = \frac{H^2}{L^2}\frac{\partial^2 T^*}{\partial x^{*2}} + \frac{\partial^2 T^*}{\partial y^{*2}} + 4y^{*2} \tag{4.5-15}$$

$$T^* = 0 \quad \text{at} \quad x^* = 0 \tag{4.5-16}$$

$$T^* = f(y^*) \quad \text{at} \quad x^* = 1 \tag{4.5-17}$$

$$T^* = 0 \quad \text{at} \quad y^* = \pm 1 \tag{4.5-18}$$

$$\frac{\partial T^*}{\partial y^*} = 0 \quad \text{at} \quad y^* = 0 \tag{4.5-19}$$

Now let us explore possible simplifications of the describing equations (step 8). The criterion for ignoring axial conduction is

$$\frac{H^2}{L^2} \ll 1 \Rightarrow \text{axial conduction can be ignored} \tag{4.5-20}$$

that is, the aspect ratio cannot be too large. Note, however, that the length L was arbitrary in that L could denote any value of the axial coordinate in the principal direction of flow. This is the principle of *local scaling* whereby we scale the problem for some fixed but arbitrary value of some coordinate, usually that in the principal direction of flow.

To ignore axial convection of heat, the dimensionless group multiplying the first term in equation (4.5-15) must be very small; that is,

$$\frac{\rho C_p U_m H^2}{kL} = \frac{U_m H}{\alpha}\frac{H}{L} = \frac{U_m H}{\nu}\frac{\nu}{\alpha}\frac{H}{L} = \text{Re}\cdot\text{Pr}\frac{H}{L} = \text{Pe}_t\frac{H}{L} \ll 1 \tag{4.5-21}$$

where $\alpha \equiv k/\rho C_p$ is the thermal diffusivity, $\nu \equiv \mu/\rho$ is the kinematic viscosity, $\text{Re} \equiv \rho U_m H/\mu$ is the Reynolds number, $\text{Pr} \equiv \nu/\alpha$ is the Prandtl number, and $\text{Pe}_t \equiv U_m H/\alpha$ is the Peclet number for heat transfer. The Reynolds number is a measure of the ratio of the convective to viscous transport of momentum. The Prandtl number is a measure of the ratio of the viscous transport of momentum to heat conduction. Hence, the *Peclet number* is a measure of the ratio of heat convection to heat conduction. We see that the criterion for ignoring axial heat convection is that the product of the Peclet number and the aspect ratio must be very small. The Peclet number in heat transfer has a role analogous to that of the Reynolds number in fluid dynamics; that is, when it is small, it justifies ignoring axial convective transport. We will see in the next example problem that when it is large, it justifies a boundary-layer approximation. Note that ignoring convective transport in the energy equation in this example is analogous to ignoring convective transport in the equations of motion, which is the basis for the creeping-flow approximation.[9]

If the conditions in equations (4.5-20) and (4.5-21) are satisfied, equations (4.5-15) through (4.5-19) reduce to

$$0 = \frac{d^2T^*}{dy^{*2}} + 4y^{*2} \tag{4.5-22}$$

$$T^* = 0 \quad \text{at} \quad y^* = \pm 1 \tag{4.5-23}$$

$$\frac{\partial T^*}{\partial y^*} = 0 \quad \text{at} \quad y^* = 0 \tag{4.5-24}$$

The solution to this simplified set of describing equations is straightforward and given by

$$T^* = \frac{1}{3}(1 - y^{*4}) \tag{4.5-25}$$

We see from this solution that our dimensionless temperature is bounded of o(1), which confirms that our scaling analysis is correct.

4.6 BOUNDARY-LAYER OR LARGE PECLET NUMBER APPROXIMATION

In the preceding example we saw that a low Peclet number justified ignoring the convective transport of thermal energy, which is analogous to a low Reynolds number justifying neglecting convective transport of momentum in the creeping-flow approximation considered in Chapter 3. In this example we consider the other end of the Peclet number spectrum by exploring the implications of a high Peclet number on heat transfer. Consider the steady-state laminar uniform (plug) flow of a Newtonian fluid with constant physical properties and temperature T_∞ intercepting a stationary, semi-infinitely long infinitely wide horizontal flat plate maintained at a constant temperature T_0 such that $T_0 > T_\infty$, as shown in Figure 4.6-1. Gravitational and viscous heating effects can be assumed to be negligible. To fully understand this example, it would be helpful to review Section 3.4, in which the boundary-layer approximation in fluid dynamics was considered. As stated in Section 3.4, boundary-layer flows are examples of *region-of-influence scaling*, for which we determine the thickness of a region wherein some effect is confined, in this case the influence of the heated flat plate that is propagated by conduction. This example also illustrates the principle of *local scaling*, in which we carry out the scaling at some arbitrary but fixed value of the axial coordinate.

[9]The creeping-flow approximation was considered in Section 3.3 and Example Problem 3.E.2.

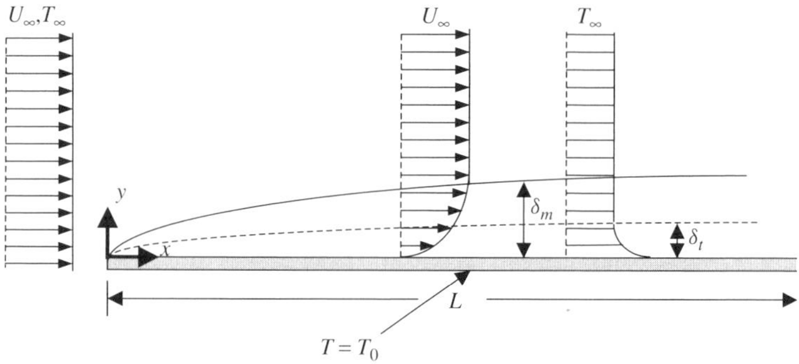

Figure 4.6-1 Steady-state laminar uniform flow of a Newtonian fluid with constant physical properties and temperature T_∞ intercepting a stationary, semi-infinitely long infinitely wide horizontal flat plate maintained at a constant temperature T_0 such that $T_0 > T_\infty$; the solid line shows the hypothetical momentum boundary-layer thickness δ_m, whereas the dash-dotted line shows the hypothetical thermal boundary-layer thickness δ_t for Pr > 1, where Pr is the Prandtl number.

The continuity equation given by equation (C.1-1), equations of motion given by equations (D.1-10) and (D.1-11), and thermal energy equation given by equation (F.1-2) in the Appendices simplify to the following for the assumed flow conditions (step 1):

$$\rho u_x \frac{\partial u_x}{\partial x} + \rho u_y \frac{\partial u_x}{\partial y} = -\frac{\partial P}{\partial x} + \mu \left(\frac{\partial^2 u_x}{\partial x^2} + \frac{\partial^2 u_x}{\partial y^2} \right) \tag{4.6-1}$$

$$\rho u_x \frac{\partial u_y}{\partial x} + \rho u_y \frac{\partial u_y}{\partial y} = -\frac{\partial P}{\partial y} + \mu \left(\frac{\partial^2 u_y}{\partial x^2} + \frac{\partial^2 u_y}{\partial y^2} \right) \tag{4.6-2}$$

$$\frac{\partial u_x}{\partial x} + \frac{\partial u_y}{\partial y} = 0 \tag{4.6-3}$$

$$\rho C_p u_x \frac{\partial T}{\partial x} + \rho C_p u_y \frac{\partial T}{\partial y} = k \left(\frac{\partial^2 T}{\partial x^2} + \frac{\partial^2 T}{\partial y^2} \right) \tag{4.6-4}$$

The corresponding boundary conditions for this flow are given by

$$u_x = U_\infty, \quad u_y = 0, \quad T = T_\infty \quad \text{at} \quad x = 0 \tag{4.6-5}$$

$$u_x = f_1(y), \quad u_y = f_2(y), \quad T = f_3(y) \quad \text{at} \quad x = L \tag{4.6-6}$$

$$u_x = 0, \quad u_y = 0, \quad T = T_0 \quad \text{at} \quad y = 0 \tag{4.6-7}$$

$$u_x = U_\infty, \quad u_y = 0, \quad T = T_\infty \quad \text{at} \quad y = \infty \tag{4.6-8}$$

where $f_1(y)$, $f_2(y)$, and $f_3(y)$ are unspecified functions. The boundary conditions for the equations of motion were discussed in Section 3.4. The boundary conditions on the energy equation given by equations (4.6-5) and (4.6-7) are the prescribed temperature of the entering fluid and at the plate, respectively. Equation (4.6-6) merely states that to solve this elliptic energy equation, a downstream boundary condition must be specified; this condition might not be known in practice. Equation (4.6-8) states that the temperature becomes equal to that of the entering fluid infinitely far above the flat plate.

Define the following dimensionless dependent and independent variables (steps 2, 3, and 4):

$$u_x^* \equiv \frac{u_x}{u_{xs}}; \qquad u_y^* \equiv \frac{u_y}{u_{ys}}; \qquad P^* \equiv \frac{P}{P_s};$$
$$T^* = \frac{T - T_r}{T_s}; \qquad x^* \equiv \frac{x}{x_s}; \qquad y_m^* \equiv \frac{y}{y_{ms}}; \qquad y_t^* \equiv \frac{y}{y_{ts}} \tag{4.6-9}$$

Note that we have allowed for different y-length scales for the energy equation and equations of motion; that is, the temperature might experience a characteristic change of $O(1)$ over a different length scale than the velocities. Introduce these dimensionless variables into the describing equations and divide each equation through by the dimensional coefficient of one term that should be retained to maintain physical significance (steps 5 and 6):

$$u_x^*\frac{\partial u_x^*}{\partial x^*} + \frac{u_{ys}x_s}{u_{xs}y_{ms}}u_y^*\frac{\partial u_x^*}{\partial y_m^*} = -\frac{P_s}{\rho u_{xs}^2}\frac{\partial P^*}{\partial x^*} + \frac{\mu}{\rho u_{xs}x_s}\frac{\partial^2 u_x^*}{\partial x^{*2}} + \frac{\mu x_s}{\rho u_{xs}y_{ms}^2}\frac{\partial^2 u_x^*}{\partial y_m^{*2}} \tag{4.6-10}$$

$$u_x^*\frac{\partial u_y^*}{\partial x^*} + \frac{u_{ys}x_s}{u_{xs}y_{ms}}u_y^*\frac{\partial u_y^*}{\partial y_m^*} = -\frac{P_s x_s}{\rho u_{xs}u_{ys}y_{ms}}\frac{\partial P^*}{\partial y_m^*} + \frac{\mu}{\rho u_{xs}x_s}\frac{\partial^2 u_y^*}{\partial x^{*2}} + \frac{\mu x_s}{\rho u_{xs}y_{ms}^2}\frac{\partial^2 u_y^*}{\partial y_m^{*2}} \tag{4.6-11}$$

$$\frac{\partial u_x^*}{\partial x^*} + \frac{u_{ys}x_s}{u_{xs}y_{ms}}\frac{\partial u_y^*}{\partial y_m^*} = 0 \tag{4.6-12}$$

$$\frac{u_{xs}y_{ts}}{u_{ys}x_s}u_x^*\frac{\partial T^*}{\partial x^*} + u_y^*\frac{\partial T^*}{\partial y_t^*} = \frac{k y_{ts}}{\rho C_p u_{ys}x_s^2}\frac{\partial^2 T^*}{\partial x^{*2}} + \frac{k}{\rho C_p u_{ys}y_{ts}}\frac{\partial^2 T^*}{\partial y_t^{*2}} \tag{4.6-13}$$

$$u_x^* = \frac{U_\infty}{u_{xs}}, \quad u_y^* = 0, \quad T^* = \frac{T_\infty - T_r}{T_s} \quad \text{at} \quad x^* = 0 \tag{4.6-14}$$

$$u_x^* = f_1^*(y_m^*), \quad u_y^* = f_2^*(y_m^*), \quad T^* = f_3(y_t^*) \quad \text{at} \quad x^* = \frac{L}{x_s} \tag{4.6-15}$$

$$u_x^* = 0, \quad u_y^* = 0, \quad T^* = \frac{T_0 - T_r}{T_s} \quad \text{at} \quad y_m^* = y_t^* = 0 \tag{4.6-16}$$

$$u_x^* = \frac{U_\infty}{u_{xs}}, \quad u_y^* = 0, \quad T^* = \frac{T_\infty - T_r}{T_s} \quad \text{at} \quad y_m^* = y_t^* = \infty \tag{4.6-17}$$

Note that we have divided equations (4.6-10) and (4.6-11) by the dimensional coefficient of the axial convection term since we are considering a high Reynolds number flow and high Peclet number heat transfer for which the convection terms must be retained. However, we have divided equation (4.6-13) by the dimensional coefficient of the transverse convection term. The reason for doing this is not obvious at this point. However, we will see that this is the principal convective term in the energy equation for $\mathrm{Pr} > 1$.

For a high Reynolds number flow for which the action of viscosity is confined to the vicinity of the boundaries, the y-length scale for the velocities will be the thickness of the momentum boundary layer or region of influence δ_m; that is, we say that $y_{ms} = \delta_m$. The scale factors for the velocity components, pressure, and axial coordinate are determined in exactly the same manner as was described in detail in Section 3.4 and are given by (step 7)

$$u_{xs} = U_\infty; \quad x_s = L; \quad u_{ys} = \frac{\delta_m}{L} U_\infty; \quad P_s = \rho U_\infty^2 \frac{\delta_m^2}{L^2} \tag{4.6-18}$$

In view of the fact that the principal viscous term in equation (4.6-10) has to be important at least within some small region in the vicinity of the flat plate, we set the dimensionless group in front of this term equal to 1 to ensure that this term is of the same size as the convection terms that are being retained. This yields the following equation for the thickness of the region of influence or momentum boundary layer:

$$\delta_m^2 = \frac{\mu L}{\rho U_\infty} = \frac{L^2}{\mathrm{Re}} \Rightarrow \delta_m = \frac{L}{\sqrt{\mathrm{Re}}} \tag{4.6-19}$$

where Re is the local Reynolds number based on using L as the characteristic length. Note that L is arbitrary in that it can be any fixed value of the axial length coordinate; that is, our scaling was done for an arbitrary length L of a semi-infinitely long flat plate; this is what is meant by the concept of *local scaling*. The reference and scale factors for the temperature are determined by setting the appropriate group in equations (4.6-14) or (4.6-17) equal to zero and in equation (4.6-16) equal to 1 to obtain

$$T_r = T_\infty; \qquad T_s = T_0 - T_\infty \tag{4.6-20}$$

In view of the fact that the principal conduction term in equation (4.6-13) has to be important at least within some small region in the vicinity of the flat plate, we set the dimensionless group in front of this term equal to 1 to ensure that this term is of the same size as the convection terms that are being retained. This yields the following equation for the thickness of the region of influence or thermal boundary layer:

$$\delta_t = \frac{kL}{\rho C_p \delta_m U_\infty} = \frac{1}{\sqrt{\text{Pr}}} \frac{L}{\sqrt{\text{Pe}_t}} = \frac{1}{\text{Pr}} \frac{L}{\sqrt{\text{Re}}} = \frac{\delta_m}{\text{Pr}} \tag{4.6-21}$$

where $\text{Pe}_t \equiv U_\infty L/\alpha$ is the local Peclet number for heat transfer based on using L as the characteristic length. A comparison of equations (4.6-19) and (4.6-21) again indicates that in heat transfer the Peclet number plays a role analogous to that of the Reynolds number in fluid dynamics. Note that for liquids $\text{Pr} > 1$; hence, the thermal boundary-layer thickness will be less than the momentum boundary-layer thickness. For gases $\text{Pr} \cong 1$; hence, the thermal and momentum boundary layers have nearly the same thickness. However, for liquid metals, $\text{Pr} < 1$; hence, the thermal boundary-layer thickness will be greater than the momentum boundary-layer thickness. The general behavior of $\delta_t(x)$ and $\delta_m(x)$ for the case when $\text{Pr} > 1$ is shown in Figure 4.6-1.

If we now rewrite our dimensionless describing equations in terms of the scales defined by equations (4.6-18) through (4.6-21), we obtain

$$u_x^* \frac{\partial u_x^*}{\partial x^*} + u_y^* \frac{\partial u_x^*}{\partial y_m^*} = -\frac{1}{\text{Re}} \frac{\partial P^*}{\partial x^*} + \frac{1}{\text{Re}} \frac{\partial^2 u_x^*}{\partial x^{*2}} + \frac{\partial^2 u_x^*}{\partial y_m^{*2}} \tag{4.6-22}$$

$$u_x^* \frac{\partial u_y^*}{\partial x^*} + u_y^* \frac{\partial u_y^*}{\partial y_m^*} = -\frac{\partial P^*}{\partial y^*} + \frac{1}{\text{Re}} \frac{\partial^2 u_y^*}{\partial x^{*2}} + \frac{\partial^2 u_y^*}{\partial y_m^{*2}} \tag{4.6-23}$$

$$\frac{\partial u_x^*}{\partial x^*} + \frac{\partial u_y^*}{\partial y_m^*} = 0 \tag{4.6-24}$$

$$\frac{1}{\text{Pr}} u_x^* \frac{\partial T^*}{\partial x^*} + u_y^* \frac{\partial T^*}{\partial y_t^*} = \frac{1}{\text{Pr} \cdot \text{Pe}_t} \frac{\partial^2 T^*}{\partial x^{*2}} + \frac{\partial^2 T^*}{\partial y_t^{*2}} \tag{4.6-25}$$

$$u_x^* = 1, \quad u_y^* = 0, \quad T^* = 0 \quad \text{at} \quad x^* = 0 \tag{4.6-26}$$

$$u_x^* = f_1^*(y_m^*), \quad u_y^* = f_2^*(y_m^*), \quad T^* = f_3(y_t^*) \quad \text{at} \quad x^* = 1 \tag{4.6-27}$$

$$u_x^* = 0, \quad u_y^* = 0, \quad T^* = 1 \quad \text{at} \quad y_m^* = y_t^* = 0 \tag{4.6-28}$$

$$u_x^* = 1, \quad u_y^* = 0, \quad T^* = 0 \quad \text{at} \quad y_m^* = y_t^* = \infty \tag{4.6-29}$$

The system of equations above is difficult to solve for several reasons. First, elliptic differential equations are involved that require specifying some downstream boundary conditions that in practice are generally not known. Second, as discussed in Section 3.4, equations (4.6-22) and (4.6-23) are coupled due to the pressure. Third, the energy equation is coupled to the solution for thc equations of motion through the two velocity components. However, if the fluid properties are not temperature-dependent, this coupling is unidirectional in that the equations of motion can be solved independent of the energy equation. In view of these complications, we seek to explore the conditions required to eliminate these two complications.

Note that in the limit of $\text{Re} \gg 1$ and $\text{Pe}_t \gg 1$, the system of equations above reduces to (step 8)

$$u_x^* \frac{\partial u_x^*}{\partial x^*} + u_y^* \frac{\partial u_x^*}{\partial y_m^*} = \frac{\partial^2 u_x^*}{\partial y_m^{*2}} \tag{4.6-30}$$

$$\frac{\partial u_x^*}{\partial x^*} + \frac{\partial u_y^*}{\partial y_m^*} = 0 \tag{4.6-31}$$

$$\frac{1}{\text{Pr}} u_x^* \frac{\partial T^*}{\partial x^*} + u_y^* \frac{\partial T^*}{\partial y_t^*} = \frac{\partial^2 T^*}{\partial y_t^{*2}} \tag{4.6-32}$$

$$u_x^* = 1, \quad u_y^* = 0, \quad T^* = 0 \qquad \text{at} \quad x^* = 0 \tag{4.6-33}$$

$$u_x^* = 0, \quad u_y^* = 0, \quad T^* = 1 \qquad \text{at} \quad y_m^* = y_t^* = 0 \tag{4.6-34}$$

$$u_x^* = 1, \quad T^* = 0 \qquad \text{at} \quad y_m^* = y_t^* = \infty \tag{4.6-35}$$

Equations (4.6-30) through (4.6-35) are the classical boundary-layer equations for flow over a heated flat plate. Note that by showing that the pressure term in equation (4.6-22) is negligible in the limit of high Reynolds number, we have eliminated the coupling between x- and y-components of the equations of motion. Moreover, we have shown that the axial viscous term in equation (4.6-22) and the axial heat-conduction term in equation (4.6-25) are negligible in the limit of very high Reynolds and Peclet numbers, respectively, and thereby have converted the elliptic into parabolic differential equations that do not require a downstream boundary condition. Note that the dimensionless y-coordinate in the equations of motion is defined differently from that in the energy equation. If these equations are recast in terms of dimensional variables and a stream function and similarity variable are introduced, they can be transformed into a set of nonlinear ordinary differential equations solved that can be solved via approximate analytical techniques or numerically.[10]

Note that the criterion for the applicability of the hydrodynamic boundary-layer approximation is

$$\text{Re} \equiv \frac{\rho U_\infty L}{\mu} = \frac{U_\infty L}{\nu} \gg 1 \Rightarrow \text{hydrodynamic boundary layer} \tag{4.6-36}$$

whereas the criterion for the applicability of the thermal boundary-layer approximation is

$$\text{Pe}_t = \frac{\rho C_p U_\infty L}{k} = \frac{U_\infty L}{\alpha} \gg 1 \Rightarrow \text{thermal boundary layer} \tag{4.6-37}$$

[10] See, for example, Bird et al., *Transport Phenomena*, 2nd ed., pp. 388–390.

Since L is merely some fixed value of the axial coordinate x, the criteria above always break down in the vicinity of the leading edge of the flat plate. Hence, if one is seeking to determine an integral quantity such as the total drag or heat flux along the flat plate, the error will not be significant if equations (4.6-36) and (4.6-37) are satisfied over most of the plate. However, the error incurred by invoking the boundary-layer approximation can be quite large in the vicinity of the leading edge of the plate for point quantities such as the local velocity components, shear stress, temperature, or heat flux. Note for 90% of the flat plate to satisfy the condition that $\text{Re} \geq O(100)$, the Reynolds number at the end of the plate must be 1000. Since the Peclet number is the product of the Reynolds number and the Prandtl number, equation (4.6-36) is more limiting than equation (4.6-37) for fluids other than liquid metals.

Note that scaling analysis suggests how a solution to the coupled heat- and momentum-transfer problem can be developed that applies from the leading edge of the plate to any arbitrary downstream distance. Recall that the coupled describing equations are difficult to solve, owing to the presence of the axial diffusion terms in both the thermal energy equation and the equations of motion. These terms require specifying downstream boundary conditions that in practice are usually not known. However, the parabolic boundary-layer equations suggested by scaling can be solved either numerically or via approximate analytical methods downstream from the leading edge of the plate. The resulting solutions for the temperature and velocity profiles then can be used as downstream boundary conditions on the full elliptic describing equations that must be solved in the vicinity of the leading edge of the flat plate. Hence, we see that scaling not only provides a systematic method for simplifying the describing equations, but also suggests a strategy for solving them.

4.7 HEAT TRANSFER WITH PHASE CHANGE

Heat transfer is very often involved in problems wherein phase change occurs, owing to the need to supply or remove the latent heat associated with the transition from one phase to another. Figure 4.7-1 shows a schematic of melting ice within porous soil that was initially at its freezing temperature T_f and then was subjected to a higher constant temperature T_0 at the ground surface. We will assume that the heat transfer is one-dimensional and purely conductive and that the physical properties are constant.[11] This example will illustrate scaling of a moving boundary problem; that is, the melting front is a boundary that moves progressively downward into the frozen soil as heat is conducted upward to the warm ground surface. We will again explore how this problem can be simplified. We use this problem to illustrate the forgiving nature of scaling by making a naïve mistake in the way we

[11]Note that the melting of ice can induce free convection heat transfer arising from the density gradients that can be generated, due to the fact that water has a density maximum at 4°C; that is, unfrozen water adjacent to melting ice is less dense than the water immediately above it, which can give rise to free convection; however, this is not likely to occur in most soils, due to their low permeability to flow.

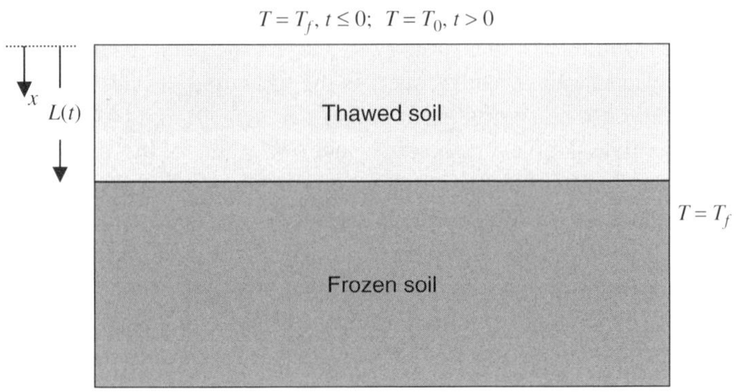

Figure 4.7-1 Unsteady-state one-dimensional heat transfer due to the imposition of a temperature T_0 at the surface of frozen water-saturated porous soil whose initial temperature was T_f where $T_0 > T_f$; the position of the thaw front denoted by $L(t)$ progressively penetrates farther into the frozen soil due to conductive heat transfer from the ground surface.

scale one of the derivatives. This will then lead to a contradiction that suggests that we rescale the equations to achieve o(1) scaling.

The describing equations are obtained by appropriately simplifying equation (F.1-2) in the Appendices and prescribing the requisite initial and boundary conditions (step 1):

$$\rho_u C_{pu} \frac{\partial T}{\partial t} = k_u \frac{\partial^2 T}{\partial x^2} \tag{4.7-1}$$

$$T = T_f \quad \text{at} \quad t = 0 \tag{4.7-2}$$

$$T = T_0 \quad \text{at} \quad x = 0 \quad \text{for } t > 0 \tag{4.7-3}$$

$$T = T_f \quad \text{at} \quad x = L(t) \tag{4.7-4}$$

where k_u, ρ_u, and C_{pu} are the effective thermal conductivity, mass density, and heat capacity, respectively, of the unfrozen soil; note that by *effective* we mean that these properties account for the presence of the solid soil and the unfrozen water that is contained in its pores. Equations (4.7-2) and (4.7-3) are the prescribed initial temperature and imposed temperature at the ground surface, respectively. Equation (4.7-4) states that ice and unfrozen water that meet at the freezing front are in thermodynamic equilibrium at the freezing temperature of water. Note that this boundary condition is applied at the moving interface between the ice and unfrozen water $L(t)$; hence, problems of this type are referred to as *moving boundary problems*. Since $L(t)$ is an additional unknown, it is necessary to prescribe an auxiliary condition to determine it. This is obtained via an integral energy balance as follows:

$$\frac{d}{dt}\int_0^L \rho_u C_{pu}(T - T^\circ)\, dx + \frac{d}{dt}\int_L^\infty \rho_f C_{pf}(T - T^\circ)\, dx = q_0 \tag{4.7-5}$$

where T° is an arbitrary reference temperature for the enthalpy or heat content, ρ_f and C_{pf} are the effective mass density and heat capacity, respectively, of the frozen soil, and q_0 is the heat transferred into the unfrozen soil at the ground surface. Applying Leibnitz's rule for differentiating an integral given by equation (H.1-2) in the Appendices and substituting equation (4.7-1) while recalling that the frozen ice remains at the constant temperature T_f yields

$$(\rho_u C_{pu} - \rho_f C_{pf})(T_f - T^\circ)\frac{dL}{dt} + \int_0^L k_u \frac{\partial^2 T}{\partial x^2}\,dx + \int_L^\infty k_f \frac{\partial^2 T}{\partial x^2}\,dx = q_0 \tag{4.7-6}$$

The first term in the above is the difference in heat content between the unfrozen and frozen soil; this can be related to ΔH_f, the latent heat of fusion of water. Hence, integrating equation (4.7-6) yields

$$\Delta H_f \varepsilon \rho_w \frac{dL}{dt} + k_u \left.\frac{\partial T}{\partial x}\right|_{x=L} - k_u \left.\frac{\partial T}{\partial x}\right|_{x=0} + k_f \left.\frac{\partial T}{\partial x}\right|_{x=\infty} - k_f \left.\frac{\partial T}{\partial x}\right|_{x=L} = q_0 \tag{4.7-7}$$

where ρ_w is the mass density of water, ε the porosity of the soil, and k_f the thermal conductivity of the frozen soil. The fourth and fifth terms in equation (4.7-7) are identically zero if there is no heat transfer in the frozen soil and the third term is equal to the last term. Hence, the auxiliary condition needed to determine the instantaneous location of the freezing front is given by

$$k_u \frac{\partial T}{\partial x} = -\Delta H_f \varepsilon \rho_w \frac{dL}{dt} \qquad \text{at} \quad x = L \tag{4.7-8}$$

This condition merely states that the heat conducted to the freezing front supplies the instantaneous latent heat required for melting the ice. To integrate equation (4.7-8), it is necessary to specify an initial condition on L; this is given by

$$L = 0 \qquad \text{at} \quad t = 0 \tag{4.7-9}$$

Note that whenever boundary conditions must be applied at a location whose position is unknown and dependent on the solution to the particular describing equations, it is necessary to use some type of integral balance to obtain an additional condition to determine the location of this boundary. In fluid dynamics this occurs for flows involving free surfaces such as were considered in Section 3.7 and Example Problem 3.E-8 and requires using an integral mass balance, which is called the *kinematic surface condition*. In heat transfer this occurs in problems such as the one considered here involving phase change and requires an integral energy balance. In some heat-transfer problems involving phase change such as evaporation, mass loss is also involved. In the latter moving boundary problems it is necessary to include both an integral energy and an integral mass balance. One of these is used as a boundary condition on the energy equation, and the other constitutes the auxiliary equation used to locate the position of the moving boundary.

Define the following dimensionless dependent and independent variables (steps 2, 3, and 4):

$$T^* \equiv \frac{T - T_r}{T_s}; \qquad x^* \equiv \frac{x}{x_s}; \qquad t^* \equiv \frac{t}{t_s}; \qquad L^* \equiv \frac{L}{L_s} \tag{4.7-10}$$

Introduce these dimensionless variables into the describing equations and divide each equation through by the dimensional coefficient of one term that should be retained to maintain physical significance (steps 5 and 6):

$$\frac{x_s^2}{\alpha_u t_s}\frac{\partial T^*}{\partial t^*} = \frac{\partial^2 T^*}{\partial x^{*2}} \tag{4.7-11}$$

$$T^* = \frac{T_f - T_r}{T_s}, \quad L^* = 0 \quad \text{at} \quad t^* = 0 \tag{4.7-12}$$

$$T^* = \frac{T_0 - T_r}{T_s} \quad \text{at} \quad x^* = 0 \quad \text{for} \quad t^* > 0 \tag{4.7-13}$$

$$T^* = \frac{T_f - T_r}{T_s} \quad \text{at} \quad x^* = \frac{L_s}{x_s}L^* \tag{4.7-14}$$

$$\left.\frac{\partial T^*}{\partial x^*}\right|_{x^*=L/x_s} = -\frac{\Delta H_f \rho_w \varepsilon x_s L_s}{k_u T_s t_s}\frac{dL^*}{dt^*} \quad \text{at} \quad x^* = \frac{L_s}{x_s}L^* \tag{4.7-15}$$

$$L^* = 0 \quad \text{at} \quad t^* = 0 \tag{4.7-16}$$

where $\alpha_u = k_u/\rho_u C_{pu}$ is the thermal diffusivity of the unfrozen soil.

We can bound the dimensionless temperature to be o(1) by setting the dimensionless group in equation (4.7-12) or (4.7-14) equal to zero to determine the reference temperature and by setting the dimensionless group in equation (4.7-13) equal to 1 to determine the temperature scale (step 7); that is,

$$\frac{T_f - T_r}{T_s} = 0 \Rightarrow T_r = T_f; \qquad \frac{T_0 - T_f}{T_s} = 1 \Rightarrow T_s = T_0 - T_f \tag{4.7-17}$$

We can bound the dimensionless spatial coordinate to be o(1) by setting the dimensionless group in equation (4.7-14) or (4.7-15) equal to 1; that is,

$$\frac{L_s}{x_s} = 1 \Rightarrow x_s = L_s \tag{4.7-18}$$

The time scale will again be the observation time t_o since this is inherently an unsteady-state problem. Since the two remaining terms in equation (4.7-15) must balance each other, to ensure that each term is o(1), we must set the dimensionless group in this equation equal to 1; this then provides the scale factor for the freezing penetration front; that is,

$$\frac{\Delta H_f \rho_w \varepsilon x_s L_s}{k_u T_s t_s} = \frac{\Delta H_f \rho_w \varepsilon L_s^2}{k_u (T_0 - T_f) t_o} = 1 \Rightarrow L_s = \left[\frac{k_u (T_0 - T_f) t_o}{\Delta H_f \rho_w \varepsilon} \right]^{1/2} \tag{4.7-19}$$

If we now rewrite our dimensionless describing equations in terms of the scales defined by equations (4.7-16) through (4.7-19), we obtain

$$\frac{k_u (T_0 - T_f)}{\Delta H_f \rho_w \varepsilon \alpha_u} \left(\frac{\partial T^*}{\partial t^*} - \frac{x^*}{L^*} \frac{dL^*}{dt^*} \frac{\partial T^*}{\partial x^*} \right) = \frac{\partial^2 T^*}{\partial x^{*2}} \tag{4.7-20}$$

$$T^* = 0 \quad \text{at} \quad t^* = 0 \tag{4.7-21}$$

$$T^* = 1 \quad \text{at} \quad x^* = 0 \quad \text{for} \quad t^* > 0 \tag{4.7-22}$$

$$T^* = 0 \quad \text{at} \quad x^* = 1 \tag{4.7-23}$$

$$\frac{dT^*}{dx^*} = -\frac{dL^*}{dt^*} \quad \text{at} \quad x^* = 1 \tag{4.7-24}$$

Note that an additional term now appears in equation (4.7-20) because of the transformation to a dimensionless spatial coordinate that is scaled with the instantaneous depth of the thawed layer. This is referred to as a *pseudo-convection term* since it involves a velocity multiplied by a spatial derivative in the same direction as that of the velocity. Pseudo-convection terms will always arise when one transforms from a stationary coordinate system to one for which either the reference or scale factor is a function of time.

Now let us assess the conditions under which the dimensionless describing equations can be simplified (step 8). We detect an immediate problem in equation (4.7-20) in that the relative importance of the transformed unsteady-state term is independent of the observation time t_o. Recall from the problem considered in Section 4.3 that the unsteady-state term should be multiplied by the inverse Fourier number, which is equal to the ratio of the observation time to the characteristic heat conduction time. For very large Fourier numbers we would anticipate that quasi-steady-state heat transfer should apply. Hence, we have obtained an unreasonable result and the forgiving nature of scaling has indicated a contradiction: namely, that quasi-steady-state conditions can never be achieved. Another contradiction inherent in this scaling is that the dimensionless thaw penetration depth is always equal to 1 since $L^* = L/L_s = L/x_s = 1$. Therefore, we need to rescale the problem; we will know that we have scaled correctly when the relevant terms are bounded of $\circ(1)$ and no contradictions occur.

We suspect that our error was introduced by scaling dL/dt with L_s/t_s. Let us rescale the describing equations by introducing a scale factor $\dot{L}_s$ for dL/dt to ensure that we bound this derivative to be $\circ(1)$:

$$\left(\frac{dL}{dt} \right)^* \equiv \frac{1}{\dot{L}_s} \frac{dL}{dt} \tag{4.7-25}$$

The other dimensionless variables are the same as defined by equation (4.7-10). Introduce these dimensionless variables into the describing equations and divide each equation through by the dimensional coefficient of one term that should be retained to maintain physical significance:

$$\frac{x_s^2}{\alpha_u t_s}\frac{\partial T^*}{\partial t^*} = \frac{\partial^2 T^*}{\partial x^{*2}} \tag{4.7-26}$$

$$T^* = \frac{T_f - T_r}{T_s}, \quad L^* = 0 \quad \text{at} \quad t^* = 0 \tag{4.7-27}$$

$$T^* = \frac{T_0 - T_r}{T_s} \quad \text{at} \quad x^* = 0 \quad \text{for} \quad t^* > 0 \tag{4.7-28}$$

$$T^* = \frac{T_f - T_r}{T_s} \quad \text{at} \quad x^* = \frac{L}{x_s} \tag{4.7-29}$$

$$\frac{\partial T^*}{\partial x^*} = -\frac{\Delta H_f \rho_w \varepsilon x_s \dot{L}_s}{k_u T_s}\left(\frac{dL}{dt}\right)^* \quad \text{at} \quad x^* = \frac{L}{x_s} \tag{4.7-30}$$

Our reference and scale factors for the temperature, the time, and the spatial coordinate remain the same as before. Since the two terms in equation (4.7-30) must balance each other, to ensure that each term is o(1), we must set the dimensionless group in this equation equal to 1; this then provides the scale factor for the melting front velocity; that is,

$$\frac{\Delta H_f \rho_w \varepsilon x_s \dot{L}_s}{k_u T_s} = \frac{\Delta H_f \rho_w \varepsilon L \dot{L}_s}{k_u (T_0 - T_f)} = 1 \Rightarrow \dot{L}_s = \frac{k_u (T_0 - T_f)}{\Delta H_f \rho_w \varepsilon L} \tag{4.7-31}$$

We see that L_s never appears explicitly in our dimensionless describing equations. Hence, L can be nondimensionalized with any relevant length scale such as the maximum thaw depth L_m. If we now rewrite our dimensionless describing equations in terms of the scales defined by equations (4.7-17), (4.7-18), and (4.7-31), we obtain

$$\frac{1}{\text{Fo}_t}\frac{\partial T^*}{\partial t^*} - \frac{\rho_u C_{pu}(T_0 - T_f)}{\Delta H_f \rho_w \varepsilon} x^* \left(\frac{dL}{dt}\right)^* \frac{\partial T^*}{\partial x^*} = \frac{\partial^2 T^*}{\partial x^{*2}} \tag{4.7-32}$$

$$T^* = 0 \quad \text{at} \quad t^* = 0 \tag{4.7-33}$$

$$T^* = 1 \quad \text{at} \quad x^* = 0 \quad \text{for} \quad t^* > 0 \tag{4.7-34}$$

$$T^* = 0 \quad \text{at} \quad x^* = 1 \tag{4.7-35}$$

$$\frac{dT^*}{dx^*} = -\left(\frac{dL}{dt}\right)^* \quad \text{at} \quad x^* = 1 \tag{4.7-36}$$

where $\text{Fo}_t \equiv \alpha_u t_o / L^2$ is the Fourier number for heat transfer. Note again that an additional pseudo-convection term appears in equation (4.7-32), due to the transformation from $T(x, t)$ to $T^*(x^*, t^*)$, in which $x^* = x/L(t)$. The dimensionless group $\rho_u C_{pu}(T_0 - T_f)/\Delta H_f \rho_w \varepsilon$ multiplying the pseudo-convection term is a ratio of the sensible heat to latent heat effects.

Now let us assess the conditions under which these dimensionless describing equations can be simplified (step 8). We see that the relative importance of the unsteady-state term in equation (4.7-32) is determined by the magnitude of the Fourier number. The two terms in this equation must balance each other. We have scaled $\partial T^*/\partial t^*$ to be o(1). However, we are not certain that $\partial^2 T^*/\partial x^{*2}$ is o(1). The fact that we have scaled $\partial T^*/\partial x^*$ to be o(1) does not necessarily ensure that $\partial^2 T^*/\partial x^{*2}$ is o(1). If $\text{Fo}_t = \text{O}(1)$, both the unsteady-state term and the conduction term will be o(1). This condition implies that

$$\frac{1}{\text{Fo}_t} = \frac{L^2}{\alpha_u t_o} = 1 \Rightarrow L = \sqrt{\alpha_u t_o} \qquad \text{for short contact times} \tag{4.7-37}$$

The question might arise as to whether $L^2/\alpha_u t_o$ can ever be much greater than 1, corresponding to very short contact times. However, this would lead to a contradiction since the dimensionless unsteady-state term should be of the same order as the dimensionless heat-conduction term. Hence, we conclude that for very short contact times, $L^2 \propto t_o$, to ensure that $L^2/\alpha_u t_o$ remains bounded as $t_o \to 0$; that is, scaling analysis permits us to infer the time dependence of the thaw penetration for short contact times.

Now let us consider the case when $\text{Fo}_t = \alpha_u t_o/L^2 \gg 1$, corresponding to very long contact times. When this condition prevails, the unsteady-state term in equation (4.7-32) can be ignored and quasi-steady-state prevails. The time dependence now enters implicitly through both the pseudoconvection term and the condition applied at the moving boundary given by equation (4.7-36). The resulting quasi-steady-state describing equations can be solved analytically. However, if in addition the dimensionless group multiplying the pseudo-convection term is small, that is, $\rho_u C_{pu}(T_0 - T_f)/\Delta H_f \rho_w \varepsilon \ll 1$, further simplification is possible. In the latter case, equation (4.7-32) predicts a linear temperature profile given by

$$T^* = 1 - x^* \tag{4.7-38}$$

When equation (4.7-38) is substituted into equation (4.7-36) and the result is cast into dimensional form and integrated, one obtains

$$\frac{dL}{dt} = \frac{k_u(T_0 - T_f)}{\Delta H_f \rho_w \varepsilon L} = L_{ts} \tag{4.7-39}$$

That is, for long contact times we find that the solution to the describing equations agrees identically with the scale factor for the thawing front velocity given by equation (4.7-31). This should not be surprising since if scaling is done properly, it should give estimates that are within an O(1) factor of those obtained by solving the describing equations. One can integrate equation (4.7-39) to obtain an equation for L as a function of t; that is,

$$L^2 = \frac{2k_u(T_0 - T_f)}{\Delta H_f \rho_w \varepsilon}(t_o - t_i) + L_i^2 \qquad \text{for long contact times} \tag{4.7-40}$$

where L_i is an integration constant; note that one cannot apply the initial condition that $L = 0$ since equation (4.7-40) does not apply at short times. However, this initial condition can be estimated from the short-time solution given by equation (4.7-37). Note also that although both equations (4.7-37) and (4.7-40) predict that L^2 will increase linearly with t_o, the short-contact time thaw-penetration rate is faster than that for the long-contact time.

In summary, if $\mathrm{Fo}_t \gg 1$, this unsteady-state moving boundary heat-transfer problem can be considered to be quasi-steady-state; that is, the describing equations can be simplified by ignoring the unsteady-state term in the thermal energy equation. For quasi-steady-state conditions the time dependence enters through the boundary condition at the moving boundary whose location is time-dependent.

4.8 TEMPERATURE-DEPENDENT PHYSICAL PROPERTIES

In Section 3.9 we used scaling analysis to determine when the incompressible flow assumption could be made for a fluid whose density was pressure-dependent. Here we consider a related coupled fluid-dynamics and heat-transfer problem in which we will use scaling to determine when the temperature-dependent shear viscosity can be assumed to be constant. Note that the manner in which scaling analysis is used to assess when the temperature-dependence of the viscosity can be ignored in this problem can be applied to assessing when the dependence of any other physical or transport property on some state variable such as temperature, pressure, or concentration can be ignored.

Figure 4.8-1 shows a schematic of the steady-state pressure-driven flow of an incompressible Newtonian liquid between two infinitely wide parallel flat plates, each of which is maintained at T_0, which is also the initial temperature of the liquid. The shear flow causes significant viscous heating that can possibly cause a progressive decrease in the liquid viscosity whose temperature dependence is given by

$$\mu = Ae^{B/T} \tag{4.8-1}$$

where A and B are positive constants. This in turn implies a possible developing flow due to the influence of the decrease in viscosity on the velocity profile. However, we will invoke the lubrication-flow[12] and low Peclet number approximations and in addition ignore axial conduction.[13] We use scaling analysis to assess when the temperature dependence of the viscosity can be ignored.

Appropriate simplification of the equations of motion given by equations (D.1-10) and (D.1-11) in the Appendices and the thermal energy equation given by

[12]Scaling analysis was applied to justify the lubrication-flow approximation in Section 3.3.

[13]Scaling analysis was applied to justify the low Peclet number approximation, ignoring the axial conduction in Section 4.5.

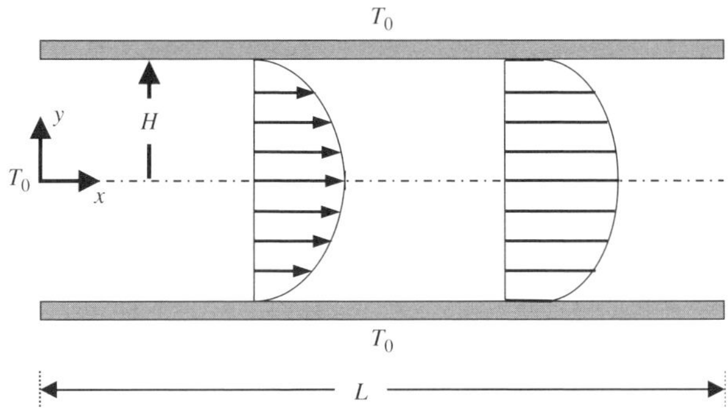

Figure 4.8-1 Steady-state pressure-driven lubrication flow of an incompressible Newtonian liquid between two infinitely wide parallel flat plates, each of which is maintained at T_0, which is also the initial temperature of the liquid; this shear flow causes viscous heating that can result in a progressive decrease in the liquid viscosity; representative velocity and temperature profiles are shown in this figure.

equation (F.1-2), and specification of the required boundary conditions yields the following set of describing equations (step 1):

$$0 = \frac{\Delta P}{L} + \frac{d}{dy}\left(\mu \frac{du_x}{dy}\right) \tag{4.8-2}$$

$$0 = k\frac{d^2T}{dy^2} + \mu\left(\frac{du_x}{dy}\right)^2 \tag{4.8-3}$$

$$\frac{du_x}{dy} = 0, \quad \frac{dT}{dy} = 0 \quad \text{at} \quad y = 0 \tag{4.8-4}$$

$$u_x = 0, \quad T = T_0 \quad \text{at} \quad y = \pm H \tag{4.8-5}$$

Since we seek to assess when the temperature dependence of the viscosity can be ignored, we need consider only small departures of the temperature from the initial temperature T_0. Hence, it is convenient to expand equation (4.8-1) in a Taylor series about T_0 at which the viscosity is μ_0:

$$\mu = \mu_0 - \frac{B\mu_0}{T_0^2}(T - T_0) \; + \; O(T - T_0)^2 \tag{4.8-6}$$

Since we need consider only the first-order effects of temperature to assess whether there is any significant change in the viscosity, truncate equation (4.8-6) after the second term on the right-hand side and substitute it into equations (4.8-2) and (4.8-3):

$$0 = \frac{\Delta P}{L} + \frac{d}{dy}\left\{\left[\mu_0 - \frac{B\mu_0}{T_0^2}(T - T_0)\right]\frac{du_x}{dy}\right\} \tag{4.8-7}$$

$$0 = k\frac{d^2T}{dy^2} + \left[\mu_0 - \frac{B\mu_0}{T_0^2}(T - T_0)\right]\left(\frac{du_x}{dy}\right)^2 \tag{4.8-8}$$

Introduce the following scale and reference factors (steps 2, 3, and 4):

$$u_x^* \equiv \frac{u_x}{u_s}; \qquad T^* \equiv \frac{T - T_r}{T_s}; \qquad x^* \equiv \frac{x}{x_s}; \qquad y^* \equiv \frac{y}{y_s} \tag{4.8-9}$$

Substitute these dimensionless variables into the describing equations and divide each equation through by the dimensional coefficient of one term that should be retained in order to maintain physical significance (steps 5 and 6):

$$0 = \frac{\Delta P}{L}\frac{y_s^2}{\mu_0 u_s} + \frac{d}{dy^*}\left\{\left[1 - \frac{BT_s}{T_0^2}\left(T^* + \frac{T_r - T_0}{T_s}\right)\right]\frac{du_x^*}{dy^*}\right\} \tag{4.8-10}$$

$$0 = \frac{d^2T^*}{dy^{*2}} + \frac{\mu_0 u_s^2}{kT_s}\left[1 - \frac{BT_s}{T_0^2}\left(T^* + \frac{T_r - T_0}{T_s}\right)\right]\left(\frac{du_x^*}{dy^*}\right)^2 \tag{4.8-11}$$

$$\frac{du_x^*}{dy^*} = 0, \quad \frac{dT^*}{dy^*} = 0 \quad \text{at} \quad y^* = 0 \tag{4.8-12}$$

$$u_x^* = 0, \quad T^* = \frac{T_0 - T_r}{T_s} \quad \text{at} \quad y^* = \pm\frac{H}{y_s} \tag{4.8-13}$$

When set equal to zero and 1, respectively, the dimensionless groups in equations (4.8-13) provide the following reference and scale factors (step 7):

$$\frac{T_0 - T_r}{T_s} = 0 \Rightarrow T_r = T_0, \qquad \frac{H}{y_s} = 1 \Rightarrow y_s = H \tag{4.8-14}$$

Since the pressure term must balance the principal viscous term for this lubrication flow, the dimensionless pressure term in equation (4.8-10) must be set equal to 1 to obtain the velocity scale:

$$\frac{\Delta P}{L}\frac{y_s^2}{\mu_0 u_s} = 1 \Rightarrow u_s = \frac{\Delta P}{L}\frac{H^2}{\mu_0} \tag{4.8-15}$$

This is a reasonable velocity scale since it is equal to the average velocity for fully developed laminar flow between two parallel flat plates. Since the viscous dissipation must be balanced by the heat conduction to the parallel flat plates, the dimensionless dissipation term in equation (4.8-11) must be set equal to 1 to obtain the temperature scale:

$$\frac{\mu_0 u_s^2}{kT_s} = \frac{\mu_0}{kT_s}\left(\frac{\Delta P}{L}\frac{H^2}{\mu_0}\right)^2 = 1 \Rightarrow T_s = \frac{H^4}{k\mu_0}\left(\frac{\Delta P}{L}\right)^2 \tag{4.8-16}$$

When the reference and scale factors defined by equations (4.8-14), (4.8-15), and (4.8-16) are substituted into our dimensionless describing equations, we obtain

$$0 = 1 + \frac{d}{dy^*}\left\{\left[1 - \frac{B}{T_0^2}\frac{H^4}{k\mu_0}\left(\frac{\Delta P}{L}\right)^2 T^*\right]\frac{du_x^*}{dy^*}\right\} \tag{4.8-17}$$

$$0 = \frac{d^2T^*}{dy^{*2}} + \left[1 - \frac{B}{T_0^2}\frac{H^4}{k\mu_0}\left(\frac{\Delta P}{L}\right)^2 T^*\right]\left(\frac{du_x^*}{dy^*}\right)^2 \tag{4.8-18}$$

$$\frac{du_x^*}{dy^*} = 0, \quad \frac{dT^*}{dy^*} = 0 \quad \text{at} \quad y^* = 0 \tag{4.8-19}$$

$$u_x^* = 0, \quad T^* = 0 \quad \text{at} \quad y^* = \pm 1 \tag{4.8-20}$$

We see that the criterion for ignoring the temperature dependence of the viscosity is given by (step 8)

$$\frac{B}{T_0^2}\frac{H^4}{k\mu_0}\left(\frac{\Delta P}{L}\right)^2 \ll 1 \tag{4.8-21}$$

If the criterion given by equation (4.8-21) is satisfied, equations (4.8-17) through (4.8-20) reduce to exactly the same equations that were considered in Section 4.5; that is, equations (4.5-22) through (4.5-24) for which the solution for the velocity profile is given by equation (4.5-1) and for which the temperature profile is given by equation (4.5-25).

4.9 THERMALLY DRIVEN FREE CONVECTION: BOUSSINESQ APPROXIMATION

Thus far we have considered problems that have involved either pure conduction or heat transfer with forced convection, that is, flow that is caused by some external driving force, such as a pump, fan, moving boundary, gravitational field, or other body force. Here we consider an example of *thermal free convection*, convection caused internal to the system, owing to a temperature gradient that creates unstable density variations. Note that unstable density variations can also be caused by concentration gradients, in which case it is referred to as *solutal free convection*. Free convection can also arise due to surface-tension gradients at an interface.

Consider a fluid with density ρ and viscosity μ that is confined between two vertical parallel plates of vertical height L separated by a distance $2H$, as shown in Figure 4.9-1. We assume that the space between the plates is capped at the top and the bottom but that $L \gg H$. The vertical plate at $y = -H$ is maintained at a constant temperature T_1, whereas the vertical plate at $y = +H$ is maintained at a constant temperature T_2, where $T_1 > T_2$. Because the density of

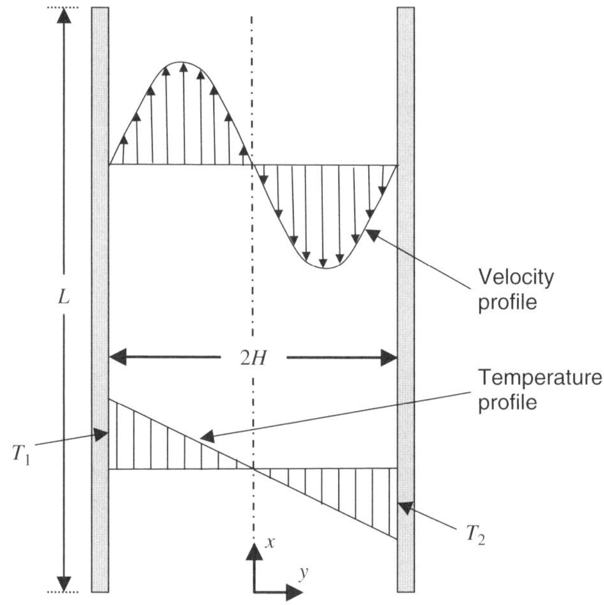

Figure 4.9-1 Steady-state fully developed buoyancy-induced free convection of a fluid confined between two vertical plates maintained at temperatures T_1 and T_2, where $T_1 > T_2$, showing representative temperature and velocity profiles.

a fluid decreases with increasing temperature, there will be a tendency for the fluid near the hot wall to flow upward and for the fluid near the cold wall to flow downward. As long as the plates are not too close together or the fluid is not too viscous, a steady-state free-convection flow can be generated for which the mass flow upward is equal to the mass flow downward.[14] We use scaling analysis to determine the conditions that permit the equations describing free-convection heat transfer to be simplified. We ignore end effects at the top and bottom of the vertical parallel plates as well as heat generation due to viscous dissipation.

Appropriate simplification of the equations of motion given by equations (D.1-10) and (D.1-11) and the thermal energy equation given by equation (F.1-2) in

[14]To determine if free-convection flow will be generated, it is necessary to carry out a stability analysis; the latter leads to the conditions required for free convection to be initiated expressed in terms of a critical value of the thermal Grashof number, $\mathrm{Gr}_t \equiv H^3 g\beta_t\,\Delta\rho/\nu^2$, or thermal Rayleigh number, $\mathrm{Ra}_t \equiv \mathrm{Gr}_t \cdot \mathrm{Pr}$, where $\Delta\rho$ in the present case is the difference in density between the fluid at the cold and hot plates; note that if the plates are too closely spaced or if the viscosity is sufficiently high, the Grashof number can be below the critical value for the inception of free convection; the reader interested in more information on stability theory applied to free convection is referred to standard references such as P. G. Drazin and W. H. Reid, *Hydrodynamic Stability*, Cambridge University Press, Cambridge, England, 1981, or J. S. Turner, *Buoyancy Effects in Fluids*, Cambridge University Press, Cambridge, England, 1973.

the Appendices and specification of the required boundary conditions yield the following set of describing equations (step 1):

$$0 = -\frac{\partial P}{\partial x} + \mu \frac{d^2 u_x}{dy^2} - \rho g \tag{4.9-1}$$

$$0 = -\frac{\partial P}{\partial y} \tag{4.9-2}$$

$$0 = \frac{d^2 T}{dy^2} \tag{4.9-3}$$

$$u_x = 0, \quad T = T_1 \quad \text{at} \quad y = -H \tag{4.9-4}$$

$$u_x = 0, \quad T = T_2 \quad \text{at} \quad y = H \tag{4.9-5}$$

Note that axial derivatives of both velocity and temperature are assumed to be zero under the assumptions of fully developed flow and no end effects. Since the density is temperature dependent, we need an appropriate equation of state. Here we consider small density variations and hence represent the density via a Taylor series expansion about the density $\overline{\rho}$ at the average temperature between the two plates, $\overline{T} \equiv (T_1 + T_2)/2$, given by

$$\rho = \rho|_{\overline{T}} + \left.\frac{\partial \rho}{\partial T}\right|_{\overline{T}} (T - \overline{T}) + \frac{1}{2} \left.\frac{\partial^2 \rho}{\partial T^2}\right|_{\overline{T}} (T - \overline{T})^2 + \cdots \tag{4.9-6}$$

$$\rho = \overline{\rho} - \overline{\rho}\beta_t (T - \overline{T}) + \overline{\rho}\gamma_t (T - \overline{T})^2 + \cdots \tag{4.9-7}$$

where β_t is the coefficient of volume expansion and γ_t is a positive constant. When equation (4.9-7) is substituted into equation (4.9-1) and truncated at the third term in the expansion, we obtain

$$0 = -\frac{\partial P}{\partial x} + \mu \frac{d^2 u_x}{dy^2} - \overline{\rho} g + \overline{\rho} g \beta_t (T - \overline{T}) - \overline{\rho} g \gamma_t (T - \overline{T})^2 \tag{4.9-8}$$

Equation (4.9-3) can be integrated directly subject to the boundary conditions given by equations (4.9-4) and (4.9-5) to obtain the following temperature profile:

$$T = \frac{T_1 + T_2}{2} - \frac{T_1 - T_2}{2} \frac{y}{H} = \overline{T} - \frac{\Delta T}{2} \frac{y}{H} \tag{4.9-9}$$

Moreover, equation (4.9-2) in combination with equation (4.9-1) implies that $\partial P/\partial x = dP/dx =$ a constant. Hence, we can evaluate dP/dx at any lateral position between the two vertical plates. It is convenient to evaluate the pressure gradient at the centerline between the two plates where $T = \overline{T}$; therefore, $dP/dx = -\overline{\rho} g$. With this simplification, equation (4.9-8) takes the form

$$0 = \mu \frac{d^2 u_x}{dy^2} + \overline{\rho} g \beta_t (T - \overline{T}) - \overline{\rho} g \gamma_t (T - \overline{T})^2 \tag{4.9-10}$$

Introduce the following scale and reference factors (steps 2, 3, and 4):

$$u_x^* \equiv \frac{u_x}{u_s}; \qquad T^* \equiv \frac{T - T_r}{T_s}; \qquad y^* \equiv \frac{y}{y_s} \tag{4.9-11}$$

Substitute these dimensionless variables into the describing equations and divide each equation through by the dimensional coefficient of one term that should be retained to maintain physical significance (steps 5 and 6):

$$0 = \frac{d^2 u_x^*}{dy^{*2}} + \frac{\overline{\rho} g \beta_t y_s^2 T_s}{\mu u_s}\left(T^* + \frac{T_r - \overline{T}}{T_s}\right) - \frac{\overline{\rho} g \gamma_t y_s^2 T_s^2}{\mu u_s}\left(T^* + \frac{T_r - \overline{T}}{T_s}\right)^2 \tag{4.9-12}$$

$$0 = \frac{d^2 T^*}{dy^{*2}} \tag{4.9-13}$$

$$u_x^* = 0, \quad T^* = \frac{T_1 - T_r}{T_s} \quad \text{at} \quad y^* = -\frac{H}{y_s} \tag{4.9-14}$$

$$u_x^* = 0, \quad T^* = \frac{T_2 - T_r}{T_s} \quad \text{at} \quad y^* = \frac{H}{y_s} \tag{4.9-15}$$

The dimensionless groups in equations (4.9-14) and (4.9-15) when set equal to 1 and zero, respectively, provide the following reference and scale factors (step 7):

$$\frac{T_2 - T_r}{T_s} = 0 \Rightarrow T_r = T_2, \quad \frac{T_1 - T_r}{T_s} = \frac{T_1 - T_2}{T_s} = 1 \Rightarrow T_s = T_1 - T_2 \tag{4.9-16}$$

In addition, the dimensionless groups in these equations provide the characteristic length scale

$$\frac{H}{y_s} = 1 \Rightarrow y_s = H \tag{4.9-17}$$

Since what causes this flow (i.e., the leading-order gravitational body force term) must balance the viscous resisting force, the corresponding dimensionless group that provides a measure of this ratio in equation (4.9-12) must be equal to 1, which provides the characteristic scale for the velocity:

$$\frac{\overline{\rho} g \beta_t y_s^2 T_s}{\mu u_s} = 1 \Rightarrow u_s = \frac{\overline{\rho} g \beta_t y_s^2 T_s}{\mu} = \frac{\overline{\rho} g \beta_t H^2 (T_1 - T_2)}{\mu} \tag{4.9-18}$$

If the reference and scale factors defined by equations (4.9-16) through (4.9-18) are substituted into our dimensionless describing equations, we obtain

$$0 = \frac{d^2 u_x^*}{dy^{*2}} + \left(T^* - \frac{1}{2}\right) - \frac{\gamma_t (T_1 - T_2)}{\beta_t}\left(T^* - \frac{1}{2}\right)^2 \tag{4.9-19}$$

$$0 = \frac{d^2 T^*}{dy^{*2}} \tag{4.9-20}$$

$$u_x^* = 0, \quad T^* = 1 \quad \text{at} \quad y^* = -1 \tag{4.9-21}$$

$$u_x^* = 0, \quad T^* = 0 \quad \text{at} \quad y^* = 1 \tag{4.9-22}$$

We see that the criterion for ignoring the higher-order temperature dependence of the density is given by (step 8)

$$\frac{\gamma_t(T_1 - T_2)}{\beta_t} \ll 1 \tag{4.9-23}$$

This simplification, which considers only the leading-order effects of the temperature on the density, is referred to as the *Boussinesq approximation*.

4.10 DIMENSIONAL ANALYSIS CORRELATION FOR COOKING A TURKEY

In dimensional analysis we seek to determine the dimensionless groups required to correlate data or to scale a process up or down. These dimensionless groups can always be determined using o(1) scaling analysis since this procedure leads to the minimum parametric representation for a set of describing equations. However, the preceding sections indicated that carrying out an o(1) scaling analysis can be somewhat complicated and time consuming. In contrast, the scaling analysis approach to dimensional analysis illustrated in this section is much easier and quicker to implement. Note, however, that it does not provide as much information as does o(1) scaling analysis for achieving the minimum parametric representation. In particular, it does not lead to groups whose magnitude can be used to assess the relative importance of particular terms in the describing equations. It also does not identify regions of influence or boundary layers whose identification can in some cases reduce the number of dimensionless groups. This first example of the use of scaling for dimensional analysis in heat-transfer applications will provide more details on the steps involved. We will also compare the results of scaling analysis to the results obtained from using the Pi theorem, to underscore the advantages of using the former to achieve the minimum parametric representation. The steps referred to here are those outlined in Section 2.4 for the scaling approach to dimensional analysis; these differ from those used in Sections 4.2 through 4.9, since no attempt is made to achieve o(1) scaling.

This first example of the use of scaling analysis for dimensional analysis in heat transfer will consider developing a correlation for determining the cooking time of a turkey. In particular, we seek to determine how long it will take to cook the 28-lb (12.7-kg) turkey shown in Figure 4.10-1. Cookbooks do not provide equations to determine the cooking time. Rather, they usually provide discrete data for the required cooking time t_c as a function of the mass of the turkey M, such as shown in the table that accompanies Figure 4.10-1. A problem arises in that this table does not indicate in any precise way how much time is required to cook a 28-lb turkey. A crude way to estimate this time might be to do some type of extrapolation from

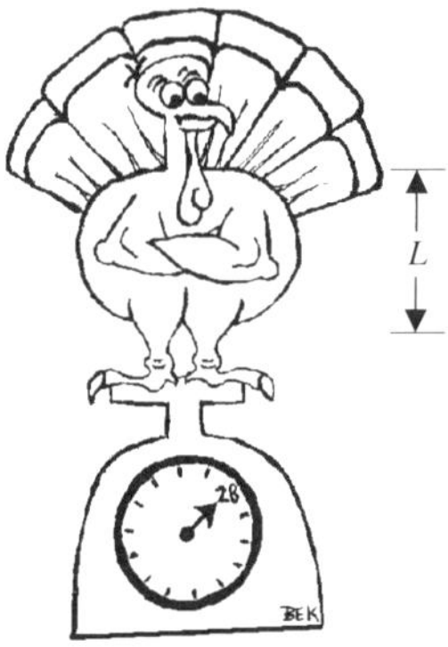

Ready to cook weight M (lb)	Total cooking time t_c (hr)
5 to 8	3 to 3½
8 to 12	3½ to 4½
12 to 16	4½ to 5½
16 to 20	5½ to 6½
20 to 24	6½ to 7

Figure 4.10-1 Schematic of a very large turkey of a characteristic length L along with cookbook data for the cooking time t_c as a function of the mass of the turkey M. (Data from General Mills, *Betty Crocker's Cookbook*, Golden Press, New York, 1972.)

the data given in this table. However, a better way is to use these data and scaling analysis to determine the underlying correlation between the weight of the turkey and the cooking time.

The time it takes to fully cook the turkey is that required to bring the center of the bird from its initial temperature T_0 up to the temperature T_1 specified by the cookbook (typically, 165°F or 73.9°C) by placing it in a preheated oven maintained at a temperature T_2 (typically 325°F or 163°C). It is reasonable to assume that the heat transfer is controlled by conduction in the turkey. To determine the time required to reach this temperature, in principle we would have to solve the unsteady-state heat-conduction equation. This is complicated by the fact that the center of the turkey is usually filled with dressing that has physical properties that differ from those of the turkey itself. A further complication is that appropriate boundary conditions need to be specified on the surface of the turkey and at the interface between the turkey and the dressing along with appropriate conditions at the center of the stuffed turkey. Indeed, this would be a complicated problem to solve numerically! However, we will see that with the aid of the data given in the cookbook, we can determine the cooking time without having to solve the describing equations.

We begin by writing the equations we would solve for the cooking time, if indeed we could solve these equations (step 1 in the scaling procedure for dimensional analysis); namely, the unsteady-state heat-conduction equations in both the turkey proper and in the dressing in its interior, which are given in generalized vector notation by equation (B.4-2) in the Appendices:

$$\frac{\partial T}{\partial t} = \alpha_T \nabla^2 T \qquad \text{(applicable in the turkey)} \tag{4.10-1}$$

$$\frac{\partial T}{\partial t} = \alpha_D \nabla^2 T \qquad \text{(applicable in the dressing)} \tag{4.10-2}$$

where ∇^2 denotes the Laplacian operator and $\alpha_T \equiv k_T/\rho_T C_{PT}$ and $\alpha_D \equiv k_D/\rho_D C_{PD}$ are the thermal diffusivities of the turkey and dressing, respectively, in which k_i, ρ_i, and C_{Pi} are the thermal conductivity, density, and heat capacity of medium i, respectively. Note that we have chosen to write the thermal energy equation in generalized notation that is not specific to any particular coordinate system; this is convenient to do in dimensional analysis since there is no need to scale specific spatial derivatives to be o(1). The initial and boundary conditions are given by

$$T = T_0 \quad \text{at} \quad t = 0 \tag{4.10-3}$$

$$T = T_2 \quad \text{at the surface of the turkey} \tag{4.10-4}$$

$$T|_+ = T|_- \quad \text{at the interface between the turkey and dressing} \tag{4.10-5}$$

$$k_T \nabla T|_+ = k_D \nabla T|_- \quad \text{at the interface between the turkey and dressing} \tag{4.10-6}$$

$$\vec{n} \cdot \nabla T = 0 \quad \text{along the plane of symmetry in the turkey} \tag{4.10-7}$$

where + and − denote the turkey and dressing side of the interface, respectively, and $\vec{n}$ is a unit vector normal to the plane of symmetry in the turkey.

To solve equations (4.10-1) through (4.10-7), we would need to describe mathematically the surface of the turkey and its interface with the dressing; this would be prohibitively difficult to do in practice. However, in dimensional analysis this challenging task can be avoided by recognizing that it is reasonable to assume that all turkeys are geometrically similar. By this we mean that although their mass might differ, their geometry is more or less the same. Assume that it takes p geometric parameters to characterize the shape of a turkey. Recall that one can form $p - 1$ dimensionless ratios from p quantities having one dimension (length in this case). For geometrically similar turkeys, these $p - 1$ dimensionless geometric ratios will be the same. Hence, one needs to include in the dimensional analysis only one geometric quantity such as some characteristic body dimension along with the quantities that characterize the heat transfer. This arbitrary characteristic length is chosen to be the maximum body width L as shown in Figure 4.10-1.

Steps 2 and 3 in the scaling procedure for dimensional analysis involve defining arbitrary scale factors for all the dependent and independent variables and reference factors for those not naturally referenced to zero. Hence, we introduce the following dimensionless variables:

$$T^* \equiv \frac{T - T_r}{T_s}; \qquad \nabla^* \equiv L\nabla; \qquad \nabla^{2*} \equiv L^2\nabla^2; \qquad t^* \equiv \frac{t}{t_s} \tag{4.10-8}$$

Note that we have chosen L for our length scale since we arbitrarily chose it to form the $p - 1$ geometric ratios that define the surface of the turkey and its interface with the dressing.

Steps 4 and 5 involve introducing these dimensionless variables into the describing equations and dividing through by the dimensional coefficient of one term in each equation. In dimensional analysis it makes no difference which term you choose. These steps yield the following dimensionless describing equations:

$$\frac{\partial T^*}{\partial t^*} = \frac{\alpha_T t_s}{L^2} \nabla^{*2} T^* \qquad \text{(applicable in the turkey)} \tag{4.10-9}$$

$$\frac{\partial T^*}{\partial t^*} = \frac{\alpha_D t_s}{L^2} \nabla^{*2} T^* \qquad \text{(applicable in the dressing)} \tag{4.10-10}$$

$$T^* = \frac{T_0 - T_r}{T_s} \qquad \text{at} \quad t^* = 0 \tag{4.10-11}$$

$$T^* = \frac{T_2 - T_r}{T_s} \qquad \text{at the surface of the turkey} \tag{4.10-12}$$

$$T^*|_+ = T^*|_- \qquad \text{at the interface between the turkey and dressing} \tag{4.10-13}$$

$$\nabla^* T^*|_+ = \frac{k_D}{k_T} \nabla^* T^*|_- \qquad \text{at the interface between the turkey and dressing} \tag{4.10-14}$$

$$\vec{n} \cdot \nabla^* T^* = 0 \qquad \text{along the plane of symmetry in the turkey} \tag{4.10-15}$$

Step 6 involves setting the various groups equal to 1 or zero to determine the scale and reference factors, respectively. It makes no difference in dimensional analysis which groups we set equal to 1 since there is no attempt to achieve $\text{o}(1)$ scaling. We need to do the latter only if we are seeking to simplify the equations by dropping one or more of the terms. In dimensional analysis we are seeking to determine the minimum parametric representation. Let us set equation (4.10-11) equal to zero to determine the reference temperature and set equation (4.10-12) equal to 1 to determine the temperature scale. Finally, let us set the dimensionless group in equation (4.10-9) equal to 1 to obtain the time scale; we could equally well have set the dimensionless group in equation (4.10-10) equal to 1 to determine this time scale. These choices then yield the following minimum parametric representation of the describing equations; that is, in terms of the minimum number of dimensionless groups:

$$\frac{\partial T^*}{\partial t^*} = \nabla^{*2} T^* \qquad \text{(applicable in the turkey)} \tag{4.10-16}$$

$$\frac{\partial T^*}{\partial t^*} = \frac{\alpha_D}{\alpha_T} \nabla^{*2} T^* \qquad \text{(applicable in the dressing)} \tag{4.10-17}$$

$$T^* = 0 \qquad \text{at} \quad t^* = 0 \tag{4.10-18}$$

$$T^* = 1 \qquad \text{at the surface of the turkey} \tag{4.10-19}$$

$$T^*|_+ = T^*|_- \qquad \text{at the interface between the turkey and dressing} \tag{4.10-20}$$

$$\nabla^* T^*|_+ = \frac{k_D}{k_T} \nabla^* T^*|_- \qquad \text{at the interface between the turkey and dressing} \tag{4.10-21}$$

$$\vec{n} \cdot \nabla^* T^* = 0 \qquad \text{along the plane of symmetry in the turkey} \tag{4.10-22}$$

The solution to equations (4.10-16) through (4.10-22) for the dimensionless temperature as a function of the dimensionless time then will be of the form

$$T^* = f_1\left(x^*, y^*, z^*, t^*, \frac{\alpha_D}{\alpha_T}, \frac{k_D}{k_T}\right) \tag{4.10-23}$$

However, we seek the particular value of the dimensionless time t_c^* at which the center of the turkey reaches the temperature T_1. The center of the turkey is located at some specific values of the dimensionless coordinates x^*, y^*, and z^*. Hence, our correlation for the dimensionless cooking time is given by

$$T_1^* = \frac{T_1 - T_0}{T_2 - T_0} = f_2\left(t_c^*, \frac{\alpha_D}{\alpha_T}, \frac{k_D}{k_T}\right) \tag{4.10-24}$$

An equivalent statement is

$$t_c^* = \frac{\alpha_T t_c}{L^2} = f_3\left(\frac{T_1 - T_0}{T_2 - T_0}, \frac{\alpha_D}{\alpha_T}, \frac{k_D}{k_T}\right) \tag{4.10-25}$$

Hence, for specified cooking conditions and geometrically similar turkeys and dressing with specified physical and transport properties, we conclude that $t_c \propto L^2$.

It is reasonable to assume for geometrically similar turkeys that the characteristic length L will be proportional to the mass of the turkey; that is,

$$L = AM^B \tag{4.10-26}$$

When equation (4.10-26) is substituted into equation (4.10-25), we obtain the following equation that can be used to correlate the data given in Table 4.10-1 for the cooking time as a function of turkey weight:

$$t_c = \frac{L^2}{\alpha_T} f'\left(\frac{T_1 - T_0}{T_2 - T_0}, \frac{\alpha_D}{\alpha_T}, \frac{k_D}{k_T}\right) = A\frac{M^{2B}}{\alpha_T} f'\left(\frac{T_1 - T_0}{T_2 - T_0}, \frac{\alpha_D}{\alpha_T}, \frac{k_D}{k_T}\right) = A'M^{B'} \tag{4.10-27}$$

where A' and B' are empirical constants that will be determined by fitting the cooking-time data given in the cookbook as a function of the mass of the turkey. Figure 4.10-2 shows a plot of the cooking time t_c as a function of the mass M of the turkey. The data points represent average values of the cooking time and turkey mass for each table entry in Figure 4.10-1; the corresponding error bars then represent the maximum deviation of the values in Figure 4.10-1 about these average values for each of the five data points. The trend line in Figure 4.10-2 fits the data with a regression coefficient $R^2 = 0.994$ and is given by the equation

$$t_c = M^{0.61} \tag{4.10-28}$$

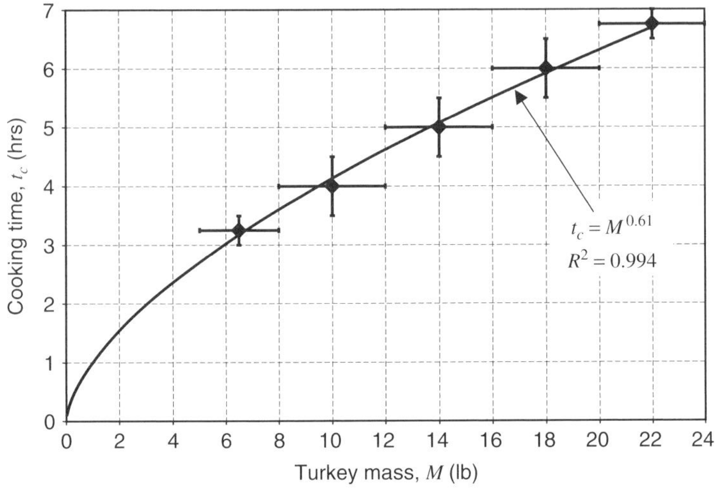

Figure 4.10-2 Cooking time in hours as a function of turkey mass in pounds. The solid line shows the correlation suggested by using the scaling approach to dimensional analysis. (Data from General Mills, *Betty Crocker's Cookbook*, Golden Press, New York, 1972.)

Note that if the turkey body were perfectly spherical with a diameter L, the exponent in equation (4.10-28) would be 0.67. Equation (4.10-28) now permits determining the time required to cook a 28-lb turkey, which is found to be 7.6 hours.

Equation (4.10-25) indicates that the cooking time can be correlated in terms of four dimensionless groups and $p - 1$ geometric ratios required to specify the shape of a turkey. It is of interest to explore the possible consequences of using the Pi theorem to develop a correlation for the cooking time. The Pi theorem approach would require first somehow identifying the quantities that enter into the correlation; these would include t_c, α_T, α_D, k_T, k_D, L, T_0, T_2, and T_1. Note that the quantity L can be expressed in terms of the mass M of the turkey via a relation of the form of equation (4.10-26). These nine quantities are expressed in terms of four units: length, time, energy, and temperature (note that energy is a fundamental unit in the system of thermodynamics when no exchange is involved between internal and mechanical energy). Hence, we might conclude that five dimensionless groups are required to correlate the cooking time (i.e., $n - m = 9 - 4 = 5$) in addition to $p - 1$ geometric aspect ratios when, in fact, our scaling analysis indicated that only four dimensional groups are needed in addition to the $p - 1$ aspect ratios. A more enlightened use of the Pi theorem cleverly might recognize that the quantities T_0, T_2, and T_1 can be combined into the single variable $(T_1 - T_0)/(T_2 - T_0)$, which then leads to the conclusion that only three dimensionless groups ($n - m = 7 - 4 = 3$) plus the $p - 1$ geometric aspect ratios are required. However, this conclusion drawn from the Pi theorem is error since our scaling analysis indicates that the minimum parametric representation involves four dimensionless groups. This error arises owing to a breakdown of the Pi theorem associated with the quantities

α_T, α_D, k_T, and k_D; the Pi theorem would indicate zero groups for these four quantities (i.e., $n - m = 4 - 4 = 0$) when in fact they lead to two dimensionless groups. This example clearly indicates the pitfalls of using the Pi theorem for dimensional analysis. Scaling analysis provides a systematic method for avoiding these problems associated with the Pi theorem.

4.11 SUMMARY

The example in Section 4.2 provided an introduction to the step-by-step procedure for scaling analysis in heat transfer. Scaling was used in this problem to assess the criterion for ignoring edge effects so that the heat transfer could be considered to be one-dimensional when predicting the temperature sufficiently far removed from the sidewalls or integral quantities such as the total heat-transfer rate. This problem involved the introduction of both reference and scale factors since the temperature was not naturally referenced to zero. It also introduced region-of-influence scaling to determine the thickness of the zone near the sidewalls, wherein the effects of lateral heat transfer could never be ignored when predicting quantities such as the temperature or heat flux at or near the sidewalls. This example provided a means for estimating the error incurred when the assumptions suggested by scaling analysis are invoked since an analytical solution was available for this heat-transfer problem. Demanding that a quantity be $O(0.1)$ in order to ignore some term in the describing equations typically results in an error of 40 to 50%; demanding that it be $O(0.01)$ reduces the error to less than 10%. However, the error that is encountered also depends on the quantity being considered; for example, point quantities within a region of influence might incur very large errors even when the relevant dimensionless group is very small.

In Section 4.3 we applied scaling analysis to unsteady-state one-dimensional heat conduction in a flat solid slab. This example led to two time scales, the observation time and the characteristic conduction time, whose ratio is the Fourier number for heat transfer. If the Fourier number is very large, the process can be assumed to be steady-state, whereas if it is very small, the unsteady-state heat transfer is confined to a region of influence or thermal boundary layer. We referred to these as the film theory and penetration theory approximations, respectively, although this terminology is generally used only for the analogous approximations in mass-transfer modeling.

Unsteady-state convective heat transfer from a solid sphere was considered in Section 4.4. A lumped-parameter boundary condition involving an appropriate heat-transfer coefficient was used to describe the heat transfer in the fluid phase. It was necessary to introduce a separate scale for the temperature gradient in this problem since the temperature did not go through a characteristic change over the characteristic length. Scaling this problem introduced the Biot number, which is the dimensionless ratio of the heat-conduction resistance in the solid sphere to the convective heat-transfer resistance in the surrounding fluid. The criterion for assuming steady-state in this problem involved an interrelationship between the

Fourier and Biot numbers; the observation time required to achieve steady-state was found to decrease with increasing Biot number. Scaling analysis provided a systematic method for arriving at the simplified equations appropriate to the small Biot number approximation whereby the temperature can be assumed to be uniform within the conducting object. Most heat-transfer textbooks do not provide any rigorous justification for the low Biot number approximation.

In Section 4.5 we considered fully developed laminar flow between two flat plates with heat generation due to viscous dissipation. The presence of both the transverse and axial conduction terms made the describing equations elliptic. This complicated the solution since the required downstream boundary condition often is unknown. The concept of local scaling in heat transfer was introduced in this problem, whereby one considers the describing equations within a domain defined by some arbitrary distance in the principal direction of flow that is assumed to be constant during the scaling analysis. In contrast to the preceding three examples, there was no explicit temperature scale in this problem; rather, the temperature scale was determined by balancing the viscous dissipation and transverse heat-conduction terms. Scaling analysis led to two important dimensionless groups in heat transfer: the Peclet and Prandtl numbers. The former is a measure of the ratio of the convection to conduction of heat, whereas the latter is a measure of the viscous transport of momentum to the conductive transfer of heat. The Peclet number has a role in heat transfer that is analogous to that of the Reynolds number in fluid dynamics. For example, we found that the convective heat transfer could be ignored if the Peclet number was very small; this is analogous to the low Reynolds number or creeping-flow approximation in fluid dynamics. We also found that the complicating effects of axial heat conduction could be ignored if the width-to-length aspect ratio was very small. The combination of small Peclet number and small aspect ratio in heat transfer is analogous to the lubrication-flow approximation in fluid dynamics.

Scaling analysis was applied to the complementary problem of high Peclet number, coupled heat and momentum transfer in Section 4.6. The problem considered here was heat transfer from a hot flat plate to the developing flow over this surface. In this problem the transverse derivative of the temperature and axial velocity were scaled with different characteristic lengths: the thermal and momentum boundary-layer thicknesses, respectively. The relative thickness of the two boundary layers depended on the Prandtl number. For liquids whose Prandtl number is much greater than 1, the thermal was thinner than the momentum boundary layer. For liquid metals whose Prandtl number is less than 1, the thermal is thicker than the momentum boundary layer. For gases whose Prandtl number is nearly 1, the two boundary layers have essentially the same thickness. We found that the boundary-layer approximation is reasonable when the Peclet and Reynolds numbers based on the local axial length scale become large (i.e., $\text{Pe}_l \equiv U_\infty L/\alpha = \text{Re} \cdot \text{Pr} \gg 1$ and $\text{Re} \equiv U_\infty \rho L/\mu \gg 1$). Note that for ordinary liquids, thermal boundary-layer analysis might apply, whereas the momentum boundary-layer analysis might not. Note also that the boundary-layer approximation must break down in the vicinity of the leading edge, where L becomes small. Hence, this problem involved both a transverse and an axial region of influence; boundary-layer theory is applicable within the former and beyond the latter. This upstream limitation of boundary-layer theory, which emerges from scaling analysis, is not mentioned in some transport and fluid mechanics textbooks. This problem also illustrated how scaling analysis can suggest a strategy for solving the describing equations by using the solution to the parabolic boundary-layer equations to specify the downstream boundary conditions on the elliptic equations that must be solved near the leading edge.

Scaling analysis was applied to heat transfer associated with phase change in Section 4.7. Conductive heat transfer causes a melting front to penetrate progressively into frozen porous media. This provided an introduction to moving boundary problems, those for which boundary conditions must be applied at surfaces whose location in turn had to be determined by solving the describing equations. Since the location of the moving boundary constituted an additional unknown, an auxiliary condition was needed. This was obtained by an integral energy balance over the domain of interest. An analogy was drawn between this auxiliary condition in heat transfer and the kinematic condition in fluid dynamics in that both are based on integral balances done to derive an additional equation to determine the location of an unspecified boundary. The forgiving nature of scaling analysis was illustrated by intentionally scaling one of the dependent variables incorrectly and thereby arriving at a contradiction: that the unsteady-state term was not multiplied by the reciprocal of the Fourier number. A proper analysis required that the time derivative of the moving boundary location be scaled with a velocity scale rather than with the ratio of the characteristic length divided by the time scale. We then found that for sufficiently large Fourier numbers, quasi-steady-state thermal penetration of the melting front could be assumed. This problem also introduced the pseudo-convection term that arises when the scaling introduces either a scale or a reference factor that is time-dependent.

In Section 4.8 we illustrated how scaling analysis could be used to determine when it is reasonable to ignore the temperature dependence of the viscosity for a problem involving laminar flow between two flat plates with transverse heat conduction and viscous dissipation. Since we sought to determine only if any significant variation in the viscosity occurred, it was sufficient to expand the equation describing the temperature dependence of the viscosity in a Taylor series and retain only the first-order correction. Scaling then identified the condition required to ignore this first-order correction. The scaling procedure used here for assessing when the temperature dependence of the viscosity can be ignored can be applied to assess the dependence of any physical or transport property on state variables such as temperature, pressure, or concentration.

In Section 4.9 we applied scaling to a free-convection heat-transfer problem; that is, to a problem wherein the driving force for flow was internal to the system, in this case due to density variations created by temperature gradients. Scaling was employed to determine when the temperature dependence of the density could be represented by the first two terms in a Taylor series expansion about the density at the average temperature. This is the basis of the classical Boussinesq approximation in the analysis of free convection.

Scaling was applied to dimensional analysis in Section 4.10. In contrast to o(1) scaling analysis, the scaling approach to dimensional analysis merely seeks to arrive at the minimum parametric representation of the problem; that is, to obtain a set of dimensionless describing equations in terms of the minimum number of dimensionless parameters. The goal of this problem was to determine the relationship between the required cooking time and turkey mass based on discrete data taken from a cookbook. This example illustrated the advantages of the scaling analysis methodology for dimensional analysis relative to using the conventional Pi theorem approach in that the latter did not lead to the minimum parametric representation. Scaling analysis led to a single dimensionless group that involved the two quantities that we sought to interrelate. By using this dimensionless group in conjunction with the discrete cookbook data it was possible to develop a fully predictive equation that related the required cooking time to the mass of the turkey.

4.E EXAMPLE PROBLEMS

4.E.1 Steady-State Heat Conduction in a Rectangular Fin

A solid metallic flat fin with a constant thermal conductivity k, length L, width W, and height H such that $H \ll L < W$ is attached at one end to a surface that is maintained at a constant temperature T_0. The convective heat-transfer flux q_0 from the surfaces of the fin to the ambient air is describing via the lumped-parameter approach and given by

$$q_0 = h(T - T_\infty) \tag{4.E.1-1}$$

where h is the heat-transfer coefficient and T_∞ is the temperature of the ambient air far removed from the fin. A schematic of the fin is shown in Figure 4.E.1-1. We use scaling analysis to explore what simplifications are possible in describing heat transfer in this fin.

The describing equations obtained by considering only the conduction terms in equation (F.1-2) in the Appendices and the necessary boundary conditions are given by (step 1)

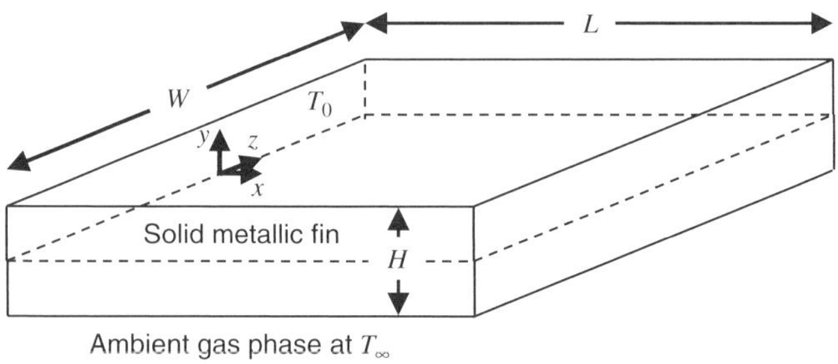

Figure 4.E.1-1 Flat solid metallic fin that has a constant thermal conductivity and length L, width W, and height H such that $H \ll L < W$; the convective heat-transfer flux q_0 from the surfaces of the fin to the ambient air is described by $h(T - T_\infty)$.

$$0 = \frac{\partial^2 T}{\partial x^2} + \frac{\partial^2 T}{\partial y^2} + \frac{\partial^2 T}{\partial z^2} \tag{4.E.1-2}$$

$$T = T_0 \quad \text{at} \quad x = 0, \quad 0 \le y \le H, \quad 0 \le z \le W \tag{4.E.1-3}$$

$$-k\frac{\partial T}{\partial x} = h(T - T_\infty) \quad \text{at} \quad x = L, \quad 0 \le y \le \frac{H}{2}, \quad 0 \le z \le \frac{W}{2} \tag{4.E.1-4}$$

$$\frac{\partial T}{\partial y} = 0 \quad \text{at} \quad y = 0, \quad 0 \le x \le L, \quad 0 \le z \le \frac{W}{2} \tag{4.E.1-5}$$

$$-k\frac{\partial T}{\partial y} = h(T - T_\infty) \quad \text{at} \quad y = \frac{H}{2}, \quad 0 \le x \le L, \quad 0 \le z \le \frac{W}{2} \tag{4.E.1-6}$$

$$\frac{\partial T}{\partial z} = 0 \quad \text{at} \quad z = 0, \quad 0 \le x \le L, \quad 0 \le y \le \frac{H}{2} \tag{4.E.1-7}$$

$$-k\frac{\partial T}{\partial z} = h(T - T_\infty) \quad \text{at} \quad z = \frac{W}{2}, \quad 0 \le x \le L, \quad 0 \le y \le \frac{H}{2} \tag{4.E.1-8}$$

Because of the planar symmetry, we have considered the heat transfer in only the upper half-width of the fin.

Introduce the following scale and reference factors (steps 2, 3, and 4):

$$T^* \equiv \frac{T - T_r}{T_s}; \qquad x^* \equiv \frac{x}{x_s}; \qquad y^* \equiv \frac{y}{y_s}; \qquad z^* \equiv \frac{z}{z_s};$$
$$\left(\frac{\partial T}{\partial x}\right)^* \equiv \frac{1}{T_{xs}}\frac{\partial T}{\partial x}; \qquad \left(\frac{\partial T}{\partial y}\right)^* \equiv \frac{1}{T_{ys}}\frac{\partial T}{\partial y}; \qquad \left(\frac{\partial T}{\partial z}\right)^* \equiv \frac{1}{T_{zs}}\frac{\partial T}{\partial z} \tag{4.E.1-9}$$

Note that we have allowed for separate scales for the temperature gradients in the x-, y-, and z-directions since there is no reason to assume that the temperature will change from its maximum to its minimum value over the length, thickness, or width of the fin, respectively. Substitute these dimensionless variables into the describing equations to obtain (steps 5 and 6)

$$0 = \frac{\partial}{\partial x^*}\left(\frac{\partial T}{\partial x}\right)^* + \frac{x_s T_{ys}}{y_s T_{xs}}\frac{\partial}{\partial y^*}\left(\frac{\partial T}{\partial y}\right)^* + \frac{x_s T_{zs}}{z_s T_{xs}}\frac{\partial}{\partial z^*}\left(\frac{\partial T}{\partial z}\right)^* \tag{4.E.1-10}$$

$$T^* = \frac{T_0 - T_r}{T_s} \quad \text{at} \quad x^* = 0, \quad 0 \le y^* \le \frac{H}{y_s}, \quad 0 \le z^* \le \frac{W}{z_s} \tag{4.E.1-11}$$

$$\left(\frac{\partial T}{\partial x}\right)^* = -\frac{hT_s}{kT_{xs}}\left(T^* + \frac{T_r - T_\infty}{T_s}\right) \quad \text{at} \quad x^* = \frac{L}{x_s}, \quad 0 \le y^* \le \frac{H}{y_s},$$
$$0 \le z^* \le \frac{W}{z_s} \tag{4.E.1-12}$$

$$\left(\frac{\partial T}{\partial y}\right)^* = 0 \quad \text{at} \quad y^* = 0, \quad 0 \le x^* \le \frac{L}{x_s}, \quad 0 \le z^* \le \frac{W}{z_s} \tag{4.E.1-13}$$

$$\left(\frac{\partial T}{\partial y}\right)^* = -\frac{hT_s}{kT_{ys}}\left(T^* + \frac{T_r - T_\infty}{T_s}\right) \quad \text{at} \quad y^* = \frac{H}{y_s}, \quad 0 \le x^* \le \frac{L}{x_s}, \quad 0 \le z^* \le \frac{W}{z_s} \tag{4.E.1-14}$$

$$\left(\frac{\partial T}{\partial z}\right)^* = 0 \quad \text{at} \quad z^* = 0, \quad 0 \le x^* \le \frac{L}{x_s}, \quad 0 \le y^* \le \frac{H}{y_s} \tag{4.E.1-15}$$

$$\left(\frac{\partial T}{\partial z}\right)^* = -\frac{hT_s}{kT_{zs}}\left(T^* + \frac{T_r - T_\infty}{T_s}\right) \quad \text{at} \quad z^* = \frac{W}{z_s}, \quad 0 \le x^* \le \frac{L}{x_s}, \quad 0 \le y^* \le \frac{H}{y_s} \tag{4.E.1-16}$$

Inspection of the dimensionless describing equations indicates that the temperature can be bounded to be $\circ(1)$ if we set $T_r = T_\infty$ and $T_s = T_0 - T_\infty$. The dimensionless spatial coordinates can be bounded to be $\circ(1)$ if we set $x_s = L$, $y_s = H$, and $z_s = W$. The dimensionless groups in equations (4.E.1-14) and (4.E.1-16) provide the following scales for the temperature gradients in the y- and z-directions, respectively (step 7):

$$T_{ys} = T_{zs} = \frac{h(T_0 - T_\infty)}{k} = \frac{hH}{k}\frac{(T_0 - T_\infty)}{H} = \text{Bi}_t\frac{T_0 - T_\infty}{H} \tag{4.E.1-17}$$

where $\text{Bi}_t \equiv hH/k$ is the Biot number for heat transfer based on the smallest length dimension H. However, it would not be appropriate to determine T_{xs} from the dimensionless group in equation (4.E.1-12) since the small amount of heat leaving the tip of the fin does not cause the axial temperature gradient; rather, we expect that the axial temperature gradient is caused by the heat being transferred from the side of the fin.

When the scale and reference factors are substituted into equations (4.E.1-10) through (4.E.1-16), we obtain the following dimensionless describing equations:

$$0 = \frac{\partial}{\partial x^*}\left(\frac{\partial T}{\partial x}\right)^* + \text{Bi}_t\frac{L(T_0 - T_\infty)}{H^2 T_{xs}}\frac{\partial}{\partial y^*}\left(\frac{\partial T}{\partial y}\right)^* + \text{Bi}_t\frac{L(T_0 - T_\infty)}{WHT_{xs}}\frac{\partial}{\partial z^*}\left(\frac{\partial T}{\partial z}\right)^* \tag{4.E.1-18}$$

$$T^* = 1 \quad \text{at} \quad x^* = 0, \quad 0 \le y^* \le 1, \quad 0 \le z^* \le 1 \tag{4.E.1-19}$$

$$\left(\frac{\partial T}{\partial x}\right)^* = -\frac{h(T_0 - T_\infty)}{kT_{xs}}T^* \quad \text{at} \quad x^* = 1, \quad 0 \le y^* \le 1, \quad 0 \le z^* \le 1 \tag{4.E.1-20}$$

$$\left(\frac{\partial T}{\partial y}\right)^* = 0 \quad \text{at} \quad y^* = 0, \quad 0 \le x^* \le 1, \quad 0 \le z^* \le 1 \tag{4.E.1-21}$$

$$\left(\frac{\partial T}{\partial y}\right)^* = -T^* \qquad \text{at} \quad y^* = 1, \quad 0 \le x^* \le 1, \quad 0 \le z^* \le 1 \tag{4.E.1-22}$$

$$\left(\frac{\partial T}{\partial z}\right)^* = 0 \qquad \text{at} \quad z^* = 0, \quad 0 \le x^* \le 1, \quad 0 \le y^* \le 1 \tag{4.E.1-23}$$

$$\left(\frac{\partial T}{\partial z}\right)^* = -T^* \qquad \text{at} \quad z^* = 1, \quad 0 \le x^* \le 1, \quad 0 \le y^* \le 1 \tag{4.E.1-24}$$

The axial conduction must be balanced by at least one of the other two terms in equation (4.E.1-18). However, since $H \ll W$, it is clear that conduction in the y-direction is far more important than that in the z-direction. Hence, we set the dimensionless group multiplying the conduction term in the y-direction equal to 1 to obtain T_{xs} as

$$\mathrm{Bi}_t \frac{L(T_0 - T_\infty)}{H^2 T_{xs}} = 1 \Rightarrow T_{xs} = \mathrm{Bi}_t \frac{L(T_0 - T_\infty)}{H^2} \tag{4.E.1-25}$$

Substituting this value for T_{xs} into equation (4.E.1-18) then yields

$$0 = \frac{\partial}{\partial x^*}\left(\frac{\partial T}{\partial x}\right)^* + \frac{\partial}{\partial y^*}\left(\frac{\partial T}{\partial y}\right)^* + \frac{H}{W}\frac{\partial}{\partial z^*}\left(\frac{\partial T}{\partial z}\right)^* \tag{4.E.1-26}$$

Inspection of equation (4.E.1-26) indicates that conduction in the z-direction can be ignored if $H/W \ll 1$, which is the case for a thin wide fin such as that being considered here (step 8). Furthermore, if we invoke the low Biot number approximation developed in Section 4.4 (i.e., $\mathrm{Bi}_t \ll 1$), equation (4.E.1-17) indicates that the temperature gradient in the y-direction will be essentially zero, which implies that the temperature will be constant across the thickness of the fin. Analogously to the procedure used in Section 4.4, let us integrate equation (4.E.1-26) across the cross-sectional area of the fin:

$$\int_{-1}^{1} \frac{\partial}{\partial x^*}\left(\frac{\partial T}{\partial x}\right)^* W\,dy^* + \int_{-1}^{1} \frac{\partial}{\partial y^*}\left(\frac{\partial T}{\partial y}\right)^* W\,dy^* = 0 \tag{4.E.1-27}$$

$$\frac{\partial}{\partial x^*}\left(\frac{\partial T}{\partial x}\right)^* \int_{-1}^{1} dy^* + \int_{T^*}^{-T^*} \partial\left(\frac{\partial T}{\partial y}\right)^* = 0 \tag{4.E.1-28}$$

$$\frac{\partial}{\partial x^*}\left(\frac{\partial T}{\partial x}\right)^* - T^* = 0 \tag{4.E.1-29}$$

Equation (4.E.1-29) is subject to the boundary conditions given by equations (4.E.1-19) and (4.E.1-20), given by

$$T^* = 1 \qquad \text{at} \quad x^* = 0 \tag{4.E.1-30}$$

$$\left(\frac{\partial T}{\partial x}\right)^* = -\frac{H}{L}T^* \cong 0 \qquad \text{at} \quad x^* = 1 \tag{4.E.1-31}$$

The solution to this simplified system of equations is straightforward and given by

$$T^* = \frac{e^{\Pi x^*}}{1 + e^{-2\Pi}} \left(e^{-2\Pi x^*} + e^{-2\Pi} \right), \quad \text{where } \Pi \equiv \sqrt{\text{Bi}_t} \frac{L}{H} \tag{4.E.1-32}$$

The solution given by equation (4.E.1-32) is applicable for small Biot numbers for which the temperature across the thickness of the fin can be assumed to be constant and for which heat transfer along the sides and tip of the fin can be neglected. These are reasonable assumptions for thin fins made from a highly conducting metal.

4.E.2 Unsteady-State Resistance Heating in a Wire

Consider a long solid wire having radius R as shown in Figure 4.E.2-1, whose initial temperature is T_0. At time $t = 0$ an alternating current begins to flow through the wire that causes electrical resistance heating whose volumetric energy generation rate G_e is given by

$$G_e = G_0 \cos \omega t \qquad \text{(energy generation rate per unit volume)} \tag{4.E.2-1}$$

where ω is the angular frequency in radians per second. The temperature at the surface of the wire is held constant at the initial temperature T_0. We use scaling to explore how the describing equations might be simplified.

The appropriately simplified form of the thermal energy equation given by equation (F.2-2) in the Appendices and the required initial and boundary conditions are given by (step 1)

$$\rho C_p \frac{\partial T}{\partial t} = k \frac{1}{r} \frac{d}{dr} \left(r \frac{dT}{dr} \right) + G_0 \cos \omega t \tag{4.E.2-2}$$

$$T = T_0 \quad \text{at} \quad t = 0 \tag{4.E.2-3}$$

$$T \text{ is bounded or } \frac{\partial T}{\partial r} = 0 \quad \text{at} \quad r = 0 \tag{4.E.2-4}$$

$$T = T_0 \quad \text{at} \quad r = R \tag{4.E.2-5}$$

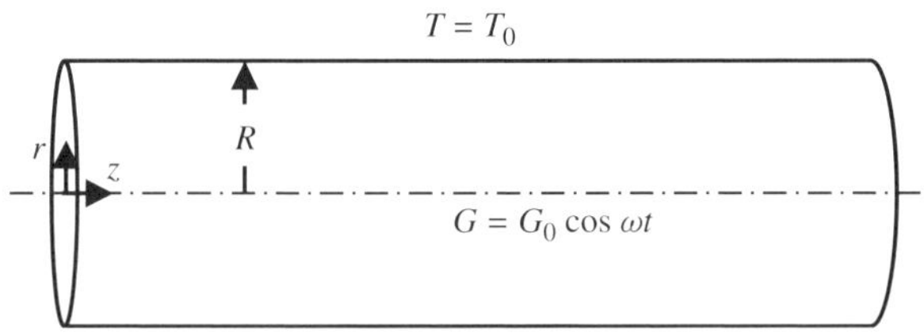

Figure 4.E.2-1 Unsteady-state heat conduction in a solid cylinder of radius R due to electrical heat generation at a volumetric rate given by $G_e = G_0 \cos \omega t$.

The boundary condition given by equation (4.E.2-4) merely states that the temperature cannot be infinite or is an extremum (a maximum in this case) at $r = 0$; this type of boundary condition is often invoked in problems involving cylindrical or spherical coordinates. Introduce the following scale and reference factors (steps 2, 3, and 4):

$$T^* \equiv \frac{T - T_r}{T_s}; \qquad r^* \equiv \frac{r}{r_s}; \qquad t^* \equiv \frac{t}{t_s} \tag{4.E.2-6}$$

Substitute these dimensionless variables into the describing equations and divide through by the dimensional coefficient of one term in each of the equations to obtain (steps 5 and 6)

$$\frac{r_s^2}{\alpha t_s}\frac{\partial T^*}{\partial t^*} = \frac{1}{r^*}\frac{d}{dr^*}\left(r^*\frac{dT^*}{dr^*}\right) + \frac{G_0 r_s^2}{kT_s}\cos\omega t_s t^* \tag{4.E.2-7}$$

$$T^* = \frac{T_0 - T_r}{T_s} \quad \text{at} \quad t^* = 0 \tag{4.E.2-8}$$

$$T^* \text{ is bounded or } \frac{\partial T^*}{\partial r^*} = 0 \quad \text{at} \quad r^* = 0 \tag{4.E.2-9}$$

$$T^* = \frac{T_0 - T_r}{T_s} \quad \text{at} \quad r^* = \frac{R}{r_s} \tag{4.E.2-10}$$

Setting the appropriate dimensionless groups in equation (4.E.2-8) and (4.E.2-10) equal to zero and 1, respectively, gives $T_r = T_0$ and $r_s = R$ (step 7). Since the heat generation must be balanced by the radial heat conduction, we set the dimensionless group in equation (4.E.2-7) equal to 1 to obtain $T_s = G_0 R^2/k$. There are three relevant time scales whose relative values determine the simplifications that are possible for this problem (step 8):

$t_s = t_o$ time scale corresponding to the observation time

$t_s = t_c = \dfrac{R^2}{\alpha}$ time scale characterizing the heat conduction (4.E.2-11)

$t_s = t_p = \dfrac{2\pi}{\omega}$ time scale characterizing the periodic rate of heat generation

Note that we use the reciprocal of the cyclic rather than the angular frequency to properly characterize the time scale for the periodic rate of heat generation.

Consider first the case where $t_s = t_o$, for which equation (4.E.2-7) simplifies to

$$\frac{1}{\text{Fo}_t}\frac{\partial T^*}{\partial t^*} = \frac{1}{r^*}\frac{d}{dr^*}\left(r^*\frac{dT^*}{dr^*}\right) + \cos\omega t_o t^* \tag{4.E.2-12}$$

where $\text{Fo}_t \equiv R^2/\alpha t_o$ is the Fourier number for heat transfer. It is not possible for the Fourier number to be much less than 1 since the conduction term has to be retained at all times. If $\mathrm{Fo}_t \cong 1$, $t_o = R^2/\alpha$ and equation (4.E.2-12) can be written as

$$\frac{\partial T^*}{\partial t^*} = \frac{1}{r^*}\frac{d}{dr^*}\left(r^*\frac{dT^*}{dr^*}\right) + \cos\frac{\omega R^2}{\alpha}t^* \tag{4.E.2-13}$$

Equation (4.E.2-13) can be simplified further if $\omega R^2/\alpha \ll 1$. This implies that the heat conduction occurs much faster than the periodic heat generation; that is, the heat conduction smoothes out the variations in temperature due to the time-dependence of the heat generation, and equation (4.E.2-13) simplifies to

$$\frac{\partial T^*}{\partial t^*} = \frac{1}{r^*}\frac{d}{dr^*}\left(r^*\frac{dT^*}{dr^*}\right) + 1 \tag{4.E.2-14}$$

This equation describes the transient heat conduction period for the special case of the conduction time scale being much smaller than the characteristic time scale for the periodic heat generation.

If $\mathrm{Fo}_t \gg 1$, the transient heat conduction effects have died out; however, the problem is still possibly unsteady state, due to the time-dependent heat-generation term. For this case the characteristic time scale is $t_s = t_p = 2\pi/\omega$ and equation (4.E.2-7) assumes the form

$$\frac{\omega R^2}{2\pi\alpha}\frac{\partial T^*}{\partial t^*} = \frac{1}{r^*}\frac{d}{dr^*}\left(r^*\frac{dT^*}{dr^*}\right) + \cos\frac{\omega R^2}{2\pi\alpha}t^* \tag{4.E.2-15}$$

Equation (4.E.2-15), which describes unsteady-state heat conduction due to time-dependent heat generation, can be simplified further if $\omega R^2/2\pi\alpha \ll 1$, whereby it reduces to

$$0 = \frac{1}{r^*}\frac{d}{dr^*}\left(r^*\frac{dT^*}{dr^*}\right) + 1 \tag{4.E.2-16}$$

which describes steady-state heat conduction due to constant heat generation.

4.E.3 Convective Heat Transfer with Injection Through Permeable Walls

Consider the steady-state fully developed laminar flow of a viscous Newtonian fluid with constant physical properties between two infinitely wide permeable parallel plates separated by a distance H as shown in Figure 4.E.3-1. There is constant injection of the same fluid at velocity V through the permeable plate at $y = H$ and constant withdrawal of the same incompressible fluid through the permeable plate at $y = 0$. The continuity equation can be used to prove that the injection and withdrawal velocities at the upper and lower plates must be equal for the flow to be fully developed. The upstream (entering) temperature of the fluid is T_1. The temperature of the upper plate at $y = H$ is also T_1, whereas the temperature of the lower plate at $y = 0$ is T_0, such that $T_1 > T_0$. Since the flow is assumed to be

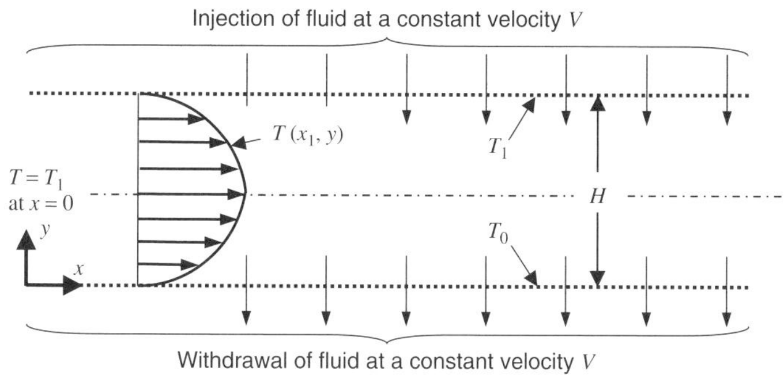

Figure 4.E.3-1 Steady-state fully developed laminar flow of a viscous Newtonian fluid with constant physical properties between two infinitely wide permeable parallel plates separated by a distance H; the same fluid is injected and withdrawn at a constant velocity V at the upper and lower plates, respectively; the fluid enters at T_1; the upper and lower plates are maintained at temperatures T_1 and T_0, respectively, where $T_1 > T_0$.

laminar and fully developed, the velocity profile is given by

$$u_x = U_m\left[1-\left(1-\frac{2y}{H}\right)^2\right] \tag{4.E.3-1}$$

where U_m is the maximum velocity. We use scaling to explore simplifying approximations that might be invoked for this heat-transfer problem.

We begin by appropriately simplifying the energy equation given by equation (F.1-2) in the Appendices (step 1). Note that we must include axial convection, transverse convection, axial conduction, transverse conduction, and viscous dissipation.

$$\rho C_p u_x \frac{\partial T}{\partial x} - \rho C_p V \frac{\partial T}{\partial y} = k\frac{\partial^2 T}{\partial x^2} + k\frac{\partial^2 T}{\partial y^2} + \mu\left(\frac{du_x}{dy}\right)^2 \tag{4.E.3-2}$$

When equation (4.E.3-1) is substituted into the above, we obtain the following set of describing equations:

$$\rho C_p U_m\left[1-\left(1-\frac{2y}{H}\right)^2\right]\frac{\partial T}{\partial x} - \rho C_p V\frac{\partial T}{\partial y} = k\frac{\partial^2 T}{\partial x^2} + k\frac{\partial^2 T}{\partial y^2} + \frac{16\mu U_m^2}{H^2}\left(1-\frac{2y}{H}\right)^2 \tag{4.E.3-3}$$

$$T = T_1 \quad \text{at} \quad x = 0 \tag{4.E.3-4}$$

$$T = f(y) \quad \text{at} \quad x = L \tag{4.E.3-5}$$

$$T = T_1 \quad \text{at} \quad y = H \tag{4.E.3-6}$$

$$T = T_0 \quad \text{at} \quad y = 0 \tag{4.E.3-7}$$

Equation (4.E.3-5) prescribes the downstream boundary condition in terms of the function $f(y)$, which might be unknown. We use o(1) scaling to explore when the axial conduction, viscous heat generation, and axial convection terms can be neglected. Introduce the following dimensionless variables involving unspecified scale and reference factors (2, 3, and 4):

$$T^* \equiv \frac{T - T_r}{T_s}; \qquad x^* \equiv \frac{x}{x_s}; \qquad y^* \equiv \frac{y}{y_s} \tag{4.E.3-8}$$

Substitute these dimensionless variables into the describing equations and divide through by the coefficient of one term in each equation (steps 5 and 6):

$$\frac{\rho C_p U_m y_s^2}{k x_s}\left[1 - \left(1 - 2\frac{y_s}{H}y^*\right)^2\right]\frac{\partial T^*}{\partial x^*} - \frac{\rho C_p V y_s}{k}\frac{\partial T^*}{\partial y^*}$$
$$= \frac{y_s^2}{x_s^2}\frac{\partial^2 T^*}{\partial x^{*2}} + \frac{\partial^2 T^*}{\partial y^{*2}} + \frac{16\mu U_m^2 y_s^2}{k T_s H^2}\left(1 - 2\frac{y_s}{H}y^*\right)^2 \tag{4.E.3-9}$$

$$T^* = \frac{T_1 - T_r}{T_s} \quad \text{at} \quad x^* = 0 \tag{4.E.3-10}$$

$$T^* = f(y^*) \quad \text{at} \quad x^* = \frac{L}{x_s} \tag{4.E.3-11}$$

$$T^* = \frac{T_1 - T_r}{T_s} \quad \text{at} \quad y^* = \frac{H}{y_s} \tag{4.E.3-12}$$

$$T^* = \frac{T_0 - T_r}{T_s} \quad \text{at} \quad y^* = 0 \tag{4.E.3-13}$$

We have divided by the coefficient of the conduction term in the y-direction since this term must be retained to satisfy the boundary conditions at the upper and lower plates.

We can bound the dimensionless temperature to be o(1) by setting the dimensionless group in equation (4.E.3-13) equal to zero, thereby determining the reference temperature, and by setting the dimensionless group in equation (4.E.3-10) equal to 1 to determine the temperature scale (step 7). We can bound the dimensionless axial and transverse coordinates to be o(1) by setting the dimensionless groups in equations (4.E.3-11) and (4.E.3-12) equal to 1. Hence, we obtain the following scale and reference factors:

$$T_r = T_0; \qquad T_s = T_1 - T_0; \qquad x_s = L; \qquad y_s = H \tag{4.E.3-14}$$

Substitution of these scale and reference factors into equations (4.E.3-9) through (4.E.3-13) yields

$$\text{Pe}_t \frac{H}{L}\left[1 - (1 - 2y^*)^2\right]\frac{\partial T^*}{\partial x^*} - \frac{\rho C_p V H}{k}\frac{\partial T^*}{\partial y^*}$$
$$= \frac{H^2}{L^2}\frac{\partial^2 T^*}{\partial x^{*2}} + \frac{\partial^2 T^*}{\partial y^{*2}} + \frac{16\mu U_m^2}{k(T_1 - T_0)}(1 - 2y^*)^2 \tag{4.E.3-15}$$

$$T^* = 1 \quad \text{at} \quad x^* = 0 \tag{4.E.3-16}$$

$$T^* = f(y^*) \quad \text{at} \quad x^* = 1 \tag{4.E.3-17}$$

$$T^* = 1 \quad \text{at} \quad y^* = 1 \tag{4.E.3-18}$$

$$T^* = 0 \quad \text{at} \quad y^* = 0 \tag{4.E.3-19}$$

where $\text{Pe}_t \equiv HU_m/\alpha$ is the Peclet number for heat transfer.

The dimensionless describing equations given by equations (4.E.3-15) through (4.E.3-19) can be simplified if any of the following conditions are satisfied (step 8):

$$\text{Pe}_t \frac{H}{L} \ll 1 \Rightarrow \text{axial heat convection can be ignored} \tag{4.E.3-20}$$

$$\frac{H^2}{L^2} \ll 1 \Rightarrow \text{axial heat conduction can be ignored} \tag{4.E.3-21}$$

$$\frac{16\mu U_m^2}{k(T_1 - T_0)} \ll 1 \Rightarrow \text{viscous heat generation can be ignored} \tag{4.E.3-22}$$

Note in particular that if the condition defined by equation (4.E.3-21) is satisfied, the elliptic thermal energy equation is reduced to a parabolic differential equation, thereby obviating the need to satisfy any downstream boundary condition on the temperature.

Let us now assume that the conditions defined by equations (4.E.3-20) through (4.E.3-22) are satisfied; that is, axial heat conduction and convection as well as viscous dissipation can be ignored. Hence, only the effects of transverse heat conduction and convection remain in our describing equations. If the transverse injection and withdrawal of fluid is sufficiently large, one might anticipate that there will be a region of influence near the lower plate where essentially all the heat transfer is occurring. We use scaling analysis to determine the thickness of this region. In this case the transverse length scale will not be H since the dimensionless temperature will experience a change of $\text{O}(1)$ over a much shorter distance. Hence, we must rescale the describing equations to determine the thickness δ_t of the region of influence or thermal boundary layer near the lower plate. Our simplified dimensional describing equations now are given by

$$-\rho C_p V \frac{\partial T}{\partial y} = k \frac{\partial^2 T}{\partial y^2} \tag{4.E.3-23}$$

$$T = T_1 \quad \text{at} \quad y = H \tag{4.E.3-24}$$

$$T = T_0 \quad \text{at} \quad y = 0 \tag{4.E.3-25}$$

If we again introduce the dimensionless variables defined by equations (4.E.3-8), we obtain the following set of dimensionless describing equations:

$$-\frac{\rho C_p V y_s}{k}\frac{\partial T^*}{\partial y^*} = \frac{\partial^2 T^*}{\partial y^{*2}} \tag{4.E.3-26}$$

$$T^* = \frac{T_1 - T_r}{T_s} \quad \text{at} \quad y^* = \frac{H}{y_s} \tag{4.E.3-27}$$

$$T^* = \frac{T_0 - T_r}{T_s} \quad \text{at} \quad y^* = 0 \tag{4.E.3-28}$$

Our reference and scale factors for the temperature will be the same as before. Since the two terms in equation (4.E.3-26) must be o(1) to balance each other, the dimensionless group multiplying the first term must be equal to 1; this then determines the transverse length scale as follows:

$$\frac{\rho C_p V y_s}{k} = 1 \Rightarrow y_s \equiv \delta_t = \frac{k}{\rho C_p V} = \frac{\alpha}{V} \tag{4.E.3-29}$$

where we have identified the transverse length scale with the thickness of the region of influence δ_t. We see that δ_t decreases with increasing injection velocity and decreasing thermal diffusivity α.

If the injection velocity is sufficiently large, δ_t will be very small so that $H/\delta_t \to \infty$. Hence, our scaled dimensionless describing equations reduce to

$$-\frac{\partial T^*}{\partial y^*} = \frac{\partial^2 T^*}{\partial y^{*2}} \tag{4.E.3-30}$$

$$T^* = 1 \quad \text{at} \quad y^* \to \infty \tag{4.E.3-31}$$

$$T^* = 0 \quad \text{at} \quad y^* = 0 \tag{4.E.3-32}$$

These equations admit an analytical solution given by

$$T^* = 1 - e^{-y^*} \tag{4.E.3-33}$$

Note that this solution indicates that T^* is bounded of o(1), as it should be if our scaling analysis was carried out properly.

4.E.4 Steady-State Heat Transfer to Falling Film Flow

Consider the fully developed laminar flow of a Newtonian liquid film of thickness H and constant physical properties down a solid vertical wall maintained at a temperature T_1, as shown in Figure 4.E.4-1. The temperature of the liquid at $x = 0$ is T_0. Assume that negligible heat is transferred to the adjacent gas phase along the length of the liquid film and that viscous heat generation can be ignored. We seek

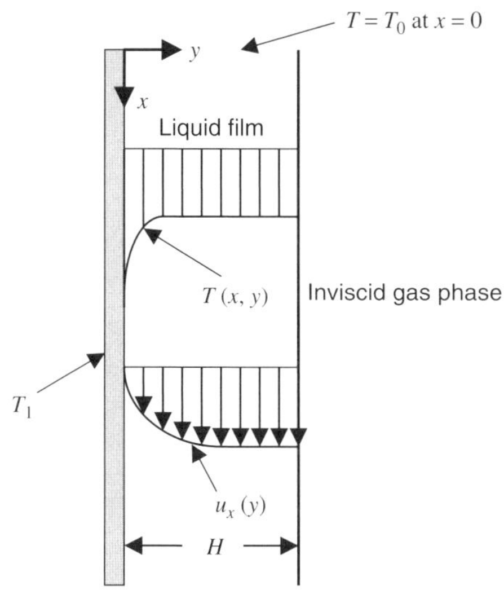

Figure 4.E.4-1 Steady-state heat transfer from a vertical plate at temperature T_1 to a liquid film in fully developed laminar flow of initial temperature T_0, thickness H, and constant physical properties.

to determine how the describing equations can be simplified. The velocity profile for fully developed laminar film flow is given by

$$u_x = U_m \left[\frac{2y}{H} - \left(\frac{y}{H} \right)^2 \right] \tag{4.E.4-1}$$

where U_m is the maximum fluid velocity, that is, the velocity at the liquid–gas interface.

The thermal energy equation given by the appropriately simplified form of equation (F.1-2) in the Appendices, along with the required boundary conditions, are given by (step 1)

$$\frac{U_m}{\alpha} \left[\frac{2y}{H} - \left(\frac{y}{H} \right)^2 \right] \frac{\partial T}{\partial x} = \frac{\partial^2 T}{\partial x^2} + \frac{\partial^2 T}{\partial y^2} \tag{4.E.4-2}$$

$$T = T_0 \quad \text{at} \quad x = 0, \quad 0 \le y \le H \tag{4.E.4-3}$$

$$T = f(y) \quad \text{at} \quad x = L, \quad 0 \le y \le H \tag{4.E.4-4}$$

$$T = T_1 \quad \text{at} \quad y = 0, \quad 0 \le x \le L \tag{4.E.4-5}$$

$$\frac{\partial T}{\partial y} = 0 \quad \text{at} \quad y = H, \quad 0 \le x \le L \tag{4.E.4-6}$$

where $\alpha = k/\rho C_p$ is the thermal diffusivity. Equation (4.E.4-2) is an elliptic differential equation that requires a downstream boundary condition, which we have indicated formally by equation (4.E.4-4); in practice, the inability to specify this downstream boundary condition precludes solving problems of this type even numerically. Introduce the following scale and reference factors (steps 2, 3, and 4):

$$T^* \equiv \frac{T - T_r}{T_s}; \qquad x^* \equiv \frac{x}{x_s}; \qquad y^* \equiv \frac{y}{y_s} \tag{4.E.4-7}$$

The resulting dimensionless describing equations are given by (steps 5 and 6)

$$\frac{\rho C_p U_m y_s^3}{k H x_s}\left(2y^* - \frac{y_s}{H} y^{*2}\right)\frac{\partial T^*}{\partial x^*} = \frac{y_s^2}{x_s^2}\frac{\partial^2 T^*}{\partial x^{*2}} + \frac{\partial^2 T^*}{\partial y^{*2}} \tag{4.E.4-8}$$

$$T^* = \frac{T_0 - T_r}{T_s} \quad \text{at} \quad x^* = 0, \quad 0 \le y^* \le \frac{H}{y_s} \tag{4.E.4-9}$$

$$T^* = \frac{f(y^*) - T_r}{T_s} \quad \text{at} \quad x^* = \frac{L}{x_s}, \quad 0 \le y^* \le \frac{H}{y_s} \tag{4.E.4-10}$$

$$T^* = \frac{T_1 - T_r}{T_s} \quad \text{at} \quad y^* = 0, \quad 0 \le x^* \le \frac{L}{x_s} \tag{4.E.4-11}$$

$$\frac{\partial T^*}{\partial y^*} = 0 \quad \text{at} \quad y^* = \frac{H}{y_s}, \quad 0 \le x^* \le \frac{L}{x_s} \tag{4.E.4-12}$$

The dimensionless temperature can be bounded to be o(1) by setting the dimensionless groups in equations (4.E.4-9) and (4.E.4-11) equal to zero and 1, respectively, to obtain $T = T_0$ and $T_s = T_1 - T_0$ (step 7). The streamwise spatial coordinate can be bounded to be o(1) by setting the appropriate group appearing in equations (4.E.4-10) through (4.E.4-12) to obtain $x_s = L$. There are two choices for bounding the cross-stream spatial coordinate to be o(1): by setting the appropriate group in any one of equations (4.E.4-9), (4.E.4-10), and (4.E.4-12) equal to 1, or by setting the dimensionless group multiplying the convection term in equation (4.E.4-8) equal to 1. However, the convection term has to balance the principal conduction term in equation (4.E.4-8); this yields the following equation for y_s:

$$\frac{\rho C_p U_m y_s^3}{k H x_s} = \frac{\rho C_p U_m y_s^3}{k H L} \Rightarrow \frac{y_s}{H} \equiv \frac{\delta_t}{H} = \left(\frac{kL}{\rho C_p U_m H^2}\right)^{1/3} = \left(\frac{1}{\text{Pe}_t}\frac{L}{H}\right)^{1/3} \tag{4.E.4-13}$$

We see that $y_s = \delta_t$ is a region of influence near the hot wall whose thickness increases slowly with axial length and is inversely proportional to the thermal Peclet number, $\text{Pe}_t \equiv U_m H/\alpha$. Sufficiently far downstream, δ_t will become equal to the film thickness H.

For large Peclet numbers the ratio δ_t/H will be quite small. This permits significant simplification of the describing equations (step 8). In particular, the aspect ratio δ_t^2/L^2 will be quite small, which permits ignoring the axial conduction term, thereby avoiding the complication of having to specify a downstream boundary condition. In addition, for sufficiently large Peclet numbers, the quadratic term in the equation for the velocity profile can be ignored since it is only the velocity in the vicinity of the solid wall that is important for a thin region of influence. Finally, since $H/\delta_t \gg 1$ for large Peclet numbers, the boundary condition by equation (4.E.4-12) can be applied at $y^* \to \infty$. The resulting simplified dimensionless describing equations are given by

$$2y^* \frac{\partial T^*}{\partial x^*} = \frac{\partial^2 T^*}{\partial y^{*2}} \tag{4.E.4-14}$$

$$T^* = 0 \quad \text{at} \quad x^* = 0, \quad 0 \le y^* \le \infty \tag{4.E.4-15}$$

$$T^* = 1 \quad \text{at} \quad y^* = 0, \quad 0 \le x^* \le 1 \tag{4.E.4-16}$$

$$\frac{\partial T^*}{\partial y^*} = 0 \Rightarrow T^* = 0 \quad \text{as} \quad y^* \to \infty, \quad 0 \le x^* \le 1 \tag{4.E.4-17}$$

This system of equations admits a solution via a similarity solution or combination of variables, as is shown in standard references.[15]

4.E.5 Unsteady-State Heat Transfer from a Sphere at Large Biot Numbers

In Section 4.4 we considered heat transfer from a solid sphere initially at temperature T_0 of radius R falling at its terminal velocity through a liquid at temperature T_∞ for which the convective heat transfer was characterized via a heat-transfer coefficient, as shown in Figure 4.4-1. We considered a scaling appropriate to a small Biot number for which all the resistance to heat transfer was in the external liquid. This implied that the temperature gradient in the sphere was negligible and the temperature was spatially uniform. Here we apply scaling analysis to the complementary case of a large Biot number corresponding to rapid heat transfer in the external liquid. To have continuity of heat flux at the surface of the sphere, the temperature gradient in the sphere will occur over a region of influence or thermal boundary layer whose thickness δ_t is the appropriate radial length scale.

The dimensional describing equations will be the same as those in Section 4.4 (step 1):

$$\frac{\partial T}{\partial t} = \alpha \frac{1}{r^2} \frac{\partial}{\partial r}\left(r^2 \frac{\partial T}{\partial r}\right) \tag{4.E.5-1}$$

$$T = T_0 \quad \text{at} \quad t \le 0, \quad 0 \le r \le R \tag{4.E.5-2}$$

$$\frac{\partial T}{\partial r} = 0 \quad \text{at} \quad r = 0, \quad t > 0 \tag{4.E.5-3}$$

$$k \frac{\partial T}{\partial r} = h(T_\infty - T) \quad \text{at} \quad r = R, \quad t > 0 \tag{4.E.5-4}$$

[15] R. B. Bird, W. E. Stewart, and E. L. Lightfoot, *Transport Phenomena*, Wiley, New York, 1960, pp. 349–350.

It is convenient to reference the coordinate system to the surface of the sphere where the thermal boundary layer is located. Hence, we define the following dimensionless variables (steps 2, 3, and 4):

$$T^* \equiv \frac{T - T_r}{T_s}; \qquad r^* \equiv \frac{R - r}{\delta_t}; \qquad t^* \equiv \frac{t}{t_s} \tag{4.E.5-5}$$

Introduce these dimensionless variables into the describing equations and divide through by the coefficient of one term in each equation that should be retained (steps 5 and 6):

$$\frac{\delta_t^2}{\alpha t_s}\frac{\partial T^*}{\partial t^*} = \frac{1}{[R/\delta_t - r^*]^2}\frac{\partial}{\partial r^*}\left[\left(\frac{R}{\delta_t} - r^*\right)^2 \frac{\partial T^*}{\partial r^*}\right] \tag{4.E.5-6}$$

$$T^* = \frac{T_0 - T_r}{T_s} \quad \text{at} \quad t^* \leq 0, \quad 0 \leq r^* \leq \frac{R}{\delta_t} \tag{4.E.5-7}$$

$$\frac{\partial T^*}{\partial r^*} = 0 \quad \text{at} \quad r^* = \frac{R}{\delta_t}, \quad t^* > 0 \tag{4.E.5-8}$$

$$\frac{\partial T^*}{\partial r^*} = -\frac{h\delta_t}{k}\left(\frac{T_\infty - T_r}{T_s} - T^*\right) \quad \text{at} \quad r^* = 0, \quad t^* > 0 \tag{4.E.5-9}$$

When set equal to zero and 1, respectively, the dimensionless groups in equations (4.E.5-7) and (4.E.5-9), indicate that $T_r = T_0$ and $T_s = T_\infty - T_0$ (step 7). Since this is inherently unsteady state, the appropriate time scale is the observation time t_o. The fact that the unsteady-state and radial heat conduction terms in equation (4.E.5-6) must be of the same order provides an estimate for δ_t:

$$\frac{\delta_t^2}{\alpha t_s} = 1 \Rightarrow \delta_t = \sqrt{\alpha t_o} = R\sqrt{\text{Fo}_t} \tag{4.E.5-10}$$

where $\text{Fo}_t \equiv \alpha t_o / R^2$ is the thermal Fourier number. Note that the thermal boundary layer penetrates progressively farther into the sphere in time. Eventually, it will penetrate to the center of the sphere when $\text{Fo}_t \cong 1$, corresponding to an observation time $t_o \cong R^2/\alpha$.

When these scale and reference factors are substituted into equations (4.E.5-6) through (4.E.5-9), we obtain the following dimensionless describing equations:

$$\frac{\partial T^*}{\partial t^*} = \frac{1}{\left[\left(\frac{1}{\sqrt{\text{Fo}_t}}\right) - r^*\right]^2}\frac{\partial}{\partial r^*}\left[\left(\frac{1}{\sqrt{\text{Fo}_t}} - r^*\right)^2 \frac{\partial T^*}{\partial r^*}\right] \tag{4.E.5-11}$$

$$T^* = 0 \quad \text{at} \quad t^* \leq 0, \quad 0 \leq r^* \leq \frac{1}{\sqrt{\text{Fo}_t}} \tag{4.E.5-12}$$

$$\frac{\partial T^*}{\partial r^*} = 0 \qquad \text{at} \quad r^* = \frac{1}{\sqrt{\text{Fo}_t}}, \quad t^* > 0 \tag{4.E.5-13}$$

$$\frac{\partial T^*}{\partial r^*} = -\text{Bi}_t\sqrt{\text{Fo}_t}(1 - T^*) \qquad \text{at} \quad r^* = 0, \quad t^* > 0 \tag{4.E.5-14}$$

where $\text{Bi}_t \equiv hR/k$ is the thermal Biot number.

Now let us consider how this system of describing equations can be simplified (step 8). We note that the curvature effects can be ignored when $\sqrt{\text{Fo}_t} \ll 1$, corresponding to short contact times for which the thermal boundary layer will be thin in comparison to the radius of the sphere. If $\sqrt{\text{Fo}_t} \ll 1$, the boundary condition given by equation (4.E.5-13) can be applied at infinity. A zero conductive heat flux far from the surface of the sphere implies no change in the temperature outside the thermal boundary layer. Hence, equation (4.E.5-13) can be replaced by the condition that $T^* = 0$ as $r^* \to \infty$. Equation (4.E.5-14) indicates that as $\text{Bi}_t \to \infty$, $T^* \to 1$ to ensure that $\partial T^*/\partial r^*$ remains bounded of $\text{o}(1)$. This implies that the surface temperature of the sphere is at T_∞ and that there is essentially no temperature gradient in the liquid. Hence, for $\sqrt{\text{Fo}_t} \ll 1$ and large Biot numbers, the describing equations simplify to

$$\frac{\partial T^*}{\partial t^*} = \frac{\partial^2 T^*}{\partial r^{*2}} \tag{4.E.5-15}$$

$$T^* = 0 \qquad \text{at} \quad t^* \leq 0, \quad 0 \leq r^* < \infty \tag{4.E.5-16}$$

$$T^* = 0 \qquad \text{at} \quad r^* \to \infty, \quad t^* > 0 \tag{4.E.5-17}$$

$$T^* = 1 \qquad \text{at} \quad r^* = 0, \quad t^* > 0 \tag{4.E.5-18}$$

This simplified set of describing equations can be solved via standard methods such as combination of variables.

4.E.6 Evaporative Cooling of a Liquid Film

An infinitely wide film of an incompressible volatile pure (i.e., single component) liquid has an initial thickness of L_0 and is resting on a solid boundary that is maintained at a constant temperature T_0. Initially, the entire liquid film is at this temperature. At time $t = 0$, evaporation from the film begins that causes the film thickness to decrease, thus implying that this is a moving boundary problem. The surrounding gas phase is assumed to transfer negligible heat to the liquid film. Hence, the latent heat of vaporization (evaporation) must be supplied entirely by heat conduction from the liquid film. This, of course, causes heat transfer within the liquid film. The evaporative mass flux at the free surface n_G is given by

$$n_G = k_G^{\bullet} p_L^{\circ} \qquad \text{(units of mass/area} \cdot \text{time)} \tag{4.E.6-1}$$

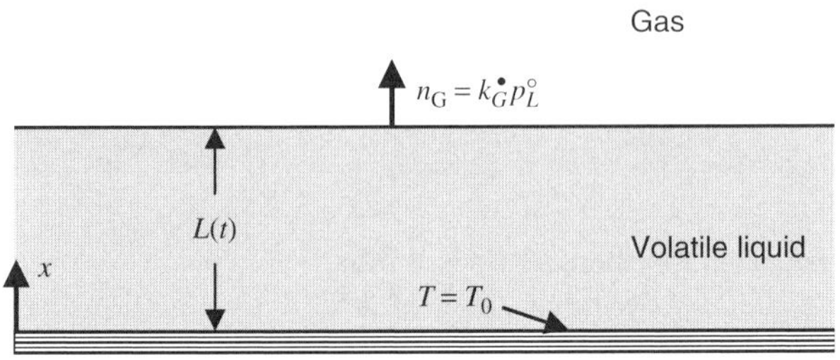

Figure 4.E.6-1 Evaporative cooling of an infinitely wide planar film of an incompressible volatile liquid of vapor pressure p_L° and initial temperature T_0, into an adjacent insoluble gas phase whose mass-transfer coefficient is $k_G^\bullet$; the gas phase is assumed to transfer negligible heat to the liquid.

in which $k_G^\bullet$ is the constant mass-transfer coefficient in the ambient gas phase and p_L° is the temperature-dependent vapor pressure of the liquid at the instantaneous temperature of the liquid–gas interface. Note that equation (4.E.6-1) assumes that the bulk of the ambient gas phase does not contain any of the evaporating component. For moderate departures of the temperature from T_0, the temperature dependence of the vapor pressure can be approximated by

$$p_L^\circ = p_{L0}^\circ + \beta(T - T_0) \tag{4.E.6-2}$$

where p_{L0}° is the vapor pressure at T_0 and β is a constant. We seek to develop a model for this evaporative cooling, film-thinning process, whose essential features are shown in Figure 4.E.6-1. In particular, we explore conditions for which the describing equations can be simplified.

The appropriately simplified form of equation (F.1-2) in the Appendices along with the initial condition and boundary condition at $x = 0$ are given by (step 1)

$$\rho_L C_{pL} \frac{\partial T}{\partial t} = k_L \frac{\partial^2 T}{\partial x^2} \tag{4.E.6-3}$$

$$T = T_0 \quad \text{at} \quad t = 0 \tag{4.E.6-4}$$

$$T = T_0 \quad \text{at} \quad x = 0 \tag{4.E.6-5}$$

where k_L, ρ_L, and C_{pL} are the thermal conductivity, density, and heat capacity of the liquid, respectively. The boundary condition at the moving upper interface is obtained from an integral energy balance as follows:

$$\frac{d}{dt}\int_0^L \rho_L C_{pL}(T - T^\circ)\,dx + \frac{d}{dt}\int_L^\infty \rho_G C_{pG}(T - T^\circ)\,dx = q_0 \tag{4.E.6-6}$$

where T° is an arbitrary reference temperature for the enthalpy or heat content and q_0 is the heat transferred into the liquid film at $x = 0$. Applying Leibnitz's rule for differentiating an integral given by equation (H.1-2) in the Appendices and substituting equation (4.E.6-3) yields

$$\rho_L C_{pL}(T_L - T^\circ)\frac{dL}{dt} + \int_0^L \rho_L C_{pL}\frac{\partial T}{\partial t}\,dx - \rho_G C_{pG}(T_L - T^\circ)\frac{dL}{dt} + \int_L^\infty \rho_G C_{pG}\frac{\partial T}{\partial t}\,dx = q_0 \tag{4.E.6-7}$$

$$(\rho_L C_{pL} - \rho_G C_{pG})(T_L - T^\circ)\frac{dL}{dt} + \int_0^L k_L \frac{\partial^2 T}{\partial x^2}\,dx + \int_L^\infty k_G \frac{\partial^2 T}{\partial x^2}\,dx = q_0 \tag{4.E.6-8}$$

$$(\rho_L C_{pL} - \rho_G C_{pG})(T_L - T^\circ)\frac{dL}{dt} + k_L \left.\frac{\partial T}{\partial x}\right|_{x=L} - k_L \left.\frac{\partial T}{\partial x}\right|_{x=0} + k_G \left.\frac{\partial T}{\partial x}\right|_{x=\infty} - k_G \left.\frac{\partial T}{\partial x}\right|_{x=L} = q_0 \tag{4.E.6-9}$$

The first term in equation (4.E.6-9) is the difference in heat content between the gas and the liquid, which is proportional to ΔH_v, the latent heat of vaporization (energy/mass). The fifth term in this equation is assumed to be zero, whereas the third term is equal to the last term. Hence, our boundary condition at the moving free interface simplifies to

$$k_L \frac{\partial T}{\partial x} = \Delta H_v \rho_L \frac{dL}{dt} \quad \text{at} \quad x = L \tag{4.E.6-10}$$

Equation (4.E.6-10) merely states that the heat conducted to the moving interface is equal to that required to vaporize the liquid.

An auxiliary condition is still needed to determine the location of the moving interface. Since mass is lost at this moving boundary, this auxiliary condition is obtained from an integral mass balance given by

$$\frac{d}{dt}\int_0^L \rho_L\,dx + \frac{d}{dt}\int_L^\infty \rho_G\,dx = 0 \tag{4.E.6-11}$$

Applying Leibnitz's rule for differentiating an integral given by equation (H.1-2) in the Appendices and substituting the one-dimensional form of the continuity equation given by equation (C.1-1) in the Appendices yields

$$\rho_L \frac{dL}{dt} - \int_0^L \frac{\partial \rho_L u_x}{\partial x}\,dx - \rho_G \frac{dL}{dt} - \int_L^\infty \frac{\partial \rho_G u_x}{\partial x}\,dx = 0 \tag{4.E.6-12}$$

$$(\rho_L - \rho_G)\frac{dL}{dt} - \rho_L u_x|_{x=L} + \rho_L u_x|_{x=0} - \rho_G u_x|_{x=\infty} + \rho_G u_x|_{x=L} = 0 \tag{4.E.6-13}$$

where u_x denotes the velocity in the x-direction. Due to the incompressibility of the liquid, the second and third terms in equation (4.E.6-13) cancel. The fourth term is identically zero, and the last term is equal to the evaporative mass-transfer flux $n_G = k_G^{\bullet}[p_{L0}^{\circ} + \beta(T - T_0)]$. Hence, the auxiliary condition for determining the location of the moving boundary simplifies to

$$(\rho_L - \rho_G)\frac{dL}{dt} \cong \rho_L \frac{dL}{dt} = n_G = k_G^{\bullet}\left[p_{L0}^{\circ} + \beta(T - T_0)\right] \qquad \text{at} \quad x = L \tag{4.E.6-14}$$

Equation (4.E.6-14) merely states that the rate of mass loss in the liquid film is equal to the rate of evaporation at the free surface. This equation requires an initial condition given by

$$L = L_0 \qquad \text{at} \quad t = 0 \tag{4.E.6-15}$$

We anticipate that for very short times, the heat transfer will be confined to a thin region of influence near the moving interface. Hence, to explore the full spectrum of possible simplifications of the describing equations, it is convenient to carry out a coordinate transformation whereby we locate the origin of our spatial coordinate at the liquid–gas interface. Define a new spatial coordinate as follows:

$$\tilde{x} \equiv L(t) - x \tag{4.E.6-16}$$

The describing equations in this new coordinate system are given by

$$\frac{\partial T}{\partial t} + \frac{dL}{dt}\frac{\partial T}{\partial \tilde{x}} = \alpha_L \frac{\partial^2 T}{\partial \tilde{x}^2} \tag{4.E.6-17}$$

$$T = T_0 \qquad \text{at} \quad t = 0 \tag{4.E.6-18}$$

$$T = T_0 \qquad \text{at} \quad \tilde{x} = L \tag{4.E.6-19}$$

$$-k_L \frac{\partial T}{\partial \tilde{x}} = \Delta H_v \, \rho_L \frac{dL}{dt} \qquad \text{at} \quad \tilde{x} = 0 \tag{4.E.6-20}$$

$$\rho_L \frac{dL}{dt} = k_G^{\bullet}[p_{L0}^{\circ} - \beta(T - T_0)] \qquad \text{at} \quad \tilde{x} = 0 \tag{4.E.6-21}$$

$$L = L_0 \qquad \text{at} \quad t = 0 \tag{4.E.6-22}$$

Note that the coordinate transformation introduces a pseudo-convection term in the energy equation.

We first seek to determine the criteria for ignoring the temperature dependence of the vapor pressure and assuming that this evaporative cooling process is at quasi-steady-state. Hence, introduce the following scale and reference factors (steps 2, 3, and 4):

$$T^* = \frac{T - T_r}{T_s}; \qquad \tilde{x}^* \equiv \frac{\tilde{x}}{\tilde{x}_s}; \qquad t^* \equiv \frac{t}{t_s}; \qquad L^* \equiv \frac{L}{L_s}; \qquad \left(\frac{dL}{dt}\right)^* \equiv \frac{1}{\dot{L}_s}\frac{dL}{dt} \tag{4.E.6-23}$$

Note that we have given dL/dt its own scale since there is no reason for it to scale as $\tilde{x}_s/t_s$. Substitute these dimensionless variables into equations (4.E.6-17) through (4.E.6-22) to obtain the following set of dimensionless describing equations (steps 5 and 6):

$$\frac{\tilde{x}_s^2}{\alpha_L t_s}\frac{\partial T^*}{\partial t^*} + \frac{\dot{L}_s \tilde{x}_s}{\alpha_L}\left(\frac{dL}{dt}\right)^* \frac{\partial T^*}{\partial \tilde{x}^*} = \frac{\partial^2 T^*}{\partial \tilde{x}^{*2}} \tag{4.E.6-24}$$

$$T^* = \frac{T_0 - T_r}{T_s} \quad \text{at} \quad t^* = 0 \tag{4.E.6-25}$$

$$T^* = \frac{T_0 - T_r}{T_s} \quad \text{at} \quad \tilde{x}^* = \frac{L}{\tilde{x}_s} \tag{4.E.6-26}$$

$$-\frac{\partial T^*}{\partial \tilde{x}^*} = \frac{\Delta H_v \rho_L \tilde{x}_s \dot{L}_s}{k_L T_s}\left(\frac{dL}{dt}\right)^* \quad \text{at} \quad \tilde{x}^* = 0 \tag{4.E.6-27}$$

$$\left(\frac{dL}{dt}\right)^* = \frac{k_G^{\bullet}\left[p_{L0}^{\circ} - \beta\left(T_s T^* + T_r - T_0\right)\right]}{\rho_L \dot{L}_s} \quad \text{at} \quad \tilde{x}^* = 0 \tag{4.E.6-28}$$

$$L^* = \frac{L_0}{L_s} \quad \text{at} \quad t^* = 0 \tag{4.E.6-29}$$

The dimensionless temperature can be referenced to zero by setting the group in either equation (4.E.6-25) or (4.E.6-26) equal to zero. For sufficiently long times such that the heat conduction penetrates through nearly the entire liquid film, the dimensionless spatial coordinate can be bounded between zero and 1 by setting the appropriate group in equation (4.E.6-26) equal to 1. The instantaneous liquid film thickness can bounded between zero and 1 by setting the dimensionless group in equation (4.E.6-29) equal to 1. Since this is an unsteady-state problem, the time scale is the observation time t_o. We can ensure that the dimensionless temperature derivative is o(1) by setting the dimensionless group in equation (4.E.6-27) equal to 1; this also determines the temperature scale factor. We can ensure that the dimensionless velocity of the free surface is o(1) by setting the dimensionless group corresponding to the leading-order term in equation (4.E.6-28) equal to 1; this also determines the scale for the free surface velocity. This results in the following dimensionless variables:

$$T^* = \frac{k_L(T - T_0)}{\Delta H_v k_G^{\bullet} p_{L0}^{\circ} L}; \qquad \tilde{x}^* = \frac{\tilde{x}}{L}; \qquad t^* = \frac{t}{t_o}; \qquad L^* = \frac{L}{L_0};$$

$$\left(\frac{dL}{dt}\right)^* \equiv \frac{\rho_L}{k_G^{\bullet} p_{L0}^{\circ}}\frac{dL}{dt} \tag{4.E.6-30}$$

These variables result in the following set of dimensionless describing equations:

$$\frac{1}{\mathrm{Fo}_t}\frac{\partial T^*}{\partial t^*} + \frac{k_G^{\bullet} p_{L0}^{\circ} L}{\alpha_L \rho_L}\left(\frac{dL}{dt}\right)^* \frac{\partial T^*}{\partial \tilde{x}^*} = \frac{\partial^2 T^*}{\partial \tilde{x}^{*2}} \tag{4.E.6-31}$$

$$T^* = 0 \quad \text{at} \quad t^* = 0 \tag{4.E.6-32}$$

$$T^* = 0 \quad \text{at} \quad \tilde{x}^* = 1 \tag{4.E.6-33}$$

$$-\frac{\partial T^*}{\partial \tilde{x}^*} = \left(\frac{dL}{dt}\right)^* \quad \text{at} \quad \tilde{x}^* = 0 \tag{4.E.6-34}$$

$$\left(\frac{dL}{dt}\right)^* = 1 - \frac{k_G^{\bullet} \beta \Delta H_v L}{k_L} T^* \quad \text{at} \quad \tilde{x}^* = 0 \tag{4.E.6-35}$$

$$L^* = 1 \quad \text{at} \quad t^* = 0 \tag{4.E.6-36}$$

where $\mathrm{Fo}_t \equiv \alpha_L t_o / L^2$ is the Fourier number for heat transfer.

Equation (4.E.6-35) indicates that the temperature dependence of the vapor pressure can be ignored if the following criterion is satisfied (step 8):

$$\frac{k_G^{\bullet} \beta\, \Delta H_v\, L}{k_L} \ll 1 \tag{4.E.6-37}$$

Note that any factors that reduce the evaporative cooling favor satisfy this criterion; these include a reduced gas-phase mass-transfer coefficient, a smaller heat of vaporization, and a higher thermal conductivity. Note also that this criterion becomes progressively easier to satisfy as time increases, owing to the presence of L in the numerator. Equation (4.E.6-31) indicates that quasi-steady-state will prevail if the following criterion is satisfied:

$$\frac{1}{\mathrm{Fo}_t} = \frac{L^2}{\alpha_L t_o} \ll 1 \tag{4.E.6-38}$$

If quasi-steady-state applies, equations (4.E.6-31) through (4.E.6-35) can be solved analytically for both constant as well as variable vapor pressure. However, the initial condition given by equation (4.E.6-36) cannot be applied, due to the long-time constraint implied by equation (4.E.6-38).

The quasi-steady-state approximation applies at very long observation times. It would be of value to explore whether a complementary short-time approximation can be developed as well. We anticipate that a short-time solution would apply when the heat transfer is confined to a thin region of influence δ_t near the upper moving interface. The dimensionless temperature goes through a change of O(1) over the thickness δ_t; therefore, to bound our spatial coordinate to be o(1), we choose our length scale factor $\tilde{x}_s = \delta_t$. Hence, our dimensionless describing equations assume the following form:

$$\frac{\delta_t^2}{\alpha_L t_o}\frac{\partial T^*}{\partial t^*} + \frac{k_G^\bullet p_{L0}^\circ \delta_t}{\alpha_L \rho_L}\left(\frac{dL}{dt}\right)^*\frac{\partial T^*}{\partial \tilde{x}^*} = \frac{\partial^2 T^*}{\partial \tilde{x}^{*2}} \tag{4.E.6-39}$$

$$T^* = 0 \quad \text{at} \quad t^* = 0 \tag{4.E.6-40}$$

$$T^* = 0 \quad \text{at} \quad \tilde{x}^* = \frac{L}{\delta_t} \tag{4.E.6-41}$$

$$-\frac{\partial T^*}{\partial \tilde{x}^*} = \left(\frac{dL}{dt}\right)^* \quad \text{at} \quad \tilde{x}^* = 0 \tag{4.E.6-42}$$

$$\left(\frac{dL}{dt}\right)^* = 1 - \frac{k_G^\bullet \beta\,\Delta H_v\,\delta_t}{k_L}T^* \quad \text{at} \quad \tilde{x}^* = 0 \tag{4.E.6-43}$$

$$L^* = 1 \quad \text{at} \quad t^* = 0 \tag{4.E.6-44}$$

Since we are considering simplifications appropriate to very short observation times, this is inherently an unsteady-state problem. This implies that the first and third terms in equation (4.E.6-39) must be retained; this provides a measure of the thickness of the region of influence given by

$$\frac{\delta_t^2}{\alpha_L t_o} = 1 \Rightarrow \delta_t = \sqrt{\alpha_L t_o} \tag{4.E.6-45}$$

Since for very short times $\delta_t \ll L$, the boundary condition given by equation (4.E.6-41) can be applied at infinity. This permits obtaining an analytical solution using the method of combination of variables if the temperature dependence of the vapor pressure can be ignored. For this short observation time scaling, the criterion for ignoring the temperature dependence of the vapor pressure is given by

$$\frac{k_G^\bullet \beta\,\Delta H_v\,\delta_t}{k_L} \ll 1 \tag{4.E.6-46}$$

Since for short times $\delta_t \ll L$, the criterion given by equation (4.E.6-46) is much easier to satisfy than that given by equation (4.E.6-37), which is applicable at longer times. Note also that the pseudo-convection term in equation (4.E.6-39) can be dropped if the following criterion is satisfied:

$$\frac{k_G^\bullet p_L^\circ \delta_t}{\alpha_L \rho_L} = \frac{k_G^\bullet p_{L0}^\circ}{\rho_L}\sqrt{\frac{t_o}{\alpha_L}} \ll 1 \tag{4.E.6-47}$$

This example illustrates one of the principal values of scaling in arriving at appropriately simplified forms of the describing equations that admit either analytical or much simpler numerical solutions. In this case we were able to simplify the describing equations so that analytical solutions could be obtained for both very short and very long observation times. These analytical solutions could be used to check the validity of a numerical solution to the complete describing equations.

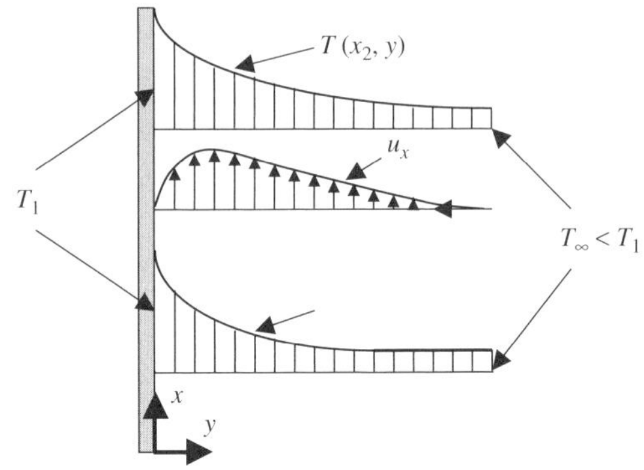

Figure 4.E.7-1 Buoyancy-induced free-convection flow next to a vertical heated plate of temperature T_1 immersed in an infinite fluid whose temperature far from the plate is T_∞; sketch shows the temperature profiles at two positions along the plate, x_1 and x_2, where $x_2 > x_1$, and the velocity profile of the component u_x parallel to the plate.

4.E.7 Free-Convection Heat Transfer Adjacent to a Vertical Heated Flat Plate

Consider a vertical flat plate of length L and temperature T_1 that is immersed in an initially quiescent fluid of temperature $T_\infty < T_1$ that can be assumed to have infinite extent in all directions, as shown in Figure 4.E.7-1. The fluid next to the heated plate will become less dense than the fluid farther removed from it. Hence, a hydrostatic pressure imbalance will occur that causes fluid near the plate to rise; in contrast to the free-convection problem considered in Section 4.9, no fluid will descend in this case, due to the assumption of infinite extent. We consider this convective flow after the transients have died out when steady-state free convection prevails. We ignore end effects at the top and bottom ends of the plate and viscous dissipation and assume constant physical properties other than the density in the gravitational body-force term in the equations of motion.

Note that this is inherently a developing flow, due to the progressive heating that occurs as the fluid moves up the plate; therefore, velocity components in both the x- and y-directions must be considered. Hence, equations (C.1-1), (D.1-10), (D.1-11), and (F.1-2) in the Appendices simplify to (step 1)

$$\rho u_x \frac{\partial u_x}{\partial x} + \rho u_y \frac{\partial u_x}{\partial y} = -\frac{\partial P}{\partial x} + \mu \left(\frac{\partial^2 u_x}{\partial x^2} + \frac{\partial^2 u_x}{\partial y^2} \right) - \rho g \tag{4.E.7-1}$$

$$\rho u_x \frac{\partial u_y}{\partial x} + \rho u_y \frac{\partial u_y}{\partial y} = -\frac{\partial P}{\partial y} + \mu \left(\frac{\partial^2 u_y}{\partial x^2} + \frac{\partial^2 u_y}{\partial y^2} \right) \tag{4.E.7-2}$$

$$\frac{\partial u_x}{\partial x} + \frac{\partial u_y}{\partial y} = 0 \tag{4.E.7-3}$$

$$\rho C_p u_x \frac{\partial T}{\partial x} + \rho C_p u_y \frac{\partial T}{\partial y} = k \left(\frac{\partial^2 T}{\partial x^2} + \frac{\partial^2 T}{\partial y^2} \right) \tag{4.E.7-4}$$

The corresponding boundary conditions are given by

$$u_x = 0, \quad u_y = 0, \quad T = T_\infty \quad \text{at} \quad x = 0 \tag{4.E.7-5}$$

$$u_x = f_1(y), \quad u_y = f_2(y), \quad T = f_3(y) \quad \text{at} \quad x = L \tag{4.E.7-6}$$

$$u_x = 0, \quad u_y = 0, \quad T = T_1 \quad \text{at} \quad y = 0 \tag{4.E.7-7}$$

$$u_x = 0, \quad u_y = 0, \quad T = T_\infty \quad \text{at} \quad y \to \infty \tag{4.E.7-8}$$

where $f_1(y)$, $f_2(y)$, and $f_3(y)$ are functions of y that often are unknown. Since the density is temperature-dependent, we need an appropriate equation of state. Here we consider small density variations and hence represent the density by means of a Taylor series expansion about the density ρ_∞ at the cold temperature T_∞ given by

$$\rho = \rho|_{T_\infty} + \left.\frac{\partial \rho}{\partial T}\right|_{T_\infty} (T - T_\infty) = \rho_\infty - \rho_\infty \beta_t (T - T_\infty) \tag{4.E.7-9}$$

where β_t is the coefficient of thermal volume expansion. In addition, it is convenient to split the pressure into dynamic, P_d, and hydrostatic, P_h, contributions as follows[16]:

$$P = P_d(x, y) + P_h(x) \tag{4.E.7-10}$$

When equations (4.E.7-9) and (4.E.7-10) are substituted into equation (4.E.7-1), we obtain

$$\rho_\infty u_x \frac{\partial u_x}{\partial x} + \rho_\infty u_y \frac{\partial u_x}{\partial y} = -\frac{\partial P_d}{\partial x} + \mu \left(\frac{\partial^2 u_x}{\partial x^2} + \frac{\partial^2 u_x}{\partial y^2} \right) + \rho_\infty \beta_t g (T - T_\infty) \tag{4.E.7-11}$$

Note that the $\rho_\infty g$ term does not appear in equation (4.E.7.11) since it cancels with the derivative of the purely hydrostatic contribution to the pressure. Note that higher-order effects of the temperature on the density are ignored in the convection terms; that is, the density appearing in the convection terms is evaluated at T_∞ and hence is denoted by ρ_∞.

Define the following dimensionless dependent and independent variables (steps 2, 3, and 4):

$$u_x^* \equiv \frac{u_x}{u_{xs}}; \quad u_y^* \equiv \frac{u_y}{u_{ys}}; \quad P^* \equiv \frac{P_d}{P_s}; \quad T^* \equiv \frac{T - T_r}{T_s};$$

$$x^* \equiv \frac{x}{x_s}; \quad y_m^* \equiv \frac{y}{y_{ms}}; \quad y_t^* \equiv \frac{y}{y_{ts}} \tag{4.E.7-12}$$

[16]Splitting the pressure into its dynamic and hydrostatic contributions is standard practice when the latter does not cause the flow; this permits eliminating the contribution of the gravitational body force from the describing equations as seen in the present example.

Note that we have allowed for different y-length scales for the energy equation and equations of motion since the temperature might experience a characteristic change of O(1) over a different length scale than the velocities. Introduce these dimensionless variables into the describing equations and divide each equation by the dimensional coefficient of one term that should be retained to maintain physical significance (steps 5 and 6):

$$\frac{u_{xs}y_{ms}^2}{\nu_\infty x_s}u_x^*\frac{\partial u_x^*}{\partial x^*}+\frac{u_{ys}y_{ms}}{\nu_\infty}u_y^*\frac{\partial u_x^*}{\partial y_m^*}=-\frac{P_s y_{ms}^2}{\mu u_{xs}x_s}\frac{\partial P^*}{\partial x^*}$$
$$+\frac{y_{ms}^2}{x_s^2}\frac{\partial^2 u_x^*}{\partial x^{*2}}+\frac{\partial^2 u_x^*}{\partial y_m^{*2}}+\frac{g\beta_t T_s y_{ms}^2}{\nu_\infty u_{xs}}\left(T^*+\frac{T_r-T_\infty}{T_s}\right) \tag{4.E.7-13}$$

$$\frac{u_{xs}y_{ms}^2}{\nu_\infty x_s}u_x^*\frac{\partial u_y^*}{\partial x^*}+\frac{u_{ys}y_{ms}}{\nu_\infty}u_y^*\frac{\partial u_y^*}{\partial y_m^*}=-\frac{P_s y_{ms}}{\mu u_{ys}}\frac{\partial P^*}{\partial y^*}+\frac{y_{ms}^2}{x_s^2}\frac{\partial^2 u_y^*}{\partial x^{*2}}+\frac{\partial^2 u_y^*}{\partial y_m^{*2}} \tag{4.E.7-14}$$

$$\frac{\partial u_x^*}{\partial x^*}+\frac{u_{ys}x_s}{u_{xs}y_{ms}}\frac{\partial u_y^*}{\partial y_m^*}=0 \tag{4.E.7-15}$$

$$\frac{u_{xs}y_{ts}}{u_{ys}x_s}u_x^*\frac{\partial T^*}{\partial x^*}+u_y^*\frac{\partial T^*}{\partial y_t^*}=\frac{\alpha_\infty y_{ts}}{u_{ys}x_s^2}\frac{\partial^2 T^*}{\partial x^{*2}}+\frac{\alpha_\infty}{u_{ys}y_{ts}}\frac{\partial^2 T^*}{\partial y_t^{*2}} \tag{4.E.7-16}$$

$$u_x^*=0,\quad u_y^*=0,\quad T^*=\frac{T_\infty-T_r}{T_s}\quad \text{at}\quad x^*=0 \tag{4.E.7-17}$$

$$u_x^*=f_1(y_m^*),\quad u_y^*=f_2(y_m^*),\quad T^*=f_3(y_t^*)\quad \text{at}\quad x^*=\frac{L}{x_s} \tag{4.E.7-18}$$

$$u_x^*=0,\quad u_y^*=0,\quad T^*=\frac{T_1-T_r}{T_s}\quad \text{at}\quad y_m^*=y_t^*=0 \tag{4.E.7-19}$$

$$u_x^*=0,\quad u_y^*=0,\quad T^*=\frac{T_\infty-T_r}{T_s}\quad \text{at}\quad y_m^*=y_t^*\to\infty \tag{4.E.7-20}$$

where $\nu_\infty \equiv \mu/\rho_\infty$ is the kinematic viscosity and $\alpha_\infty \equiv k/\rho_\infty C_p$ is the thermal diffusivity.

For this flow the effect of viscosity will be confined to a thin region near the vertical plate; hence, the y-length scale for the equations of motion will be the thickness of the momentum boundary layer or region of influence δ_m; that is, $y_{ms}=\delta_m$ (step 7). The momentum boundary-layer thickness is obtained by balancing the convection terms with the principal viscous term in equation (4.E.7-13). Similarly, the effect of heat conduction will be confined to a thin region δ_t, although not necessarily of the same thickness as that of the momentum boundary layer; hence, the y-length scale for the energy equation will be the thickness of the thermal boundary layer or region of influence δ_t; that is, $y_{ts}=\delta_t$. The thermal boundary-layer thickness is obtained by balancing the transverse heat-convection and heat-conduction terms in equation (4.E.7-16). To determine the axial velocity scale u_{xs}, we need to balance what causes the flow with the principal resistance to flow; the former is the gravitationally induced body force; the latter is the principal viscous term.

The transverse velocity scale u_{ys} is obtained from the continuity equation since this is inherently a developing flow. One might be tempted to obtain P_s from the dimensionless group multiplying the pressure term in equation (4.E.7-13). However, the pressure term in equation (4.E.7-13) does not cause the free convection flow; the latter is caused by the gravitational body-force term in this equation. However, the pressure does cause the flow in the y-direction, which is the reason why we determine its scale by setting the dimensionless group containing P_s in equation (4.E.7-14) equal to 1. The axial length scale and temperature reference and scale factors are obtained from the boundary conditions as described in Sections 3.4 and 4.6. These considerations then result in the following scale factors:

$$u_{xs} = (g\beta_t\,\Delta T\,L)^{0.5}; \qquad u_{ys} = \left(\frac{\nu_\infty^2 g\beta_t\,\Delta T}{L}\right)^{0.25}; \qquad P_s = \left(\frac{\mu^2 g\beta_t\,\Delta T}{L}\right)^{0.5};$$

$$T_s = T_1 - T_\infty \equiv \Delta T; \qquad x_s = L; \qquad y_{ms} = \delta_m = \frac{L}{\mathrm{Re}^{0.5}} = \frac{L}{\mathrm{Gr}_t^{0.25}};$$

$$y_{ts} = \delta_t = \frac{L}{\mathrm{Pr}^{0.5}\mathrm{Pe}_t^{0.5}} = \frac{L}{\mathrm{Pr}\cdot\mathrm{Gr}_t^{0.25}} = \frac{\delta_m}{\mathrm{Pr}} \tag{4.E.7-21}$$

where $\mathrm{Re} \equiv u_{xs}L/\nu_\infty$ is the Reynolds number, $\mathrm{Pe}_t \equiv u_{xs}L/\alpha_\infty = \mathrm{Re}\cdot\mathrm{Pr}$ is the Peclet number for heat transfer, $\mathrm{Pr} \equiv \nu_\infty/\alpha_\infty$ is the Prandtl number, and $\mathrm{Gr}_t \equiv L^3 g\beta_t\,\Delta T/\nu_\infty^2$ is the Grashof number for heat transfer. Note that the Grashof number is a measure of the ratio of the free convection to viscous transport of momentum; as such, it is the analog of the Reynolds number for free convection.[17] Note that the last of equations (4.E.7-21) indicates that $\delta_t < \delta_m$ for normal liquids, $\delta_t \cong \delta_m$ for gases, and $\delta_t > \delta_m$ for liquid metals.

If we now rewrite our dimensionless describing equations in terms of the scales defined by equations (4.E.7-21), we obtain

$$u_x^*\frac{\partial u_x^*}{\partial x^*} + u_y^*\frac{\partial u_x^*}{\partial y_m^*} = -\frac{1}{\mathrm{Gr}_t^{0.5}}\frac{\partial P^*}{\partial x^*} + \frac{1}{\mathrm{Gr}_t^{0.5}}\frac{\partial^2 u_x^*}{\partial x^{*2}} + \frac{\partial^2 u_x^*}{\partial y_m^{*2}} + T^* \tag{4.E.7-22}$$

$$u_x^*\frac{\partial u_y^*}{\partial x^*} + u_y^*\frac{\partial u_y^*}{\partial y_m^*} = -\frac{\partial P^*}{\partial y_m^*} + \frac{1}{\mathrm{Gr}_t^{0.5}}\frac{\partial^2 u_y^*}{\partial x^{*2}} + \frac{\partial^2 u_y^*}{\partial y_m^{*2}} \tag{4.E.7-23}$$

$$\frac{\partial u_x^*}{\partial x^*} + \frac{\partial u_y^*}{\partial y_m^*} = 0 \tag{4.E.7-24}$$

$$\frac{1}{\mathrm{Pr}}u_x^*\frac{\partial T^*}{\partial x^*} + u_y^*\frac{\partial T^*}{\partial y_T^*} = \frac{1}{\mathrm{Pr}^2\mathrm{Gr}_t^{0.5}}\frac{\partial^2 T^*}{\partial x^{*2}} + \frac{\partial^2 T^*}{\partial y_T^{*2}} \tag{4.E.7-25}$$

[17] The thermal Rayleigh number, defined as $\mathrm{Ra}_t \equiv L^3\alpha_\infty g\beta_t\,\Delta T/\nu_\infty = \mathrm{Gr}_t\cdot\mathrm{Pr}$, is another important dimensionless group that appears in free-convection problems; it is a measure of the ratio of the free convection to viscous transport of heat; as such, it is the analog of the thermal Peclet number for free convection.

$$u_x^* = 0, \quad u_y^* = 0, \quad T^* = 0 \quad \text{at} \quad x^* = 0 \tag{4.E.7-26}$$

$$u_x^* = f_1(y_m^*), \quad u_y^* = f_2(y_m^*), \quad T^* = f_3(y_t^*) \quad \text{at} \quad x^* = 1 \tag{4.E.7-27}$$

$$u_x^* = 0, \quad u_y^* = 0, \quad T^* = 1 \quad \text{at} \quad y_m^* = y_t^* = 0 \tag{4.E.7-28}$$

$$u_x^* = 0, \quad u_y^* = 0, \quad T^* = 0 \quad \text{at} \quad y_m^* = y_t^* = \infty \tag{4.E.7-29}$$

We now can consider how these scaled dimensionless describing equations can be simplified (step 8). Note that if the Grashof number is very large, such that $\text{Gr}_t^{0.5} \gg 1$, the pressure and axial viscous momentum transfer terms can be dropped from equation (4.E.7-22). The former simplification implies that the x-component is decoupled from the solution to the y-component of the equations of motion; hence, the latter equation can be ignored. Dropping the axial viscous momentum transfer term from equation (4.E.7-22) converts it from an elliptic into a parabolic differential equation; this obviates the need to specify downstream boundary conditions that in many cases are unknown. A very large Grashof number also implies that the axial heat-conduction term can be dropped from equation (4.E.7-23), which also converts it from an elliptic into a parabolic differential equation, again avoiding the need to specify a downstream boundary condition. The resulting describing equations that are applicable in the limit of very large Grashof number are given by

$$u_x^* \frac{\partial u_x^*}{\partial x^*} + u_y^* \frac{\partial u_x^*}{\partial y_m^*} = \frac{\partial^2 u_x^*}{\partial y_m^{*2}} + T^* \tag{4.E.7-30}$$

$$\frac{\partial u_x^*}{\partial x^*} + \frac{\partial u_y^*}{\partial y_m^*} = 0 \tag{4.E.7-31}$$

$$\frac{1}{\text{Pr}} u_x^* \frac{\partial T^*}{\partial x^*} + u_y^* \frac{\partial T^*}{\partial y_t^*} = \frac{\partial^2 T^*}{\partial y_t^{*2}} \tag{4.E.7-32}$$

$$u_x^* = 0, \quad u_y^* = 0, \quad T^* = 0 \quad \text{at} \quad x^* = 0 \tag{4.E.7-33}$$

$$u_x^* = 0, \quad u_y^* = 0, \quad T^* = 1 \quad \text{at} \quad y_m^* = y_t^* = 0 \tag{4.E.7-34}$$

$$u_x^* = 0, \quad u_y^* = 0, \quad T^* = 0 \quad \text{at} \quad y_m^* = y_t^* \to \infty \tag{4.E.7-35}$$

These simplified describing equations are often given in standard references with little or no justification.[18] Scaling analysis clearly provides a systematic method for developing these simplified equations and for understanding their limitations. The latter are explored further in the practice problems at the end of the chapter.

[18] See, for example, Bird et al., *Transport Phenomena*, 2nd ed., pp. 346–349; note that equations (4.E.7-30) and (4.E.7-33) differ from those in *Transport Phenomena* due to allowing for different radial length scales in the equations of motion and energy equation.

4.E.8 Dimensional Analysis Correlation for Electrical Heat Generation in a Wire

In Example Problem 4.E.2, we applied scaling analysis to unsteady-state radial heat conduction in a wire due to electrical heat generation. Assume now that we wish to develop a correlation for the instantaneous average temperature or spatially averaged temperature of the wire $\overline{T}(t)$ shown in Figure 4.E.2-1, which is defined

$$\overline{T}(t) \equiv \frac{1}{\pi R^2} \int_0^R T(r, t) 2\pi r \, dr \tag{4.E.8-1}$$

Let us recast equation (4.E.8-1) in terms of the dimensionless variables defined in Example Problem 4.E.2, given by

$$T^* \equiv \frac{T - T_0}{G_0 R^2 / k}; \qquad r^* \equiv \frac{r}{R}; \qquad t^* \equiv \frac{\alpha t}{R^2} \tag{4.E.8-2}$$

$$\overline{T}(t) \equiv \frac{G_0 R^2}{k} \int_0^1 T^*(r^*, t^*) 2r^* \, dr^* + T_0 \tag{4.E.8-3}$$

Employing the results of our scaling analysis for Example Problem 4.E.2 obviates the need to apply steps 1 through 7 in the scaling analysis procedure for dimensional analysis outlined in Chapter 2. Equation (4.E.8-3) can be rearranged in form

$$\frac{k[\overline{T}(t) - T_0]}{G_0 R^2} \equiv \overline{T}^*(t^*) = \int_0^1 T^*(r^*, t^*) 2r^* \, dr^* \tag{4.E.8-4}$$

Equation (4.E.8-4) implies that $\overline{T}^*$, the dimensionless average temperature, is a function of t^*, the dimensionless time, and any dimensionless groups that enter into the solution for $T^*(r^*, t^*)$. Equation (4.E.2-15) indicates that once the transients have died out, only one additional dimensionless group is involved in the solution for $T^*(r^*, t^*)$: namely, $\omega R^2/\alpha$. Hence, the correlation for the instantaneous average temperature will involve two dimensionless groups and the dimensionless time; that is,

$$\overline{T}^* \equiv \frac{k(\overline{T} - T_0)}{G_0 R^2} = f\left(t^*; \frac{\omega R^2}{\alpha}\right) \tag{4.E.8-5}$$

Now let us assume that we seek a correlation for $\overline{T}_m$ the maximum in the spatially averaged temperature fluctuation. In principal, this would be obtained by setting the time derivative of the spatially averaged temperature equal to zero and determining the corresponding maximum. Hence, the correlation for $\overline{T}_m^*$, the dimensionless maximum in the average temperature, will involve only two dimensionless groups; that is,

$$\overline{T}_m^* \equiv \frac{k(\overline{T}_m - T_0)}{G_0 R^2} = f\left(\frac{\omega R^2}{\alpha}\right) \tag{4.E.8-6}$$

However, if the dimensionless group $\omega R^2/\alpha \ll 1$, corresponding to very low frequency currents, thin wires, or wires having a very high thermal conductivity, the right-hand side of equation (4.E.8-6) can be expanded in a Taylor series and truncated at the first term (step 9). For this special case the dimensionless maximum spatially averaged temperature is a constant; that is,

$$\overline{T}_m^* \equiv \frac{k(\overline{T}_m - T_0)}{G_0 R^2} = \text{a constant} \qquad (4.E.8\text{-}7)$$

It is instructive to compare the results of scaling for dimensional analysis to those of the Pi theorem. A naive application of the Pi theorem for correlating the spatially averaged temperature $\overline{T}$ would indicate that four dimensionless groups were necessary; this follows from having nine dimensional quantities ($\overline{T}$, T_0, G_0, ω, t, R, k, ρ, and C_p) in five units (mass, length, time, energy, and temperature). However, scaling analysis reveals that the spatially averaged temperature can be correlated with only three dimensionless groups. The Pi theorem will yield this same result if one recognizes that the quantities k, ρ, and C_p can be combined into a single quantity α and that $\overline{T}$ and T_0 can be combined into a single quantity $\overline{T} - T_0$. This reduces the number of quantities to seven and the number of units to four, thereby indicating three dimensionless groups. Scaling analysis avoids having to invoke the subtle arguments required to ensure that the Pi theorem yields the minimum parametric representation.

4.P PRACTICE PROBLEMS

4.P.1 Steady-State Conduction in a Slab with a Specified Cooling Flux

Consider steady-state heat conduction in the solid slab considered in Section 4.2 and shown in Figure 4.2-1. The boundary conditions at $x = 0$, $x = W$, and $y = H$ remain the same; however, the constant-temperature boundary condition at $y = 0$ is replaced by a constant-heat-flux condition given by $q_y = -q_2$ where $q_2 > 0$.

(a) Use scaling analysis to determine the appropriate temperature scales.
(b) Determine the criterion for ignoring lateral heat conduction.

4.P.2 Steady-State Conduction in a Slab with a Specified Heat Flux

Consider steady-state heat conduction in the solid slab considered in Section 4.2 and shown in Figure 4.2-1. The boundary conditions at $x = 0$, $x = W$, and $y = 0$ remain the same; however, the constant-temperature boundary condition at $y = H$ is replaced by a constant-heat-flux condition given by $q_y = -q_2$ where $q_2 > 0$.

(a) Use scaling analysis to determine the appropriate temperature scales.
(b) Determine the criterion for ignoring lateral heat conduction.

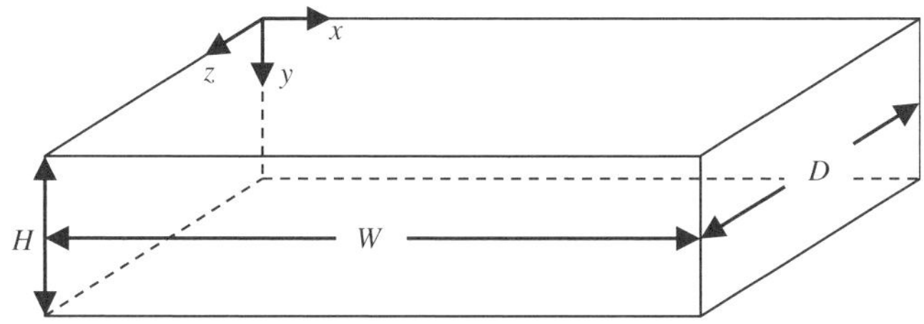

Figure 4.P.3-1 Steady-state multidimensional heat conduction in a solid rectangular parallelepiped of width W, depth D, and height H and constant physical properties.

4.P.3 Steady-State Heat Conduction in a Rectangular Parallelepiped

Consider the solid rectangular parallelepiped of width W, depth D, and height H and constant physical properties shown in Figure 4.P.3-1.

(a) Write the appropriate form of the energy equation along with the boundary conditions required.

(b) Use scaling analysis to determine the criterion for ignoring conduction in the x-direction.

(c) Use scaling analysis to determine the criterion for ignoring conduction in the z-direction.

(d) Derive an equation for the thickness of the region of influence near the sidewalls at $x = 0$ and $x = W$ wherein lateral heat conduction cannot be ignored in predicting quantities such as the temperature or heat flux near these boundaries.

(e) Derive an equation for the thickness of the region of influence near the sidewalls at $z = 0$ and $z = D$ wherein lateral heat conduction cannot be ignored in predicting quantities such as the temperature or heat flux near these boundaries.

4.P.4 Steady-State Conduction in a Cylinder with Specified Temperatures at Its Boundaries

Consider the steady-state heat conduction in a solid cylinder of radius R and constant physical properties due to a high temperature T_0 applied at $z = 0$ and a low temperature T_1 applied at $z = L$ as well as at the lateral surface as shown in Figure 4.P.4-1.

(a) Write the appropriate form of the energy equation along with the boundary conditions required.

(b) Use scaling analysis to determine the criterion for ignoring radial conduction.

(c) Derive an equation for the thickness of the region of influence near the circumferential boundary of the cylinder wherein radial heat conduction

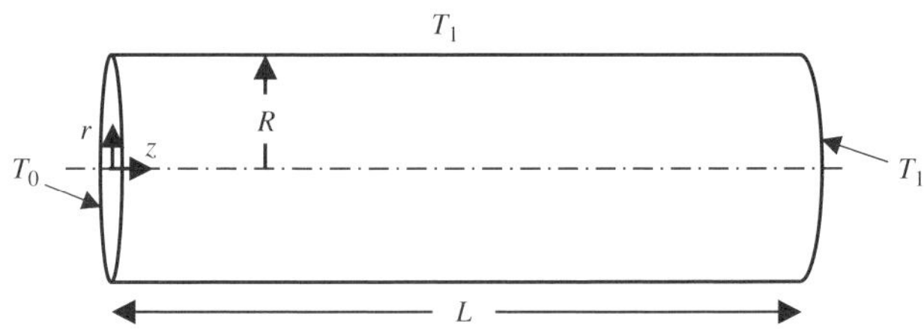

Figure 4.P.4-1 Steady-state heat conduction in a solid circular cylinder due to a high temperature T_0 applied at $z = 0$ and a low temperature T_1 applied at $z = L$ and at the circumferential boundary at $r = R$.

cannot be ignored in predicting quantities such as the temperature or heat flux near this boundary.

4.P.5 Steady-State Conduction in an Annulus with Specified Temperatures at Its Boundaries

Consider steady-state heat conduction in a solid with an annular cross-sectional area and constant physical properties due to a high temperature T_1 applied at its inner surface at R_1 and a low temperature T_2 applied at its outer surface at R_2 as well as at the two ends of the cylinder at $z = 0$ and $z = L$ as shown in Figure 4.P.5-1.

(a) Write the appropriate form of the energy equation and the boundary conditions required.

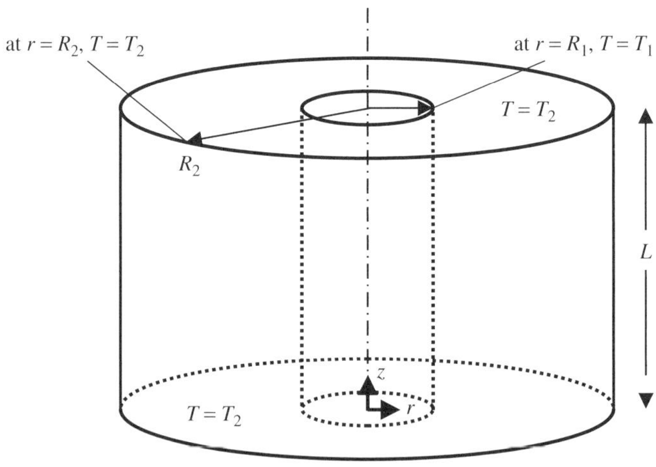

Figure 4.P.5-1 Steady-state heat conduction in a solid annulus with constant physical properties whose inner surface at R_1 is held at a high temperature T_1 and whose outer surface at R_2 and ends at $z = 0$ and $z = L$ are held at a low temperature T_2.

(b) Use scaling analysis to determine the criterion for ignoring axial conduction; be certain to include a reference scale for the dimensionless radial coordinate since it is not naturally referenced to zero for an annulus.

(c) Derive an equation for the thickness of the region of influence near the two ends of the annulus wherein axial heat conduction cannot be ignored in predicting quantities such as the temperature or heat flux near this boundary.

4.P.6 Steady-State Heat Conduction in a Circular Fin

A solid metallic circular flat fin with a constant thermal conductivity k, width H, and radius R_2 is attached to a cylindrical pipe having a radius R_1 that is maintained at a constant temperature T_1, where $R_2 - R_1 \gg H$. The convective heat-transfer flux q from the surfaces of the fin to the ambient air is described via the lumped-parameter approach and is given by

$$q = h(T - T_\infty) \tag{4.P.6-1}$$

where h is the heat-transfer coefficient and T_∞ is the temperature of the ambient air far removed from the fin. A schematic of this fin is shown in Figure 4.P.6-1.

(a) Write the appropriate form of the energy equation and the boundary conditions required.

(b) Determine the criterion for assuming that the temperature is uniform across the thickness of the fin.

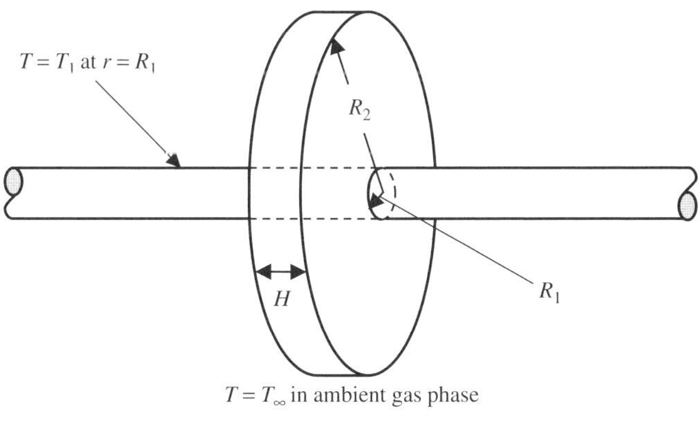

Figure 4.P.6-1 Solid flat circular metallic fin with a constant thermal conductivity k, radius R_2, and thickness H attached to a cylindrical pipe having a radius R_1 and temperature T_1; the convective heat-transfer flux from the surfaces of the fin to the ambient air is described by $q = h(T - T_\infty)$.

(c) Integrate the simplified describing equations that result from the approximation made in part (b) to obtain an equation for the axial temperature distribution within the fin.

4.P.7 Unsteady-State Axial Heat Conduction in a Solid Cylinder

A solid cylinder with constant physical properties, radius R, and length L is initially at a constant temperature T_0 as shown in Figure 4.P.7-1. At time $t = 0$ the temperature of the face of this cylinder at $z = 0$ is raised to T_1 while the face at $z = L$ is maintained at T_0. The circumferential boundary of the cylinder is perfectly insulated so that there is no heat transfer in the radial direction.

(a) Write the appropriate form of the energy equation and the initial and boundary conditions required.
(b) Scale the describing equations to determine the criterion for assuming that steady-state heat transfer is achieved.
(c) Scale the describing equations appropriate to very short contact times for which the heat conduction has not penetrated the entire length of the cylinder, and determine the criterion for the applicability of this approximation.
(d) Derive an equation for the region of influence or thermal penetration for the conditions in part (c).

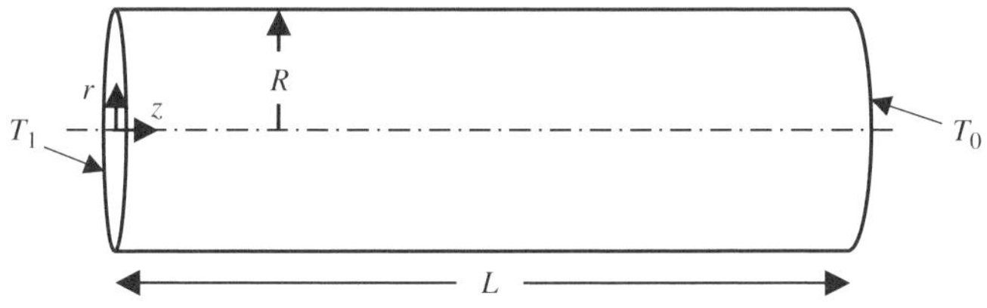

Figure 4.P.7-1 Unsteady-state axial heat conduction in a solid cylinder with constant physical properties, a perfectly insulated circumferential boundary, and an initial temperature T_0 due to a high temperature T_1 being imposed on the face at $z = 0$ while the face at $z = L$ is maintained at T_0.

4.P.8 Unsteady-State Radial Heat Conduction in a Solid Cylinder

A solid cylinder with constant physical properties, radius R, and length L is initially at a constant temperature T_0 as shown in Figure 4.P.8-1. At time $t = 0$ the temperature of the circumferential boundary of this cylinder at $r = R$ is raised to T_1. The ends of the cylinder at $z = 0$ and $z = L$ are perfectly insulated so that there is no heat transfer in the axial direction.

(a) Write the appropriate form of the energy equation and the initial and boundary conditions required.

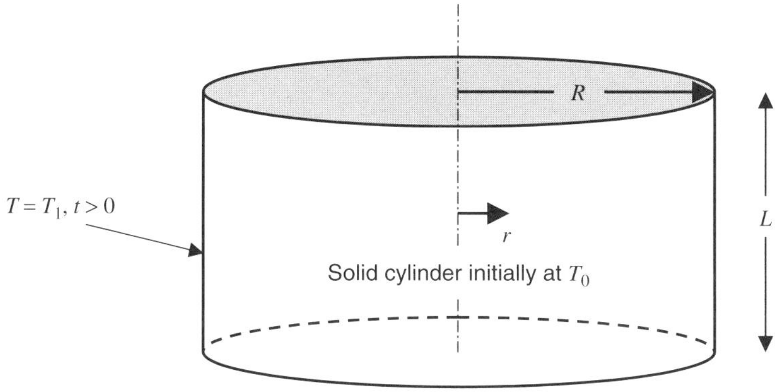

Figure 4.P.8-1 Unsteady-state axial heat conduction in a solid cylinder with constant physical properties, insulated ends, and an initial temperature T_0 due to a high temperature T_1 being imposed on the face at $r = R$.

(b) Scale the describing equations to determine the criterion for assuming that thermal equilibrium has been achieved.

(c) Scale the describing equations appropriate to very short contact times for which the thermal energy has not penetrated entirely through the cylinder and determine the criterion for the applicability of this approximation.

(d) Derive an equation for the region of influence or thermal penetration for the conditions in part (c).

4.P.9 Unsteady-State Radial Heat Conduction in a Spherical Shell

A solid spherical shell with constant physical properties, an inner radius R_1, and an outer radius R_2 is initially at a constant temperature T_0, as shown in Figure 4.P.9-1. At time $t = 0$ the temperature of the inner boundary at $r = R_1$ is raised to T_1 and convective heat transfer to a surrounding flowing fluid whose bulk temperature is maintained at T_0 occurs at the outer boundary at R_2; the heat flux at the outer boundary is described by a constant heat-transfer coefficient and is given by

$$q_r = h(T - T_0) \qquad \text{at} \quad r = R_2 \tag{4.P.9-1}$$

(a) Write the appropriate form of the energy equation and the initial and boundary conditions required.

(b) Scale the describing equations to determine the criterion for justifying that steady-state heat transfer can be assumed. Note that it is necessary to introduce a separate scale for the radial temperature gradient since this derivative does not necessarily scale with the characteristic temperature scale divided by the length scale; also introduce a reference factor for the spatial variable since it is not naturally referenced to zero.

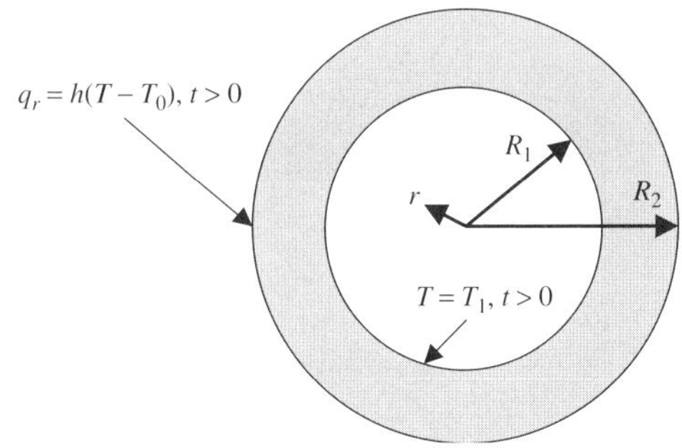

Figure 4.P.9-1 Unsteady-state heat conduction in a solid spherical shell with constant physical properties, inner radius R_1, outer radius R_2, and initial temperature T_0; at time $t = 0$ a high temperature T_1 is applied at the inner boundary, and convective heat transfer to the surrounding fluid whose bulk temperature is maintained at T_0 occurs at the outer boundary.

(c) Based on your scaling analysis in part (b), determine the criterion for ignoring curvature effects on the heat conduction within the spherical shell.

(d) Solve the simplified describing equations appropriate to parts (b) and (c) for the temperature profile in the spherical shell and determine the temperature at $r = R_2$.

(e) Scale the describing equations appropriate to very short contact times for which the thermal energy has not penetrated entirely through the spherical shell, and determine the criterion for the applicability of this approximation.

(f) Derive an equation for the region of influence or thermal penetration for the conditions in part (e).

4.P.10 Steady-State Conduction in a Cylinder with External Phase Convection

Consider steady-state heat conduction in a solid cylinder of radius R and constant physical properties due to a high temperature T_0 applied at $z = 0$ and a low temperature T_L applied at $z = L$, as shown in Figure 4.P.4-1. However, the circumferential boundary at $r = R$ is exposed to convective heat transfer in an external fluid phase; the heat flux in this external phase is described by a lumped-parameter condition given by

$$q_r = h(T - T_\infty) \tag{4.P.10-1}$$

where T_∞ is the temperature in the bulk of the external fluid phase far removed from the solid cylinder.

(a) Write the appropriate form of the energy equation and the boundary conditions required.

(b) Use scaling analysis to determine the criterion for ignoring radial conduction. Introduce a separate scale for the radial derivative of the temperature since there is no reason to assume that it will scale with the characteristic temperature scale divided by the characteristic radial length scale.

(c) If the criterion you derived in part (b) is satisfied, radial heat conduction can be ignored everywhere within the solid cylinder, even near the circumferential boundary; that is, there is no region of influence in this case near the circumferential boundary. Explain why there is no region of influence in this case, where a lumped-parameter heat-flux boundary condition is prescribed at the lateral boundaries, whereas there would be a region of influence if a constant temperature were prescribed at this boundary.

4.P.11 Unsteady-State Heat Transfer to a Sphere at Small Biot Numbers

In Section 4.4 we considered convective heat transfer to a solid sphere falling at its terminal velocity through a fluid. Scaling analysis was used to obtain a criterion for achieving steady-state, which in this case meant thermal equilibrium between the sphere and the liquid; this criterion is given by equation (4.4-17). However, if the Biot number is very small, the describing equations can be simplified and solved analytically; the resulting solution for the dimensionless temperature is given by equation (4.4-23).

(a) Use the solution obtained for the small Biot number approximation to determine the dimensionless time required for the temperature to reach 1% of its final equilibrium temperature and compare this result to the criterion we derived for achieving steady-state.

(b) The small Biot number approximation implies that the temperature within the sphere is nearly uniform. Use the solution for the small Biot number approximation and the results of the scaling analysis in Section 4.4 to estimate the dimensionless time required for the dimensionless temperature at the center of the sphere to differ from the surface temperature by less than 1.0%.

4.P.12 Unsteady-State Heat Transfer in a Solid Sphere

In Section 4.4 we considered unsteady-state heat transfer to a solid sphere that was initially at temperature T_0 due to falling at its terminal velocity through a hot liquid at temperature T_∞. We used scaling analysis to arrive at the criterion for assuming that the temperature of the sphere was essentially uniform across its radius at any time. This corresponded to the small Biot number approximation, for which the heat transfer is controlled by the external liquid phase. The exact analytical solution for this heat-transfer problem is given by

$$\frac{T - T_\infty}{T_0 - T_\infty} = \sum_{n=1}^{\infty} C_n e^{-\lambda_n^2 \mathrm{Fo}_t} \frac{R}{\lambda_n r} \sin \frac{\lambda_n r}{R} \qquad (4.P.12\text{-}1)$$

where $\mathrm{Fo}_t \equiv \alpha t/R^2$ is the Fourier number for heat transfer and C_n is a coefficient whose value is obtained from

$$C_n = \frac{4(\sin\lambda_n - \lambda_n \cos\lambda_n)}{2\lambda_n - \sin 2\lambda_n} \tag{4.P.12-2}$$

for which the λ_n are the positive roots of

$$1 - \lambda_n \cot\lambda_n = \mathrm{Bi}_t \tag{4.P.12-3}$$

where $\mathrm{Bi}_t = hR/k$ is the Biot number for heat transfer. Compare the approximate solution that we obtained for the small Biot number approximation given by equation (4.4-23) to the exact solution given by the above for $\mathrm{Bi}_t = 0.01$ and $\mathrm{Bi}_t = 0.1$; for the case of the exact solution, compute the temperature at the center of the sphere.

4.P.13 Unsteady-State Convective Heat Transfer to a Plane Wall

Consider an infinitely long solid plane wall that is initially at a high temperature T_0 with constant physical properties and thickness $2H$. This wall is cooled on both sides by a flowing liquid whose bulk temperature far removed from the wall is T_∞. The heat transfer in the cold liquid is characterized by a heat-transfer coefficient h. A schematic of this heat-transfer problem is shown in Figure 4.P.13-1.

(a) Write the appropriate form of the energy equation along with the initial and boundary conditions required.

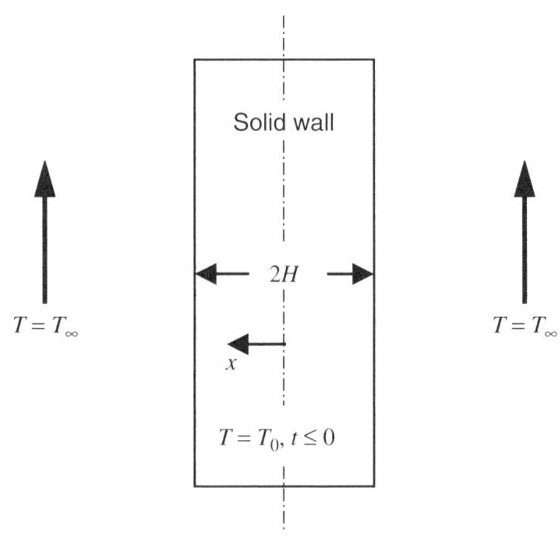

Figure 4.P.13-1 Infinitely long solid plane wall with constant physical properties, thickness $2H$, and initial temperature T_0 subject to convective cooling at its lateral boundaries by a liquid whose temperature is T_∞ far from the wall.

(b) Use scaling analysis to determine the criterion for ignoring the conductive heat-transfer resistance in the wall relative to that in the external liquid phase.
(c) Solve the simplified describing equations justified by the criterion derived in part (b).
(d) The exact analytical solution for this heat-transfer problem is given by

$$\frac{T - T_\infty}{T_0 - T_\infty} = \sum_{n=1}^{\infty} C_n e^{-\lambda_n^2 \mathrm{Fo}_t} \cos \frac{\lambda_n x}{H} \tag{4.P.13-1}$$

where $\mathrm{Fo}_t \equiv \alpha t / H^2$ is the Fourier number for heat transfer and C_n is a coefficient whose value is obtained from

$$C_n = \frac{4 \sin \lambda_n}{2\lambda_n + \sin 2\lambda_n} \tag{4.P.13-2}$$

for which the λ_n are the positive roots of the equation

$$\lambda_n \tan \lambda_n = \mathrm{Bi}_t \tag{4.P.13-3}$$

where $\mathrm{Bi}_t = hR/k$ is the Biot number for heat transfer. Compare the approximate solution that you obtained in part (c) to the exact solution given by the above for $\mathrm{Bi}_t = 0.01$ and $\mathrm{Bi}_t = 0.1$; for the case of the exact solution, compute the temperature at the center of the plane wall.

4.P.14 Unsteady-State Convective Heat Transfer to a Solid Cylinder

Consider an infinitely long solid cylinder initially at a high temperature T_0 with constant physical properties and radius R. This cylinder is cooled by immersing it in a flowing liquid whose upstream temperature is T_∞ and whose upstream velocity perpendicular to the axis of the cylinder is U_∞. The heat transfer in the cold liquid is characterized via a heat-transfer coefficient h. Correlations for the Nusselt number, the dimensionless heat-transfer coefficient, as a function of the Reynolds number for flow over a cylinder exposed to a liquid flowing at constant velocity are available in standard references.[19] A schematic of this heat-transfer problem is shown in Figure 4.P.14-1.

(a) Write the appropriate form of the energy equation along with the initial and boundary conditions required.
(b) Use scaling analysis to determine the criterion for ignoring the conductive heat-transfer resistance in the solid cylinder relative to that in the external liquid phase.

[19]See, for example, Bird et al., *Transport Phenomena*, 2nd ed., p. 440.

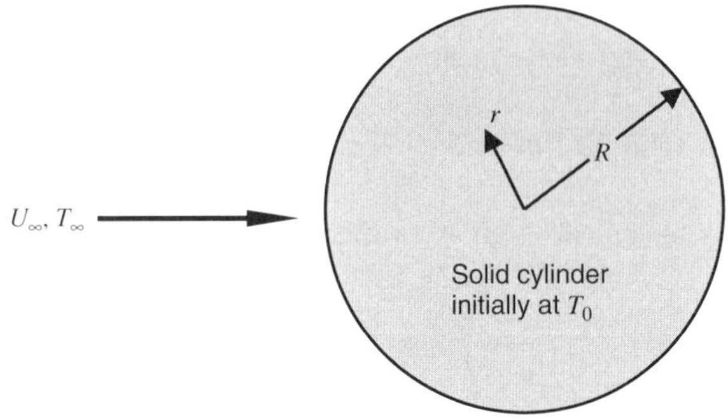

Figure 4.P.14-1 Infinitely long solid cylinder with constant physical properties, radius R, and initial temperature T_0 immersed in a flowing liquid whose upstream velocity and temperature are U_∞ and T_∞ such that $T_\infty < T_0$ and for which the convective heat transfer in the liquid phase is described by a constant heat-transfer coefficient h.

(c) Solve the simplified set of describing equations justified by the criterion that you derived in part (b).

(d) The exact analytical solution for this heat-transfer problem is given by

$$\frac{T - T_\infty}{T_0 - T_\infty} = \sum_{n=1}^{\infty} C_n e^{-\lambda_n^2 \mathrm{Fo}_t} J_0\left(\frac{\lambda_n x}{R}\right) \tag{4.P.14-1}$$

where $\mathrm{Fo}_t \equiv \alpha t/R^2$ is the Fourier number and C_n is a coefficient whose value is obtained from

$$C_n = \frac{2}{\lambda_n} \frac{J_1(\lambda_n)}{J_0^2(\lambda_n) + J_1^2(\lambda_n)} \tag{4.P.14-2}$$

where the λ_n are the positive roots of

$$\lambda_n \frac{J_1(\lambda_n)}{J_0(\lambda_n)} = \mathrm{Bi}_t \tag{4.P.14-3}$$

where $\mathrm{Bi}_t \equiv hR/k$ is the Biot number for heat transfer and J_i is the ith-order Bessel function of the first kind. Compare the approximate solution that you obtained in part (c) to the exact solution given by the above for $\mathrm{Bi}_t = 0.01$ and $\mathrm{Bi}_t = 0.1$; for the case of the exact solution, compute the temperature at the center of the cylinder.

4.P.15 Entrance Effect Limitations in Laminar Slit Flow

In Section 4.5 we considered the steady-state fully developed laminar flow of a viscous Newtonian fluid with constant physical properties between two infinitely wide parallel plates that have a separation distance $2H$, length L, and maintained at the temperature T_0, which was also the temperature of the entering fluid. We scaled the describing equations to determine the criteria for ignoring both the axial convection and axial conduction of heat.

(a) The criterion for ignoring axial heat conduction breaks down near the leading edge of the two parallel plates. Use scaling analysis to estimate the thickness of the region of influence wherein axial heat conduction cannot be ignored.

(b) Retaining the axial conduction term in the describing equations for the entrance region complicates solving this program since a downstream boundary condition is required. Describe a procedure whereby the solution to the simplified equations sufficiently far downstream from the entrance region can be used to obtain a solution for this heat-transfer problem along the entire length of the two flat plates.

4.P.16 Convective Heat Transfer for Fully Developed Laminar Flow Between Heated Parallel Flat Plates

Consider the steady-state heat transfer associated with fully developed laminar flow of a Newtonian liquid with constant physical properties and initial temperature T_0 between two parallel flat plates of length L, separated by a distance $2H$, and maintained at a temperature T_1 such that $T_1 > T_0$, as shown in Figure 4.P.16-1. However, heat generation owing to viscous dissipation can also occur. It can be assumed that the laminar flow velocity profile is fully developed and given by

$$u_x = U_m \left[1 - \left(\frac{y}{H} \right)^2 \right] \tag{4.P.16-1}$$

where U_m is the maximum fluid velocity at the centerline between the two plates.

(a) Write the appropriately simplified form of the energy equation and associated boundary conditions.

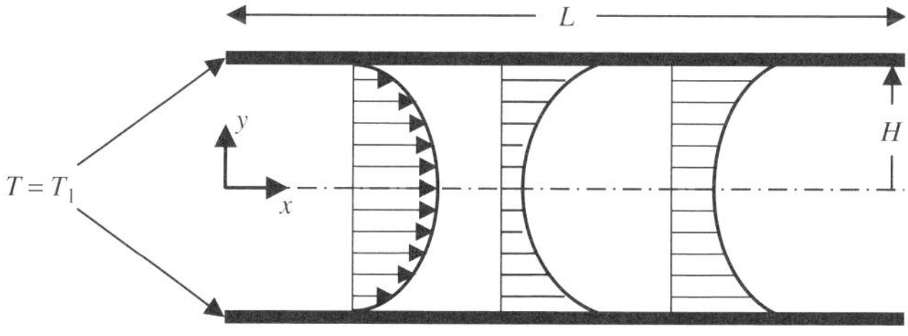

Figure 4.P.16-1 Steady-state heat transfer to fully developed flow of a Newtonian fluid with constant physical properties and initial temperature T_0 between two parallel solid flat plates of length L, separated by a distance $2H$, and maintained at temperature T_1; the figure shows the fully developed laminar flow velocity profile and sketches of the developing temperature profile at two locations.

(b) Scale the describing equations for conditions such that the predominant heating is caused by heat transfer from the two plates and for conditions such that the transverse conduction is sufficiently large so that heat penetration occurs essentially across the entire cross section between the two flat plates.
(c) Determine the criterion for ignoring the axial heat conduction.
(d) Determine the criterion for ignoring viscous heat generation.
(e) Determine the criterion for ignoring axial convection.

4.P.17 Entrance Effect Limitations in the Thermal Boundary-Layer Approximation for Falling Film Flow

In Example Problem 4.E.4 we considered a thermal boundary-layer approximation for heat transfer from a heated vertical plate to laminar film flow. We found that the heat transfer was confined essentially to a thin boundary layer near the heated wall if the Peclet number were sufficiently high. This thermal boundary-layer approximation also involved ignoring axial heat conduction.

(a) Determine the thickness of the region of influence near the leading edge of the vertical plate within which the thermal boundary-layer approximation breaks down.
(b) If the axial heat conduction cannot be ignored near the leading edge, problems are encountered in solving the describing equations, owing to the lack of a downstream boundary condition. Outline a procedure whereby a solution could be obtained to the describing equations over the full length of the vertical plate by using a solution to the thermal boundary-layer equations derived in Example Problem 4.E.4. Note that it is only necessary to describe the procedure that you would use to solve the describing equations.

4.P.18 Thermal Boundary-Layer Heat Transfer for Fully Developed Laminar Flow Between Heated Parallel Flat Plates

Consider the steady-state heat transfer associated with fully developed laminar flow of a Newtonian liquid with constant physical properties and initial temperature T_0 between two parallel flat plates separated by a distance $2H$. The two plates are maintained at the same temperature T_0 over the length defined by $0 \le x \le L_0$. However, the temperature of the two plates is raised to T_1 over the length defined by $L_0 \le x \le L_1$ as shown in Figure 4.P.18-1. It can be assumed that viscous dissipation can be ignored and that the fully developed laminar flow velocity profile is given by

$$u_x = U_m \left[1 - \left(\frac{y}{H}\right)^2\right] \tag{4.P.18-1}$$

(a) Write the appropriately simplified form of the energy equation and associated boundary conditions.

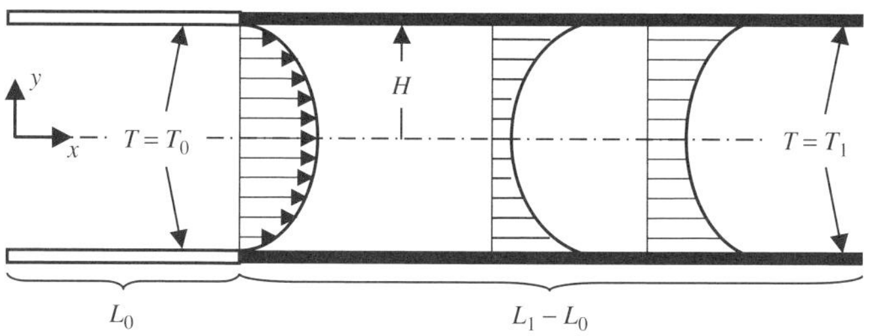

Figure 4.P.18-1 Steady-state heat transfer to fully developed flow of a Newtonian fluid with constant physical properties and initial temperature T_0 between two parallel solid flat plates separated by a distance $2H$; the two plates are maintained at temperature T_0 for $0 \leq x \leq L_0$ and maintained at temperature T_1 for $L_0 \leq x \leq L_1$; the figure shows the fully developed laminar flow velocity profile and sketches of the developing temperature profile at two locations.

(b) Scale the describing equations for conditions such that axial convection of heat is significant; note that for these conditions there will be a region of influence or thermal boundary layer δ_t near each flat plate; hence, recast the describing equations in terms of a coordinate system located on one of the two plates and provide an estimate of the thickness of this thermal boundary layer.

(c) Determine the criterion for ignoring the axial heat conduction.

(d) Determine the thickness of the region of influence near the leading edge of the heated zone wherein axial heat conduction cannot be neglected.

(e) If the thermal boundary layer is sufficiently thin, it is possible to use a simplified form of the velocity profile in the convective heat-transfer term in the energy equation; determine the criterion for employing a linear velocity profile in the region near the plates. Note that this simplification is often referred to as the *Lévêque approximation*.[20]

4.P.19 Heat Transfer from a Hot Inviscid Gas to Fully Developed Laminar Falling Film Flow

Consider the fully developed laminar flow down a solid vertical wall of a Newtonian liquid film of thickness H, constant physical properties, and initial temperature T_0. This liquid film contacts an inviscid gas phase whose temperature is T_1 such that $T_1 > T_0$. The vertical wall can be assumed to be perfectly insulated and viscous dissipation in the liquid film can be ignored. A sketch of this heat-transfer problem is shown in Figure 4.P.19-1. The velocity profile for fully developed laminar film flow is given by

$$u_x = U_m \left[1 - \left(\frac{y}{H} \right)^2 \right] \quad (4.P.19\text{-}1)$$

[20] M. A. Lévêque, *Ann. Mines*, **13**, 201 (1928).

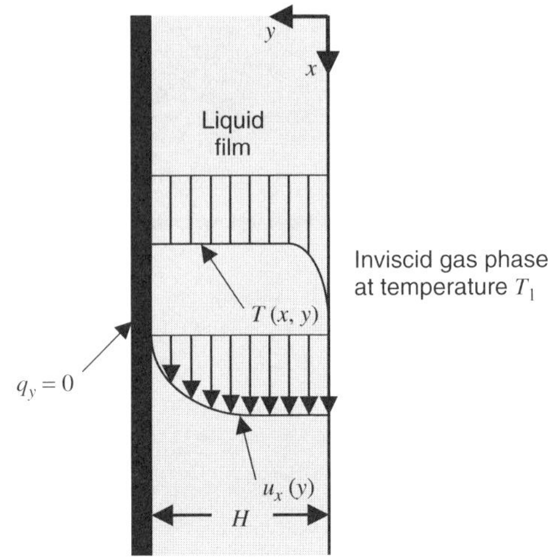

Figure 4.P.19-1 Steady-state heat transfer from a hot inviscid gas phase at temperature T_1 to a liquid film in fully developed laminar flow with an initial temperature T_0, thickness H, and constant physical properties.

(a) Write the appropriately simplified form of the energy equation and associated boundary conditions.

(b) Scale the describing equations accounting for axial convection, axial conduction, and transverse conduction of heat; note that there will be a region of influence or thermal boundary layer δ_t near the liquid–gas interface.

(c) Based on your scaling analysis in part (b), provide an estimate of the thickness of the thermal boundary layer.

(d) Determine the criterion for ignoring the axial heat conduction.

(e) Determine the thickness of the region of influence near the leading edge of the falling film flow wherein axial heat conduction cannot be neglected.

(f) If the thermal boundary layer is sufficiently thin, it is possible to use a simplified form of the velocity profile in the convective heat-transfer term in the energy equation; determine the criterion for employing a constant value of the velocity near the liquid–gas interface.

(g) Estimate the length required for thermal penetration to reach the vertical wall.

4.P.20 Thermal Boundary-Layer Development Along a Heated Flat Plate

In Section 4.6 we considered the thermal boundary-layer approximation for laminar flow over a horizontal flat plate maintained at a constant temperature. We found that the criterion for making the thermal boundary-layer approximation was a very large Peclet number. At the end of Section 4.6 it was stated that the thermal boundary-layer approximation breaks down in the vicinity of the leading edge of the flat plate.

(a) Use scaling analysis to provide an estimate of the thickness of the region of influence near the leading edge of the flat plate wherein the thermal boundary-layer approximation breaks down.

(b) Compare the thicknesses of the regions where the thermal boundary-layer and momentum boundary-layer approximations break down for different values of the Prandtl number.

4.P.21 Thermal Boundary-Layer Development with an Unheated Entry Region

In Section 4.6 we considered the thermal boundary-layer approximation for laminar flow over a horizontal flat plate maintained at a constant temperature over its entire length. Now consider the case where the plate is maintained at T_∞, the temperature of the fluid upstream from the plate over a length L_0, after which the plate is maintained at the temperature T_0, where $T_0 > T_\infty$, as shown in Figure 4.P.21-1. Note that in this case the momentum boundary layer will begin developing before the thermal boundary layer. Note also that different axial length scales will be required for the equations of motion and the energy equation. Assume that the Prandtl number Pr ≥ 1.

(a) Consider the scaling for this heat-transfer problem for conditions such that both the Reynolds number and Peclet number are large. Determine the appropriate scale and reference factors for the dependent and independent variables.

(b) Use your scaling analysis to obtain estimates for both the momentum and thermal boundary-layer thicknesses.

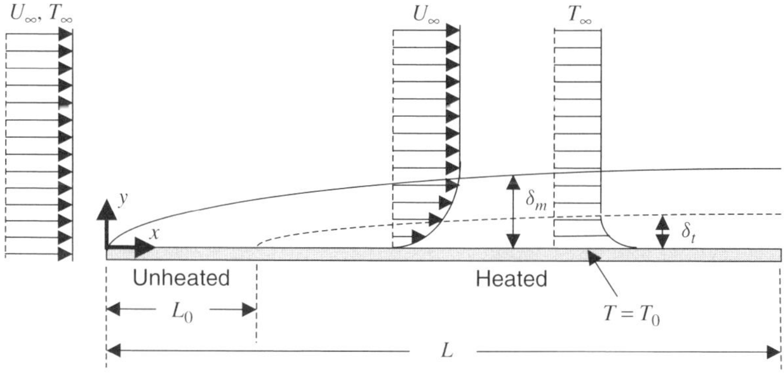

Figure 4.P.21-1 Steady-state laminar uniform flow of a Newtonian fluid with constant physical properties, temperature T_∞, and velocity U_∞ intercepting a stationary semi-infinite infinitely wide horizontal flat plate; the latter is maintained at T_∞ for $0 \leq x \leq L_0$ and at $T_0 > T_\infty$ for $L_0 < x \leq L$; the solid line shows the hypothetical momentum boundary-layer thickness δ_m, and the dashed line shows the hypothetical thermal boundary-layer thickness δ_t for Pr > 1, where Pr is the Prandtl number.

(c) Use scaling analysis to provide an estimate of the axial length of the region of influence near the leading edge of the heated region wherein the thermal boundary-layer approximation breaks down.

(d) Discuss whether the approximations made in the equations of motion or in the energy equation are more limiting with respect to ignoring the axial diffusion terms.

4.P.22 Thermal Boundary-Layer Development with Flux Condition

Consider the steady-state laminar uniform plug flow of a Newtonian liquid with constant physical properties, temperature T_∞, and velocity U_∞ intercepting a stationary semi-infinite infinitely wide horizontal impermeable flat plate as shown in Figure 4.6-1. However, a constant heat flux q_0 is maintained along the surface of the flat plate rather than a constant temperature. Gravitational and viscous heating effects can be assumed to be negligible. In this problem we use scaling to determine the criteria for making the thermal boundary-layer approximation; that is, the conditions for which axial heat conduction can be ignored and for which the heat transfer can be assumed to be confined to a thin thermal boundary layer near the plate.

(a) Write the appropriate forms of the equations of motion and thermal energy equation applicable to this boundary-layer flow; it is not necessary here to justify the forms of these equations by scaling; that is, you can begin with the dimensional momentum and thermal boundary-layer equations that resulted from the scaling done in Section 4.6.

(b) Write the boundary conditions required to solve the coupled equations of motion and thermal energy equation.

(c) In scaling the describing equations for this problem, it is necessary to introduce a scale factor for the y-derivative of the temperature due to the flux condition at the plate. This implies that the temperature scale will be different from that obtained in Section 4.6. In view of these considerations, determine the appropriate scale factors for the temperature and its y-derivative. In determining the temperature scale, keep in mind that heat convection in both the x- and y-directions must be retained for large Peclet numbers.

(d) Derive an equation for the thermal boundary-layer thickness δ_t and discuss any differences between your result and that obtained in the boundary-layer problem considered in Section 4.6.

(e) Determine the criterion for making the thermal boundary-layer approximation.

(f) Determine the thickness of the region of influence near the leading edge of the plate wherein the thermal boundary-layer approximation is not applicable.

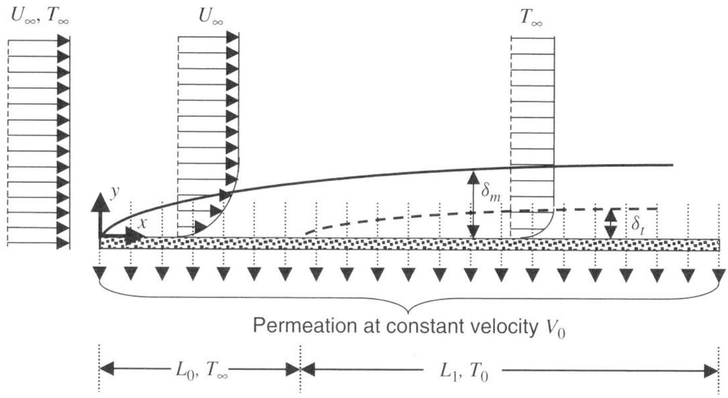

Figure 4.P.23-1 Steady-state laminar uniform flow of an incompressible viscous Newtonian liquid with constant physical properties, temperature T_∞, and velocity U_∞ intercepting a stationary semi-infinite infinitely wide horizontal permeable flat plate along which there is a constant suction velocity V_0; the surface of the flat plate is maintained at the temperature T_∞ over the distance $0 \le x \le L_0$; however, the temperature is increased to T_0 over the distance $L_0 \le x \le L_1$.

4.P.23 Thermal Boundary-Layer Development with Suction

Consider the steady-state laminar uniform plug flow of a Newtonian liquid with constant physical properties, temperature T_∞, and velocity U_∞ intercepting a stationary semi-infinite infinitely wide horizontal permeable flat plate, as shown in Figure 4.P.23-1. The surface of the flat plate is maintained at the temperature T_∞ over the distance $0 \le x \le L_0$; however, the temperature is increased to T_0 over the distance $L_0 \le x \le L_1$. Suction is applied over the entire length of the plate to cause a constant velocity V_0 normal to the plate. Gravitational and viscous heating effects can be assumed to be negligible.

(a) Write the appropriate forms of the equations of motion and thermal energy equation applicable to this boundary-layer flow; it is not necessary here to justify the forms of these equations by scaling; that is, you can begin with the dimensional momentum and thermal boundary-layer equations that resulted from the scaling done in Section 4.6.

(b) Write the boundary conditions required to solve the coupled equations of motion and thermal energy equation.

(c) We might anticipate that with boundary-layer suction such as we have in this problem, both the momentum and thermal boundary-layer thicknesses might ultimately become constant rather than grow without bound as they do for a boundary layer on a semi-infinite flat plate without suction or with blowing. Use scaling analysis to determine the criteria for obtaining both a constant momentum boundary-layer thickness as well as a constant thermal boundary-layer thickness; express your answers in terms of dimensionless groups, which must be very small.

(d) For the constant momentum and thermal boundary-layer conditions obtained in part (c), determine the temperature profile.

(e) Discuss whether the boundary-layer suction will increase or decrease the heat transfer relative to when no suction is used.

4.P.24 Evaporative Cooling of a Liquid Film with Radiative Heat Transfer

In Example Problem 4.E.6 we considered the cooling that occurs when a volatile liquid evaporates into an ambient gas phase. In our analysis of this problem we assumed that there was no significant heat transfer from the ambient gas phase to the liquid film. Consider now the effect of radiative heat transfer on this evaporative cooling process; assume that the radiative heat-transfer flux is given by

$$q_x = \sigma\varepsilon(T^4 - T_\infty^4) \tag{4.P.24-1}$$

in which σ is the Stefan–Boltzmann constant, ε the emissivity of the surface of the film, and T_∞ the temperature of the medium that is causing the radiative heat transfer.

(a) The presence of the radiative heat transfer will alter the boundary condition at the moving interface that is obtained from an integral energy balance; derive this modified boundary condition.

(b) Use scaling analysis to determine the criterion for ignoring the radiative heat transfer.

(c) Use scaling analysis to determine the criterion for ignoring the evaporative cooling relative to the radiative heating for the case when $T_\infty > T_0$.

(d) Use scaling analysis to determine the criterion to ensure that the radiative heat transfer is sufficient to maintain the liquid film at its initial temperature T_0.

4.P.25 Melting of Frozen Soil Due to Constant Radiative Heat Flux

In Section 4.7 we considered the melting of frozen soil initially at its freezing point due to a higher temperature being applied at the ground surface. Assume now that the melting is caused by a constant radiative heat flux q_0 at the ground surface rather than by a higher temperature being applied. Note that this is a reasonable condition for melting, due to exposure of the ground surface to solar radiation.

(a) Consider carefully whether this change in the boundary condition at the ground surface will change the integral energy balance that is needed to determine the instantaneous location of the freezing front.

(b) Determine the temperature scale appropriate to this modified condition at the ground surface.

(c) Determine the criterion for assuming that the melting is quasi-steady-state.

(d) What relationship between the thaw depth and time does scaling imply for very short contact times?

4.P.26 Melting of Frozen Soil Initially at Subfreezing Temperature

In our scaling analysis of the melting of frozen soil in Section 4.7, we made the somewhat unrealistic assumption that the soil was initially at its freezing point T_f. Assume now that the soil is infinitely thick and initially at a temperature T_0, where $T_0 < T_f$. This modified initial condition implies that heat transfer will be to the freezing front from above but away from it in the frozen region below. Hence, the thermal energy equation must be considered in the regions both above and below the freezing front. Assume that k_u, ρ_u, and C_{pu} are the effective thermal conductivity, mass density, and heat capacity, respectively, of the unfrozen soil and that k_f, ρ_f, and C_{pf} are the effective thermal conductivity, mass density, and heat capacity, respectively, of the frozen soil.

(a) Write the appropriate form of the thermal energy equation in both regions.
(b) Write the initial and boundary conditions required to solve the equations in part (a).
(c) Derive the auxiliary condition required to determine the location of the melting front.
(d) Scale the describing equations to determine when heat transfer to the underlying frozen soil can be neglected.
(e) Determine the criteria for assuming that the melting is quasi-steady-state; be careful to consider the implications of heat transfer to the underlying ice.
(f) Determine the thickness of the region of influence wherein the heat transfer in the frozen soil can be assumed to be confined.

4.P.27 Freezing of Water-Saturated Soil Initially Above Its Freezing Temperature

Consider water-saturated soil initially at a temperature T_∞ above its freezing temperature T_f. The ground surface then is subjected to a subfreezing temperature $T_0 < T_f$ that eventually causes a freezing front to propagate down through the soil, as shown in Figure 4.P.27-1. We consider modeling this freezing process from the instant at which the upper surface of the water-saturated soil reaches its freezing point T_f; that is, you do not need to consider the unsteady-state heat-transfer process during which the temperature at the soil surface drops to the freezing point. Note that these conditions imply that heat is transferred from the freezing front in the upward direction but is transferred to the freezing front from the unfrozen soil beneath it. Hence, the thermal energy equation must be considered in the regions both above and below the freezing front. Assume that k_u, ρ_u, and C_{pu} are the effective thermal conductivity, mass density, and heat capacity, respectively, of the unfrozen soil and that k_u, ρ_f, and C_{pf} are the effective thermal conductivity, mass density, and heat capacity, respectively, of the frozen soil.

(a) Write the appropriate form of the thermal energy equation in both regions.
(b) Write the initial and boundary conditions required to solve the equations in part (a).

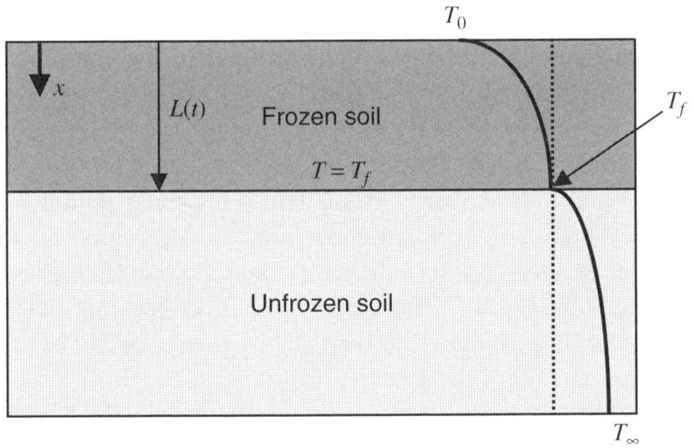

Figure 4.P.27-1 Unsteady-state one-dimensional heat transfer due to the imposition of a temperature T_0 at the surface of unfrozen water-saturated porous soil whose initial temperature was T_∞, where $T_0 < T_\infty$; the position of the freezing front denoted by $L(t)$ penetrates progressively farther into the unfrozen soil, owing to conductive heat transfer to the ground surface.

(c) Derive the auxiliary condition required to determine the location of the melting front.

(d) Scale the describing equations to determine when heat transfer from the underlying unfrozen soil can be neglected.

(e) Determine the criteria for assuming that the melting is quasi-steady-state; be careful to consider the implications of heat transfer from the underlying unfrozen soil.

(f) Determine the thickness of the region of influence wherein the heat transfer in the unfrozen soil can be assumed to be confined.

4.P.28 Freezing of Water-Saturated Soil Overlaid by Snow

Figure 4.P.28-1 shows a schematic of unsteady-state one-dimensional heat conduction involving freezing of a water-saturated soil overlaid by a layer of snow of constant thickness L_1. Initially, both the entire snow layer and unfrozen water-saturated soil are assumed to be at T_f, the freezing temperature of water. At time $t = 0$, freezing is initiated such that a freezing front $L(t)$ penetrates the soil at a rate determined by the dynamics of the heat conduction. The prevailing winds cause forced-convection heat transfer at the interface between the snow and the air that can be described by the following lumped-parameter heat-flux condition:

$$q = h(T_\infty - T) \tag{4.P.28-1}$$

where h is the heat-transfer coefficient, T_∞ the temperature of the ambient air flowing over the snow ($T_\infty < T_f$), and T the unspecified temperature at the

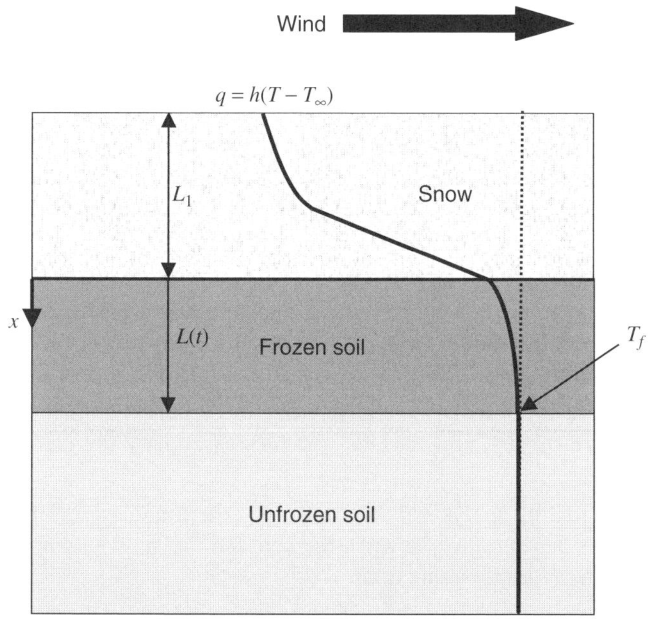

Figure 4.P.28-1 Unsteady-state one-dimensional freezing of water-saturated soil whose initial temperature was T_f overlaid by snow of thickness L_1 due to convective heat transfer at the ground surface; the position of the freezing front denoted by $L(t)$ penetrates progressively farther into the unfrozen soil, due to conductive heat transfer through the frozen soil and overlying snow.

interface between the snow and the air. The relevant properties include the following:

C_{p1}	effective heat capacity of the snow
C_{pu}	effective heat capacity of the unfrozen water-saturated soil
k_1	effective thermal conductivity of the snow
k_u	effective thermal conductivity of the unfrozen water-saturated soil
ρ_1	effective density of the snow
ρ_u	effective density of the unfrozen water-saturated soil
ρ_f	effective density of the frozen water-saturated soil
α_1	effective thermal diffusivity of the snow
α_u	effective thermal diffusivity of the unfrozen water-saturated soil
ΔH_f	latent heat of fusion of pure water
ε	porosity of the soil

Note that since snow is a very good insulator, $\alpha_1 \ll \alpha_u$.

(a) Write the appropriate forms of the thermal energy equation in both the snow and frozen soil that describe this unsteady-state heat-transfer problem; assume constant physical and transport properties.

(b) Write the appropriate initial and boundary conditions required to solve this problem.

(c) Derive the appropriate form of the auxiliary condition required to locate the instantaneous freezing front.

(d) Scale this heat-transfer problem, noting that it is necessary to introduce separate length scales for the heat transfer in the snow and frozen water-saturated soil. Note also that it is necessary to introduce a separate scale factor for the freezing-front velocity dL/dt since this does not scale with the ratio of the characteristic length divided by the characteristic time.

(e) Determine the criteria for assuming that this heat transfer is quasi-steady-state; discuss the implications of your results with respect to short and long observation times.

(f) Use your scaling analysis to determine the criterion that implies that the temperature of the snow surface becomes essentially the same as the bulk air T_∞.

(g) Use your scaling analysis to determine the criterion that implies that the temperature of the interface between the soil and the snow essentially becomes equal to the freezing temperature T_f.

(h) What is the implication of the result you obtained in part (g) for the rate of freezing-front penetration?

(i) What are the implications for the dimensional temperature if the heat-transfer coefficient goes to zero? What are the implications if it goes to infinity? Use your scaled equations to answer these questions.

(j) Use your scaling analysis to determine the criterion for neglecting the effect of the snow on the freezing process.

4.P.29 Heat Conduction in a Cylinder with Temperature-Dependent Thermal Diffusivity

Consider steady-state axial heat conduction in a long solid cylinder with a perfectly insulated lateral boundary due to a high temperature T_0 being imposed at $z = 0$ and a low temperature T_L being imposed at $z = L$, as shown in Figure 4.P.29-1. However, the thermal conductivity of the cylinder has a temperature dependence given by

$$k = k_0 - \beta(T - T_0) \tag{4.P.29-1}$$

where k_0 is the thermal conductivity evaluated at the reference temperature T_0 and β is a positive constant. The other relevant physical properties of the solid can be assumed to be constant.

(a) Write the appropriate form of the thermal energy equation and required boundary conditions for this heat-transfer problem.

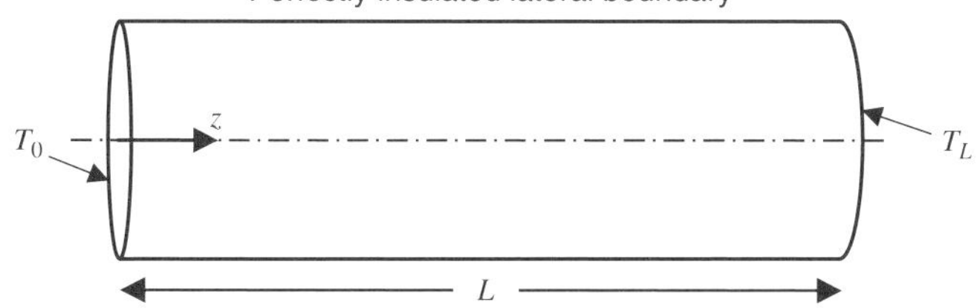

Figure 4.P.29-1 Steady-state axial conduction in a solid cylinder with a temperature-dependent thermal conductivity that is perfectly insulated along its lateral boundary.

(b) Scale the describing equations to determine when the temperature dependence of the thermal conductivity can be ignored.

4.P.30 Entry Region Effects for Free Convection Heat Transfer Adjacent to a Vertical Heated Flat Plate

In Example Problem 4.E.7 we considered steady-state free convection induced by immersing a heated vertical plate into an infinite fluid. We found that if the Grashof number is sufficiently large, boundary-layer simplifications can be made for both the equations of motion and the energy equation.

(a) The scaling for this problem was only outlined; complete the details of the scaling analysis; in particular, show how the scale factors in equation (4.E.7-21) were obtained.
(b) The boundary-layer analysis leading to the simplified set of describing equations given by equations (4.E.7-30) breaks down near the leading edge of the vertical plate. Determine the length of the region of influence near the leading edge wherein viscous and conductive transport in the axial direction cannot be ignored.
(c) Retaining the viscous and conductive transport in the axial direction in the equations of motion and energy equation implies that these equations will be elliptic and therefore require downstream boundary conditions. Indicate how the results of scaling analysis can be used to obtain a solution to this convective heat-transfer problem over the full length of the vertical plate.

4.P.31 Free Convection from a Heated Vertical Plate with Wall Suction

In Example Problem 4.E.7 we considered steady-state free convection induced by immersing a heated vertical plate into an infinite fluid. Assume now that a uniform suction velocity V_0 is applied along the heated wall. We will assume in this analysis that the Grashof number is sufficiently large so that the boundary-layer simplifications can be made for both the equations of motion and the energy equation.

(a) Use scaling analysis to determine the criterion for ignoring the effect of the wall suction.

(b) We might anticipate that ultimately, the suction will cause both the momentum and thermal boundary-layer thicknesses to become constant. Use scaling analysis to determine the criteria for obtaining a both a constant momentum boundary-layer thickness as well as a constant thermal boundary-layer thickness; express your answer in terms of dimensionless groups, which must be very small.

4.P.32 Correlation for Temperature in a Slab with Heat Generation

An infinite solid slab of thickness $2H$ and thermal conductivity k is initially at a uniform temperature T_0. There is a uniform volumetric rate of heat production within the slab given by G_e (energy/volume · time). The slab is cooled on each side by a fluid whose temperature far from the slab is given by T_∞. The heat-transfer coefficient between the slab and the fluid is given by h (energy/area · time · degree).

(a) Use the Pi theorem method to obtain the dimensionless groups needed to correlate the instantaneous temperature at the center of the slab.

(b) Use the scaling method for dimensional analysis to obtain the dimensionless groups needed to correlate the instantaneous temperature at the center of the slab. Reconcile any differences with the result you obtained in part (a).

(c) Simplify your scaling analysis result for the special case of steady-state heat transfer.

(d) Based on your result in part (c), derive an equation for the factor by which the centerline temperature will change if the generation rate is increased by 50%.

4.P.33 Correlation for Steady-State Heat Transfer from a Sphere

A highly conducting solid sphere of radius R is maintained at a constant temperature T_0 and fixed in a fluid stream having density ρ, viscosity μ, heat capacity C_p, and thermal conductivity k and whose velocity and temperature far from the sphere are U_∞ and T_∞, respectively, where $T_0 > T_\infty$. We seek to develop a correlation for the steady-state heat-transfer coefficient defined by

$$h \equiv \frac{\overline{q}}{T_0 - T_\infty} \tag{4.P.33-1}$$

where $\overline{q}$ is the heat flux averaged over the surface of the sphere.

(a) Write the appropriate forms of both the equations of motion and the energy equation and their boundary conditions for this heat-transfer problem.

(b) Use the scaling method for dimensional analysis to obtain the dimensionless groups needed to correlate the heat-transfer coefficient defined by equation (4.P.33-1).

(c) Consider how the correlation that you obtained in part (b) simplifies in the limit of zero flow, that is, $U_\infty = 0$.

(d) Compare the result that you obtained in part (b) to the standard correlation for the Nusselt number given by[21]

$$\mathrm{Nu} = 2 + 0.60\mathrm{Re}^{1/2}\mathrm{Pr}^{1/3} \tag{4.P.33-2}$$

where the Nusselt and Reynolds numbers are based on the sphere diameter.

(e) Equation (4.P.33-2) predicts that Nu = 2 in the limit of zero Reynolds number; this implies that the overall heat transfer is twice that of purely conductive heat transfer in the limit of zero Reynolds number. Is this result reasonable? Explain this result by considering the analytical solution for purely conductive steady-state heat transfer from a sphere.

4.P.34 Correlation for Hot Wire Anemometer Performance

A hot wire anemometer is a device for measuring the velocity in a flowing fluid. This device consists of a thin cylindrical wire of radius R that has a very high thermal conductivity. The hot wire anemometer determines the velocity by measuring the current required to cause sufficient electrical heat generation G_e (energy/time · volume) to maintain the wire at a constant temperature T_0 that is higher than the temperature T_∞ of the flowing fluid. The electrical heat generation that is required to maintain the wire at a constant temperature will change depending on the velocity of the flowing fluid and its relevant physical properties: its density ρ, thermal conductivity C_p, and thermal conductivity k. We seek to develop a correlation that will relate the electrical heat generation G to the velocity of the fluid.

(a) Write the appropriate forms of both the equations of motion and the energy equation and their boundary conditions for this heat-transfer problem.

(b) Use the scaling method for dimensional analysis to obtain the dimensionless groups needed to relate the electrical heat generation to the fluid velocity.

(c) Compare the result that you obtained in part (b) to one of the standard correlations for forced convection heat transfer to or from a cylinder given by[22]

$$\begin{aligned}\mathrm{Nu} = {} & (0.376\mathrm{Re}^{1/2} + 0.057\mathrm{Re}^{2/3})\mathrm{Pr}^{1/3} \\ & + 0.92\left[\ln\left(\frac{7.4055}{\mathrm{Re}}\right) + 4.18\mathrm{Re}\right]^{-1/3}\mathrm{Re}^{1/3}\mathrm{Pr}^{1/3}\end{aligned} \tag{4.P.34-1}$$

where Pr is the Prandtl number, and Nu and Re, the Nusselt and Reynolds number, are based on the cylinder diameter. In particular, show how the above is a special case of the more general result that you obtained from dimensional analysis.

[21] Bird et al., *Transport Phenomena*, 2nd ed., p. 439.
[22] Ibid., p. 440.

(d) Consider the implications of equation (4.P.34-1) in the limit of zero Reynolds number. Is your result reasonable? Provide an explanation for the strange behavior observed in this limit by considering purely conductive steady-state heat transfer from a cylinder.

4.P.35 Correlation for Unsteady-State Heat Transfer to a Sphere with Temperature-Dependent Thermal Conductivity

A solid sphere of radius R, heat capacity C_p, density ρ, and initial temperature T_0 is immersed at time $t = 0$ into a hot fluid whose temperature far from the sphere is T_∞; the heat transfer between the sphere and the fluid is described by a constant heat-transfer coefficient h. The thermal conductivity of the sphere is temperature-dependent and described by

$$k = k_0 + \beta k_0 (T - T_0) \tag{4.P.35-1}$$

where k_0 and β are constants.

(a) Use the Pi theorem method to obtain the dimensionless groups needed to correlate the instantaneous temperature at any radial position within the sphere.

(b) Use the scaling method for dimensional analysis to obtain the dimensionless groups needed to correlate the instantaneous temperature at any radial position within the sphere. Reconcile any differences with the result you obtained in part (a).

(c) Assume now that we wish to develop a correlation for the instantaneous temperature at the surface of a sphere that has a radius of 1 m by studying a sphere that has a diameter of 5 cm. How can the dimensionless correlation obtained from data taken for the 5-cm sphere be used to determine the thermal response of the 1-m sphere? Indicate any conditions that need to be satisfied with respect to the studies on the small sphere in order to do this.

4.P.36 Characterization of Home Freezer Performance

Assume that we want to characterize the performance of home freezers whose shape is that of geometrically similar rectangular parallelepipeds of height L_1, width L_2, and depth L_3. The operating cost of a freezer is directly proportional to the total heat-transfer rate (energy/time) resulting from the difference between the ambient room temperature T_∞ and that of the interior of the freezer wall. In a well-designed freezer, this heat-transfer rate is controlled by conduction through its walls, all of which have thickness H. The insulation in the freezer walls is characterized by its density ρ, heat capacity C_p, and thermal conductivity k. To assess freezer performance, a simple test is designed that involves suspending an incandescent light bulb at the center of the freezer; the test involves measuring the instantaneous temperature at a fixed point as a function of time after the light bulb is turned on. The light bulb can be assumed to be a radiating point source having a constant heating rate of $\tilde{G}$ (energy/time). Each point along a given wall in the freezer receives a heat flux that depends on the distance between the wall and the light bulb. To characterize freezer performance, we consider a correlation for the instantaneous temperature at the midpoint of the inside surface of one wall of the freezer. Clearly, larger freezers will have a smaller heat flux than smaller freezers at this same point if the same light bulb is used. Moreover, the instantaneous temperature at the midpoint of the front wall in general will be different from than at the midpoint of the sidewall.

(a) Use the Pi theorem method to obtain the dimensionless groups needed to correlate the instantaneous temperature at the midpoint of the inside surface of one wall of the freezer.

(b) Use the scaling method for dimensional analysis to obtain the dimensionless groups needed to correlate the instantaneous temperature at the midpoint of the inside surface of one wall of the freezer. Reconcile any differences with the result you obtained in part (a).

(c) Simplify your scaling analysis result for the special case of steady-state heat transfer.

(d) Two groups of investigators carry out separate tests on the same freezer. However, one group measures the temperature at the inside of the front wall, whereas the other group measures the temperature at the inside of the sidewall. How can the data from these two groups be consolidated onto one plot for freezer performance?

(e) The operating cost for a freezer depends on the total heat transfer (energy/time) from the ambient air through the freezer walls. Determine the factor by which the insulation thickness needs to be changed to ensure that a freezer that is 50% larger in its three dimensions has the same operating costs as its smaller counterpart.

5 Applications in Mass Transfer

We also assume that the disc is infinitely wide, so that the concentration is a function only of z. I find this assumption mind-boggling, but it is justified by the success of the following calculations.[1]

5.1 INTRODUCTION

The quotation above underscores the uncertainty regarding some topics in mass-transfer analysis, in this case the assumption of a uniformly accessible surface offered by the rotating disk. This problem as well as many others are discussed in this chapter, where we consider the application of scaling analysis to mass transfer. The organization of this chapter is the same as that used in Chapters 3 and 4. Clearly, it is essential to read Chapter 2 in order to understand the scaling procedure used in this chapter. Since mass transfer can occur due to both species diffusion and convection, it is also useful to read Chapter 3 to fully understand the material in this chapter. Again, the first few examples are developed in detail, as was done in Chapters 3 and 4. We again use the ordering symbols o(1) and O(1) introduced in Chapter 2. The symbol o(1) implies that the magnitude of the quantity can range between zero and more-or-less 1, whereas the symbol O(1) implies that the magnitude of the quantity is more-or-less 1 but not much less than 1.

Many of the topics considered in this chapter, such as film theory, penetration theory, and boundary-layer analysis are quite similar to those considered for heat transfer in Chapter 4. However, mass transfer is uniquely different from heat transfer in that in contrast to conductive transport, diffusive transport, can cause bulk flow; that is, the diffusion of species can result in a net movement of mass, thereby causing a bulk flow velocity. This velocity can convect species to complement the diffusional transport and can also distort the concentration profiles from what they

[1]E. L. Cussler, *Diffusion: Mass Transfer in Fluid Systems*, Cambridge University Press, Cambridge, England, 1985, p. 76.

Scaling Analysis in Modeling Transport and Reaction Processes: A Systematic Approach to Model Building and the Art of Approximation, By William B. Krantz

would be for purely diffusive transport. For this reason, in the first example we will use scaling to assess when the convective transport arising from the species diffusion can be neglected. Note that no attempt is made here to provide a detailed derivation of the describing equations that are used in the scaling analysis. Hence, the material in this chapter provides a useful supplement for a foundation course in mass transfer. The reader is referred to the appendices, which summarize the species-balance equation in generalized vector–tensor notation as well as in rectangular, cylindrical, and spherical coordinates. These equations serve as the starting point for each example problem.

The goal in Sections 5.2 through 5.8 is to use scaling analysis to develop classical approximations made in mass-transfer modeling. Hence, Sections 5.2 and 5.3 use scaling to develop the film theory and penetration theory models. Although these two models are developed for a stationary liquid film, they can be applied to a variety of complex problems for which the resistance to mass transfer can be associated with one-dimensional transport through a film near one of the boundaries. Mass transfer often involves either homogeneous reactions that occur in the bulk of the system or heterogeneous reactions that occur on the boundaries; these are considered in Sections 5.4 and 5.5, respectively. When convective transport is large, the mass-transfer resistance can be confined to a thin region of influence or boundary layer; this is considered in Section 5.6. If mass transfer causes significant mass loss or gain and/or densification or expansion, moving boundaries can be involved; these are considered in Section 5.7. In Chapter 4 we used scaling analysis to determine when the temperature dependence of the physical and transport properties needs to be considered. In Section 5.8 we apply scaling analysis to simplify the describing equations when the diffusivity is concentration-dependent. Scaling is applied to solutally induced buoyancy-driven free convection in Section 5.9. Finally, the scaling analysis approach is applied to dimensional analysis in developing a correlation for the performance of a membrane–lung oxygenator in Section 5.10. Several additional worked example problems are included. In particular, these examples use scaling analysis to develop systematically the criteria for Taylor dispersion, field-flow fractionation, the uniformly accessible rotating disk, and small Thiele modulus flows. Unworked practice problems are included at the end of the chapter.

5.2 FILM THEORY APPROXIMATION

The first example is used to develop the basis for the classical film theory and penetration theory approximations for modeling complex mass-transfer problems. These two models were developed for heat-transfer applications in Section 4.3. In this chapter we develop these models in separate sections since scaling will be used not only to develop the criteria for the film theory and penetration theory approximations, but also to determine the criterion for ignoring the convective mass transfer that can be generated by diffusion. In this section the film theory model is developed, and in Section 5.3 the penetration theory approximation is considered.

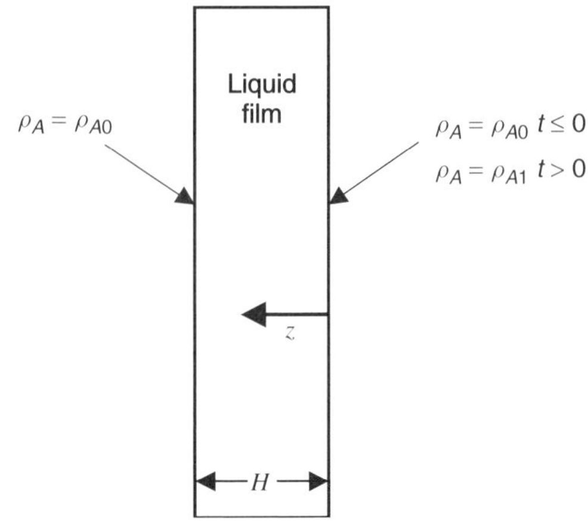

Figure 5.2-1 Unsteady-state one-dimensional binary mass transfer in an infinitely wide liquid film of thickness H due to a sudden change in concentration from ρ_{A0} to ρ_{A1} at one boundary.

Consider the liquid film shown in Figure 5.2-1 that has thickness H and consists of components A and B, whose initial mass concentrations (i.e., mass per unit volume) are ρ_{A0} and ρ_{B0}, respectively. At time $t = 0$ the concentration of component A at one boundary is increased to ρ_{A1}, while its concentration at the other boundary is maintained at ρ_{A0}. This causes diffusion of A and B since a concentration gradient in one component causes a complementary gradient in the other.

In modeling mass transfer involving n components, one can either write n species-balance equations, or $n - 1$ species-balance equations and the overall mass balance. This follows from the fact that the sum of the n species-balance equations is the same as an overall mass balance. Since it is reasonable to assume that the mass density is constant for an incompressible liquid, it is convenient here to consider the species-balance equations in terms of the mass fluxes. Hence, step 1 consists of writing the appropriately simplified continuity and species-balance equations, given by equations (C.1-1) and (G.1-1) in the Appendices as follows:

$$\frac{\partial \rho}{\partial t} = -\frac{\partial}{\partial z}(\rho u) \Rightarrow \frac{\partial u}{\partial z} = 0 \Rightarrow u = u(t) \tag{5.2-1}$$

$$\frac{\partial \rho_A}{\partial t} = -\frac{\partial n_A}{\partial z} \tag{5.2-2}$$

where the mass-average velocity u is defined by

$$u \equiv \frac{n_A + n_B}{\rho} \tag{5.2-3}$$

in which the mass fluxes n_A and n_B of components A and B, respectively, are given by

$$n_A = \rho_A u - \rho D_{AB}\frac{\partial \omega_A}{\partial z} = \rho_A u - D_{AB}\frac{\partial \rho_A}{\partial z} \tag{5.2-4}$$

$$n_B = \rho_B u - \rho D_{AB}\frac{\partial \omega_B}{\partial z} = \rho_B u - D_{AB}\frac{\partial \rho_B}{\partial z} \tag{5.2-5}$$

in which $\rho = \rho_A + \rho_B$ is the overall mass density, $\omega_A = \rho_A/\rho$ and $\omega_B = \rho_B/\rho$ the mass fractions of components A and B, respectively, and D_{AB} the binary diffusion coefficient. The integration in equation (5.2-1) follows from the fact that we are assuming an incompressible liquid. Note that we need to consider only the overall continuity equation and the species-balance equation for one of the two species since the sum of the two species-balance equations is equal to the continuity equation. Equations (5.2-4) and (5.2-5) indicate that the total mass flux is the sum of a convective flux that is proportional to the mass-average velocity and a purely diffusive flux given by the term proportional to the concentration gradient. Equation (5.2-3) shows that the convective velocity arises from the mass transfer since it is proportional to the sum of the mass fluxes.

Note that for this problem we chose to express the concentrations in terms of mass per unit volume and mass fractions. This was convenient since it is generally quite reasonable to assume that liquids have a constant mass density. A constant mass density in this case implies that the mass-average velocity u is a function only of time. We could also have expressed the concentrations in terms of moles per unit volume and mole fractions. However, in the case of mass transfer, the molar density even for liquids might not remain constant. Note, however, that molar concentrations are particularly convenient for mass transfer in gases since the molar density is constant for an ideal gas at constant temperature and pressure. Section 5.7 considers mass transfer in an ideal gas at constant temperature and pressure for which molar concentrations are used.

Equations (5.2-1) to (5.2-5) constitute five equations in five unknowns: ρ_A, ρ_B, n_A, n_B, and u. However, equations (5.2-3), (5.2-4), and (5.2-5) are not independent, that is, the sum of equations (5.2-4) and (5.2-5) is equal to equation (5.2-3). Hence, an additional equation is needed. This can be either some specified relationship between the mass fluxes or an equation of state that relates the mass density to the concentration.[2] However, when the latter is specified, it is also necessary to know the value of the velocity at one boundary since a spatial integration is required to obtain the velocity from the continuity equation. Here we specify that the ratio of the mass fluxes of the two components is a constant, that is,

$$\frac{n_B}{n_A} = \kappa, \qquad \text{a constant} \tag{5.2-6}$$

[2]Note that specifying an equation of state for the mass density does not contradict the fact that $\rho = \rho_A + \rho_B$; for example, we might specify that $\rho = \omega_A \rho_A^0 + \omega_B \rho_B^0$, where ρ_A^0 and ρ_B^0 denote pure component densities. The latter provides an independent equation from which the mass-average velocity can be determined; this is explored further in Example Problem 5.E.1.

This specification includes several special situations of interest in mass transfer. The condition $\kappa = 0$ corresponds to unimolecular mass transfer of component A in stationary phase B. However, note that unimolecular mass transfer is a misnomer in that it does not imply that component B is not diffusing. Indeed, the diffusive and convective transport of component B in equation (5.2-5) exactly balance each other for unimolecular mass transfer, so that their sum is zero. The condition $\kappa = -1$ corresponds to equimass counterdiffusion of components A and B. That is, the mass flux of component A is exactly equal in magnitude and opposite in direction to that of component B. Note that we define κ so that it is bounded of $\mathrm{o}(1)$; hence, the case of unimolecular diffusion of component B in stationary phase A is treated by inverting the ratio of mass fluxes in equation (5.2-6).

Step 1 is completed by specifying the requisite initial and boundary conditions given by

$$\rho_A = \rho_{A0} \quad \text{at} \quad t \le 0, \quad 0 \le z \le H \tag{5.2-7}$$

$$\rho_A = \rho_{A1} \quad \text{at} \quad z = 0, \quad 0 < t \le \infty \tag{5.2-8}$$

$$\rho_A = \rho_{A0} \quad \text{at} \quad z = H, \quad 0 \le t \le \infty \tag{5.2-9}$$

Equation (5.2-7) is the prescribed initial condition, whereas equations (5.2-8) and (5.2-9) are the known conditions at the two boundaries. We will use scaling analysis to explore when steady-state mass transfer can be assumed. We also use scaling to assess when the convective transport arising from the diffusion can be ignored.

We begin by defining dimensionless variables involving unspecified scale and reference factors (steps 2, 3, and 4):

$$\rho_A^* \equiv \frac{\rho_A - \rho_{Ar}}{\rho_{As}}; \qquad u^* \equiv \frac{u}{u_s}; \qquad n_A^* \equiv \frac{n_A}{n_{As}}; \qquad z^* \equiv \frac{z}{z_s}; \qquad t^* \equiv \frac{t}{t_s} \tag{5.2-10}$$

We then introduce these dimensionless variables into the describing equations and divide through by the coefficient of one term in each of these equations that we believe should be retained (steps 5 and 6):

$$\frac{\rho_{As} z_s}{n_{As} t_s} \frac{\partial \rho_A^*}{\partial t^*} = -\frac{\partial n_A^*}{\partial z^*} \tag{5.2-11}$$

$$\frac{n_{As} z_s}{\rho_{As} D_{AB}} n_A^* = \frac{u_s z_s}{D_{AB}} \rho_A^* u^* - \frac{\partial \rho_A^*}{\partial z^*} \tag{5.2-12}$$

$$u^* = \frac{(1+\kappa) n_{As}}{\rho u_s} n_A^* \tag{5.2-13}$$

$$\rho_A^* = \frac{\rho_{A0} - \rho_{Ar}}{\rho_{As}} \quad \text{at} \quad t^* \le 0, \quad 0 \le z^* \le \frac{H}{z_s} \tag{5.2-14}$$

$$\rho_A^* = \frac{\rho_{A1} - \rho_{Ar}}{\rho_{As}} \quad \text{at} \quad z^* = 0, \quad 0 < t^* \le \infty \tag{5.2-15}$$

$$\rho_A^* = \frac{\rho_{A0} - \rho_{Ar}}{\rho_{As}} \quad \text{at} \quad z^* = \frac{H}{z_s}, \quad 0 \le t^* \le \infty \tag{5.2-16}$$

Now let us proceed to determine the scale factors (step 7). The dimensionless concentration can be bounded of o(1) by setting the group containing the concentration scale and reference factors in equations (5.2-15) and (5.2-16) equal to 1 and zero, respectively, to obtain

$$\frac{\rho_{A0} - \rho_{Ar}}{\rho_{As}} = 0 \Rightarrow \rho_{Ar} = \rho_{A0}; \qquad \frac{\rho_{A1} - \rho_{Ar}}{\rho_{As}} = 1 \Rightarrow \rho_{As} = \rho_{A1} - \rho_{A0} \tag{5.2-17}$$

Since we seek to determine when steady-state conditions apply, the appropriate time scale is the observation or contact time; that is, $t_s = t_o$. The manner in which the length scale factor is determined depends on the observation time. Let us assume that the dimensionless group containing the length scale in equation (5.2-16) determines z_s. Although this bounds the dimensionless spatial coordinate to be o(1), it does not necessarily bound the dimensionless concentration gradient to be o(1). Indeed, the concentration gradient could involve a much shorter length scale during the early stages of mass transfer when the species diffusion has not penetrated very far from the boundary at $z = 0$. However, let us assume that

$$\frac{H}{z_s} = 1 \Rightarrow z_s = H \tag{5.2-18}$$

The scale factor for the mass flux is obtained by setting the appropriate dimensionless group in equation (5.2-12) equal to 1, thereby obtaining

$$\frac{n_{As} z_s}{\rho_{As} D_{AB}} = 1 \Rightarrow n_{As} = \frac{D_{AB}(\rho_{A1} - \rho_{A0})}{H} \tag{5.2-19}$$

The scale factor for the mass-average velocity is obtained by setting the dimensionless group in equation (5.2-13) equal to 1 as follows:

$$\frac{(1+\kappa) n_{As}}{\rho u_s} = 1 \Rightarrow u_s = \frac{(1+\kappa) D_{AB}(\rho_{A1} - \rho_{A0})}{\rho H} \tag{5.2-20}$$

Equation (5.2-20) indicates why we defined κ to be o(1); that is, the scale factor for the mass-average velocity would become unbounded for the case of unimolecular diffusion of component B. For the latter case we redefine κ by inverting the fluxes in equation (5.2-6) to ensure that κ is always o(1).

Substitution of the aforementioned scale and reference factors into the describing equations yields

$$\frac{1}{\mathrm{Fo}_m} \frac{\partial \rho_A^*}{\partial t^*} = -\frac{\partial n_A^*}{\partial z^*} \tag{5.2-21}$$

$$n_A^* = \Pi_1 \rho_A^* n_A^* - \frac{\partial \rho_A^*}{\partial z^*} \tag{5.2-22}$$

$$\rho_A^* = 0 \quad \text{at} \quad t^* \leq 0, \quad 0 \leq z^* \leq 1 \tag{5.2-23}$$

$$\rho_A^* = 1 \quad \text{at} \quad z^* = 0, \quad 0 \leq t^* \leq \infty \tag{5.2-24}$$

$$\rho_A^* = 0 \quad \text{at} \quad z^* = 1, \quad 0 \leq t^* \leq \infty \tag{5.2-25}$$

where

$$\text{Fo}_m \equiv \frac{D_{AB}t_o}{H^2} \quad \text{and} \quad \Pi_1 \equiv (1+\kappa)\frac{\rho_{A1} - \rho_{A0}}{\rho} \tag{5.2-26}$$

in which Fo_m is the solutal Fourier number or Fourier number for mass transfer, although this terminology is rarely used. Its physical significance is analogous to that for the Fourier number in heat transfer; namely, it is a measure of the ratio of the contact time to the characteristic time for molecular transport via either conduction or diffusion.[3] The dimensionless group Π_1 is a measure of the ratio of the convective to diffusive mass transfer and is physically bounded to be less than 1 since the maximum value of κ is zero and the maximum value of $\rho_{A1} - \rho_{A0}$ has to be less than ρ.

Now let us explore possible simplifications of the describing equations (step 8). Note that if $\text{Fo}_m \gg 1$, the unsteady-state term in equation (5.2-21) becomes insignificant; hence, the mass transfer is steady-state; that is,

$$\text{Fo}_m \equiv \frac{D_{AB}t_o}{H^2} \gg 1 \Rightarrow \text{steady-state mass transfer} \tag{5.2-27}$$

Note that a large Fourier number in this problem ensures that the mass transfer is truly steady-state, in contrast to quasi-steady-state. The latter implies that the unsteady-state term in the species-balance equation is negligible but that the problem is still unsteady state, owing to the time dependence that enters through the boundary conditions. If the following condition is satisfied, the convective contribution to the mass-transfer flux can be ignored:

$$\Pi_1 \equiv (1+\kappa)\frac{\rho_{A1} - \rho_{A0}}{\rho} \ll 1 \tag{5.2-28}$$

Note that this condition is satisfied identically for equimass counterdiffusion for which $\kappa = -1$. Note if the inequality in equation (5.2-28) is satisfied, both the convective mass flux as well as the effect of the bulk flow on the distortion of the concentration profiles can be neglected.

The steady-state describing equations that result when equation (5.2-27) is satisfied form the basis of film theory. The latter is used to model complex problems for which the resistance to mass transfer can be assumed to be confined to a thin film near one of the boundaries of the system; for example, mass transfer in turbulent pipe flow can be modeled assuming that the mass-transfer resistance is confined

[3]The Fourier number in heat transfer was introduced in Section 4.3.

to a thin film near the wall within which the turbulent eddies are damped and the transfer is by diffusion. Another valuable use of film theory is to correct mass-transfer coefficients obtained from empirical correlations for the effect of large mass-transfer fluxes [i.e., corresponding to $\Pi_1 = \mathrm{O}(1)$] when the mass transfer is not contact-time limited. That is, correlations for mass-transfer coefficients are usually obtained in the limit of very small mass-transfer fluxes (i.e., $\Pi_1 \ll 1$) since this minimizes the number of dimensionless groups required. Film theory can be used to derive an equation for the ratio of the mass-transfer coefficient at high flux to that at low flux. By multiplying the resulting equation by the mass-transfer coefficient obtained from the empirical correlation valid only at low fluxes, one can obtain a reliable estimate of the mass-transfer coefficient applicable at high fluxes. This is discussed in more detail by Bird et al.[4]

Assume now that $\mathrm{Fo}_m \gg 1$, corresponding to steady-state mass transfer. The resulting set of simplified describing equations can be solved analytically to obtain the following solution for the mass flux of component A:

$$n_A^* = -\frac{1}{\Pi_1} \ln(1 - \Pi_1) \tag{5.2-29}$$

Note that for $\Pi_1 \ll 1$, corresponding to negligible convective mass transfer, equation (5.2-29) reduces to

$$n_A^* = 1 \tag{5.2-30}$$

Let us assess the error incurred in determining n_A^* when convective mass transfer is ignored. This error is 5.1% and 0.50% for $\Pi_1 = 0.1$ and 0.01, respectively. This is typical for scaling analysis: namely, that the error incurred in making some approximation becomes negligible if the particular dimensionless group involved in the criterion is $\mathrm{o}(0.01)$.

5.3 PENETRATION THEORY APPROXIMATION

In the preceding example we sought to determine when steady-state conditions applied. Hence, we bounded z^* to be $\mathrm{o}(1)$ by setting $z_s = H$. However, this length scale is not appropriate for short contact times for which the diffusion does not penetrate across the entire thickness of the film shown in Figure 5.2-1. The appropriate length scale for short contact times is obtained by balancing the unsteady-state and diffusion terms in equation (5.2-11), that is, by setting the dimensionless group in this equation equal to 1 to obtain

$$\frac{z_s^2}{D_{AB} t_o} = 1 \Rightarrow z_s = \sqrt{D_{AB} t_o} \tag{5.3-1}$$

[4]R. B. Bird, W. E. Stewart, and E. N. Lightfoot, *Transport Phenomena*, 2nd ed., Wiley, Hoboken, NJ, 2002, pp. 704–706.

Note that z_s defines a region of influence or boundary layer wherein all the mass transfer is confined. The thickness of this region of influence increases in time until it eventually penetrates the entire film. The other scale and reference factors remain unchanged from those determined in Section 5.2. The resulting dimensionless describing equations are given by

$$\frac{\partial \rho_A^*}{\partial t^*} = -\frac{\partial n_A^*}{\partial z^*} \tag{5.3-2}$$

$$n_A^* = \Pi_1 \rho_A^* n_A^* - \frac{\partial \rho_A^*}{\partial z^*} \tag{5.3-3}$$

$$\rho_A^* = 0 \quad \text{at} \quad t^* \leq 0, \quad 0 \leq z^* \leq \frac{H}{\sqrt{D_{AB} t_o}} \tag{5.3-4}$$

$$\rho_A^* = 1 \quad \text{at} \quad z^* = 0, \quad 0 < t^* \leq \infty \tag{5.3-5}$$

$$\rho_A^* = 0 \quad \text{at} \quad z^* = \frac{H}{\sqrt{D_{AB} t_o}}, \quad 0 \leq t^* \leq \infty \tag{5.3-6}$$

where Π_1 is defined by equation (5.2-26). We again see that if the criterion given by equation (5.2-28) is satisfied, the convective contribution to the mass-transfer flux in equation (5.3-3) can be ignored. Moreover, if

$$\frac{H}{\sqrt{D_{AB} t_o}} \gg 1 \tag{5.3-7}$$

the boundary condition defined by equation (5.3-6) can be applied at infinity. The solution to this simplified set of describing equations is given in standard references.[5]

The simplified describing equations that result when equation (5.3-7) is satisfied form the basis of penetration theory. This is also used to model complex problems for which the diffusive mass transfer is contact-time limited; for example, mass transfer from a gas phase to liquid film flow down a short vertical wall. Penetration theory is also used to correct mass-transfer coefficients obtained from empirical correlations for the effect of large mass-transfer fluxes [i.e., corresponding to $\Pi_1 = O(1)$]. However, penetration theory is used to make this correction when the mass transfer is contact-time limited; that is, for short contact times.[6] The procedure for making this correction is analogous to that used for film theory; namely, an equation is derived for the ratio of the mass-transfer coefficient at high flux to that at low flux. By multiplying the mass-transfer coefficient obtained from the low flux empirical correlation by this ratio, an estimate of the mass-transfer coefficient applicable at high fluxes is obtained. This is discussed in more detail by Bird et al.[7]

[5] Ibid., pp. 613–617.

[6] It is interesting to note that the high flux correction factors obtained from film theory and penetration theory do not differ significantly even though these two models apply at opposite ends of the mass-transfer contact-time spectrum; the reason for this is that the correction factor involves the ratio of the mass-transfer coefficients.

[7] Bird et al., *Transport Phenomena*, 2nd ed., pp. 706–708.

5.4 SMALL PECLET NUMBER APPROXIMATION FOR LAMINAR FLOW WITH A HOMOGENEOUS REACTION

Consider the steady-state fully developed laminar flow of a Newtonian liquid with constant physical properties between upper and lower lateral boundaries that consist of infinitely wide parallel semipermeable membranes separated by a distance $2H$ and having length L as shown in Figure 5.4-1. The laminar flow velocity profile is given by

$$u_x = 2\overline{U}\left(1 - \frac{y^2}{H^2}\right) \tag{5.4-1}$$

where $\overline{U}$ is the average velocity. The incoming liquid feed stream consists of pure component B. This feed stream reacts with component A, which is injected continuously through the semipermeable membrane boundaries at a constant molar flux N_{Aw}. Since the injection rate of component A is sufficiently low, its concentration remains dilute. Hence, component A is the limiting reactant and the reaction rate is given by

$$R_A = k_1 c_A \tag{5.4-2}$$

where R_A is the rate of homogeneous reaction of component A (moles/volume·time) and k_1 is the reaction rate constant for a first-order reaction (time^{-1}).

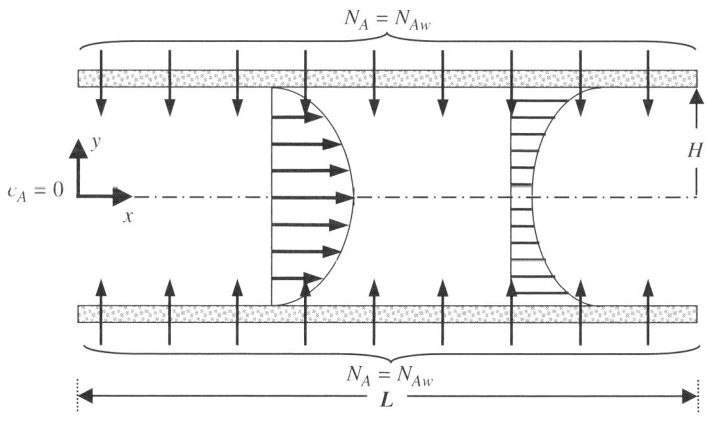

Figure 5.4-1 Laminar flow with a homogeneous chemical reaction; the liquid feed stream consists of pure component B; the feed stream undergoes an irreversible first-order homogeneous reaction with component A that is injected through the semipermeable membrane boundaries at a constant flux N_{Aw}; the injection rate is sufficiently low to ensure that the concentration of component A is dilute, thereby making it the limiting reactant; the concentration profile for component A and fully developed laminar velocity profile are shown in the figure.

The appropriate form of the species-balance equation given by equation (G.1-5) in the Appendices and corresponding boundary conditions are (step 1)

$$2\overline{U}\left(1-\frac{y^2}{H^2}\right)\frac{\partial c_A}{\partial x} = D_{AB}\frac{\partial^2 c_A}{\partial x^2} + D_{AB}\frac{\partial^2 c_A}{\partial y^2} - k_1 c_A \tag{5.4-3}$$

$$c_A = 0 \quad \text{at} \quad x = 0 \tag{5.4-4}$$

$$c_A = f(y) \quad \text{at} \quad x = L \tag{5.4-5}$$

$$D_{AB}\frac{\partial c_A}{\partial y} = -N_{Aw} \quad \text{at} \quad y = \pm H \tag{5.4-6}$$

$$\frac{\partial c_A}{\partial y} = 0 \quad \text{at} \quad y = 0 \tag{5.4-7}$$

where $f(y)$ is a function of y. Note that each term in equation (G.1-5) was divided by the molecular weight of component A in order to convert the mass concentration to molar concentration for the assumed dilute solution having constant mass and molar densities. This is a nontrivial problem to solve, due to the elliptic nature of the describing equations. The presence of the second-order axial derivative requires that a downstream boundary condition be specified. Often, these downstream conditions are not known, which precludes solving the describing equations. Clearly, one would like to know how these describing equations might be simplified to permit a tractable solution. In particular, one would like to know when the axial diffusion and convection terms might be neglected. We use o(1) scaling to determine the criteria for neglecting these terms.

Define the following dimensionless variables involving unspecified scale factors (steps 2, 3, and 4):

$$c_A^* \equiv \frac{c_A}{c_s}; \qquad x^* \equiv \frac{x}{x_s}; \qquad y^* \equiv \frac{y}{y_s} \tag{5.4-8}$$

Note that there is no need to introduce a reference factor for the concentration since it is naturally referenced to zero. Introduce these dimensionless variables into the describing equations and divide through by the coefficient of one term in each equation that should be retained (steps 5 and 6):

$$\frac{2\overline{U}y_s^2}{D_{AB}x_s}\left(1-\frac{y_s^2}{H^2}y^{*2}\right)\frac{\partial c_A^*}{\partial x^*} = \frac{y_s^2}{x_s^2}\frac{\partial^2 c_A^*}{\partial x^{*2}} + \frac{\partial^2 c_A^*}{\partial y^{*2}} - \frac{k_1 y_s^2}{D_{AB}}c_A^* \tag{5.4-9}$$

$$c_A^* = 0 \quad \text{at} \quad x^* = 0 \tag{5.4-10}$$

$$c_A^* = f(y^*) \quad \text{at} \quad x^* = \frac{L}{x_s} \tag{5.4-11}$$

$$\frac{\partial c_A^*}{\partial y^*} = -\frac{N_{Aw}y_s}{D_{AB}c_s} \quad \text{at} \quad y^* = \pm\frac{H}{y_s} \tag{5.4-12}$$

$$\frac{\partial c_A^*}{\partial y^*} = 0 \quad \text{at} \quad y^* = 0 \tag{5.4-13}$$

Step 7 involves bounding the independent and dependent dimensionless variables to be o(1). This can be done for the spatial coordinates by setting the dimensionless groups containing x_s and y_s in equations (5.4-11) and (5.4-12) equal to 1; that is,

$$\frac{L}{x_s} = 1 \Rightarrow x_s = L; \qquad \frac{H}{y_s} = 1 \Rightarrow y_s = H \tag{5.4-14}$$

Since the dimensionless concentration gradient must be bounded of o(1), we set the dimensionless group in equation (5.4-12) equal to 1; this yields the following scale factor for the concentration:

$$\frac{N_{Aw}H}{D_{AB}c_s} = 1 \Rightarrow c_s = \frac{N_{Aw}H}{D_{AB}} \tag{5.4-15}$$

Substitution of the scale and reference factors defined by equations (5.4-14) and (5.4-15) into the dimensionless describing equations given by equations (5.4-9) through (5.4-13) yields

$$2\text{Pe}_m \frac{H}{L}\left(1 - y^{*2}\right)\frac{\partial c_A^*}{\partial x^*} = \frac{H^2}{L^2}\frac{\partial^2 c_A^*}{\partial x^{*2}} + \frac{\partial^2 c_A^*}{\partial y^{*2}} - \text{Th}^2 c_A^* \tag{5.4-16}$$

$$c_A^* = 0 \quad \text{at} \quad x^* = 0 \tag{5.4-17}$$

$$c_A^* = f(y^*) \quad \text{at} \quad x^* = 1 \tag{5.4-18}$$

$$\frac{\partial c_A^*}{\partial y^*} = -1 \quad \text{at} \quad y^* = \pm 1 \tag{5.4-19}$$

$$\frac{\partial c_A^*}{\partial y^*} = 0 \quad \text{at} \quad y^* = 0 \tag{5.4-20}$$

where $\text{Pe}_m \equiv \overline{U}H/D_{AB}$ is the solutal Peclet number or Peclet number for mass transfer; note that $\text{Pe}_m = \overline{U}H/\nu \cdot \nu/D_{AB} = \text{Re} \cdot \text{Sc}$, where Re and Sc denote the Reynolds and Schmidt numbers, respectively. The Reynolds number is a measure of the ratio of the convection to the viscous or molecular transport of momentum. The Schmidt number is a measure of the ratio of the viscous or molecular transport of momentum to the diffusive or molecular transport of species. Hence, the Peclet number is a measure of the ratio of the convective to diffusive or molecular transport of species. The dimensionless group $\text{Th} \equiv \sqrt{k_1 H^2/D_{AB}}$, known as the *Thiele modulus* is a measure of the ratio of the characteristic time for diffusion relative to that for homogeneous reaction.

Now let us explore possible simplifications of the describing equations (step 8). The criterion for ignoring axial diffusion is

$$\frac{H^2}{L^2} \ll 1 \Rightarrow \text{axial diffusion can be ignored} \tag{5.4-21}$$

that is, the aspect ratio must be small. Note, however, that the length L was arbitrary in that L could denote any value of the axial coordinate in the principal direction of flow. This is the principle of *local scaling*, whereby we scale the problem for some fixed but arbitrary value of some coordinate, usually that in the principal direction of flow. When the inequality in equation (5.4-21) applies, the elliptic describing equation is reduced to a parabolic equation, thereby obviating the need to satisfy any downstream boundary condition. Recall that prescribing this downstream boundary condition often is problematic and precludes solving elliptic describing equations. Note that for distances sufficiently close to the entry region for the flow, the axial diffusion term cannot be ignored. This entry region length can be determined by assessing the criterion for when axial dispersion is significant; that is,

$$\frac{H^2}{L^2} \geq 0.1 \Rightarrow L \leq 10H \tag{5.4-22}$$

In order to ignore axial convection of species, the dimensionless group multiplying the first term in equation (5.4-16) must be very small; that is,

$$\mathrm{Pe}_m \cdot \frac{H}{L} \ll 1 \Rightarrow \text{axial convection of species can be ignored} \tag{5.4-23}$$

We see that the criterion for ignoring axial species convection is that the Peclet number be very small. The Peclet number in mass transfer has a role analogous to that of the Reynolds number in fluid dynamics; that is, when it is small, it justifies ignoring axial convective transport. We see in the next example problem that when it is large, it justifies a boundary-layer approximation. Note that ignoring convective transport in the species-balance equation in this example is analogous to ignoring convective transport in the equations of motion that are the basis of the creeping-flow approximation.[8]

If the conditions in equations (5.4-21) and (5.4-23) are satisfied, equations (5.4-16) through (5.4-20) reduce to

$$0 = \frac{\partial^2 c_A^*}{\partial y^{*2}} - \mathrm{Th}^2 c_A^* \tag{5.4-24}$$

$$\frac{\partial c_A^*}{\partial y^*} = -1 \quad \text{at} \quad y^* = \pm 1 \tag{5.4-25}$$

$$\frac{\partial c_A^*}{\partial y^*} = 0 \quad \text{at} \quad y^* = 0 \tag{5.4-26}$$

Note that equation (5.4-24) implies that the transverse diffusion of species A into the flowing liquid is balanced by its consumption, owing to the homogeneous

[8]The creeping-flow approximation was considered in Section 3.3 and Example Problem 3.E.2.

reaction. This in turn implies that $\text{Th}^2 \cong 1$. The solution to this simplified set of describing equations is straightforward and given by

$$c_A^* = \frac{1}{\text{Th}} \frac{e^{\text{Th}\cdot y} + e^{-\text{Th}\cdot y}}{e^{\text{Th}} - e^{-\text{Th}}} \tag{5.4-27}$$

Note that the dimensionless concentration predicted by equation (5.4-27) is no longer bounded of $\circ(1)$ when $\text{Th} \ll 1$. This implies that the reaction is not sufficiently fast to prevent the concentration of component A from building up due to the continuous injection through the membrane boundaries. That is, our scaling implicitly assumed that the transverse diffusion of component A was balanced by its consumption, due to the homogeneous reaction. This is no longer true if the homogeneous reaction rate becomes small. In this case the convection term rather than the reaction term must balance the transverse diffusion term; that is, the describing equations must be rescaled appropriately.

Now let us consider the case when the Thiele modulus is very large, thereby implying a very fast homogeneous reaction. This implies that component A will be consumed within a region of influence near the two membrane boundaries. In this case the transverse length scale is no longer H since the dimensionless concentration experiences a change of $\circ(1)$ over a much shorter length scale that can be determined by balancing the reaction and transverse diffusion terms in the describing equations. To achieve $\circ(1)$ scaling for the very fast reaction case, we introduce a region-of-influence scale δ_s that is a measure of the distance from the membrane boundaries over which the homogeneous reaction consumes component A entirely. Since the diffusive mass transfer and homogeneous reaction are occurring very close to the membrane boundaries, it is convenient to recast the describing equations in terms of a coordinate measured from the wall defined by $\tilde{y} = H - y$. Hence, our describing equations assume the form

$$\frac{d^2 c_A^*}{d\tilde{y}^{*2}} - \frac{k_1 \delta_s^2}{D_{AB}} c_A^* = 0 \tag{5.4-28}$$

$$\frac{dc_A^*}{d\tilde{y}^*} = -\frac{N_{Aw}\delta_s}{D_{AB}c_s} \quad \text{at} \quad \tilde{y}^* = 0 \tag{5.4-29}$$

$$\frac{dc_A^*}{d\tilde{y}^*} = 0 \quad \text{at} \quad \tilde{y}^* = \frac{H}{\delta_s} \tag{5.4-30}$$

Note that we have assumed that the Peclet number and aspect ratio are sufficiently small to permit ignoring convective and diffusive transport in the axial direction. To determine the thickness of the region of influence and to bound the dimensionless concentration gradient to be $\circ(1)$, we set the following groups equal to 1:

$$\frac{k_1 \delta_s^2}{D_{AB}} = 1 \Rightarrow \delta_s = \sqrt{\frac{D_{AB}}{k_1}} \tag{5.4-31}$$

$$\frac{N_{Aw}\delta_s}{D_{AB}c_s} = \frac{N_{Aw}}{c_s\sqrt{D_{AB}k_1}} = 1 \Rightarrow c_s = \frac{N_{Aw}}{\sqrt{D_{AB}k_1}} \tag{5.4-32}$$

We see that equation (5.4-32) provides the scale for the concentration. Since $\delta_s/H \ll 1$, the describing equations reduce to

$$\frac{d^2c_A^*}{d\tilde{y}^{*2}} - c_A^* = 0 \tag{5.4-33}$$

$$\frac{dc_A^*}{d\tilde{y}^*} = -1 \quad \text{at} \quad \tilde{y}^* = 0 \tag{5.4-34}$$

$$\frac{dc_A^*}{d\tilde{y}^*} = 0 \quad \text{at} \quad \tilde{y}^* \to \infty \tag{5.4-35}$$

The solution to these equations is given by

$$c_A^* = e^{-\tilde{y}^*} \tag{5.4-36}$$

This solution indicates that, indeed, c_A^* is bounded of $\text{o}(1)$.

5.5 SMALL DAMKÖHLER NUMBER APPROXIMATION FOR LAMINAR FLOW WITH A HETEROGENEOUS REACTION

Figure 5.5-1 shows a schematic of steady-state fully developed laminar flow of a Newtonian fluid with constant physical properties in a cylindrical tube of radius R containing a solute A having an initial concentration c_{A0} that undergoes a first-order irreversible reaction along length L. The heterogeneous reaction is assumed to be irreversible and first-order with a reaction-rate constant $\hat{k}_1$ (length/time). We use scaling analysis to simplify the describing equations; in particular, we assess the criterion for making the classical *plug-flow reactor* approximation; that is, a flow reactor in which the velocity can be assumed to be uniform at its average value $\overline{U}$ and that is surface-reaction limited.

The appropriately simplified species-balance equation given by equation (G.2-5) in the Appendices and the requisite boundary conditions are (step 1)

$$u_z\frac{\partial c_A}{\partial z} = D_{AB}\frac{\partial^2 c_A}{\partial z^2} + D_{AB}\frac{1}{r}\frac{\partial}{\partial r}\left(r\frac{\partial c_A}{\partial r}\right) \tag{5.5-1}$$

$$c_A = c_{A0} \quad \text{at} \quad z = 0 \tag{5.5-2}$$

$$c_A = f(r) \quad \text{at} \quad z = L \tag{5.5-3}$$

$$\frac{\partial c_A}{\partial r} = 0 \quad \text{at} \quad r = 0 \tag{5.5-4}$$

$$-D_{AB}\frac{\partial c_A}{\partial r} = \hat{k}_1 c_A \quad \text{at} \quad r = R \tag{5.5-5}$$

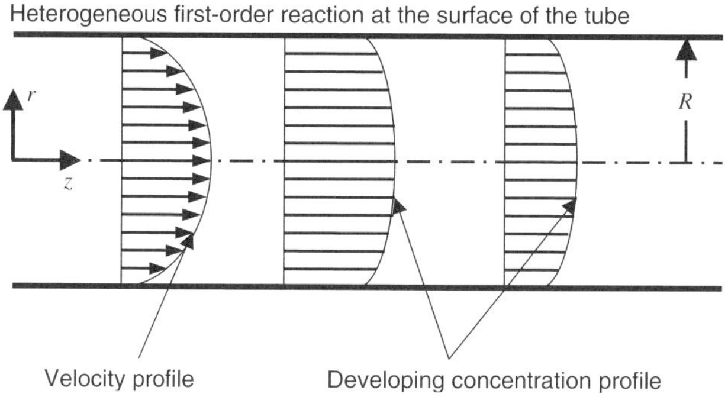

Figure 5.5-1 Steady-state fully developed laminar flow of a Newtonian fluid with constant physical properties undergoing a first-order heterogeneous reaction at the wall of a cylindrical tube having radius R; the fully developed velocity profile is shown along with the concentration profiles at two axial positions.

where D_{AB} is the binary diffusion coefficient, c_{A0} the initial concentration of the reactant A, $f(r)$ an unspecified function of r, $\hat{k}_1$ the first-order heterogeneous reaction-rate constant, and u_z the laminar flow velocity, given by

$$u_z = 2\overline{U}\left(1 - \frac{r^2}{R^2}\right) \tag{5.5-6}$$

where $\overline{U}$ is the average velocity. Note that each term in equation G.2-5 has been divided by the molecular weight of component A in order to convert the mass concentration into molar concentration for the dilute solution, which is assumed to have constant mass and molar densities. The boundary condition given by equation (5.5-3) is required because of the elliptic nature of equation (5.5-1). Since the function $f(r)$ is often unknown in practice, the describing equations cannot be solved even numerically.

Introduce the following dimensionless variables (steps 2 and 3):

$$c_A^* = \frac{c_A}{c_s}; \qquad r^* = \frac{r}{r_s}; \qquad z^* = \frac{z}{z_s} \tag{5.5-7}$$

Substitute these dimensionless variables into the describing equations and divide through by the coefficient of one term in each equation (steps 4 and 5):

$$\frac{2\overline{U}r_s^2}{D_{AB}z_s}\left[1 - \left(\frac{r_s}{R}\right)^2 r^{*2}\right]\frac{\partial c_A^*}{\partial z^*} = \frac{r_s^2}{z_s^2}\frac{\partial^2 c_A^*}{\partial z^{*2}} + \frac{1}{r^*}\frac{\partial}{\partial r^*}\left(r^*\frac{\partial c_A^*}{\partial r^*}\right) \tag{5.5-8}$$

$$c_A^* = \frac{c_{A0}}{c_s} \quad \text{at} \quad z^* = 0 \tag{5.5-9}$$

$$c_A^* = f(r^*) \quad \text{at} \quad z^* = \frac{L}{z_s} \tag{5.5-10}$$

$$\frac{\partial c_A^*}{\partial r^*} = 0 \qquad \text{at} \quad r^* = 0 \tag{5.5-11}$$

$$\frac{\partial c_A^*}{\partial r^*} = -\frac{\hat{k}_1 r_s}{D_{AB}} c_A^* \qquad \text{at} \quad r^* = \frac{R}{r_s} \tag{5.5-12}$$

The dimensionless groups in equations (5.5-9), (5.5-10), and (5.5-12) suggest the following choices for the scale factors to achieve o(1) scaling (step 7): $c_s = c_{A0}$, $z_s = L$, and $r_s = R$; note that we are employing local scaling here since L can be any specified value of z. Substitution of these scale factors into equations (5.5-8) through (5.5-12) then yields the following set of dimensionless describing equations:

$$2\text{Pe}_m \frac{R}{L}\left(1 - r^{*2}\right)\frac{\partial c_A^*}{\partial z^*} = \frac{R^2}{L^2}\frac{\partial^2 c_A^*}{\partial z^{*2}} + \frac{1}{r^*}\frac{\partial}{\partial r^*}\left(r^* \frac{\partial c_A^*}{\partial r^*}\right) \tag{5.5-13}$$

$$c_A^* = 1 \qquad \text{at} \quad z^* = 0 \tag{5.5-14}$$

$$c_A^* = f\left(r^*\right) \qquad \text{at} \quad z^* = 1 \tag{5.5-15}$$

$$\frac{\partial c_A^*}{\partial r^*} = 0 \qquad \text{at} \quad r^* = 0 \tag{5.5-16}$$

$$\frac{\partial c_A^*}{\partial r^*} = -\text{Da}^{\text{II}} c_A^* \qquad \text{at} \quad r^* = 1 \tag{5.5-17}$$

where $\text{Pe}_m \equiv \overline{U}R/D_{AB}$ is the Peclet number for mass transfer and $\text{Da}^{\text{II}} \equiv \hat{k}_1 R/D_{AB}$ is the second Damköhler number, which is a measure of the ratio of the time scale for radial diffusion to that for the heterogeneous reaction.

Now let us consider how this set of dimensionless describing equations can be simplified (step 8). If $R^2/L^2 \ll 1$, the axial diffusion term can be ignored in equation (5.5-13). If $\text{Da}^{\text{II}} \ll 1$, equation (5.5-17) implies that $\partial c_A^*/\partial r^* \ll 1$ since $c_A^* = \text{o}(1)$. This in turn implies that the concentration will not vary significantly in the radial direction; that is, $c_A^* \cong 1$ across the tube. Hence, equation (5.5-13) can be integrated as follows:

$$\int_0^1 2\text{Pe}_m \frac{R}{L}\left(1 - r^{*2}\right)\frac{\partial c_A^*}{\partial z^*} 2\pi r^* \, dr^* = \int_0^1 \frac{1}{r^*}\frac{\partial}{\partial r^*}\left(r^* \frac{\partial c_A^*}{\partial r^*}\right) 2\pi r^* \, dr^* \tag{5.5-18}$$

$$2\text{Pe}_m \frac{R}{L}\int_0^1 \left(1 - r^{*2}\right)\frac{\partial c_A^*}{\partial z^*} r^* \, dr^* = \int_0^{-\text{Da}^{\text{II}} c_A^*} \partial\left(r^* \frac{\partial c_A^*}{\partial r^*}\right) \tag{5.5-19}$$

Since the concentration is essentially uniform across the tube for very small Damköhler numbers, $\partial c_A^*/\partial z^* = dc_A^*/dz^*$; hence, equation (5.5-19) simplifies to

$$\frac{1}{2}\text{Pe}_m \frac{R}{L}\frac{dc_A^*}{dz^*} = -\text{Da}^{\text{II}} c_A^* \tag{5.5-20}$$

It is instructive to recast equation (5.5-20) into dimensional form, which is given by

$$\overline{U}\frac{dc_A}{dz} = -\frac{2\hat{k}_1 c_A}{R} \tag{5.5-21}$$

Equation (5.5.21) is equivalent to assuming that the fluid is convected down the tube at a uniform velocity $\overline{U}$ in the absence of any radial diffusion; however, the concentration changes axially, due to the heterogeneous reaction at the wall of the tube. Hence, the small Damköhler number approximation results in the classical *plug-flow reactor* assumption, which often is assumed without systematic justification.[9] Scaling provides a systematic method for arriving at this approximation.

Integrating equation (5.5-20) results in the following solution for the dimensionless concentration profile:

$$c_A^* = e^{-(2\text{Da}^{\text{II}}/\text{Pe}_m)(L/R)z^*} \tag{5.5-22}$$

If the Damköhler number is small, the exponential can be expanded in a Taylor series and truncated at two terms to obtain the following approximate solution:

$$c_A^* \cong 1 - \frac{2\text{Da}^{\text{II}}}{\text{Pe}_m}\frac{L}{R}z^* \tag{5.5-23}$$

The error in the solution given by equation (5.5-23) will be in the range 10 to 100% if $\text{Da}^{\text{II}} = \text{O}(0.1)$ and 1 to 10% if $\text{Da}^{\text{II}} = \text{O}(0.01)$.

Note that there is an analogy between the small Damköhler number approximation in modeling convective mass transfer with heterogeneous chemical reaction and the small Biot number approximation in modeling convective heat transfer from a solid particle that was considered in Section 4.4. When the Damköhler number is small, there is a negligible variation in the concentration over the cross section of the reactor; hence, the mass transfer rate is controlled by the resistance external to the fluid offered by the heterogeneous reaction. When the Biot number is small, there is a negligible variation in the temperature across the solid particle; hence, the heat transfer rate is controlled by the resistance external to the particle associated with convection to the surrounding fluid. This demonstrates another advantage of scaling: that it establishes the analogies between the various transport processes systematically.

5.6 LARGE PECLET NUMBER APPROXIMATION FOR MASS TRANSFER IN FALLING FILM FLOW

In Example Problem 4.E.4 we considered heat transfer to a liquid film flowing in fully developed laminar flow down a heated vertical wall. We found that at a

[9]By *plug flow* we mean that the velocity is uniform across the cross-section of the reactor; in the present example, for small Damköhler numbers this plug flow moves at an average velocity $\overline{U}$.

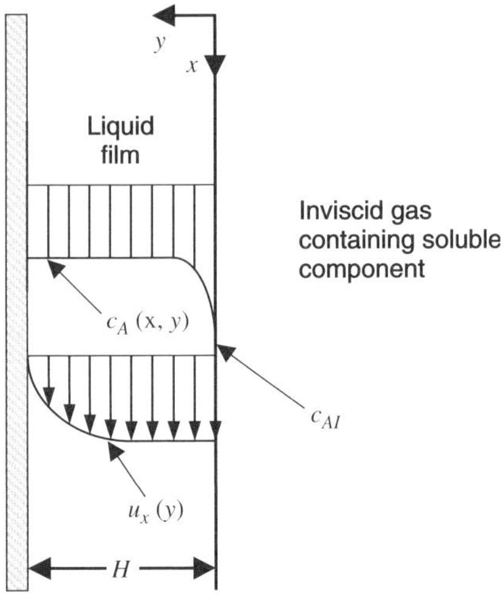

Figure 5.6-1 Steady-state gas absorption of a soluble component A from an inviscid gas phase to a liquid film in fully developed laminar flow down an impermeable vertical wall; the liquid film has initial concentration c_{A0}, interfacial concentration c_{AI}, thickness H, and constant physical properties.

sufficiently high Peclet number, the heat transfer was confined to a thin boundary layer in the vicinity of the wall. Here we consider a closely related problem: absorption of a soluble component from an inviscid gas into a liquid film in fully developed laminar flow down an impermeable solid wall as shown in Figure 5.6-1. The liquid film has thickness H and constant physical properties. It is assumed to have an initial concentration c_{A0} and an interfacial concentration c_{AI} established via equilibrium with the adjacent gas phase. The velocity profile in the liquid film is given by

$$u_x = U_m \left[1 - \left(\frac{y}{H}\right)^2\right] \tag{5.6-1}$$

where U_m is the maximum liquid velocity; namely, at the liquid–gas interface. We use scaling to explore how the describing equations can be simplified; in particular, we determine the conditions required to assume that the mass transfer is confined to a thin boundary layer near the liquid–gas interface.

The appropriate form of the species-balance equation given by equation (G.1-5) in the Appendices and corresponding boundary conditions are (step 1)

$$U_m \left[1 - \left(\frac{y}{H}\right)^2\right] \frac{\partial c_A}{\partial x} = D_{AB} \frac{\partial^2 c_A}{\partial x^2} + D_{AB} \frac{\partial^2 c_A}{\partial y^2} \tag{5.6-2}$$

$$c_A = c_{A0} \quad \text{at} \quad x = 0 \tag{5.6-3}$$

$$c_A = f(y) \quad \text{at} \quad x = L \tag{5.6-4}$$

$$c_A = c_{AI} \quad \text{at} \quad y = 0 \tag{5.6-5}$$

$$\frac{\partial c_A}{\partial y} = 0 \quad \text{at} \quad y = H \tag{5.6-6}$$

where $f(y)$ is some function of y and L is any arbitrary distance downstream. Each term in equation (G.1-5) has been divided by the molecular weight of component A to convert the mass concentrations to molar concentrations. These describing equations are similar to those encountered in Section 5.5; that is, they include an elliptic differential equation that is nontrivial to solve, owing to the need to satisfy some downstream boundary condition, which often is unknown. Hence, we use o(1) scaling to determine when this elliptic differential equation can be simplified into a parabolic equation for which a downstream boundary condition is not necessary.

Define the following dimensionless variables involving unspecified scale and reference factors (steps 2, 3, and 4):

$$c_A^* \equiv \frac{c_A - c_r}{c_s}; \qquad x^* \equiv \frac{x}{x_s}; \qquad y^* \equiv \frac{y}{y_s} \tag{5.6-7}$$

Note that a reference factor is introduced to ensure that the concentration is bounded between zero and 1. Introduce these dimensionless variables into the describing equations and divide through by the coefficient of one term in each equation that should be retained (steps 5 and 6):

$$\frac{U_m y_s^2}{D_{AB} x_s}\left[1 - \left(\frac{y_s}{H}\right)^2 y^{*2}\right]\frac{\partial c_A^*}{\partial x^*} = \frac{y_s^2}{x_s^2}\frac{\partial^2 c_A^*}{\partial x^{*2}} + \frac{\partial^2 c_A^*}{\partial y^{*2}} \tag{5.6-8}$$

$$c_A^* = \frac{c_{A0} - c_r}{c_s} \quad \text{at} \quad x^* = 0 \tag{5.6-9}$$

$$c_A^* = f(y^*) \quad \text{at} \quad x^* = \frac{L}{x_s} \tag{5.6-10}$$

$$c_A^* = \frac{c_{AI} - c_r}{c_s} \quad \text{at} \quad y^* = 0 \tag{5.6-11}$$

$$\frac{\partial c_A^*}{\partial y^*} = 0 \quad \text{at} \quad y^* = \frac{H}{y_s} \tag{5.6-12}$$

Step 7 involves bounding the independent and dependent dimensionless variables to be o(1). This is done for the dimensionless concentration by setting the dimensionless groups in equations (5.6-9) and (5.6-11) equal to zero and 1, respectively, to obtain $c_r = c_{A0}$ and $c_s = c_{AI} - c_{A0}$. The length scale for the axial spatial coordinate is obtained by setting the dimensionless group in equation (5.6-10) equal to 1, thereby obtaining $x_s = L$. We seek to determine the thickness of the region of influence near the liquid–gas interface wherein the mass transfer is confined. This region of influence occurs because the flowing liquid convects species downstream before it can diffuse through the entire thickness of the film. Hence, the convection term must be of the same magnitude as the cross-stream diffusion term in equation (5.6-8). This then implies that we demand the following, which leads to an estimate of the region-of-influence thickness δ_s:

$$\frac{U_m y_s^2}{D_{AB} L} = \frac{U_m \delta_s^2}{D_{AB} L} = 1 \Rightarrow \frac{\delta_s}{H} = \sqrt{\frac{1}{\text{Pe}_m} \frac{L}{H}} \tag{5.6-13}$$

We see that δ_s increases with the downstream distance L and is inversely proportional to the solutal Peclet number, $\text{Pe}_m \equiv U_m H / D_{AB}$, which is a measure of the ratio of the convection to diffusion of species. Sufficiently far downstream, δ_s will become equal to the film thickness H. Substitution of these scale and reference factors into equations (5.6-8) through (5.6-12) yields the following set of dimensionless describing equations:

$$\left(1 - \frac{1}{\text{Pe}_m} \frac{L}{H} y^{*2}\right) \frac{\partial c_A^*}{\partial x^*} = \frac{1}{\text{Pe}_m} \frac{H}{L} \frac{\partial^2 c_A^*}{\partial x^{*2}} + \frac{\partial^2 c_A^*}{\partial y^{*2}} \tag{5.6-14}$$

$$c_A^* = 0 \quad \text{at} \quad x^* = 0 \tag{5.6-15}$$

$$c_A^* = f\left(y^*\right) \quad \text{at} \quad x^* = 1 \tag{5.6-16}$$

$$c_A^* = 1 \quad \text{at} \quad y^* = 0 \tag{5.6-17}$$

$$\frac{\partial c_A^*}{\partial y^*} = 0 \quad \text{at} \quad y^* = \left(\text{Pe}_m \frac{H}{L}\right)^{\frac{1}{2}} \tag{5.6-18}$$

For large Peclet numbers the ratio δ_s / H will be quite small. This permits significant simplification of the describing equations (step 8). In particular, $H/(L \cdot \text{Pe}_m)$ will be quite small away from the leading edge of the liquid film, which permits ignoring the axial diffusion term, thereby avoiding the complication of having to specify a downstream boundary condition. In addition, for sufficiently large Peclet numbers, the quadratic term in the equation for the velocity profile can be ignored since it is only the interfacial velocity that is important for a thin region of influence. Finally, for large Peclet numbers the boundary condition by equation (5.6-18) can be applied at $y^* = \infty$ except when L becomes large. The resulting simplified dimensionless describing equations are

$$\frac{\partial c_A^*}{\partial x^*} = \frac{\partial^2 c_A^*}{\partial y^{*2}} \tag{5.6-19}$$

$$c_A^* = 0 \quad \text{at} \quad x^* = 0 \tag{5.6-20}$$

$$c_A^* = 1 \quad \text{at} \quad y^* = 0 \tag{5.6-21}$$

$$\frac{\partial c_A^*}{\partial y^*} = 0 \quad \text{at} \quad y^* \to \infty \tag{5.6-22}$$

This system of equations admits a solution via a similarity solution or combination of variables. The resulting solution will provide accurate predictions of the concentration profiles and mass-transfer rates for sufficiently long films for which ignoring the leading edge effects is a reasonable assumption.[10]

5.7 QUASI-STEADY-STATE APPROXIMATION FOR MASS TRANSFER DUE TO EVAPORATION

A volatile liquid composed of pure component A having constant physical properties and an initial depth L_0 is contained in a cylindrical tube of radius R and height H. At time $t = 0$ this liquid is exposed to the gas volume in this tube, which is filled initially with an insoluble gas composed of pure component B. Pure component B is also blown continuously over the top of the tube, thereby causing one-dimensional binary diffusion of A in B, as shown in Figure 5.7-1. This is inherently an unsteady-state problem both because it is an initial value problem and because the liquid level will drop in time continuously, due to the evaporative mass loss. We use scaling to assess the following: when quasi-steady-state can be assumed; when convective mass transfer can be ignored, and when one can neglect the pseudo-convection term that arises from the transformation from a fixed to a moving coordinate system.

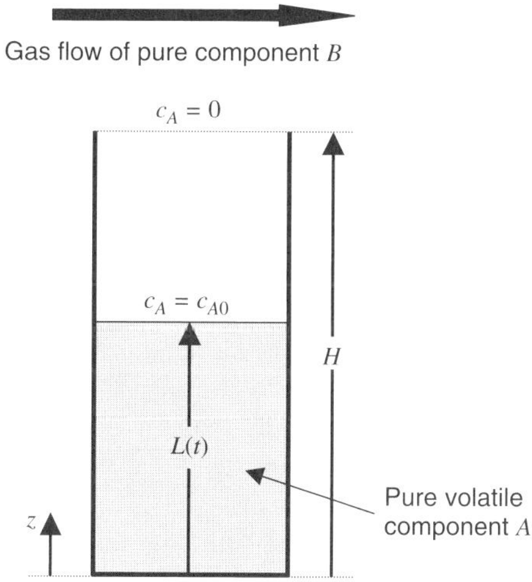

Figure 5.7-1 Unsteady-state evaporation of pure liquid A into an insoluble gas B; x_{A0}, the gas-phase mole fraction of component A, is dictated by thermodynamic equilibrium between the liquid and gas phases at the prevailing temperature and pressure; a gas stream of pure component B is blown over the top of the tube to maintain the composition of component A at zero.

[10]The conditions under which these approximations break down are considered in Practice Problem 5.P.5.

Since we are considering gas mass transfer, it is advantageous to use molar concentrations and molar fluxes rather than mass concentrations and mass fluxes, owing to the fact that the overall molar density rather than the mass density is constant at fixed temperature and pressure.[11] Step 1 consists of writing the appropriately simplified continuity and species-balance equations, given by equations (C.2-4) and (G.2-6) in the Appendices, as:

$$\frac{\partial c}{\partial t} = -\frac{\partial}{\partial z}(c\hat{u}) \Rightarrow \frac{\partial \hat{u}}{\partial z} = 0 \Rightarrow \hat{u} = \hat{u}(t) \tag{5.7-1}$$

$$\frac{\partial c_A}{\partial t} = -\frac{\partial N_A}{\partial z} \tag{5.7-2}$$

$$\frac{\partial c_B}{\partial t} = -\frac{\partial N_B}{\partial z} \tag{5.7-3}$$

where the molar-average velocity $\hat{u}$ is defined by

$$\hat{u} \equiv \frac{N_A + N_B}{c} \tag{5.7-4}$$

in which N_A and N_B are the molar fluxes of components A and B, respectively, given by

$$N_A = -cD_{AB}\frac{\partial x_A}{\partial z} + c_A\hat{u} = -D_{AB}\frac{\partial c_A}{\partial z} + c_A\hat{u} \tag{5.7-5}$$

$$N_B = -cD_{AB}\frac{\partial x_B}{\partial z} + c_B\hat{u} = -D_{AB}\frac{\partial c_B}{\partial z} + c_B\hat{u} \tag{5.7-6}$$

in which $c = c_A + c_B$ is the overall molar density, $x_A = c_A/c$ and $x_B = c_B/c$ are the mole fractions of components A and B, respectively, and D_{AB} is the binary diffusion coefficient. The integration in equation (5.7-1) follows from the fact that we are assuming an ideal gas at constant temperature and pressure. Note that only one of equations (5.7-2) and (5.7-3) needs be considered since the sum of these two equations results in equation (5.7-1). We again see that equations (5.7-5) and (5.7-6) indicate that the total molar flux is the sum of a purely diffusive flux and a convective flux that is proportional to the molar average velocity. Equation (5.7-4) shows that the convective velocity arises from the mass transfer since it is proportional to the sum of the molar fluxes.

Equations (5.7-1) through (5.7-6) constitute six equations in five unknowns: c_A, c_B, N_A, N_B, and $\hat{u}$. However, only four of these equations are independent; thus, an additional equation is needed. In this case this comes from the fact that component B is insoluble in liquid component A; hence, it follows that

$$\hat{u}(t) = \frac{N_A + N_B}{c} = \frac{1}{c}N_A \qquad \text{at} \quad z = L \tag{5.7-7}$$

[11]For moderate pressures the ideal gas law is applicable; hence, $c = P/RT$ = a constant for specified P and T.

The requisite initial and boundary conditions are given by

$$c_A = 0 \quad \text{at} \quad t \le 0, \quad L \le z \le H \tag{5.7-8}$$

$$c_A = c_{A0} \quad \text{at} \quad z = L(t), \quad 0 < t \le \infty \tag{5.7-9}$$

$$c_A = 0 \quad \text{at} \quad z = H, \quad 0 \le t \le \infty \tag{5.7-10}$$

Equation (5.7-8) specifies the composition of component A in the gas phase at the interface that is determined by thermodynamic equilibrium at the prevailing temperature and pressure. This boundary condition is applied at the moving interface $L(t)$ between the liquid and gas phases. Problems of this type are referred to as *moving boundary problems*. Since $L(t)$ is an additional unknown, it is necessary to prescribe an auxiliary condition to determine it. This is obtained via an integral mass balance on component A as follows:

$$\frac{d}{dt}\int_0^L \frac{\rho}{M_A}\,dz + \frac{d}{dt}\int_L^H c_A\,dz = -N_A|_{z=H} \tag{5.7-11}$$

where ρ is the mass density of pure liquid component A and M_A is the molecular weight of component A. Applying Leibnitz's rule for differentiating an integral given by equation (H.1-2) in the Appendices and substituting equation (5.7-2) yields

$$\left(\frac{\rho}{M_A} - c_A|_{z=L}\right)\frac{dL}{dt} + N_A|_{z=L} = 0 \tag{5.7-12}$$

In arriving at equation (5.7-12) we have used the fact that the liquid density is constant and that component B is insoluble in the pure liquid A. However, since $c_{A0} \ll \rho/M_A$, equation (5.7-12) simplifies to the following auxiliary condition to determine the instantaneous location of the liquid–gas interface:

$$\frac{dL}{dt} = -\frac{M_A}{\rho}N_A \quad \text{at} \quad z = L(t) \tag{5.7-13}$$

This condition merely states that the rate that the interface recedes is proportional to the rate at which component A is transferred to the gas phase. To integrate equation (5.7-13), it is necessary to specify an initial condition for L; this is given by

$$L = L_0 \quad \text{at} \quad t = 0 \tag{5.7-14}$$

Define the following dimensionless variables (steps 2, 3, and 4):

$$c_A^* \equiv \frac{c_A}{c_{As}}; \quad \hat{u}^* \equiv \frac{\hat{u}}{u_s}; \quad N_A^* \equiv \frac{N_A}{N_{As}}; \quad L^* \equiv \frac{L}{L_s};$$
$$\left(\frac{dL}{dt}\right)^* \equiv \frac{1}{L_s}\frac{dL}{dt}; \quad z^* \equiv \frac{z - z_r}{z_s}; \quad t^* \equiv \frac{t}{t_s} \tag{5.7-15}$$

Note that we have introduced a reference factor for the spatial coordinate since it is not naturally referenced to zero in the gas phase where the diffusion is occurring. However, there is no need to introduce a reference factor for the molar concentration since it is naturally referenced to zero. We have also introduced scale factors for the interface location and velocity since there is no reason to expect that these will scale with z_s.

Introduce these dimensionless variables into the describing equations and divide through by the coefficient of one term in each of these equations that should be retained (steps 5 and 6) to obtain the following:

$$\frac{c_{As} z_s}{N_{As} t_s} \frac{\partial c_A^*}{\partial t^*} = -\frac{\partial N_A^*}{\partial z^*} \tag{5.7-16}$$

$$\frac{N_{As} z_s}{D_{AB} c_{As}} N_A^* = -\frac{\partial c_A^*}{\partial z^*} + \frac{u_s z_s}{D_{AB}} c_A^* \hat{u}^* \tag{5.7-17}$$

$$\hat{u}^*(t) = \frac{N_{As}}{c u_s} N_A^* \quad \text{at} \quad z^* = \frac{L - z_r}{z_s} \tag{5.7-18}$$

$$c_A^* = 0, \quad L^* = \frac{L_0}{L_s} \quad \text{at} \quad t^* \le 0, \quad \frac{L - z_r}{z_s} \le z^* \le \frac{H - z_r}{z_s} \tag{5.7-19}$$

$$c_A^* = \frac{c_{A0}}{c_{As}} \quad \text{at} \quad z^* = \frac{L - z_r}{z_s}, \quad 0 < t^* \le \infty \tag{5.7-20}$$

$$c_A^* = 0 \quad \text{at} \quad z^* = \frac{H - z_r}{z_s}, \quad 0 \le t^* \le \infty \tag{5.7-21}$$

$$\left(\frac{dL}{dt}\right)^* = -\frac{M_A N_{As}}{\rho \dot{L}_s} N_A^* \quad \text{at} \quad z^* = \frac{L - z_r}{z_s} \tag{5.7-22}$$

Now let us proceed to determine the scale factors (step 7). The dimensionless concentration can be bounded of $o(1)$ by setting the group containing the concentration scale factor in equation (5.7-20) equal to 1, thereby obtaining $c_{As} = c_{A0}$. Since we seek to determine when steady-state conditions apply, the appropriate time scale is the observation time; that is, $t_s = t_o$. The dimensionless spatial coordinate can be bounded to be $o(1)$ by setting the dimensionless groups in equations (5.7-20) and (5.7-21) equal to zero and 1, respectively, to obtain $z_r = L$ and $z_s = H - L$. The scale factor for the molar flux is obtained by setting the appropriate dimensionless group in equation (5.7-17) equal to 1, thereby obtaining $N_{As} = D_{AB} c_{A0}/(H - L)$. The scale factor for the molar average velocity is obtained by setting the dimensionless group in equation (5.2-18) equal to 1, thereby obtaining $u_s = D_{AB} x_{A0}/(H - L)$. The scale factor for the liquid-layer depth and its time rate of change are obtained by setting the dimensionless groups in equations (5.7-19) and (5.7-22) equal to 1, thereby obtaining $L = L_0$ and $\dot{L}_s = M_A D_{AB} c_{A0}/\rho(H - L)$, respectively.

Substitution of these scale and reference factors yields the following set of dimensionless describing equations:

$$\frac{1}{\text{Fo}_m}\frac{\partial c_A^*}{\partial t^*} + \frac{M_A c_{A0}}{\rho}\left(z^* - 1\right)\left(\frac{dL}{dt}\right)^* \frac{\partial c_A^*}{\partial z^*} = -\frac{\partial N_A^*}{\partial z^*} \tag{5.7-23}$$

$$N_A^* = -\frac{\partial c_A^*}{\partial z^*} + x_{A0} c_A^* \hat{u}^* \tag{5.7-24}$$

$$\hat{u}^*(t) = N_A^* \quad \text{at} \quad z^* = 0 \tag{5.7-25}$$

$$c_A^* = 0, \quad L^* = 1 \quad \text{at} \quad t^* \leq 0 \tag{5.7-26}$$

$$c_A = 1 \quad \text{at} \quad z^* = 0, \quad 0 \leq t^* \leq \infty \tag{5.7-27}$$

$$c_A^* = 0 \quad \text{at} \quad z^* = 1, \quad 0 \leq t^* \leq \infty \tag{5.7-28}$$

$$\left(\frac{dL}{dt}\right)^* = -N_A^* \quad \text{at} \quad z^* = 0 \tag{5.7-29}$$

where $\text{Fo}_m \equiv D_{AB}t_o/(H - L)^2$ is the solutal Fourier number or Fourier number for mass transfer. Note that the additional term appearing in equation (5.7-23) arises due to the transformation from a stationary to a moving coordinate system implied by our new z^* variable. This additional term is referred to as a *pseudo-convection term*. The latter will arise in a moving boundary problem if the scaling involves a transformation from a stationary to a moving coordinate system.

Now let us explore possible simplifications of the describing equations (step 8). Note that if $\text{Fo}_m \gg 1$, the unsteady-state term in equation (5.7-23) becomes insignificant; hence, the mass transfer is quasi-steady-state; that is,

$$\text{Fo}_m \equiv \frac{D_{AB}t_o}{(H - L)^2} \gg 1 \Rightarrow \text{quasi-steady-state mass transfer} \tag{5.7-30}$$

Quasi-steady-state implies that the unsteady-state term in the species-balance equation is negligible but that the problem is still unsteady-state, due to the time-dependence that enters through the boundary condition applied at the moving liquid–gas interface. If $x_{A0} \ll 1$, corresponding to a very dilute gas-phase concentration, the convective contribution to the mass-transfer flux in equation (5.7-24) can be ignored. In addition, if $M_A c_{A0}/\rho \ll 1$, corresponding to the gas phase having a much smaller molar density than the liquid, the pseudo-convection term can be ignored. When the latter condition and that given by equation (5.7-30) are satisfied, the solution to this problem can be obtained analytically and is available in standard references.[12]

5.8 MEMBRANE PERMEATION WITH NONCONSTANT DIFFUSIVITY

A dense (i.e., nonporous) membrane of thickness H fabricated from pure polymer B is used to separate component A from a liquid feed solution that establishes a

[12] Bird et al., *Transport Phenomena*, 2nd ed., pp. 545–549.

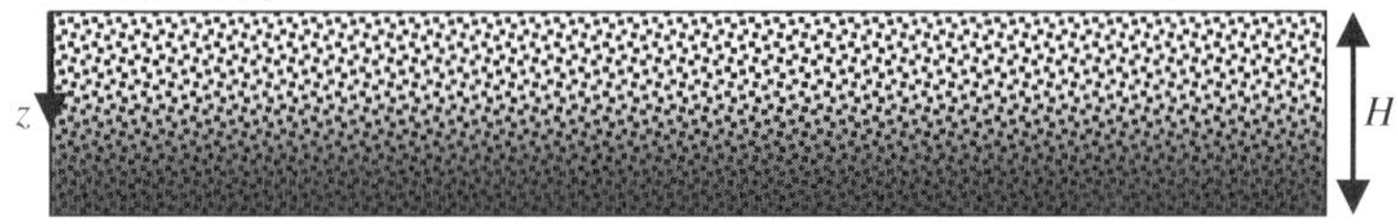

Figure 5.8-1 Diffusion of component A through a dense polymeric membrane of component B of thickness H; component A is a weak plasticizing agent that causes membrane swelling, which in turn causes the diffusion coefficient to increase; this is shown schematically by the lighter shading in the regions where swelling is occurring.

concentration ρ_{Af} (mass per unit volume) in the membrane on the feed side; the other component in the feed solution is insoluble in the polymeric membrane.[13] The concentration of component A on the permeate product side of the membrane is maintained at zero. A schematic of this mass-transfer process is shown in Figure 5.8-1, where the origin of the coordinate system has been located at the feed side. Component A is a weak plasticizing agent for the polymer and causes it to swell, thereby increasing the diffusion coefficient or diffusivity. Since this swelling is proportional to the concentration of component A, it causes a change in the diffusivity across the membrane, whose dependence on the concentration is given by

$$D_{AB} = D_0 e^{\beta \rho_A} \tag{5.8-1}$$

where β is a positive constant and D_0 is the diffusion coefficient at infinite dilution for which $\rho_A = 0$. We use scaling to ascertain the following: when the concentration dependence of the diffusion coefficient can be neglected; how we can ascertain that membrane swelling is occurring; and the effective thickness of the membrane wherein the resistance to diffusion is confined.

If the species-balance equation given by equation (G.1-1) in the Appendices is appropriately simplified, we obtain (step 1)

$$\frac{dn_A}{dz} = 0 \Rightarrow n_A = K \tag{5.8-2}$$

where K is an integration constant and n_A is the mass flux of component A given by

$$n_A = K = -D_{AB}\frac{d\rho_A}{dz} + \rho_A u \cong -D_{AB}\frac{d\rho_A}{dz} = -D_0 e^{\beta\rho_A}\frac{d\rho_A}{dz} \tag{5.8-3}$$

The solubility of most solutes in polymeric materials is small, so that the bulk flow term involving the mass-average velocity u can be ignored. The corresponding

[13] Mass rather than molar concentrations are used for polymer systems, due to their very high molecular weight relative to the diffusing solute, which would result in extremely small mole fractions.

boundary conditions are given by

$$\rho_A = \rho_{A0} \quad \text{at} \quad z = 0 \tag{5.8-4}$$

$$\rho_A = 0 \quad \text{at} \quad z = H \tag{5.8-5}$$

Introduce the following dimensionless variables (steps 2, 3, and 4):

$$\rho_A^* \equiv \frac{\rho_A}{\rho_{As}}; \qquad z^* \equiv \frac{z}{z_s} \tag{5.8-6}$$

Substitute these dimensionless variables into equations (5.8-3) through (5.8-5) (steps 5 and 6):

$$1 = -\frac{D_0 \rho_{As}}{K z_s} e^{\beta \rho_{As} \rho_A^*} \frac{d\rho_A^*}{dz^*} \tag{5.8-7}$$

$$\rho_A^* = \frac{\rho_{A0}}{\rho_{As}} \quad \text{at} \quad z^* = 0 \tag{5.8-8}$$

$$\rho_A^* = 0 \quad \text{at} \quad z^* = \frac{H}{z_s} \tag{5.8-9}$$

Now let us proceed to determine the reference and scale factors (step 7). Equation (5.8-8) permits bounding ρ_A^* to be $\circ(1)$ by choosing $\rho_{As} = \rho_{A0}$. We have two choices for bounding z^* to be $\circ(1)$: namely, by setting the dimensionless groups either in equation (5.8-7) or (5.8-9) equal to 1; the proper choice depends on the situation for which we are scaling. If we are scaling to determine when the membrane swelling effect can be neglected so that the diffusivity can be assumed to be constant, the resistance to diffusion is offered by the entire membrane thickness. Hence, setting $z_s = H$ provides the appropriate length scale. However, if membrane swelling is appreciable so that the resistance to diffusion is confined to a thin region of influence near $z = H$, the appropriate length scale is obtained from equation (5.8-7); that is, $z_s = D_0 \rho_{A0}/K$.

Let us first determine the criterion for ignoring the effect of membrane swelling on the mass transfer (step 8). After substituting the scale factors, our dimensionless describing equations assume the form

$$1 = -\frac{D_0 \rho_{A0}}{KH} e^{\beta \rho_{A0} \rho_A^*} \frac{d\rho_A^*}{dz^*} = -\frac{D_0 \rho_{A0}}{KH}(1 + \beta \rho_{A0} \rho_A^*) \frac{d\rho_A^*}{dz^*} \tag{5.8-10}$$

$$\rho_A^* = 1 \quad \text{at} \quad z^* = 0 \tag{5.8-11}$$

$$\rho_A^* = 0 \quad \text{at} \quad z^* = 1 \tag{5.8-12}$$

Note that to assess the effect of membrane swelling on the diffusion coefficient, we have expanded the equation for the concentration-dependent diffusivity in a Taylor series about its minimum value at $\rho_A = 0$. This procedure of considering only the leading order behavior is standard practice when using scaling to assess the importance of the concentration or temperature dependence of a physical or transport property. This does not limit the results of our scaling analysis since if the criterion that we determine for assuming constant diffusivity is not satisfied for a weak concentration dependence, it most certainly will not be satisfied for a strong dependence. Hence, if the following criterion is satisfied, we can assume that the diffusivity is constant:

$$\Pi_1 \equiv \beta \rho_{A0} \ll 1 \tag{5.8-13}$$

If this criterion is satisfied, the solution to the approximate describing equations for the mass concentration and flux is given by

$$\rho_A^* = 1 - z^* \quad \text{and} \quad \frac{n_{A0} H}{D_0 \rho_{A0}} = \frac{KH}{D_0 \rho_{A0}} = 1 \tag{5.8-14}$$

where n_{A0} denotes the constant mass flux in the absence of any membrane swelling; this is the same as the mass flux that would occur if the entire membrane had the diffusivity corresponding to $\rho_A = 0$. Note that the solution for K could have been obtained merely by setting the dimensionless group in the approximate form of equation (5.8-10) equal to 1.

Now let us consider when the swelling effect is large. For this case the diffusivity increases markedly so that the concentration gradient in equation (5.8-7) is large only near the boundary at $z = H$, where the concentration is nearly zero. For this condition to prevail, equation (5.8-7) indicates that the following criterion must apply:

$$\Pi_1 \equiv \beta \rho_{A0} \gg 1 \tag{5.8-15}$$

The dimensionless group in equation (5.8-7) then implies that the thickness of the region of influence or characteristic length scale for the dimensionless concentration to experience a change of $\circ(1)$ is given by $z_s \equiv \delta_s = D_0 \rho_{A0}/K = D_0 \rho_{A0}/n_A$, where n_A is the mass flux in the presence of significant membrane swelling; note that $n_A > n_{A0}$. Hence, if one knows D_0, the diffusivity for mass transfer of component A through polymer B at infinite dilution, one can determine if swelling is occurring merely by measuring the mass flux for a specified feed concentration. If the measured n_A exceeds n_{A0}, it indicates that swelling is occurring.

It is instructive to compare the results of our scaling analysis with the predictions of the analytical solution for the mass concentration and flux for the exact set of describing equations that is given by

$$\rho_A^* = \frac{1}{\Pi_1} \ln\left[z^* + e^{\Pi_1}(1 - z^*)\right] \quad \text{and} \quad \frac{n_A H}{D_0 \rho_{A0}} = \frac{1}{\Pi_1}\left(e^{\Pi_1} - 1\right) \tag{5.8-16}$$

When the solution for the mass flux in equation (5.8-16) is substituted into the equation for δ_s, the thickness of the region of influence wherein the concentration

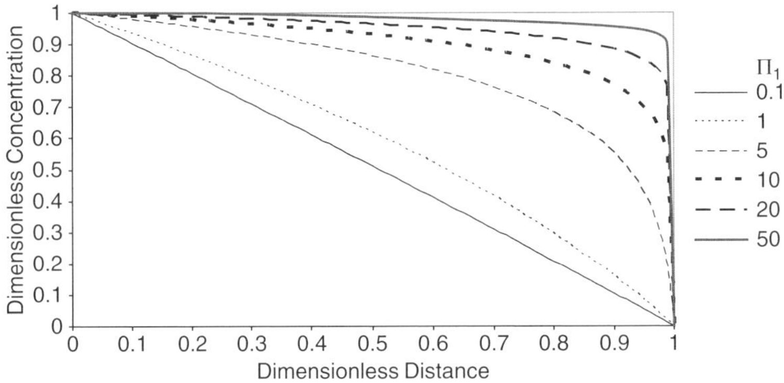

Figure 5.8-2 Dimensionless concentration ρ_A as a function of dimensionless distance z^* for various values of $\Pi_1 \equiv \beta\rho_{A0}$ for diffusion through a polymeric membrane subject to swelling; for large values of Π_1 the resistance to mass transfer is confined to a thin region of influence or boundary layer.

gradient is O(1), we obtain

$$\delta_s = \frac{\Pi_1}{e^{\Pi_1} - 1} \tag{5.8-17}$$

Figure 5.8-2 plots the predictions of equation (5.8-16) for ρ_A^* as a function of z^* for various values of Π_1. We see that the exact solution reduces to that given by equation (5.8-14) for the case of negligible swelling when $\Pi_1 \leq 0.1$ and that the concentration gradient is confined to a thin boundary layer or region of influence when $\Pi_1 \geq 50$, corresponding to pronounced swelling. These trends are consistent with the criteria emanating from our scaling analysis given by equations (5.8-13) and (5.8-15).

5.9 SOLUTALLY DRIVEN FREE CONVECTION DUE TO EVAPOTRANSPIRATION FROM A VERTICAL CYLINDER

Consider the annular region between an inner permeable vertical cylinder of radius R_1 and length L and an outer impermeable cylindrical shell of radius R_2 as shown in Figure 5.9-1. The permeable inner cylinder transpires water vapor (evapotranspiration) into the annular region between the two concentric cylinders. The concentration of water vapor in the air adjacent to the inner cylinder, c_{A0}, is greater than that in the ambient air, $c_{A\infty}$. Since water vapor is less dense than air, a hydrostatic pressure imbalance is generated that causes air near the inner cylinder to rise, thereby generating free convection. We consider the steady-state free convection that prevails after the transients have died out. We ignore viscous dissipation and end effects and assume constant physical properties other than the density in the gravitational body-force term in the equations of motion. We consider a local

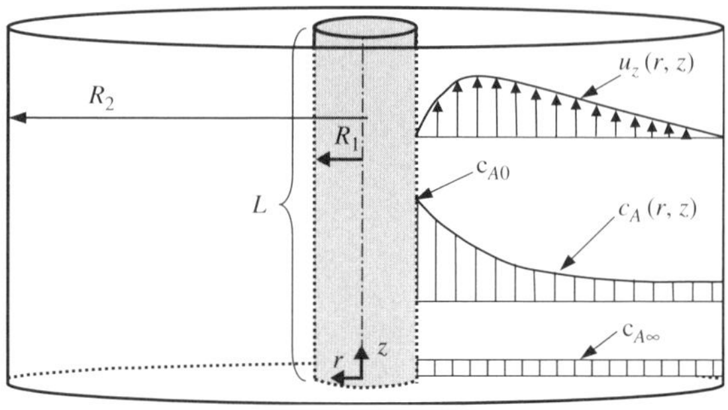

Figure 5.9-1 Buoyancy-induced free-convection flow in the annular gap between two concentric cylinders of radii R_1 and R_2 due to evapotranspiration through the inner cylinder that causes the concentration c_{A0} of the water to be greater than that of the ambient air, $c_{A\infty}$; the sketch shows the developing concentration and axial velocity profiles.

scaling for which L denotes any arbitrary length along the cylinder. We use scaling analysis to determine how the describing equations for free convection can be simplified and to develop a criterion for ignoring the effect of the outer cylindrical boundary on the solutally driven free convection.[14]

This is inherently a developing flow, due to the progressive evapotranspiration that occurs as the humid air moves up the cylinder; therefore, velocity components in both the r- and z-directions must be considered. In this analysis we assume constant physical properties except for the density that appears in the buoyancy term in the equations of motion; elsewhere the density is assumed to be constant. Hence, equations (D.2-10), (D.2-12), (C.2-2), and (G.2-5) in the Appendices simplify to (step 1)

$$\rho u_r \frac{\partial u_z}{\partial r} + \rho u_z \frac{\partial u_z}{\partial z} = -\frac{\partial P}{\partial z} + \mu \frac{1}{r}\frac{\partial}{\partial r}\left(r \frac{\partial u_z}{\partial r}\right) + \mu \frac{\partial^2 u_z}{\partial z^2} - \rho g \tag{5.9-1}$$

$$\rho u_r \frac{\partial u_r}{\partial r} + \rho u_z \frac{\partial u_r}{\partial z} = -\frac{\partial P}{\partial r} + \mu \frac{\partial}{\partial r}\left[\frac{1}{r}\frac{\partial}{\partial r}(r u_r)\right] + \mu \frac{\partial^2 u_r}{\partial z^2} \tag{5.9-2}$$

$$\frac{1}{r}\frac{\partial}{\partial r}(r u_r) + \frac{\partial u_z}{\partial z} = 0 \tag{5.9-3}$$

$$u_r \frac{\partial c_A}{\partial r} + u_z \frac{\partial c_A}{\partial z} = D_{AB}\frac{1}{r}\frac{\partial}{\partial r}\left(r \frac{\partial c_A}{\partial r}\right) + D_{AB}\frac{\partial^2 c_A}{\partial z^2} \tag{5.9-4}$$

[14]This provides a good model for evapotranspiration from *Phycomyces*, a large single-celled sporangiophore that has a cylindrical "stem" and a spherical "head". This organism has the interesting property that it "senses" the presence of objects near it and responds by growing away from them. If a cylindrical shell is placed concentric with the axis of *Phycomyces*, it will grow faster. If the cylindrical shell is placed eccentric with the axis of *Phycomyces*, it will grow away from the closer boundary. This interesting behavior, referred to as the *avoidance phenomenon*, has been shown to be caused by the influence of lateral boundaries on the free-convection boundary layer that is created due to evapotranspiration of water vapor from *Phycomyces*. A free-convection model for the avoidance phenomenon is developed in J. J. Pellegrino, R. L. Sani, and R. I. Gamow, *J. Theor. Bio.*, **105**(1), 77–90 (1983).

Note that each term in equation (G.2-5) has been divided by the molecular weight, which is assumed to be constant for the dilute solutions assumed. The corresponding boundary conditions are given by

$$u_r = 0, \quad u_z = 0, \quad c_A = c_{A0} \quad \text{at} \quad r = R_1 \tag{5.9-5}$$

$$u_r = 0, \quad u_z = 0, \quad \frac{\partial c_A}{\partial r} = 0 \quad \text{at} \quad r = R_2 \tag{5.9-6}$$

$$u_r = 0, \quad u_z = 0, \quad c_A = c_{A\infty} \quad \text{at} \quad z = 0 \tag{5.9-7}$$

$$u_r = f_1(r), \quad u_z = f_2(r), \quad c_A = f_3(r) \quad \text{at} \quad z = L \tag{5.9-8}$$

where $f_1(r)$, $f_2(r)$, and $f_3(r)$ are unspecified functions of r that often are unknown. Equation (5.9-5) constitutes the no-slip and impermeable boundary conditions at the inner boundary. The evapotranspiration through the inner boundary will contribute to the radial component of velocity, which is assumed to be a negligible effect in this scaling analysis.[15] Equation (5.9-6) states that the velocities as well as the molar flux are zero at the impermeable outer boundary. Equation (5.9-7) states that the fluid is quiescent and at ambient conditions at the bottom of the annular gap. Equation (5.9-8) is a formal statement that downstream boundary conditions are required for these coupled elliptic equations.

Since the density is concentration-dependent, we need an appropriate equation of state. Here we consider small variations and hence will represent the density via a truncated Taylor series expansion about its value ρ_∞ at the ambient concentration $c_{A\infty}$:

$$\rho = \rho|_{c_{A\infty}} + \left.\frac{\partial \rho}{\partial c_A}\right|_{c_{A\infty}} (c_A - c_{A\infty}) = \rho_\infty - \rho_\infty \beta_s (c_A - c_{A\infty}) \tag{5.9-9}$$

where β_s is the coefficient of solutal volume expansion. It is convenient to split the pressure into dynamic, P_d, and hydrostatic, P_h, contributions[16]:

$$P = P_d(r, z) + P_h(z) \tag{5.9-10}$$

When equations (5.9-9) and (5.9-10) are substituted into equation (5.9-1), we obtain

$$\rho_\infty u_r \frac{\partial u_z}{\partial r} + \rho_\infty u_z \frac{\partial u_z}{\partial z} = -\frac{\partial P_d}{\partial z} + \mu \frac{1}{r}\frac{\partial}{\partial r}\left(r \frac{\partial u_z}{\partial r}\right) + \mu \frac{\partial^2 u_z}{\partial z^2} + \rho_\infty \beta_s g (c_A - c_{A\infty}) \tag{5.9-11}$$

Note that the $\rho_\infty g$ term does not appear in equation (5.9-11) because it cancels with the derivative of the purely hydrostatic contribution to the pressure.

[15]Scaling analysis is used to develop the criterion for making this assumption in Practice Problem 5.P.23.

[16]Splitting the pressure into its dynamic and hydrostatic contributions was discussed in connection with Example Problem 4.E.7.

Define the following dimensionless dependent and independent variables (steps 2, 3, and 4):

$$u_z^* \equiv \frac{u_z}{u_{zs}}; \quad u_r^* \equiv \frac{u_r}{u_{rs}}; \quad P^* \equiv \frac{P_d}{P_s}; \quad c_A^* \equiv \frac{c_A - c_r}{c_s};$$
$$z^* \equiv \frac{z}{z_s}; \quad r_m^* \equiv \frac{r - r_r}{r_{ms}}; \quad r_c^* \equiv \frac{r - r_r}{r_{cs}} \tag{5.9-12}$$

Note that we have allowed for different radial length scales for the species-balance equation and equations of motion since the concentration might experience a characteristic change of O(1) over a different length scale than that of the velocities. Introduce these dimensionless variables into the describing equations and divide each equation by the dimensional coefficient of one term that should be retained in order to maintain physical significance (steps 5 and 6):

$$\frac{u_{rs} r_{ms}}{\nu_\infty} u_r^* \frac{\partial u_z^*}{\partial r_m^*} + \frac{u_{zs} r_{ms}^2}{\nu_\infty z_s} u_z^* \frac{\partial u_z^*}{\partial z^*} = -\frac{P_s r_{ms}^2}{\mu u_{zs} z_s} \frac{\partial P^*}{\partial z^*}$$
$$+ \frac{1}{r_m^* + r_r/r_{ms}} \frac{\partial}{\partial r_m^*} \left[\left(r_m^* + \frac{r_r}{r_{ms}} \right) \frac{\partial u_z^*}{\partial r_m^*} \right] + \frac{r_{ms}^2}{z_s^2} \frac{\partial^2 u_z^*}{\partial z^{*2}} + \frac{g \beta_s c_s r_{ms}^2}{\nu_\infty u_{zs}} \left(c_A^* + \frac{c_r - c_{A\infty}}{c_s} \right) \tag{5.9-13}$$

$$\frac{u_{rs} r_{ms}}{\nu_\infty} u_r^* \frac{\partial u_r^*}{\partial r_m^*} + \frac{u_{zs} r_{ms}^2}{\nu_\infty z_s} u_z^* \frac{\partial u_r^*}{\partial z^*} = -\frac{P_s r_{ms}}{\mu u_{rs}} \frac{\partial P^*}{\partial r_m^*}$$
$$+ \frac{\partial}{\partial r_m^*} \left\{ \frac{1}{r_m^* + r_r/r_{ms}} \frac{\partial}{\partial r_m^*} \left[\left(r_m^* + \frac{r_r}{r_{ms}} \right) u_r^* \right] \right\} + \frac{r_{ms}^2}{z_s^2} \frac{\partial^2 u_r^*}{\partial z^{*2}} \tag{5.9-14}$$

$$\frac{1}{r_m^* + r_r/r_{ms}} \frac{\partial}{\partial r_m^*} \left[\left(r_m^* + \frac{r_r}{r_{ms}} \right) u_r^* \right] + \frac{u_{zs} r_{ms}}{u_{rs} z_s} \frac{\partial u_z^*}{\partial z^*} = 0 \tag{5.9-15}$$

$$\frac{u_{rs} r_{cs}}{D_{AB}} u_r^* \frac{\partial c_A^*}{\partial r_c^*} + \frac{u_{zs} r_{cs}^2}{D_{AB} z_s} u_z^* \frac{\partial c_A^*}{\partial z^*} = \frac{1}{r_c^* + r_r/r_{cs}} \frac{\partial}{\partial r_c^*} \left[\left(r_c^* + \frac{r_r}{r_{cs}} \right) \frac{\partial c_A^*}{\partial r_c^*} \right] + \frac{r_{cs}^2}{z_s^2} \frac{\partial^2 c_A^*}{\partial z^{*2}} \tag{5.9-16}$$

$$u_r^* = 0, \quad u_z^* = 0 \quad \text{at} \quad r_m^* = \frac{R_1 - r_r}{r_{ms}}$$
$$c_A^* = \frac{c_{A0} - c_r}{c_s} \quad \text{at} \quad r_c^* = \frac{R_1 - r_r}{r_{cs}} \tag{5.9-17}$$

$$u_r^* = 0, \quad u_z^* = 0 \quad \text{at} \quad r_m^* = \frac{R_2 - r_r}{r_{ms}}$$
$$\frac{\partial c_A^*}{\partial r_c^*} = 0 \quad \text{at} \quad r_c^* = \frac{R_2 - r_r}{r_{cs}} \tag{5.9-18}$$

$$u_z^* = 0, \quad u_r^* = 0, \quad c_A^* = \frac{c_{A\infty} - c_r}{c_s} \quad \text{at} \quad z^* = 0 \tag{5.9-19}$$

$$u_z^* = f_1(r_m^*), \quad u_r^* = f_2(r_m^*), \quad c_A^* = f_3(r_c^*) \quad \text{at} \quad z^* = \frac{L}{z_s} \tag{5.9-20}$$

where $\nu_\infty \equiv \mu/\rho_\infty$ is the kinematic viscosity. Note that the boundary conditions given by equation (5.9-5) are not applied at the same values of the dimensionless radial coordinate owing to the different radial scale factors for the equations of motion and the species-balance equation; the same comment applies to equation (5.9-6).

The effect of viscosity for this flow will be confined to a thin region near the vertical cylinder; hence, the radial length scale for the equations of motion will be the thickness of the momentum boundary layer or region of influence δ_m; that is, $r_{ms} = \delta_m$. The momentum boundary-layer thickness is obtained by balancing the convection terms with the principal viscous term in equation (5.9-13). Similarly, the effect of species diffusion will also be confined to a thin region δ_s, although not necessarily of the same thickness as that of the momentum boundary layer; hence, the radial length scale for the species-balance equation will be the thickness of the solutal boundary layer or region of influence δ_s; that is, $r_{cs} = \delta_s$. The solutal boundary-layer thickness is obtained by balancing the radial species convection and diffusion terms in equation (5.9-16). To determine the axial velocity scale u_{zs}, we need to balance what causes the flow with the principal resistance to flow; the former is the gravitationally induced body force, whereas the latter is the principal viscous term. The transverse velocity scale u_{rs} is obtained from the continuity equation since this is inherently a developing flow. One might be tempted to obtain P_s from the dimensionless group multiplying the pressure term in equation (5.9-13). However, the pressure term in equation (5.9-13) does not cause the free-convection flow; the latter is caused by the gravitational body-force term in this equation. The pressure does cause the flow in the r-direction, which is the reason why we determine its scale by setting the dimensionless group containing P_s in equation (5.9-14) equal to 1. The reference and scale factors for the axial length and concentration are obtained from the boundary conditions to ensure that these variables are bounded of $o(1)$. These considerations then result in the following scale factors (step 7):

$$u_{zs} = (g\beta_s\, \Delta c_A\, L)^{0.5}; \qquad u_{rs} = \left(\frac{\nu_\infty^2 g\beta_s\, \Delta c_A}{L}\right)^{0.25};$$

$$P_s = \left(\frac{\mu^2 g\beta_s\, \Delta c_A}{L}\right)^{0.5}; \qquad c_s = c_{A0} - c_{A\infty} \equiv \Delta c_A; \qquad z_s = L;$$

$$r_{ms} = \delta_m = \frac{L}{\text{Re}^{0.5}} = \frac{L}{\text{Gr}_m^{0.25}}; \qquad r_{cs} = \delta_s = \frac{L}{\text{Sc}^{0.5}\text{Pe}_m^{0.5}} = \frac{L}{\text{Sc}\cdot\text{Gr}_m^{0.25}} = \frac{\delta_m}{\text{Sc}} \tag{5.9-21}$$

where $\text{Re} \equiv u_{zs}L/\nu_\infty$ the Reynolds number, $\text{Pe}_m \equiv u_{zs}L/D_{AB} = \text{Re}\cdot\text{Sc}$ the solutal Peclet number, $\text{Sc} \equiv \nu_\infty/D_{AB}$ the Schmidt number, and $\text{Gr}_m \equiv L^3 g\beta_s\, \Delta c_A/\nu_\infty^2$ the solutal Grashof number. Note that the Grashof number is a measure of the ratio of the free convection to viscous transport of momentum; as such, it is the analog of the Reynolds number for free convection.[17] Note that the last of equations (5.9-21) indicates that $\delta_s < \delta_m$ for liquids and $\delta_s \cong \delta_m$ for gases.

If we now rewrite our dimensionless describing equations in terms of the scales defined by equations (5.9-21), we obtain

$$u_r^* \frac{\partial u_z^*}{\partial r_m^*} + u_z^* \frac{\partial u_z^*}{\partial z^*} = -\frac{1}{\mathrm{Gr}_m^{0.5}} \frac{\partial P^*}{\partial z^*} + \frac{1}{r_m^* + R_1/\delta_m} \frac{\partial}{\partial r_m^*} \left[\left(r_m^* + \frac{R_1}{\delta_m} \right) \frac{\partial u_z^*}{\partial r_m^*} \right] + \frac{1}{\mathrm{Gr}_m^{0.5}} \frac{\partial^2 u_z^*}{\partial z^{*2}} + c_A^* \tag{5.9-22}$$

$$u_r^* \frac{\partial u_r^*}{\partial r_m^*} + u_z^* \frac{\partial u_r^*}{\partial z^*} = -\frac{\partial P^*}{\partial r_m^*} + \frac{\partial}{\partial r_m^*} \left\{ \frac{1}{r_m^* + R_1/\delta_m} \frac{\partial}{\partial r_m^*} \left[\left(r_m^* + \frac{R_1}{\delta_m} \right) u_r^* \right] \right\} + \frac{1}{\mathrm{Gr}_m^{0.5}} \frac{\partial^2 u_r^*}{\partial z^{*2}} \tag{5.9-23}$$

$$\frac{1}{r_m^* + R_1/\delta_m} \frac{\partial}{\partial r_m^*} \left[\left(r_m^* + \frac{R_1}{\delta_m} \right) u_r^* \right] + \frac{\partial u_z^*}{\partial z^*} = 0 \tag{5.9-24}$$

$$u_r^* \frac{\partial c_A^*}{\partial r_c^*} + \frac{1}{\mathrm{Sc}} u_z^* \frac{\partial c_A^*}{\partial z^*} = \frac{1}{r_c^* + (R_1/\delta_m)\mathrm{Sc}} \frac{\partial}{\partial r_c^*} \left[\left(r_c^* + \frac{R_1}{\delta_m}\mathrm{Sc} \right) \frac{\partial c_A^*}{\partial r_c^*} \right] + \frac{1}{\mathrm{Sc}^2 \mathrm{Gr}_m^{0.5}} \frac{\partial^2 c_A^*}{\partial z^{*2}} \tag{5.9-25}$$

$$u_r^* = 0, \quad u_z^* = 0, \quad c_A^* = 1 \quad \text{at} \quad r_m^* = r_c^* = 0 \tag{5.9-26}$$

$$u_r^* = 0, \quad u_z^* = 0 \quad \text{at} \quad r_m^* = \frac{R_2 - R_1}{\delta_m}; \qquad \frac{\partial c_A^*}{\partial r_c^*} = 0 \quad \text{at} \quad r_c^* = \frac{R_2 - R_1}{\delta_s} \tag{5.9-27}$$

$$u_z^* = 0, \quad u_r^* = 0, \quad c_A^* = 0 \quad \text{at} \quad z^* = 0 \tag{5.9-28}$$

$$u_z^* = f_1(r_m^*), \quad u_r^* = f_2(r_m^*), \quad c_A^* = f_3(r_c^*) \quad \text{at} \quad z^* = 1 \tag{5.9-29}$$

We can now consider how these scaled dimensionless describing equations can be simplified (step 8). Note that if the Grashof number is very large such that $\mathrm{Gr}_m^{0.5} \gg 1$, the pressure and axial viscous momentum transfer terms can be dropped from equation (5.9-22). The former simplification implies that the z-component is decoupled from the solution to the r-component of the equations of motion; hence,

[17]The solutal Rayleigh number, defined as $\mathrm{Ra}_m \equiv L^3 D_{AB} g \beta_s \Delta c_A / \nu_\infty = \mathrm{Gr}_m \cdot \mathrm{Pr}$, is another important dimensionless group that appears in free-convection problems; it is a measure of the ratio of the free convection to viscous transport of heat; as such, it is the analog of the solutal Peclet number for free convection.

the latter equation can be ignored. Dropping the axial viscous momentum transfer term from equation (5.9-22) converts it from an elliptic into a parabolic differential equation; this obviates the need to specify downstream boundary conditions, which in many cases are unknown. A very large Grashof number also implies that the axial species-diffusion term can be dropped from equation (5.9-25), which also converts it from an elliptic into a parabolic differential equation, again avoiding the need to specify a downstream boundary condition. If in addition, $R_1 \gg \delta_m$ and $R_1 \gg \delta_s$, the curvature effects can be neglected. Finally, if $R_2 - R_1 \gg \delta_m$ and $R_2 - R_1 \gg \delta_s$, the boundary conditions given by equation (5.9-27) can be applied at infinity. If the aforementioned criteria are satisfied, the resulting simplified describing equations are given by

$$u_r^* \frac{\partial u_z^*}{\partial r_m^*} + u_z^* \frac{\partial u_z^*}{\partial z^*} = \frac{\partial^2 u_z^*}{\partial r_m^{2*}} + c_A^* \tag{5.9-30}$$

$$\frac{\partial u_r^*}{\partial r_m^*} + \frac{\partial u_z^*}{\partial z^*} = 0 \tag{5.9-31}$$

$$u_r^* \frac{\partial c_A^*}{\partial r_c^*} + \frac{1}{\text{Sc}} u_z^* \frac{\partial c_A^*}{\partial z^*} = \frac{\partial^2 c_A^*}{\partial r_c^{*2}} \tag{5.9-32}$$

$$u_r^* = 0, \quad u_z^* = 0, \quad c_A^* = 1 \quad \text{at} \quad r_m^* = r_c^* = 0 \tag{5.9-33}$$

$$u_r^* = 0, \quad u_z^* = 0, \quad \frac{\partial c_A^*}{\partial r_c^*} = 0 \quad \text{at} \quad r_m^* = r_c^* = \infty \tag{5.9-34}$$

$$u_z^* = 0, \quad u_r^* = 0, \quad c_A^* = 0 \quad \text{at} \quad z^* = 0 \tag{5.9-35}$$

In particular, one sees that the outer cylindrical boundary will have a negligible effect on the free convection if $\delta_m/R_1 = L/(R_1 \text{Gr}_m^{0.25}) \ll 1$. Note that this criterion will break down for sufficiently long cylinders owing to the thickening of the free-convection boundary layer associated with progressively more evapotranspiration along the cylinder.

5.10 DIMENSIONAL ANALYSIS FOR A MEMBRANE–LUNG OXYGENATOR

Section 2.4 we discussed the scaling analysis procedure for dimensional analysis and its advantages relative to the Pi theorem. Here we apply this procedure to develop a correlation for the performance of a membrane–lung oxygenator. The steps referred to here are those outlined in Section 2.4; these differ from those used in Sections 5.2 through 5.9 since no attempt is made to achieve $\circ(1)$ scaling. Note that $\circ(1)$ scaling analysis, which was illustrated in Sections 5.2 through 5.9, always leads to the minimum parametric representation for a set of describing equations; hence, it can always be used to identify the appropriate dimensionless groups. However, carrying out an $\circ(1)$ scaling analysis can be somewhat complicated and time-consuming. In contrast, the scaling analysis approach to dimensional analysis illustrated in this section is much easier and quicker to implement. However, it does not provide as much information as does o(1) scaling analysis for achieving the minimum parametric representation. In particular, it does not lead to groups whose magnitude can be used to assess the relative importance of particular terms in the describing equations. It also does not identify regions of influence or boundary layers, whose identification in some cases can reduce the number of dimensionless groups.

The oxygenator of interest here involves a bundle of cylindrical hollow-fiber membranes encased in a tubular housing. Because the hollow-fiber membranes are both microporous and hydrophobic, they provide a means for mass transfer of oxygen and carbon dioxide to and from the blood, respectively, while preventing direct contact between the gas and liquid. The mass transfer is controlled on the blood side because of the inability of the oxygen-absorbing hemoglobin "particles" to closely approach the inner surface of the membrane. Improvement in oxygenator performance has focused on various means to reduce the resistance to mass transfer on the blood side of the membrane. One very effective way to accomplish this is to oscillate the hollow fibers relative to the blood flow to increase the oxygen concentration gradients adjacent to the membrane.[18] We use the scaling analysis approach to dimensional analysis to determine the dimensionless groups required to correlate the effects of oscillating the hollow fibers on the performance of the oxygenator. It is sufficient here to consider the effect of oscillations on the oxygen mass transfer to the blood flow in a single hollow-fiber membrane of radius R and length L, as shown in Figure 5.10-1.

Step 1 in the scaling procedure for dimensional analysis consists of writing the appropriate describing equations for the oxygen mass transfer to the blood, which will be assumed to be in fully developed periodically pulsed laminar flow.[19] The

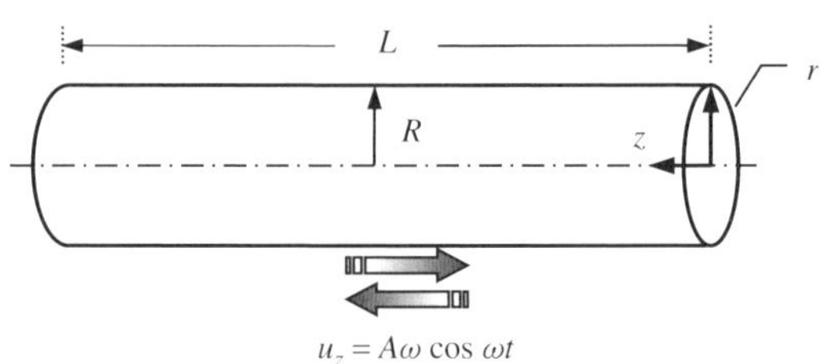

Figure 5.10-1 Single hollow fiber of radius R and length L in a membrane–lung oxygenator; axial oscillations having amplitude A and angular frequency ω are used to increase the concentration gradients at the interior wall, where the resistance to mass transfer is concentrated.

[18]R. R. Bilodeau, R. J. Elgas, W. B. Krantz, and M. E. Voorhees, U.S. patent 5,626,759, issued May 6, 1997.

[19]Note that this is an uncommon example of a fully developed unsteady-state flow; that is, the axial velocity does not change in the axial direction at any instant of time, yet it is a function of time due to the oscillating wall.

mass-transfer coefficient $k_L^{\bullet}$ provides a convenient measure of the effect of the oscillating wall on the oxygen mass transfer and is defined in terms of $\overline{N}_{Aw}$, the time-average molar flux of oxygen at the inner wall of the hollow-fiber membrane, as[20]

$$k_L^{\bullet} \equiv \frac{\overline{N}_{Aw}}{\Delta c_{\mathrm{lm}}} \tag{5.10-1}$$

where Δc_{lm} is the log-mean concentration driving force, defined as

$$\Delta c_{\mathrm{lm}} \equiv \frac{(c_{Aw} - c_{AL}) - (c_{Aw} - c_{A0})}{\ln{(c_{Aw} - c_{AL})/(c_{Aw} - c_{A0})}} \tag{5.10-2}$$

in which c_{Aw} is the equilibrium oxygen concentration at the blood side of the membrane, and c_{A0} and c_{AL} are the average concentrations in the blood at $z = 0$ and at $z = L$, respectively, and

$$\overline{N}_{Aw} \equiv \frac{\int_0^{2\pi/\omega} N_{Aw}\, dt}{\int_0^{2\pi/\omega} dt} = \frac{\omega}{2\pi} \int_0^{2\pi/\omega} N_{Aw}\, dt \tag{5.10-3}$$

The molar flux N_{Aw} appearing in equation (5.10-3) is equal to the local molar flux averaged over the length of the hollow-fiber membrane and is defined

$$N_{Aw} \equiv \frac{\int_0^L D_{AB}\, (\partial c_A/\partial r)|_{r=R}\, dz}{\int_0^L dz} = \frac{1}{L} \int_0^L D_{AB} \left.\frac{\partial c_A}{\partial r}\right|_{r=R} dz \tag{5.10-4}$$

where D_{AB} is the binary diffusion coefficient for oxygen in blood. Note that the bulk flow contribution to the molar flux has been ignored because the solutions are dilute. When equations (5.10-3) and (5.10-4) are substituted into equation (5.10-1), we obtain

$$k_L^{\bullet} = \frac{\omega D_{AB}}{2\pi L\, \Delta c_{\mathrm{lm}}} \int_0^{2\pi/\omega} \int_0^L \left.\frac{\partial c_A}{\partial r}\right|_{r=R} dz\, dt \tag{5.10-5}$$

Both c_{AL} appearing in Δc_{lm} and c_A in equation (5.10-5) are obtained from a solution to the axisymmetric form of the convective diffusion equation in cylindrical coordinates given by equation (G.2-5) in the Appendices

$$\frac{\partial c_A}{\partial t} + u_z \frac{\partial c_A}{\partial z} = \frac{D_{AB}}{r} \frac{\partial}{\partial r}\left(r \frac{\partial c_A}{\partial r}\right) \tag{5.10-6}$$

in which u_z is the local axial fluid velocity. In arriving at equation (5.10-6) each term in equation (G.2-5) has been divided by the molecular weight, which is

[20]Note that the mass-transfer coefficient is defined in terms of a flux divided by a driving force; as such, it can be expressed in several different ways, depending on the units chosen for these quantities as well as whether the total flux or just the diffusive flux is used; the various definitions of the mass-transfer coefficient are discussed in standard references such as Bird et al., *Transport Phenomena*, 2nd ed., pp. 672–675.

constant for the assumed dilute solution. An analytical solution for u_z for fully developed laminar flow subject to axial harmonic pulsations of the tube wall has been developed and is of the form[21]

$$\frac{u_z}{\overline{U}} = f\left(\frac{r}{R}, \omega t, \frac{\nu}{\omega R^2}, \frac{A\omega}{\overline{U}}\right) \tag{5.10-7}$$

where $\overline{U}$ is the average fluid velocity in the hollow fiber and ν is the kinematic viscosity of blood. The dimensional analysis correlation given by equation (5.10-7) is also developed in Practice Problem 3.P.38. The boundary and periodic solution conditions are given by

$$\frac{\partial c_A}{\partial r} = 0 \quad \text{at} \quad r = 0, \quad 0 \le z \le L \tag{5.10-8}$$

$$c_A = c_{Aw} \quad \text{at} \quad r = R, \quad 0 \le z \le L \tag{5.10-9}$$

$$c_A = c_{A0} \quad \text{at} \quad z = 0 \tag{5.10-10}$$

$$c_A|_t = c_A|_{t+2\pi/\omega} \tag{5.10-11}$$

Steps 2 and 3 in the scaling procedure for dimensional analysis involve defining arbitrary scale factors for all the dependent and independent variables and reference factors for those not naturally referenced to zero. Hence, we introduce the following dimensionless variables:

$$c_A^* \equiv \frac{c_A - c_{Ar}}{c_{As}}; \qquad u^* \equiv \frac{u_z}{u_s}; \qquad r^* \equiv \frac{r}{r_s}; \qquad z^* \equiv \frac{z}{z_s}; \qquad t^* \equiv \frac{t}{t_s} \tag{5.10-12}$$

Steps 4 and 5 involve introducing these dimensionless variables into the describing equations and dividing through by the dimensional coefficient of one term in each equation. In the scaling analysis procedure for dimensional analysis, in contrast to o(1) scaling analysis, it makes no difference which term is chosen in this step. These steps then yield the following dimensionless describing equations:

$$\frac{2\pi k_L^{\bullet} L\, \Delta c_{\text{lm}}^* r_s}{\omega D_{AB} z_s t_s} = \int_0^{2\pi/\omega t_s} \int_0^{L/z_s} \left.\frac{\partial c_A^*}{\partial r^*}\right|_{r^*=R/r_s} dz^*\, dt^* \tag{5.10-13}$$

$$\Delta c_{\text{lm}}^* \equiv \frac{-c_{AL}^* + (c_{A0} - c_{Ar})/c_{As}}{\ln\left(\dfrac{c_{Aw} - c_{Ar}}{c_{Aw} - c_{A0}} - \dfrac{c_{As}}{c_{Aw} - c_{A0}} c_{AL}^*\right)} \tag{5.10-14}$$

$$\frac{r_s^2}{D_{AB} t_s} \frac{\partial c_A^*}{\partial t^*} + \frac{u_s r_s^2}{D_{AB} z_s} u_z^* \frac{\partial c_A^*}{\partial z^*} = \frac{1}{r^*} \frac{\partial}{\partial r^*}\left(r^* \frac{\partial c_A^*}{\partial r^*}\right) \tag{5.10-15}$$

[21] W. B. Krantz, R. R. Bilodeau, M. E. Voorhees, and R. J. Elgas, *J. Membrane Sci.*, **124**, 283–299 (1997).

$$u_z^* = \frac{\overline{U}}{u_s} f\left(\frac{r_s}{R}r^*, \omega t_s t^*, \frac{\nu}{\omega R^2}, \frac{A\omega}{\overline{U}}\right) \tag{5.10-16}$$

$$\frac{\partial c_A^*}{\partial r^*} = 0 \quad \text{at} \quad r^* = 0, \quad 0 \le z^* \le \frac{L}{z_s} \tag{5.10-17}$$

$$c_A^* = \frac{c_{Aw} - c_{Ar}}{c_{As}} \quad \text{at} \quad r^* = \frac{R}{r_s}, \quad 0 \le z^* \le \frac{L}{z_s} \tag{5.10-18}$$

$$c_A^* = \frac{c_{A0} - c_{Ar}}{c_{As}} \quad \text{at} \quad z^* = 0 \tag{5.10-19}$$

$$c_A^*|_{t^*} = c_A^*|_{t^*+2\pi/\omega t_s} \tag{5.10-20}$$

Step 6 involves setting the various groups equal to 1 or zero to determine the scale and reference factors, respectively. Since there is no attempt to achieve o(1) in the scaling approach for dimensional analysis, it makes no difference what groups are chosen in this step. Let us set equation (5.10-19) equal to zero to determine the reference concentration $c_{Ar} = c_{A0}$ and the appropriate group in equation (5.10-18) equal to 1 to determine the concentration scale $c_{As} = c_{Aw} - c_{A0}$. Setting the remaining dimensionless groups in equation (5.10-18) equal to 1 determines the radial and axial length scales $r_s = R$ and $z_s = L$, respectively; note, however, that there is no assurance that these length factors will scale the corresponding derivatives with respect to these spatial coordinates to be of o(1). The dimensionless group in equation (5.10-16) provides the velocity scale $u_s = \overline{U}$. Either the dimensionless group in equation (5.10-15) or in equation (5.10-20) can be set equal to 1 to determine the time scale; let us arbitrarily choose the latter, thereby obtaining $t_s = 2\pi/\omega$. These choices then yield the following minimum parametric representation of the describing equations; that is, in terms of the minimum number of dimensionless groups:

$$\text{Sh} = \frac{c_{AL}^*}{\ln(1 - c_{AL}^*)} \int_0^1 \int_0^1 \left.\frac{\partial c_A^*}{\partial r^*}\right|_{r^*=1} dz^*\, dt^* \tag{5.10-21}$$

$$(\Delta c_{\text{lm}}^*)_{\text{v}} \equiv \frac{-c_{AL}^*}{\ln(1 - c_{AL}^*)} \tag{5.10-22}$$

$$\frac{\omega R^2}{2\pi D_{AB}} \frac{\partial c_A^*}{\partial t^*} + \frac{2\text{Gz}}{\pi} u_z^* \frac{\partial c_A^*}{\partial z^*} = \frac{1}{r^*} \frac{\partial}{\partial r^*}\left(r^* \frac{\partial c_A^*}{\partial r^*}\right) \tag{5.10-23}$$

$$u_z^* = f\left(r^*, t^*, \frac{\nu}{\omega R^2}, \frac{A\omega}{\overline{U}}\right) \tag{5.10-24}$$

$$\frac{\partial c_A^*}{\partial r^*} = 0 \quad \text{at} \quad r^* = 0, \quad 0 \le z^* \le 1 \tag{5.10-25}$$

$$c_A^* = 1 \quad \text{at} \quad r^* = 1, \quad 0 \le z^* \le 1 \tag{5.10-26}$$

$$c_A^* = 0 \quad \text{at} \quad z^* = 0 \tag{5.10-27}$$

$$c_A^*|_{t^*} = c_A^*|_{t^*+1} \tag{5.10-28}$$

where $(\Delta c^*_{\text{lm}})_v$ denotes Δc^*_{lm} in the presence of axial oscillations or vibrations and where

$$\text{Sh} \equiv \frac{k^\bullet_L R}{D_{AB}} \quad \text{is the Sherwood number} \tag{5.10-29}$$

$$\text{Gz} \equiv \frac{\pi \overline{U} R^2}{2 D_{AB} L} = \frac{\pi R \text{Pe}_m}{2L} \quad \text{is the Graetz number} \tag{5.10-30}$$

The Sherwood number provides a measure of the overall mass transfer to that by diffusion alone; as such, it is analogous to the Nusselt number in heat transfer.[22] The Graetz number provides a measure of the relative convection to diffusional transport of species; as can be seen from equation (5.10-30), it is proportional to the solutal Peclet number, which is also a measure of this same ratio. However, the Graetz number takes into account the finite contact time for mass transfer via the ratio R/L. Equation (5.10-21) implies that the Sherwood number is a function of the dimensionless groups involved in determining c^*_{AL} and $\partial c^*_A/\partial r^*\big|_{r^*=1}$ and hence will be functions of only the dimensionless groups involved in solving equation (5.10-23). These include the two dimensionless groups appearing in this equation as well as the two involved in determining u^* as indicated by equation (5.10-24). Hence, we conclude that membrane–lung oxygenator performance can be correlated in terms of five dimensionless groups; that is,

$$\text{Sh} = f\left(\text{Gz}, \frac{\omega R^2}{D_{AB}}, \frac{\omega R^2}{\nu}, \frac{A\omega}{\overline{U}}\right) \tag{5.10-31}$$

Note that a naive application of the Pi theorem would imply that eight dimensionless groups would be required (i.e., $n = 11$ and $m = 3$).

The five dimensionless groups in the correlation for membrane–lung oxygenator performance given by equation (5.10-31) are not unique. Step 7 involves isolating certain quantities by the procedure indicated formally by equation (2.4-2). Here it is convenient to isolate the angular frequency into just two groups by the following operation:

$$\frac{\omega R^2/D_{AB}}{\omega R^2/\nu} = \frac{\nu}{D_{AB}} = \text{Sc} \tag{5.10-32}$$

where Sc is the Schmidt number. Hence, our modified correlation for the Sherwood number is given by

$$\text{Sh} = f\left(\text{Gz,Sc}, \frac{\omega R^2}{\nu}, \frac{A\omega}{\overline{U}}\right) \tag{5.10-33}$$

Recasting our correlation in terms of the Schmidt number is particularly convenient for designing a membrane blood oxygenator since the Schmidt number is fixed for

[22]The dimensionless group corresponding to the Sherwood number defined here is sometimes referred to as the Nusselt number for mass transfer.

this system. Hence, one needs to consider only three variable groups in correlating performance data for a membrane blood oxygenator. Note that the angular frequency could be isolated into just one dimensionless group. However, the resulting group would still contain the variable amplitude and hence would not be fixed by specifying the blood–oxygen system.

Step 8 in the scaling procedure for dimensional analysis involves exploring the correlation for various limiting values of the dimensionless groups. For large Sc (i.e., for blood) we can use the expansion in Sc^{-1} suggested by equation (2.4-3) truncated to one term to conclude that the oxygenator performance can be correlated in terms of only four dimensionless groups; that is,

$$\mathrm{Sh} = f\left(\mathrm{Gz}, \frac{\omega R^2}{\nu}, \frac{A\omega}{\overline{U}}\right) \qquad \text{for } \mathrm{Sc}^{-1} \to 0 \tag{5.10-34}$$

The dimensionless group $\omega R^2/\nu$ is the ratio of the characteristic time for viscous penetration to that for the periodic wall oscillations. One might anticipate that the wall oscillations will not have much effect on the Sherwood number for very small values of $\omega R^2/\nu$ since this implies negligible wall motion. We would also not expect much effect on the Sherwood number for very large values of $\omega R^2/\nu$, since this implies negligible time for the wall oscillations to penetrate the fluid. The dimensionless group $A\omega/\overline{U}$ is the ratio of the wall velocity to the average velocity of the fluid. If $A\omega/\overline{U}$ is very small, the wall oscillations are insignificant relative to the fluid velocity, in which case they will have a negligible effect on the Sherwood number. If $A\omega/\overline{U}$ is very large, it implies a very slow flow for which the oxygen transport is not limited by contact time; hence, the oscillations will also have a negligible effect for large values of this group. These arguments suggest that the effects of the oscillations on the Sherwood number will exhibit a maximum with respect to the values of $\omega R^2/\nu$ and $A\omega/\overline{U}$; that is, the effect of wall oscillations on the mass transfer involves a "tuned" response whereby the maximum effect will be achieved only over a relatively narrow range of oscillation amplitudes and frequencies.

The design of a membrane–lung oxygenator will be revisited in Chapter 7. This same problem will be analyzed in Section 7.2 using o(1) scaling analysis to achieve the minimum parametric representation rather than using the scaling approach to dimensional analysis. We will see that o(1) scaling analysis yields far more information about the design of a membrane–lung oxygenator. In particular, it will show that the oxygenator performance can be correlated in terms of four dimensionless groups rather than five. Moreover, o(1) scaling analysis permits predicting the optimum frequency, within a multiplicative constant of O(1), required to achieve maximum enhancement of the oxygen mass transfer to the blood.

5.11 SUMMARY

In Section 5.2, we provided an introduction to the step-by-step procedure for applying scaling analysis in mass transfer. We considered binary diffusion in a stagnant film of liquid subject to an instantaneous change in the concentration at one of its boundaries. Scaling was used to determine the criterion for applying the film theory model. This criterion requires that the Fourier number or ratio of the contact to diffusion time be very large. Scaling was also used to assess when the bulk-flow contribution that arises from a net mass-transfer flux can be neglected. This criterion requires that either the mass fluxes be nearly equal in magnitude or that the relative concentration change across the film be small. When this bulk-flow contribution is potentially important, it is necessary either to know some relationship between the mass-transfer fluxes or to have an equation of state for the mass density. In this problem the ratio of the mass fluxes was specified.[23] Since this problem could be solved analytically, it was possible to estimate the error incurred when the bulk-flow effect was ignored. This was seen to be less than 0.1% when the criterion given by equation (5.2-28) was o(0.01). Note that the film theory model developed in this section is frequently used to correct mass-transfer coefficients obtained from literature correlations for the effects of bulk flow.[24] Interestingly, the ratio of the mass-transfer coefficient in the presence to that in the absence of bulk-flow effects is not a strong function of the actual mass-transfer configuration. For this reason film theory can be used to correct for the bulk-flow effects even when the model is not applicable to the particular mass-transfer problem.

In Section 5.3 we considered the same physical situation as in Section 5.2. However, short-contact-time or small Fourier number scaling was done whereby the diffusion and unsteady-state terms were balanced in the species-balance equation. This led to a length scale that was identified with a region of influence or solutal boundary layer within which all the mass transfer is confined. The resulting simplified describing equations provide the basis for penetration theory. The latter is the short-contact-time complement to film theory. Penetration theory provides a better model for correcting mass-transfer coefficients obtained from the literature for bulk-flow effects when the contact time or Fourier number is small.

Mass transfer in fully developed laminar flow between permeable parallel membrane walls in the presence of a homogeneous chemical reaction was considered in Section 5.4. The presence of both the transverse and axial diffusion terms made the describing equations elliptic. This complicated the solution since the required downstream boundary condition is often unknown. The concept of local scaling in mass transfer was introduced in this problem, whereby one considers the describing equations within a domain defined by some arbitrary distance in the principal direction of flow that is assumed to be constant during the scaling analysis. In contrast to the preceding two examples, there was no explicit concentration scale in this problem. Hence, the concentration scale was determined by balancing the mass flux through the permeable membrane walls and transverse diffusion terms. Scaling analysis led to three important dimensionless groups: the Peclet number for mass transfer, Schmidt number, and Thiele modulus. The Peclet number for mass transfer is a measure of the convection to diffusion of mass. The Schmidt

[23]Note that Example Problem 5.E.1 considers scaling to determine when the bulk-flow effect can be neglected when an equation of state for the mass density is known.

[24]The interested reader is referred to Bird et al., *Transport Phenomena*, pp 704–706.

number is a measure of the viscous transport of momentum to the diffusive transfer of mass. The Thiele modulus is a measure of the characteristic time for diffusion relative to that for homogeneous reaction. The Peclet number has a role in mass transfer analogous to that of the Reynolds number in fluid dynamics. For example, we found that the convective mass transfer could be ignored if the Peclet number was very small; this is analogous to the low Reynolds number or creeping-flow approximation in fluid dynamics. We also found that the complicating effects of axial diffusion could be ignored if the aspect ratio was very small. The combination of small Peclet number and small aspect ratio in mass transfer is analogous to the lubrication-flow approximation in fluid dynamics. We found that for very large Thiele moduli, the homogeneous chemical reaction was confined to a thin region of influence or boundary layer near the permeable membrane walls.

In Section 5.5 we considered a complementary problem to that discussed in Section 5.4: mass transfer with a heterogeneous chemical reaction at the wall for fully developed laminar tube flow. In this case, scaling was used to determine when the classical plug-flow reactor approximation can be made; that is, when one can ignore any radial concentration gradients and thereby represent the axial convection using the average rather than the local velocity. It was necessary to introduce separate scales for both the axial and radial concentration gradients since neither of these scaled with the characteristic length divided by the characteristic time. This problem introduced the second Damköhler number, which is a measure of the time scale for radial diffusion to that for the heterogeneous reaction. When this dimensionless group was small, the plug-flow reactor approximation can be made. The small Damköhler number approximation in mass transfer is analogous to the small Biot number approximation in heat transfer in that both imply negligible resistance to transfer within the control volume relative to that at some boundary.

The problem considered in Section 5.6 involved mass transfer to falling film flow. Scaling was used to determine when the mass transfer could be assumed to be confined to a thin boundary layer or region of influence near the interface. For large Peclet numbers the mass transfer will be confined to a boundary layer that is sufficiently thin to assume that the film is infinitely thick. Moreover, if the product of the Peclet number and the length-to-thickness aspect ratio is large, axial diffusion can be neglected. These two approximations greatly simplify the describing equations; in particular, they obviate the need to apply a downstream boundary condition, which in many cases is not known.

In Section 5.7 we used scaling to assess when the quasi-steady-state (QSS) approximation can be made for the evaporation of a pure volatile liquid into an insoluble gas in a cylindrical tube. Quasi-steady-state implies that time does not enter explicitly in the describing differential equations, but implicitly through one or more boundary conditions. This is a moving boundary problem for which an auxiliary condition is required to locate the interface. The velocity that arises because of the diffusive mass transfer was determined in this problem from the additional condition that the gas was insoluble in the liquid. A proper scaling analysis required introducing a reference factor for the independent variable since it was not naturally referenced to zero. Moreover, it was necessary to introduce a separate scale factor for the interface velocity since it does not necessarily scale with the ratio of the characteristic length divided by the characteristic time. Since this problem involved a transformation from a stationary to a moving coordinate system, a pseudo-convection term was generated. Scaling analysis indicated that QSS can be assumed if the Fourier number is very large. Moreover, it provided criteria for when the convective mass transfer and pseudoconvection term could be neglected.

In Section 5.8 we applied scaling analysis to permeation through a polymeric membrane whose swelling caused the diffusivity to be concentration-dependent. For this reason it was necessary to scale the diffusivity as well. Two length scales were possible depending on how markedly the diffusivity changed with concentration. If the resistance to mass transfer was distributed through the entire cross-section of the membrane, the appropriate length scale was its thickness. However, if the diffusivity decreased markedly with concentration near one of the boundaries, the resistance to mass transfer was confined to a thin region of influence whose thickness was the appropriate length scale. To determine whether the diffusivity could be assumed to be constant, the equation describing its concentration-dependence was expanded in a Taylor series in which only the first-order correction was retained in the scaling analysis. Scaling then identified the condition required to ignore this first-order correction. For this problem it was possible to compare the solution to the complete describing equations with the simplified form of these equations for both negligible swelling and significant swelling; this confirmed that the simplified equation emanating from scaling provided accurate solutions when the appropriate criteria for the validity of these approximation were satisfied.

In Section 5.9 we applied scaling to a free-convection mass-transfer problem, that is, to a problem wherein the driving force for flow was internal to the system, in this case due to density variations created by concentration gradients. Scaling was employed to arrive at the free-convection boundary-layer equations and to determine when curvature effects could be neglected. This problem introduced the solutal Grashof and Rayleigh numbers, which are the free convection analogues of the Reynolds and Peclet numbers; that is, the former is a measure of the ratio of the free convection to viscous transport of momentum, whereas the latter is a measure of the free convection to diffusive transport of species.

Scaling was applied to dimensional analysis in Section 5.10. In contrast to o(1) scaling analysis, the scaling approach to dimensional analysis merely seeks to arrive at the minimum parametric representation of the problem; that is, to obtain a set of describing equations in terms of the minimum number of dimensionless parameters. The scaling approach to dimensional analysis was applied here to a novel membrane–lung oxygenator that employed axial oscillations to enhance the mass transfer. Scaling analysis was used to determine the dimensionless groups required to correlate the effects of the oscillations on the performance of the oxygenator. This problem introduced the Sherwood number, a dimensionless group that is a measure of the ratio of the overall mass transfer to that by diffusion alone, and the Graetz number. The latter is closely related to the Peclet number since it is a measure of the ratio of the convection to diffusion of species. However, the Graetz number includes an aspect ratio that accounts for the effect of a limited contact time. This example illustrated the advantages of the scaling analysis methodology for dimensional analysis relative to using the conventional Pi theorem approach in that the latter did not lead to the minimum parametric representation.

5.E EXAMPLE PROBLEMS

5.E.1 Evaporative Casting of a Polymer Film

Consider a binary mixture of volatile solvent A and nonvolatile polymer B that has initial mass fractions ω_{A0} and ω_{B0}, respectively, cast as a thin planar film having an initial thickness L_0 on an impermeable plate as shown in Figure 5.E.1-1. At time $t = 0$, the volatile solvent begins to evaporate into the gas phase, thereby causing the film to thin so that its instantaneous thickness is $L(t)$; evaporative cooling effects will be assumed to be negligible in this analysis. The evaporative mass transfer is described via a lumped-parameter approach with an appropriate mass-transfer coefficient. The overall mass density of the polymer film is assumed to be given by

$$\rho = \omega_A \rho_A^0 + \omega_B \rho_B^0 \Rightarrow \rho = \rho_B^0 + \Delta\rho_{AB}^0 \omega_A \tag{5.E.1-1}$$

where ρ_i^0 is the pure component mass density of component i and $\Delta\rho_{AB}^0 \equiv \rho_A^0 - \rho_B^0$; note that an alternative equation of state for the density could be used in this scaling analysis. We use scaling to determine criteria for when the following approximations can be made: Convective mass transfer arising from densification can be neglected; convective mass-transfer effects on the film thinning can

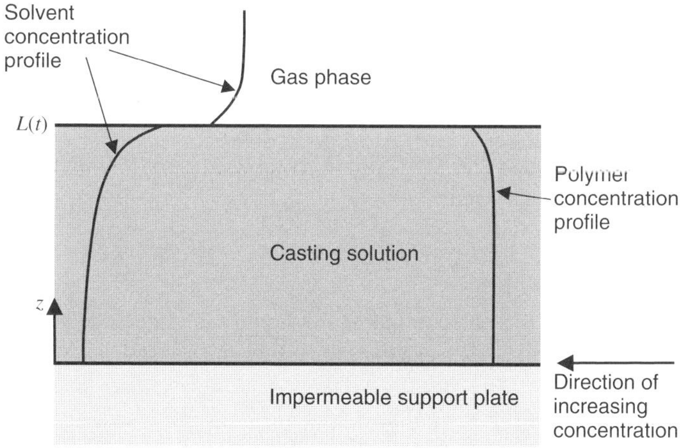

Figure 5.E.1-1 Representative concentration profiles during the evaporative casting of a dense film from a solution of a volatile solvent and nonvolatile polymer on an impermeable plate; the instantaneous thickness of the planar film is $L(t)$.

be ignored; the density can be assumed to be constant; quasi-steady-state applies; and the mass transfer can be assumed to be gas-phase controlled. This example illustrates how to handle a mass-transfer problem involving a nonconstant density and to scale properties such as the mass density that are functions of the dependent variable; it also provides another example of a moving boundary problem.

The appropriately simplified forms of the continuity and species-balance equations given by (C.1-1) and (G.1-1) in the Appendices are (step 1)

$$\frac{\partial \rho}{\partial t} = -\frac{\partial}{\partial z}(\rho u) \Rightarrow \Delta\rho^0_{AB}\frac{\partial \omega_A}{\partial t} = -\frac{\partial}{\partial z}(\rho u) \tag{5.E.1-2}$$

$$\frac{\partial \rho_A}{\partial t} = \frac{\partial(\omega_A \rho)}{\partial t} = -\frac{\partial n_A}{\partial z} \tag{5.E.1-3}$$

where

$$n_A = -\rho D_{AB}\frac{\partial \omega_A}{\partial z} + \omega_A \rho u \tag{5.E.1-4}$$

in which u is the mass-average velocity. Equations (5.E.1-2) and (5.E.1-3) can be combined to obtain

$$\frac{\partial}{\partial z}(\rho u) = \frac{\Delta\rho^0_{AB}/\rho}{1+\omega_A \Delta\rho^0_{AB}/\rho}\frac{\partial n_A}{\partial z} \tag{5.E.1-5}$$

Equation (5.E.1-5) can, in turn, be integrated to obtain the following explicit equation for the mass-average velocity u:

$$u = \frac{1}{\rho}\int_0^z \frac{\Delta\rho^0_{AB}/\rho}{1+\omega_A \Delta\rho^0_{AB}/\rho}\frac{\partial n_A}{\partial z}\,dz \tag{5.E.1-6}$$

$$u = \frac{1}{\rho}\int_0^{n_A} \frac{\Delta\rho^0_{AB}/\rho}{1+\omega_A \Delta\rho^0_{AB}/\rho}\,dn_A \tag{5.E.1-7}$$

in which the following boundary conditions corresponding to an impermeable lower solid boundary have been used:

$$u = 0, \quad n_A = 0 \quad \text{at} \quad z = 0 \tag{5.E.1-8}$$

A boundary condition is also required at the liquid–gas interface; this is obtained by an integral species balance over the entire film thickness:

$$\frac{d}{dt}\int_0^{L(t)} (\omega_A \rho)\,dz + \frac{d}{dt}\int_{L(t)}^{\infty} (\tilde{\omega}_A \tilde{\rho})\,dz = 0 \quad \text{at} \quad z = L(t) \tag{5.E.1-9}$$

where $\tilde{\omega}_A$ and $\tilde{\rho}$ denote the mass fraction of component A and mass density in the gas phase. Applying Leibnitz's rule for differentiating an integral given by equation (H.1-2) in the Appendices and substituting equations (5.E.1-3) and (5.E.1-4) for the casting solution and similar equations for the gas phase yields

$$\rho D_{AB}\frac{\partial \omega_A}{\partial z} = \omega_A \rho u - k_G^{\bullet} K_A \omega_A - \omega_A \rho \frac{dL}{dt} \qquad \text{at} \quad z = L(t), \quad 0 \le t \le \infty \tag{5.E.1-10}$$

In arriving at equation (5.E.1-10), the mass flux of component A in the gas phase at the liquid–gas interface was represented via a lumped-parameter approach in terms of the mass-transfer coefficient $k_G^{\bullet}$. This introduces the distribution coefficient K_A, which incorporates the effects of nonideal solution behavior and other concentration-dependent factors needed to interrelate the liquid and gas concentrations at the interface. Equation (5.E.1-10) also assumes that the bulk of the gas phase contains none of the evaporating solvent and that the mass density of the gas phase is much less than that of the casting solution.

Since this is a moving boundary problem, owing to the mass loss and densification, an auxiliary condition is required to locate the instantaneous position of the interface. This is determined by an integral mass balance as follows:

$$\frac{d}{dt}\int_0^{L(t)} \rho\, dz = -k_G^{\bullet} K_A \omega_A \qquad \text{at} \quad z = L(t) \tag{5.E.1-11}$$

Applying Leibnitz's rule for differentiating an integral given by equation (H.1-2) in the Appendices and substituting equation (5.E.1-2) yields the following auxiliary condition to determine $L(t)$:

$$\rho \frac{dL}{dt} = \rho u - k_G^{\bullet} K_A \omega_A \qquad \text{at} \quad z = L(t) \tag{5.E.1-12}$$

In arriving at equation (5.E.1-12) we have used the boundary condition that $u = 0$ at $z = 0$, corresponding to an impermeable boundary. The initial conditions required to solve Equations (5.E.1-3), (5.E.1-10), and (5.E-12) are given by

$$\omega_A = \omega_{A0}, \quad L = L_0 \qquad \text{at} \quad t = 0, \qquad 0 \le z \le L(t) \tag{5.E.1-13}$$

Equation (5.E.1-12) can be used to simplify equation (5.E.1-10) to yield the following form of the boundary condition at the moving interface:

$$\rho D_{AB}\frac{\partial \omega_A}{\partial z} = -k_G^{\bullet} K_A \omega_A (1 - \omega_A) \qquad \text{at} \quad z = L(t), \qquad 0 \le t \le \infty \tag{5.E.1-14}$$

Introduce the following dimensionless dependent and independent variables (steps 2, 3, and 4):

$$\omega_A^* \equiv \frac{\omega_A}{\omega_{As}}; \qquad \rho^* \equiv \frac{\rho}{\rho_s}; \qquad u^* \equiv \frac{u}{u_s}; \qquad n_A^* \equiv \frac{n_A}{n_{As}};$$

$$K_A^* \equiv \frac{K_A}{K_{As}}; \qquad z^* \equiv \frac{z}{z_s}; \qquad t^* \equiv \frac{t}{t_s}; \qquad L^* \equiv \frac{L}{L_s}; \qquad \left(\frac{dL}{dt}\right)^* \equiv \frac{1}{L_s}\frac{dL}{dt} \tag{5.E.1-15}$$

Note that we have introduced a scale factor for ω_A even though it is dimensionless with a maximum possible value of 1 because we seek to bound it of o(1). Indeed, the maximum concentration ω_{A0} could be much less than 1. However, it is not necessary to introduce a reference factor for ω_A since its minimum value is zero, corresponding to complete evaporation of the volatile component. We have introduced a separate scale for dL/dt, the velocity of the interface since there is no reason why this should scale as z_s/t_s.[25] However, we have not introduced a separate scale for the derivative of the concentration since we are considering a longer time scaling for which the concentration will undergo its characteristic change over the instantaneous thickness of the film, not over some boundary layer near the upper interface.[26] Scale factors are also introduced for the mass density ρ and distribution coefficient K_A since they depend on the concentration.

Introduce these dimensionless variables into equations ((5.E.1-1)), (5.E.1-3), (5.E.1-4), (5.E.1-5), (5.E.1-7), (5.E.1-8), (5.E.1-12), and (5.E.1-14), and divide each equation through by the dimensional coefficient of one term that should be retained in order to maintain physical significance to obtain the following set of dimensionless describing equations (steps 5 and 6):

$$\rho^* = \frac{\rho_B^0}{\rho_s} + \frac{\Delta\rho_{AB}^0 \omega_{As}}{\rho_s}\omega_A^* \tag{5.E.1-16}$$

$$\frac{z_s \omega_{As}\rho_s}{n_{As} t_s}\frac{\partial(\omega_A^* \rho^*)}{\partial t^*} = -\frac{\partial n_A^*}{\partial z^*} \tag{5.E.1-17}$$

$$n_A^* = \frac{\omega_{As}\rho_s u_s}{n_{As}}\omega_A^* \rho^* u^* - \frac{D_{AB}\omega_{As}\rho_s}{n_{As} z_s}\rho^* \frac{\partial \omega_A^*}{\partial z^*} \tag{5.E.1-18}$$

$$u^* = \frac{n_{As}\,\Delta\rho_{AB}^0}{u_s \rho_s^2}\frac{1}{\rho^*}\int_0^{z^*}\frac{1/\rho^*}{1+(\Delta\rho_{AB}^0\,\omega_{As}/\rho_s)(\omega_A^*/\rho^*)}\frac{\partial n_A^*}{\partial z^*}dz^* \tag{5.E.1-19}$$

$$\omega_A^* = \frac{\omega_{A0}}{\omega_{As}}, \quad L^* = \frac{L_0}{L_s} \quad \text{at} \quad t^* = 0, \quad 0 \le z^* \le \frac{L}{z_s} \tag{5.E.1-20}$$

$$n_A^* = 0, \quad u^* = 0 \quad \text{at} \quad z^* = 0, \quad 0 \le t^* \le \infty \tag{5.E.1-21}$$

$$\rho^* \frac{\partial \omega_A^*}{\partial z^*} = -\frac{k_G^\bullet K_{As} z_s}{D_{AB}\rho_s} K_A^* \omega_A^* (1 - \omega_{As}\omega_A^*) \quad \text{at} \quad z^* = \frac{L}{z_s}, \quad 0 \le t^* \le \infty \tag{5.E.1-22}$$

$$\rho^* \left(\frac{dL}{dt}\right)^* = \frac{u_s}{\dot{L}_s}\rho^* u^* - \frac{k_G^\bullet K_{As}\omega_{As}}{\dot{L}_s \rho_s} K_A^* \omega_A^* \quad \text{at} \quad z^* = \frac{L}{z_s} \tag{5.E.1-23}$$

[25]The consequences of scaling the interface velocity with z_s/t_s were discussed and illustrated in Section 4.7.

[26]However, for a short-time scaling where the concentration change is confined to a region of influence near the upper surface, one must introduce a separate scale for the spatial derivative of the concentration, which is determined from the mass-flux boundary condition given by equation (5.E.1–14); this short-time scaling is considered in Practice Problem 5.P.20.

The following considerations dictate determining our scale factors (step 7). Assessing when quasi-steady-state can be assumed necessarily implies a scaling analysis for longer contact times, for which the effect of the evaporation will have penetrated through the entire liquid film. This consideration is important since it implies that the characteristic length is the entire thickness of the liquid film rather than some region of influence near the liquid–gas interface. Hence, to bound the dimensionless spatial coordinate and film thickness to be $\circ(1)$, we set the appropriate dimensionless groups in equation (5.E.1-20) equal to 1 to obtain $z_s = L(t)$ and $L_s = L_0$. Note that z and L scale differently since for longer contact times, z ranges between 0 and $L(t)$, whereas $L(t)$ ranges between L_0 initially to some smaller value at the end of the evaporation process. The scale for the mass fraction comes from equation (5.E.1-20) and is given by $\omega_{As} = \omega_{A0}$. The density scale comes from equation (5.E.1-16) and is given by $\rho_s = \rho_B^0$. Since we seek to determine when the quasi-steady-state assumption is applicable, our time scale is the observation time; that is, $t_s = t_o$. The scale for the species mass flux comes from balancing the principal terms in equation (5.E.1-18) to obtain $n_{As} = D_{AB}\omega_{A0}\rho_B^0/L$. The scale for the front velocity comes from balancing the principal terms in equation (5.E.1-23), which yields $\dot{L}_s = k_G^{\bullet} K_{A0}\omega_{A0}/\rho_B^0$. A proper scale for K_{As} that bounds K_A^* to be $\mathrm{O}(1)$ is the initial value of K_A denoted here by K_{A0} for which the concentration is known. The scale for the mass-average velocity comes from equation (5.E.1-19), which yields $u_s = D_{AB}\omega_{A0}\Delta\rho_{AB}^0/\rho_B^0 L$.

These scale factors then result in the following set of dimensionless describing equations, which will permit us to assess when quasi-steady and negligible convection can be assumed (step 8):

$$\rho^* = 1 + \frac{\Delta\rho_{AB}^0}{\rho_B^0}\omega_{A0}\omega_A^* \tag{5.E.1-24}$$

$$\frac{1}{\mathrm{Fo}_m}\frac{\partial(\omega_A^*\rho^*)}{\partial t^*} - \mathrm{Bi}_m\omega_{A0}z^*\left(\frac{dL}{dt}\right)^*\frac{\partial(\omega_A^*\rho^*)}{\partial z^*} = -\frac{\partial n_A^*}{\partial z^*} \tag{5.E.1-25}$$

$$n_A^* = \frac{\Delta\rho_{AB}^0}{\rho_B^0}\omega_{A0}\omega_A^*\rho^* u^* - \rho^*\frac{\partial\omega_A^*}{\partial z^*} \tag{5.E.1-26}$$

$$u^* = \frac{1}{\rho^*}\int_0^{z^*}\frac{1/\rho^*}{1+\omega_{A0}(\Delta\rho_{AB}^0/\rho_B^0)(\omega_A^*/\rho^*)}\frac{\partial n_A^*}{\partial z^*}dz^* \tag{5.E.1-27}$$

$$\omega_A = 1, \quad L^* = 1 \quad \text{at} \quad t^* = 0, \quad 0 \le z^* \le 1 \tag{5.E.1-28}$$

$$n_A^* = 0, \quad u^* = 0, \quad \text{at} \quad z^* = 0, \quad 0 \le t^* \le \infty \tag{5.E.1-29}$$

$$\rho^*\frac{\partial\omega_A^*}{\partial z^*} = -\mathrm{Bi}_m K_A^*\omega_A^*(1-\omega_{A0}\omega_A^*) \quad \text{at} \quad z^* = 1, \quad 0 \le t^* \le \infty \tag{5.E.1-30}$$

$$\rho^*\left(\frac{dL}{dt}\right)^* = \frac{1}{\mathrm{Bi}_m}\frac{\Delta\rho_{AB}^0}{\rho_B^0}\rho^* u^* - K_A^*\omega_A^* \quad \text{at} \quad z^* = 1 \tag{5.E.1-31}$$

where $\text{Fo}_m \equiv D_{AB}t_o/L^2$ and $\text{Bi}_m \equiv k_G^{\bullet} K_{As} L/D_{AB}\rho_B^0$ are the solutal Fourier and solutal Biot numbers, respectively, for mass transfer. Note that an additional pseudo-convection term now appears in equation (5.E.1-25) because of the transformation from a fixed coordinate system to one that is referenced to the moving interface between the liquid and gas phases. Pseudo-convection terms will always arise when one transforms from a stationary coordinate system to a moving system. Equation (5.E.1-26) indicates that the convection arising from densification will have a negligible effect on the mass-transfer flux if the following criterion is satisfied:

$$\frac{\Delta\rho_{AB}^0}{\rho_B^0}\omega_{A0} \ll 1 \Rightarrow \frac{\Delta\rho_{AB}^0}{\rho_B^0}\omega_{A0} = \text{o}(0.01) \qquad (5.\text{E}.1\text{-}32)$$

This criterion indicates that convective transport can be ignored when the densities of the two components are nearly the same (i.e., $\Delta\rho_{AB}^0 \cong 0$) or for very dilute solutions. Equation (5.E.1-31) indicates that the convection will have a negligible effect on the film thinning if the following criterion is satisfied:

$$\frac{1}{\text{Bi}_m}\frac{\Delta\rho_{AB}^0}{\rho_B^0} \ll 1 \Rightarrow \frac{1}{\text{Bi}_m}\frac{\Delta\rho_{AB}^0}{\rho_B^0} = \text{o}(0.01) \qquad (5.\text{E}.1\text{-}33)$$

This criterion is also satisfied when the density difference between the two components is small or when the Biot number is very large, which implies a negligible resistance to mass transfer in the gas phase relative to that in the liquid. Quasi-steady-state mass transfer can be assumed if the following criterion is satisfied:

$$\text{Fo}_m \gg 1 \qquad (5.\text{E}.1\text{-}34)$$

corresponding to sufficiently long contact times. Note that quasi-steady-state implies that the time dependence enters implicitly through the boundary condition at the upper surface rather than explicitly via the unsteady-state term in equation (5.E.1-25). The concentration dependence of the density given by equation (5.E.1-24) can be ignored if the following criterion is satisfied:

$$\frac{\Delta\rho_{AB}^0}{\rho_B^0}\omega_{A0} \ll 1 \Rightarrow \frac{\Delta\rho_{AB}^0}{\rho_B^0}\omega_{A0} = \text{o}(1) \qquad (5.\text{E}.1\text{-}35)$$

Equation (5.E.1-30) indicates that an additional simplification is possible when the Biot number is very small: namely, that

$$\frac{\partial\omega_A^*}{\partial z^*} \ll 1 \quad \text{if} \quad \text{Bi}_m \ll 1 \qquad (5.\text{E}.1\text{-}36)$$

This simplified problem, which is the small Biot number approximation for mass transfer, is solved in a manner similar to that shown in detail in Section 4.4 for an analogous problem in heat transfer.

5.E.2 Taylor Dispersion

Sir Geoffrey Taylor developed a widely used approximation for the manner in which a pulse or step change in concentration disperses as it is convected downstream; this phenomenon, which is called *Taylor dispersion*, causes the sharp boundary at the leading edge of the pulse or step change to become diffuse, due to the combined action of species diffusion and differential convection arising from the velocity profile.[27] Sir Geoffrey Taylor developed this approximation via very clever intuitive arguments. In this example we develop the Taylor dispersion approximation using systematic scaling analysis.

Consider steady-state fully developed laminar flow of a Newtonian fluid composed of pure component B having constant physical properties in a cylindrical tube having radius R, as shown in Figure 5.E.2-1. At time $t = 0$, the fluid is changed instantaneously to pure component A while maintaining all other aspects of the flow. Although the interface between fluid A and B is initially planar, it will begin to disperse as a result of convection and molecular diffusion. Convective transport will cause component A near the center of the tube to move farther downstream than that near the wall, due to the parabolic velocity profile. Superimposed on this convection of component A are both radial and axial diffusion; the former will cause diffusion of component A between the center of the tube and the wall. This combined species diffusion and convection, which is due to a nonuniform velocity profile, is referred to as *Taylor dispersion*.

We begin by writing the appropriately simplified species-balance equation given by equation (G.2-5) in the Appendices (step 1):

$$\frac{\partial c_A}{\partial t} + u_z \frac{\partial c_A}{\partial z} = D_{AB} \frac{\partial^2 c_A}{\partial z^2} + D_{AB} \frac{1}{r} \frac{\partial}{\partial r}\left(r \frac{\partial c_A}{\partial r}\right) \tag{5.E.2-1}$$

where the axial velocity is given by

$$u_z = 2\overline{U}\left(1 - \frac{r^2}{R^2}\right) \tag{5.E.2-2}$$

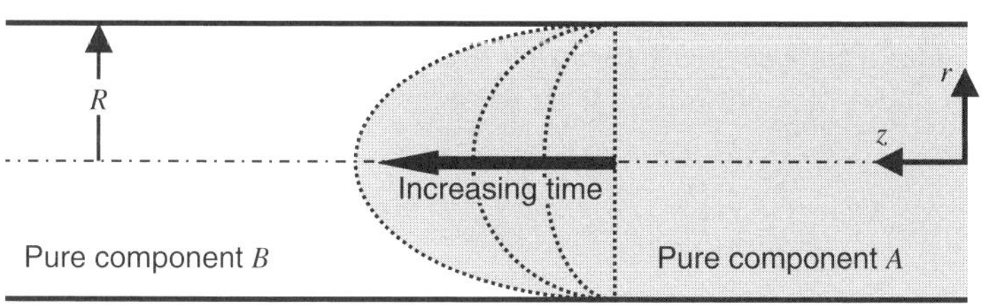

Figure 5.E.2-1 Fully developed laminar flow in a cylindrical tube of radius R showing the displacement of liquid B by the continuous injection of immiscible liquid A due to Taylor dispersion that arises from the combined effects of diffusion and convection via a nonuniform velocity profile.

[27] G. I. Taylor, *Proc. R. Soc. London*, **225A**, 473–477 (1954).

in which $\overline{U}$ is the average velocity. Note that each term in equation (G.2-5) has been divided by the molecular weight of component A in arriving at equation (5.E.2-1). The center of the dispersion zone containing both components A and B will be convected at the average velocity. Hence, we define a new axial coordinate $\tilde{z} \equiv z - \overline{U}t$ and transform equations (5.E.2-1) and (5.E.2-2) to a convected coordinate system as follows:

$$\frac{\partial c_A}{\partial t} + \overline{U}\left(1 - 2\frac{r^2}{R^2}\right)\frac{\partial c_A}{\partial \tilde{z}} = D_{AB}\frac{\partial^2 c_A}{\partial \tilde{z}^2} + D_{AB}\frac{1}{r}\frac{\partial}{\partial r}\left(r\frac{\partial c_A}{\partial r}\right) \tag{5.E.2-3}$$

where the time derivative is now evaluated at constant $\tilde{z}$ rather than at constant z; that is, the time derivative of the concentration is evaluated at a position relative to the convected dispersion zone. Note that this coordinate transformation modifies the convection term in the species-balance equation. The corresponding initial and boundary conditions are given by

$$c_A = 0 \quad \text{at} \quad t = 0 \tag{5.E.2-4}$$

$$c_A \text{ is bounded} \quad \text{at} \quad r = 0 \tag{5.E.2-5}$$

$$\frac{\partial c_A}{\partial r} = 0 \quad \text{at} \quad r = R \tag{5.E.2-6}$$

$$c_A = c_{A0} \quad \text{at} \quad \tilde{z} = -\overline{U}t \tag{5.E.2-7}$$

$$c_A = 0 \quad \text{at} \quad \tilde{z} \to \infty \tag{5.E.2-8}$$

The boundary condition given by equation (5.E.2-5) merely states that the concentration remains finite at the centerline; this is a common condition applied at a point or axis of symmetry. Equation (5.E.2-6) states that the walls of the tube are impermeable.

Define the following dimensionless variables (steps 2, 3, and 4):

$$c_A^* \equiv \frac{c_A}{c_s}; \qquad r^* \equiv \frac{r}{r_s}; \qquad \tilde{z}^* \equiv \frac{\tilde{z}}{\tilde{z}_s}; \qquad t^* \equiv \frac{t}{t_s} \tag{5.E.2-9}$$

Introduce these dimensionless variables into the describing equations and divide through by the coefficient of one term in each equation (steps 5 and 6):

$$\frac{r_s^2}{D_{AB}t_s}\frac{\partial c_A^*}{\partial t^*} + \frac{\overline{U}r_s^2}{D_{AB}\tilde{z}_s}\left(1 - 2\frac{r_s^2}{R^2}r^{*2}\right)\frac{\partial c_A^*}{\partial \tilde{z}^*} = \frac{r_s^2}{x_s^2}\frac{\partial^2 c_A^*}{\partial \tilde{z}^{*2}} + \frac{1}{r^*}\frac{\partial}{\partial r^*}\left(r^*\frac{\partial c_A^*}{\partial r^*}\right) \tag{5.E.2-10}$$

$$c_A^* = 0 \quad \text{at} \quad t^* = 0 \tag{5.E.2-11}$$

$$c_A^* \text{ is bounded} \quad \text{at} \quad r^* = 0 \tag{5.E.2-12}$$

$$\frac{\partial c_A^*}{\partial r^*} = 0 \quad \text{at} \quad r^* = \frac{R}{r_s} \tag{5.E.2-13}$$

$$c_A^* = \frac{c_{A0}}{c_s} \quad \text{at} \quad \tilde{z}^* = -\frac{\overline{U} t_s}{\tilde{z}_s} t^* \tag{5.E.2-14}$$

$$c_A^* = 0 \quad \text{at} \quad \tilde{z}^* \to \infty \tag{5.E.2-15}$$

Now let us determine the scale factors (step 7). The dimensionless concentration and radial coordinate can be bounded to be o(1) by the following choice of scale factors that emanate from the dimensionless groups in equations (5.E.2-13) and (5.E.2-14): $c_s = c_{A0}$ and $r_s = R$. Since this is inherently unsteady-state mass transfer, we choose the observation time as our time scale; that is, $t_s = t_o$. Since the principal terms in the dispersion process are axial convection and radial diffusion, the dimensionless group multiplying the convection term in equation (5.E.2-10) is set equal to 1, which provides the axial length scale; that is, $\tilde{z}_s = \overline{U} R^2 / D_{AB}$. Our dimensionless describing equations then assume the form

$$\frac{1}{\text{Fo}_m} \frac{\partial c_A^*}{\partial t^*} + \left(1 - 2r^{*2}\right) \frac{\partial c_A^*}{\partial \tilde{z}^*} = \frac{1}{\text{Pe}_m^2} \frac{\partial^2 c_A^*}{\partial \tilde{z}^{*2}} + \frac{1}{r^*} \frac{\partial}{\partial r^*} \left(r^* \frac{\partial c_A^*}{\partial r^*} \right) \tag{5.E.2-16}$$

$$c_A^* = 0 \quad \text{at} \quad t^* = 0 \tag{5.E.2-17}$$

$$c_A^* \text{ is bounded} \quad \text{at} \quad r^* = 0 \tag{5.E.2-18}$$

$$\frac{\partial c_A^*}{\partial r^*} = 0 \quad \text{at} \quad r^* = 1 \tag{5.E.2-19}$$

$$c_A^* = 1 \quad \text{at} \quad \tilde{z}^* = -\text{Fo}_m t^* \tag{5.E.2-20}$$

$$c_A^* = 0 \quad \text{at} \quad \tilde{z}^* \to \infty \tag{5.E.2-21}$$

where

$$\text{Fo}_m \equiv \frac{D_{AB} t_o}{R^2} \tag{5.E.2-22}$$

is the solutal Fourier number or Fourier number for mass transfer, which is a measure of the ratio of the observation time to the characteristic diffusion time, and

$$\text{Pe}_m \equiv \frac{\overline{U} R}{D_{AB}} = \frac{\overline{U} R}{\nu} \frac{\nu}{D_{AB}} = \text{Re} \cdot \text{Sc} \tag{5.E.2-23}$$

is the solutal Peclet number or Peclet number for mass transfer, which is a measure of the convection to diffusion of species, Re is the Reynolds number, which is a measure of the convection to viscous transport of momentum, and Sc is the Schmidt number, which is a measure of the viscous transport of momentum to diffusive transport of species.

If the following conditions apply, these describing equations can be reduced to those considered by Sir Geoffrey Taylor in his classical development of Taylor dispersion (step 8; see footnote 27):

$$\text{Fo}_m \gg 1 \Rightarrow \text{quasi-steady-state can be assumed} \tag{5.E.2-24}$$

$$\mathrm{Pe}_m^2 \gg 1 \Rightarrow \text{axial diffusion can be ignored} \tag{5.E.2-25}$$

Note that ignoring this unsteady-state term in equation (5.E.2-16) term does not mean that the dispersion is not time-dependent; it merely means that it is not time-dependent in a convected coordinate system; the time dependence enters through the axial coordinate in the convected coordinate system. Sir Geoffrey developed an approximate solution to the resulting simplified system of describing equations by assuming that $\partial c_A^*/-\partial \tilde{z}^*$ was constant (see footnote 27). He used intuitive arguments to conclude that his solution was a reasonable approximation to the solution for the full set of describing equations if $\mathrm{Fo}_m \gg 0.25$ and $\mathrm{Pe}_m \gg 6.9$. These conditions agree well with those given equations (5.E.2-24) and (5.E.2-25). The full set of describing equations were solved numerically by Gill et al.; their solution confirms the criteria given by equations (5.E.2-24) and (5.E.2-25) for the applicability of Sir Geoffrey's approximate solution.[28]

5.E.3 Convective Diffusion in a Tapered Pore

Consider steady-state binary gas-phase diffusion at constant pressure and temperature through a pore having a nonconstant circular cross-sectional area whose radius R is given by $R = R_0 - \beta\sqrt{z}$, where β is a constant. Figure 5.E.3-1 shows a cross-sectional view of this model pore along a plane that cuts through its axis of symmetry. The binary gas mixture at the mouth of the pore is assumed to be dilute in the diffusing component, whose concentration is maintained at a constant value c_{A0} (moles/volume). The concentration of the diffusing component is maintained at zero at the other end of the pore, where $z = L$. The diffusion coefficient may be

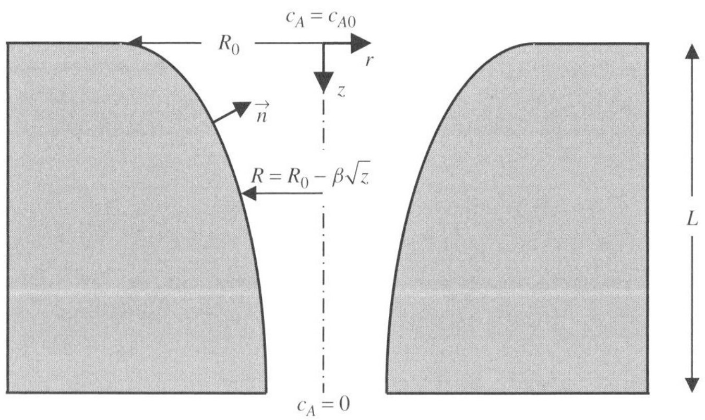

Figure 5.E.3-1 Binary diffusion of a dilute gas at constant temperature and pressure through a tapered pore having length L and a circular cross-section with a radius given by $R = R_0 - \beta\sqrt{z}$.

[28] V. Ananthakrishnan, W. N. Gill, and A. J. Barduhn, *A.I.Ch.E. J.*, **11**(6), 1063–1072 (1965).

assumed to be constant. Scaling analysis is used to determine when radial diffusion can be ignored relative to axial diffusion.

The appropriately simplified species-balance equation given by equation (G.2-10) in the Appendices and corresponding boundary conditions are given by

$$0 = D_{AB}\frac{\partial^2 c_A}{\partial z^2} + D_{AB}\frac{1}{r}\frac{\partial}{\partial r}\left(r\frac{\partial c_A}{\partial r}\right) \tag{5.E.3-1}$$

$$c_A = c_{A0} \quad \text{at} \quad z = 0 \tag{5.E.3-2}$$

$$c_A = 0 \quad \text{at} \quad z = L \tag{5.E.3-3}$$

$$\frac{\partial c_A}{\partial r} = 0 \quad \text{at} \quad r = 0 \quad \text{for } 0 \le z \le L \tag{5.E.3-4}$$

$$\vec{N}_A \cdot \vec{n} = 0 \quad \text{at} \quad r = R(z) = R_0 - \beta\sqrt{z}, \quad 0 \le z \le L \tag{5.E.3-5}$$

where $\vec{n}$ is the unit normal vector to the pore wall as shown in Figure 5.E.3-1 (step 1).

Define the following dimensionless variables (steps 2, 3, and 4):

$$c_A^* \equiv \frac{c_A}{c_s}; \qquad N_A^* \equiv \frac{N_A}{N_{As}}; \qquad z^* \equiv \frac{z}{z_s}; \qquad r^* \equiv \frac{r}{r_s} \tag{5.E.3-6}$$

A scale factor is introduced for the mass-transfer flux N_A, although this is a formality since it will not be necessary to determine this scale factor to assess when radial diffusion can be ignored.

Introduce these dimensionless variables into the describing equations and divide through by the coefficient of one term in each equation (steps 5 and 6):

$$0 = \frac{\partial^2 c_A^*}{\partial z^{*2}} + \frac{z_s^2}{r_s^2}\frac{1}{r^*}\frac{\partial}{\partial r^*}\left(r^*\frac{\partial c_A^*}{\partial r^*}\right) \tag{5.E.3-7}$$

$$c_A^* = \frac{c_{A0}}{c_s} \quad \text{at} \quad z^* = 0 \tag{5.E.3-8}$$

$$c_A^* = 0 \quad \text{at} \quad z^* = \frac{L}{z_s} \tag{5.E.3-9}$$

$$\frac{\partial c_A^*}{\partial r^*} = 0 \quad \text{at} \quad r^* = 0 \quad \text{for } 0 \le z^* \le \frac{L}{z_s} \tag{5.E.3-10}$$

$$\vec{N}_A^* \cdot \vec{n} = 0 \quad \text{at} \quad r^* = \frac{R_0}{r_s} - \frac{\beta\sqrt{z_s}}{r_s}\sqrt{z^*}, \quad 0 \le z^* \le \frac{L}{z_s} \tag{5.E.3-11}$$

The following considerations determine the scale factors (step 7). The dimensionless concentration and axial coordinate can be bounded to be o(1) by the following choice of scale factors, which emanate from the dimensionless groups in equations (5.E.3-8) and (5.E.3-9): $c_s = c_{A0}$ and $z_s = L$. Note that to ensure that

r^* is o(1), we want the largest possible value for r_s; this corresponds to setting the appropriate dimensionless group in equation (5.E.3-11) equal to 1 to obtain $r_s = R_0$.

Our dimensionless describing equations then assume the form

$$0 = \frac{\partial^2 c_A^*}{\partial z^{*2}} + \left(\frac{L}{R_0}\right)^2 \frac{1}{r^*}\frac{\partial}{\partial r^*}\left(r^* \frac{\partial c_A^*}{\partial r^*}\right) \tag{5.E.3-12}$$

$$c_A^* = 1 \quad \text{at} \quad z^* = 0 \tag{5.E.3-13}$$

$$c_A^* = 0 \quad \text{at} \quad z^* = 1 \tag{5.E.3-14}$$

$$\frac{\partial c_A^*}{\partial r^*} = 0 \quad \text{at} \quad r^* = 0 \quad \text{for } 0 \leq z^* \leq 1 \tag{5.E.3-15}$$

$$\vec{N}_A^* \cdot \vec{n} = 0 \quad \text{at} \quad r^* = 1 - \frac{\beta\sqrt{L}}{R_0}\sqrt{z^*}, \quad 0 \leq z^* \leq 1 \tag{5.E.3-16}$$

Equation (5.E.3-12) then indicates that to ignore radial relative to axial diffusion the following criterion must be satisfied (step 8):

$$\left(\frac{L}{R_0}\right)^2 \ll 1 \tag{5.E.3-17}$$

Note that this criterion will always break down for sufficiently long pores. This limitation is explored further in Practice Problem 5.P.3.

5.E.4 Dissolution of a Spherical Capsule

Consider a solid spherical capsule of a pure material A having an initial radius R_0. Assume that this capsule is ingested into the stomach, where it progressively dissolves while undergoing a first-order reaction with the stomach fluid B given by $R_A = k_1 c_A$ (moles/volume·time), in which k_1 is the reaction-rate constant. Equilibrium is assumed at the interface between the capsule and stomach fluid, at which the concentration of A is c_{A0} (moles/volume).[29] The solution will be assumed to be sufficiently dilute so that any bulk flow arising from the mass transfer is negligible; that is, the fluid phase is assumed to be quiescent so that the mass transfer is purely diffusive. A schematic of this dissolution process is shown in Figure 5.E.4-1. We use scaling to explore how the describing equations can be simplified.

[29]Note that in practice it would be difficult to measure this equilibrium concentration if component A reacts in the liquid phase with component B. However, it would be possible to infer the equilibrium concentration from measurements of the dissolution process if the kinetic constant for the homogeneous reaction were known.

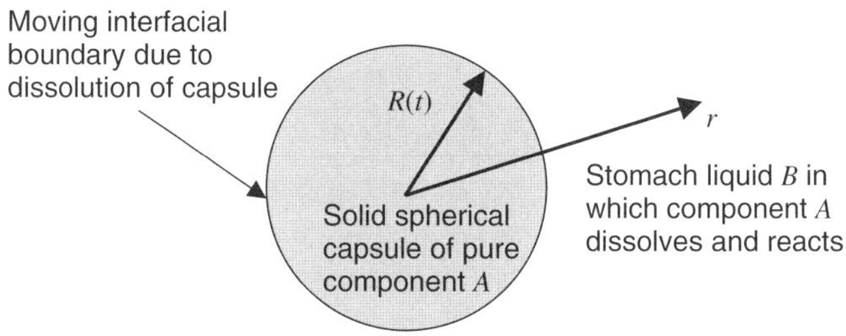

Figure 5.E.4-1 Solid spherical capsule of pure component A dissolving in stomach liquid B, accompanied by a first-order homogeneous chemical reaction.

The appropriately simplified form of the species-balance equation in spherical coordinates given by equation (G.3-10) in the Appendices and the corresponding initial and boundary conditions are (step 1)

$$\frac{\partial c_A}{\partial t} = D_{AB}\left[\frac{1}{r^2}\frac{\partial}{\partial r}\left(r^2\frac{\partial c_A}{\partial r}\right)\right] - k_1 c_A \tag{5.E.4-1}$$

$$c_A = 0, \quad R = R_0 \quad \text{at} \quad t = 0 \tag{5.E.4-2}$$

$$c_A = c_{A0} \quad \text{at} \quad r = R(t) \tag{5.E.4-3}$$

$$c_A = 0 \quad \text{as} \quad r \to \infty \tag{5.E.4-4}$$

where c_A is the molar concentration (moles/volume) and R_0 is the initial radius of the capsule. Since the dissolution of the spherical capsule implies that this is a moving boundary problem, an auxiliary condition is needed to locate the instantaneous position of the interface, denoted here by $R(t)$. This comes from an integral species balance on the spherical capsule and the stomach fluid, shown in detail here to illustrate how the homogeneous reaction term is handled:

$$\frac{d}{dt}\int_0^R \frac{\rho_A^0}{M_A} 4\pi r^2\, dr + \frac{d}{dt}\int_R^\infty c_A 4\pi r^2\, dr = -\int_R^\infty k_1 c_A 4\pi r^2\, dr \tag{5.E.4-5}$$

where M_A and ρ_A^0 are the molecular weight and solid mass density of pure component A, respectively. The last term in equation (5.E.4-5) accounts for loss of species A due to the homogeneous chemical reaction. Application of Leibnitz's rule for differentiating an integral given by equation (H.1-2) in the Appendices and substitution of equation (5.E.4-1) into equation (5.E.4-5) then yields

$$\left(\frac{\rho_A^0}{M_A} - c_A\right) R^2 \frac{dR}{dt} - D_{AB} R^2 \frac{\partial c_A}{\partial r} = 0 \quad \text{at} \quad r = R \tag{5.E.4-6}$$

Note that for dilute solutions, $c_A \ll \rho_A^0/M_A$. The initial condition needed to solve equation (5.E.4-6) is given in equation (5.E.4-2).

Define the following dimensionless variables (steps 2, 3, and 4):

$$c_A^* \equiv \frac{c_A}{c_s}; \qquad R^* \equiv \frac{R}{R_s}; \qquad \left(\frac{dR}{dt}\right)^* \equiv \frac{1}{\dot{R}_s}\frac{dR}{dt}; \qquad r^* \equiv \frac{r - r_r}{r_s}; \qquad t^* \equiv \frac{t}{t_s} \tag{5.E.4-7}$$

Note that we have introduced a reference factor since r is not naturally referenced to zero. We also have introduced a separate scale factor, $\dot{R}_s$, for the dissolution velocity of the spherical capsule dR/dt since there is no reason to assume that this will scale as r_s/t_s.

Introduce these dimensionless variables into the describing equations and divide through by the dimensional coefficient of one term in each equation (steps 5 and 6):

$$\frac{r_s^2}{D_{AB}t_s}\frac{\partial c_A^*}{\partial t^*} = \frac{1}{(r^* + r_r/r_s)^2}\frac{\partial}{\partial r^*}\left[(r^* + r_r/r_s)^2\frac{\partial c_A^*}{\partial r^*}\right] - \frac{k_1 r_s^2}{D_{AB}}c_A^* \tag{5.E.4-8}$$

$$c_A^* = 0, \quad R^* = \frac{R_0}{R_s} \quad \text{at} \quad t^* = 0 \tag{5.E.4-9}$$

$$c_A^* = \frac{c_{A0}}{c_s} \quad \text{at} \quad r^* = \frac{R - r_r}{r_s} \tag{5.E.4-10}$$

$$c_A^* = 0 \quad \text{as} \quad r^* \to \infty \tag{5.E.4-11}$$

$$\left(\frac{dR}{dt}\right)^* - \frac{M_A D_{AB} c_s}{\rho_A^0 \dot{R}_s r_s}\frac{\partial c_A^*}{\partial r^*} = 0 \quad \text{at} \quad r^* = \frac{R - r_r}{r_s} \tag{5.E.4-12}$$

The boundary condition given by equation (5.E.4-10) indicates that we can bound c_A^* to be o(1) by setting $c_s = c_{A0}$ and can reference r^* to zero by choosing $r_r = R$. The proper scale factor for the spatial coordinate depends on the conditions for which we are scaling this problem. Let us first consider the case where the homogeneous chemical reaction is sufficiently slow such that quasi-steady-state is never possible. Hence, the proper time scale is the observation time t_o. For this case the diffusion term must always balance the unsteady-state term in equation (5.E.4-8); the proper scale for the spatial coordinate is $r_s = \sqrt{D_{AB}t_o}$. The dimensionless sphere size can be bounded to be o(1) by setting the dimensionless group in equation (5.E.4-9) to 1 to obtain $R_s = R_0$. Since the two terms in equation (5.E.4-12) must balance, the scale for the dissolution velocity of the capsule is found to be $\dot{R}_s = M_A c_{A0}\sqrt{D_{AB}}/\rho_A^0\sqrt{t_o}$.

These scale and reference factors then result in the following dimensionless describing equations (step 7):

$$\frac{\partial c_A^*}{\partial t^*} - \frac{M_A c_{A0}}{\rho_A^0}\left(\frac{dR}{dt}\right)^*\frac{\partial c_A^*}{\partial r^*} = \frac{1}{\left(r^* + \dfrac{R}{\sqrt{D_{AB}t_o}}\right)^2}\frac{\partial}{\partial r^*}\left[\left(r^* + \frac{R}{\sqrt{D_{AB}t_o}}\right)^2\frac{\partial c_A^*}{\partial r^*}\right] - k_1 t_o c_A^* \tag{5.E.4-13}$$

$$c_A^* = 0, \quad R^* = 1 \quad \text{at} \quad t^* = 0 \tag{5.E.4-14}$$

$$c_A^* = 1 \quad \text{at} \quad r^* = 0 \tag{5.E.4-15}$$

$$c_A^* = 0 \quad \text{as} \quad r^* \to \infty \tag{5.E.4-16}$$

$$\left(\frac{dR}{dt}\right)^* - \frac{\partial c_A^*}{\partial r^*} = 0 \quad \text{at} \quad r^* = 0 \tag{5.E.4-17}$$

Note that an additional term now appears in equation (5.E.4-13) because of the transformation from a fixed coordinate system to one that is referenced to the moving interface of the spherical capsule. This is another example of the fact that one has to be careful when applying the chain rule of differentiation in transforming to the dimensionless variables.

Now let us consider how equations (5.E.4-13) through (5.E.4-17) can be simplified (step 8). Note that the characteristic length $r_s = \sqrt{D_{AB}t_o}$ defines a region of influence wherein the diffusive mass transfer is essentially confined. For sufficiently short times such that $R/\sqrt{D_{AB}t_0} \gg 1$, one can ignore the curvature effects in equation (5.E.4-13). Note, however, that this approximation will break down for sufficiently long times or when the capsule size becomes very small. One can estimate when the curvature effects become important by using the scale factor $\dot{R}_s$ to obtain an approximate solution for the instantaneous location of the capsule interface applicable for short contact times:

$$\dot{R}_s \cong \frac{dR}{dt} = -\frac{M_A c_{A0}\sqrt{D_{AB}}}{\rho_A^0\sqrt{t_o}} \Rightarrow R \cong R_0 - \frac{2M_A c_{A0}\sqrt{D_{AB}t_o}}{\rho_A^0} \tag{5.E.4-18}$$

Hence, curvature effects associated with the spherical geometry will need to be considered when

$$\frac{\sqrt{D_{AB}t_0}}{R_0 - 2M_A c_{A0}\sqrt{D_{AB}t_0}/\rho_A^0} = \mathrm{O}(1) \tag{5.E.4-19}$$

Note that equation (5.E.4-18) also provides an estimate of the time required for the capsule to dissolve. A further simplification of equation (5.E.4-13) is possible if $k_1 t_0 \ll 1$, in which case the effect of the homogeneous chemical reaction can be ignored. Note, however, that this approximation will always break down for sufficiently long contact times such that $k_1 t_0 = \mathrm{O}(1)$. In arriving at this set of describing equations we already have made the dilute solution assumption; that is, $M_A c_{A0}/\rho_A^0 \ll 1$, which implies that the pseudo-convection term arising from the coordinate transformation can also be ignored.

Now let us consider conditions for which the homogeneous reaction term is always important. In this case the characteristic length scale is obtained by balancing the diffusion term with the reaction term in equation (5.E.4-8), thereby obtaining $r_s = \sqrt{D_{AB}/k_1}$. This in turn implies that $\dot{R}_s = M_A c_{A0}\sqrt{D_{AB}k_1}/\rho_A^0$. All the other scale and reference factors remain the same. Hence, our dimensionless describing equations assume the form

$$\frac{1}{k_1 t_o}\frac{\partial c_A^*}{\partial t^*} - \frac{M_A c_{A0}}{\rho_A^0}\left(\frac{dR}{dt}\right)^* \frac{\partial c_A^*}{\partial r^*}$$

$$= \frac{1}{\left(r^* + R\sqrt{k_1/D_{AB}}\right)^2}\frac{\partial}{\partial r^*}\left[\left(r^* + R\sqrt{\frac{k_1}{D_{AB}}}\right)^2 \frac{\partial c_A^*}{\partial r^*}\right] - c_A^* \quad (5.E.4\text{-}20)$$

$$c_A^* = 0, \quad R^* = 1 \quad \text{at} \quad t^* = 0 \quad (5.E.4\text{-}21)$$

$$c_A^* = 1 \quad \text{at} \quad r^* = 0 \quad (5.E.4\text{-}22)$$

$$c_A^* = 0 \quad \text{as} \quad r^* \to \infty \quad (5.E.4\text{-}23)$$

$$\left(\frac{dR}{dt}\right)^* - \frac{\partial c_A^*}{\partial r^*} = 0 \quad \text{at} \quad r^* = 0 \quad (5.E.4\text{-}24)$$

The condition for ignoring the pseudo-convection term arising from the coordinate transformation again is $M_A c_{A0}/\rho_A^0 \ll 1$. In this case the curvature effects can be ignored when $R\sqrt{k_1/D_{AB}} \gg 1$. However, this condition will break down for sufficiently long times when the capsule radius becomes very small. Again, one can estimate when the curvature effects become important by using the scale factor $\dot{R}_s$ to obtain an approximate solution for the instantaneous location of the capsule interface:

$$\dot{R}_s \cong \frac{dR}{dt} = -\frac{M_A c_{A0}\sqrt{D_{AB}k_1}}{\rho_A^0} \Rightarrow R \cong R_0 - \frac{M_A c_{A0}\sqrt{D_{AB}k_1}}{\rho_A^0} t_o \quad (5.E.4\text{-}25)$$

Hence, curvature effects associated with the spherical geometry will need to be considered when

$$\left(R_0 - \frac{M_A c_{A0}\sqrt{D_{AB}k_1}}{\rho_A^0} t_o\right)\sqrt{\frac{k_1}{D_{AB}}} = \mathrm{O}(1) \quad (5.E.4\text{-}26)$$

For fast reaction conditions it is also possible to achieve quasi-steady-state mass transfer since the unsteady-state term in equation (5.E.4-20) becomes insignificant when $k_1 t_0 \gg 1$. In the latter case all the mass transfer and accompanying chemical reaction occur within a region of influence whose constant thickness is given by $r_s = \sqrt{D_{AB}/k_1}$; for very fast reaction conditions this boundary layer can become very thin. However, unless the reaction is instantaneous, the unsteady-state term will become significant for sufficiently short observation times.

5.E.5 Mass Transfer to a Rotating Disk: Uniformly Accessible Surface

In Example Problem 3.E.4 we scaled the hydrodynamics for a circular disk rotating about its axis at a constant angular frequency ω (radians/second) in an infinite Newtonian fluid having constant physical properties. We used scaling analysis in the aforementioned example to provide an estimate for the region of influence

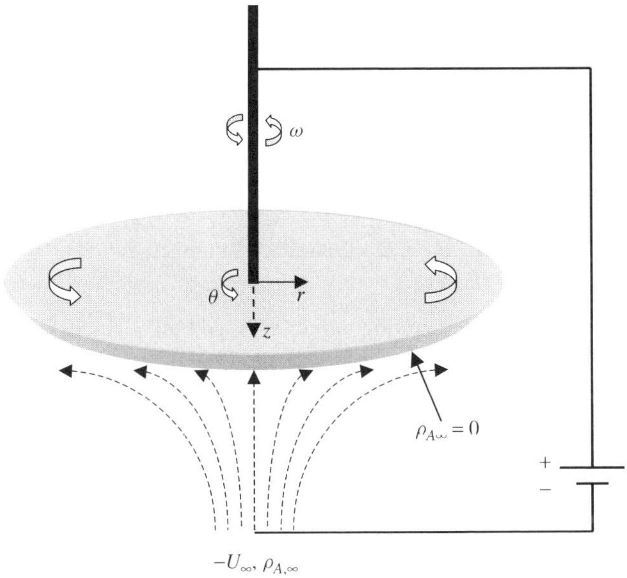

Figure 5.E.5-1 Mass transfer in laminar flow created by a flat disk rotating in an unbounded fluid; the rotational motion of the disk draws fluid toward the disk; the axial velocity and concentration infinitely far removed from the disk are U_∞ and ρ_∞, respectively.

across which the motion of the disk influences the velocity components. Assume now that the infinite fluid contains a solute that diffuses toward the rotating disk, at which it undergoes an electrochemical reaction that reduces its concentration to zero; that is, the rotating disk constitutes one electrode in an electrochemical system, as shown in Figure 5.E.5-1. We will use scaling to provide an estimate of the thickness of the mass-transfer boundary layer. A knowledge of this mass-transfer boundary-layer thickness is important since it must be much smaller than the radius of the rotating disk, to ensure that the finite container and edge effects are negligible.

We build our mass-transfer model based on the results of Example Problem 3.E.4. In theory, this mass-transfer problem involves the coupled equations of motion and species-balance equations. However, we assume that the solution is dilute and the mass-transfer rates are sufficiently small so that the solution to the equations-of-motion is decoupled from that for species balance. Hence, the velocity profiles are given by the solution to the appropriately simplified equations of motion given in Example Problem 3.E.4, which established that u_z, P, ru_r, and ru_θ are functions only of the axial coordinate z.

The rotating disk is analyzed as a uniformly accessible surface: that is, a surface for which the mass-transfer flux is not a function of the radial or angular coordinates. This concept of the uniformly accessible surface is presented without proof in standard references on mass transfer. Hence, we begin this example by developing a rigorous proof that in the absence of finite container and edge effects, the rotating disk, provides a uniformly accessible surface. To do this, consider a species balance on a control volume having arbitrary radius r and extending from the surface of the rotating disk far into the quiescent region of the fluid, where there is only an axial velocity component given by $u_z = -U_\infty$, and the concentration is ρ_∞, expressed in terms of the species mass per unit volume. A species balance on this control volume then yields

$$\rho_{A\infty} U_\infty \pi r^2 = \int_0^r n_{Az}|_{z=0} \cdot 2\pi \tilde{r}\, d\tilde{r} - \int_0^\infty n_{Ar} \cdot 2\pi r\, dz \tag{5.E.5-1}$$

where $\tilde{r}$ denotes a dummy integration variable. From Fick's law of diffusion applied at the surface of the disk and at the circumferential boundary defined by the arbitrary value of r along with the results of equations (3.E.4-1), we have

$$n_{Az} = -D_{AB}\frac{\partial \rho_A}{\partial z} + \rho_A u_z \Rightarrow n_{Az}|_{z=0} = -D_{AB}\frac{\partial \rho_A}{\partial z} \tag{5.E.5-2}$$

$$n_{Ar} = -D_{AB}\frac{\partial \rho_A}{\partial r} + \rho_A u_r = -D_{AB}\frac{\partial \rho_A}{\partial r} + \rho_A r f_1(z) \tag{5.E.5-3}$$

Substituting equations (5.E.5-2) and (5.E.5-3) into equation (5.E.5-1) then yields

$$\begin{aligned}\rho_{A\infty} U_\infty \pi r^2 = &-2\pi D_{AB}\int_0^r \left.\frac{\partial \rho_A}{\partial z}\right|_{z=0} \tilde{r}\, d\tilde{r} \\ &+ 2\pi r D_{AB}\int_0^\infty \frac{\partial \rho_A}{\partial r}\, dz - 2\pi r^2 \int_0^\infty \rho_A f_1(z)\, dz\end{aligned} \tag{5.E.5-4}$$

Equation (5.E.5-4) can be rearranged into the form

$$\rho_{A\infty} U_\infty + 2\int_0^\infty \rho_A f_1(z)\, dz = \frac{2D_{AB}}{r}\left(\int_0^\infty \frac{\partial \rho_A}{\partial r}\, dz - \int_0^r \left.\frac{\partial \rho_A}{\partial z}\right|_{z=0} d\tilde{r}\right) \tag{5.E.5-5}$$

Since the left-hand side of equation (5.E.5-5) is a constant, to ensure that the right-hand side is also a constant, the only possible solution for ρ_A is of the form

$$\rho_A = f_5(z) \tag{5.E.5-6}$$

that is, the concentration is a function only of the axial coordinate. This then implies that the mass-transfer flux along the surface of the rotating disk is a constant; that is, the surface of the disk is uniformly accessible. For this reason the rotating disk is frequently used to determine kinetic constants for reacting system since the electrode current provides a direct measure of the mass-transfer flux.

We now can proceed with our scaling analysis to estimate the mass-transfer boundary-layer thickness. The relevant form of the species-balance equation given by equation (G.2-5) in the Appendices for a uniformly accessible rotating disk then is given by the following:

$$u_z \frac{d\rho_A}{dz} = D_{AB} \frac{d^2 \rho_A}{dz^2} \tag{5.E.5-7}$$

The corresponding boundary conditions are given by (step 1)

$$\rho_A = 0 \quad \text{at} \quad z = 0 \tag{5.E.5-8}$$

$$\rho_A = \rho_{A\infty} \quad \text{as} \quad z \to \infty \tag{5.E.5-9}$$

Define the following dimensionless variables involving unspecified scale factors (steps 2, 3, and 4):

$$\rho_A^* \equiv \frac{\rho_A}{\rho_s}; \qquad u_z^* \equiv \frac{u_z}{u_{zs}}; \qquad z^* \equiv \frac{z}{z_s} \tag{5.E.5-10}$$

We have used the scaling results of Example Problem 3.E.4 in defining the dimensionless velocity. Introduce these dimensionless variables into the describing equations and divide through by the coefficient of one term in each equation that should be retained (steps 5 and 6):

$$u_z^* \frac{d\rho_A^*}{dz^*} = \frac{D_{AB}}{u_{zs} z_s} \frac{d^2 \rho_A^*}{dz^{*2}} \tag{5.E.5-11}$$

$$\rho_A^* = 0 \quad \text{at} \quad z^* = 0 \tag{5.E.5-12}$$

$$\rho_A^* = \frac{\rho_{A\infty}}{\rho_s} \quad \text{as} \quad z^* \to \infty \tag{5.E.5-13}$$

In determining our scale factors (step 7) we first recognize that the velocity scale was determined in Example Problem 3.E.4 and is given by equation (3.E.4-25) as $u_{zs} = \sqrt{\omega \nu}$. Equations (5.E.5-12) and (5.E.5-13) indicate that the dimensionless concentration can be bounded to be o(1) if we set $\rho_s = \rho_{A\infty}$. Since both diffusive and convective transport cannot be neglected in this problem, the two terms in equation (5.E.5-11) must balance; this provides the length scale $z_s = D_{AB}/u_{zs}$; hence, $z_s = D_{AB}/\sqrt{\omega \nu}$.

Now let us use our scaling results to enhance our understanding of this mass-transfer problem (step 8). The length scale z_s is a measure of the region of influence or boundary-layer thickness δ_s wherein the concentration goes from its far-field value of $\rho_{A\infty}$ to zero. It is instructive to recast $z_s = \delta_s$ in terms of the momentum boundary-layer thickness δ_m determined in Example Problem 3.E.4:

$$z_s = \delta_s = \frac{D_{AB}}{\sqrt{\omega \nu}} = \frac{D_{AB}}{\nu} \delta_m = \frac{1}{\text{Sc}} \delta_m \tag{5.E.5-14}$$

where $\mathrm{Sc} \equiv \nu/D_{AB}$ is the Schmidt number, which is a measure of the ratio of the viscous transfer of momentum to the diffusive transfer of mass. Since $\mathrm{Sc} \gg 1$ for liquids, the concentration boundary layer will be much thinner than the momentum boundary layer. Hence, the limiting criterion with respect to the importance of finite container and edge effects will be determined by the momentum rather than the concentration boundary-layer thickness; that is, the radius of the rotating disk must be much greater than δ_m.

5.E.6 Field-Flow Fractionation

Field-flow fractionation is a technique for separating small particles such as proteins and viruses from a carrier fluid such as water by combining a longitudinal laminar flow with a transverse field. The latter can be a thermal gradient, centrifugal force, electrical field, or transverse flow. Here we consider the latter, which is referred to as *flow-field-flow fractionation*. A transverse flow field can be imposed by making the closely spaced parallel lateral walls of the horizontal flow channel consist of two permeable membranes. Inflow and outflow of the same carrier fluid (without particles) occurs through the upper and lower membranes, respectively. This drives the particles, which are injected as a pulse in the axially flowing fluid, toward the lower membrane. This increase in particle concentration at the lower membrane causes a counterdiffusion of particles toward the upper membrane. The opposing convective and diffusive fluxes establish a layering of the particles near the lower membrane. Larger particles that have smaller diffusivities form layers closer to the lower membrane wall. Due to the fact that the axial velocity is smaller near the membrane surface, the larger particles will be eluted or pass through the flow-field-flow fractionation device more slowly than will the smaller particles. Hence, the total volume eluted from the device correlates directly with the particle size, thereby achieving the desired separation if the channel is sufficiently long. A schematic of the flow-field-flow fractionation device is shown in Figure 5.E.6-1. An early analysis of flow-field-flow fractionation claimed that the thickness of the steady-state exponential layer formed near the lower membrane would be equal to the binary diffusion coefficient D_{AB} divided by the transverse flow velocity V.[30] Here we use scaling to justify this claim and to ascertain the criteria required for its applicability.

The species-balance equation given by equation (G.1-5) in the Appendices reduces to the following for flow-field-flow fractionation (step 1):

$$\frac{\partial c_A}{\partial t} + u_x \frac{\partial c_A}{\partial x} + u_y \frac{\partial c_A}{\partial y} = D_{AB} \frac{\partial^2 c_A}{\partial x^2} + D_{AB} \frac{\partial^2 c_A}{\partial y^2} \tag{5.E.6-1}$$

Note that each term in equation (G.1-5) has been divided by the molecular weight of component A in order to arrive at equation (5.E.6-1). Scaling can be used to show that the velocity profile will not be affected by the transverse flow if the

[30] J. C. Giddings, F. J. F. Yang, and M. N. Myers, *Science*, **193**, 1244–1245 (1976).

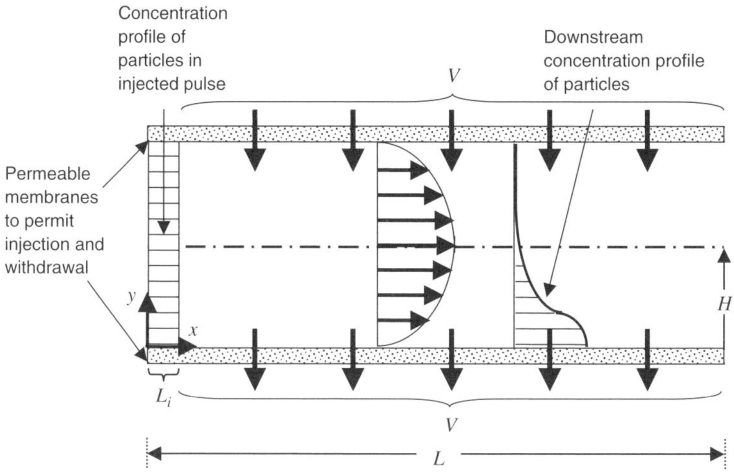

Figure 5.E.6-1 Flow-field-flow fractionation showing injection of a pulse of particles having a uniform concentration and initial width L_i into fully developed laminar flow; these particles are convected downstream due to the axial velocity profile and in the transverse direction due to injection and withdrawal of fluid through permeable membranes at the upper and lower boundaries; a balance between transverse convection and diffusion causes the particles to be concentrated in a thin layer near the lower membrane boundary.

Reynolds number based on the transverse velocity is very small: that is, if the following criterion is satisfied[31]:

$$\mathrm{Re}_V \equiv \frac{\rho H V}{\mu} \ll 1 \tag{5.E.6-2}$$

where ρ is the mass density, H the spacing between the parallel membranes, V the transverse injection velocity, and μ the shear viscosity, in which case the fully developed laminar flow velocity profile referenced to a coordinate system whose origin is located at the lower membrane boundary is given by

$$u_x = 2\overline{U}\left(\frac{2y}{H} - \frac{y^2}{H^2}\right) \tag{5.E.6-3}$$

where $\overline{U}$ is the average axial velocity. In problems such as this that involve the injection of a concentration pulse or plug, it is convenient to transform equation (5.E.6-1) into a coordinate system that moves at the average velocity. The reason for doing this is that under appropriate conditions the problem might be considered to be steady-state in a coordinate system translated at the appropriate average velocity. The appropriate average velocity is not necessarily $\overline{U}$ if, indeed, the particles are concentrated near the lower membrane boundary. We use scaling to determine the appropriate average velocity and the conditions under which

[31] See Practice Problem 3.P.32, which applies scaling analysis to justify this approximation.

steady-state can be assumed. Hence, we define a new axial coordinate $\tilde{x} \equiv x - \overline{U}_w t$, where $\overline{U}_w$ is an appropriate average velocity near the membrane boundary wherein the particles are confined. In this convected coordinate system equation (5.E.6-1) assumes the form:

$$\frac{\partial c_A}{\partial t} + (u_x - \overline{U}_w)\frac{\partial c_A}{\partial \tilde{x}} + u_y \frac{\partial c_A}{\partial y} = D_{AB}\frac{\partial^2 c_A}{\partial \tilde{x}^2} + D_{AB}\frac{\partial^2 c_A}{\partial y^2} \tag{5.E.6-4}$$

where the time derivative is now evaluated at constant $\tilde{x}$ rather than at constant x. The corresponding initial and boundary conditions are given by

$$\left.\begin{matrix} c_A = c_{A0} & \text{for} \quad 0 \le \tilde{x} \le L_i \\ c_A = 0 & \text{for} \quad L_i \le \tilde{x} \le L \end{matrix}\right\} \quad \text{at} \quad t = 0 \tag{5.E.6-5}$$

$$c_A = 0 \qquad \text{at} \quad \tilde{x} = -\overline{U}_w t \tag{5.E.6-6}$$

$$c_A = f(y) \qquad \text{at} \quad \tilde{x} = L - \overline{U}_w t \tag{5.E.6-7}$$

$$N_A = -D_{AB}\frac{\partial c_A}{\partial y} - V c_A = 0 \qquad \text{at} \quad y = 0 \tag{5.E.6-8}$$

$$N_A = -D_{AB}\frac{\partial c_A}{\partial y} - V c_A = 0 \qquad \text{at} \quad y = 2H \tag{5.E.6-9}$$

where D_{AB} is the binary diffusion coefficient. The initial condition given by equation (5.E.6-5) states that a pulse having concentration c_{A0} is injected over length L_i. Equation (5.E.6-6) states that the inlet concentration drops to zero after the initial injection of the particles. Equation (5.E.6-7) is a formal statement that a downstream boundary condition must be specified, although in practice this condition might not be known. Equations (5.E.6-8) and (5.E.6-9) are a statement that the particles cannot permeate through the membranes that constitute the lower and upper boundaries, respectively.

Define the following dimensionless variables (steps 2, 3, and 4):

$$c_A^* \equiv \frac{c_A}{c_s}; \qquad \tilde{x}^* \equiv \frac{\tilde{x}}{\tilde{x}_s}; \qquad y^* \equiv \frac{y}{y_s}; \qquad \left(\frac{\partial c_A}{\partial \tilde{x}}\right)^* \equiv \frac{1}{c_{xs}}\frac{\partial c_A}{\partial \tilde{x}} \tag{5.E.6-10}$$

Note that we have introduced a scale for the axial derivative of the concentration since it will not scale with the ratio of the concentration scale divided by the axial length scale; indeed, the characteristic value of this gradient is determined by change in concentration over the convected pulse of particles in the wall region. Introduce these dimensionless variables into the describing equations and divide through by the coefficient of one term in each equation (steps 5 and 6):

$$\begin{aligned} &\frac{y_s^2}{D_{AB} t_s}\frac{\partial c_A^*}{\partial t^*} + \frac{4\overline{U} y_s^3 c_{xs}}{D_{AB} H c_s}\left(y^* - \frac{1}{2}\frac{y_s}{H} y^{*2} - \frac{1}{4}\frac{\overline{U}_w H}{\overline{U} y_s}\right)\left(\frac{\partial c_A}{\partial \tilde{x}}\right)^* \\ &\quad - \frac{V y_s}{D_{AB}}\frac{\partial c_A^*}{\partial y^*} = \frac{y_s c_{xs}}{\tilde{x}_s c_s}\frac{\partial}{\partial \tilde{x}^*}\left(\frac{\partial c_A}{\partial \tilde{x}}\right)^* + \frac{\partial^2 c_A^*}{\partial y^{*2}} \end{aligned} \tag{5.E.6-11}$$

$$\left.\begin{array}{lll} c_A^* = \dfrac{c_{A0}}{c_s} & \text{for} & 0 \le \tilde{x}^* \le \dfrac{L_i}{x_s} \\ c_A^* = 0 & \text{for} & \dfrac{L_i}{\tilde{x}_s} \le \tilde{x}^* \le \dfrac{L}{\tilde{x}_s} \end{array}\right\} \quad \text{at} \quad t^* = 0 \tag{5.E.6-12}$$

$$c_A^* = 0 \quad \text{at} \quad \tilde{x}^* = -\frac{\overline{U}_w t_s}{\tilde{x}_s} t^* \tag{5.E.6-13}$$

$$c_A^* = f(y^*) \quad \text{at} \quad \tilde{x}^* = \frac{L}{\tilde{x}_s} - \frac{\overline{U}_w t_s}{\tilde{x}_s} t^* \tag{5.E.6-14}$$

$$\frac{\partial c_A^*}{\partial y^*} + \frac{V y_s}{D_{AB}} c_A^* = 0 \quad \text{at} \quad y^* = 0 \tag{5.E.6-15}$$

$$\frac{\partial c_A^*}{\partial y^*} + \frac{V y_s}{D_{AB}} c_A^* = 0 \quad \text{at} \quad y^* = \frac{2H}{y_s} \tag{5.E.6-16}$$

The scale factors are determined via the following considerations (step 7). The dimensionless concentration and axial coordinate can be bounded to be O(1) by setting the relevant dimensionless groups in equation (5.E.6-12) equal to 1 to obtain $c_s = c_{A0}$ and $x_s = L$. The principle of flow-field-flow fractionation is that the particles are confined within a thin layer whose thickness is determined by a balance between the transverse convection and counterdiffusion of particles. This implies that the two terms in equation (5.E.6-15) must balance each other; this yields the transverse length scale as

$$\frac{V y_s}{D_{AB}} = 1 \Rightarrow y_s = \frac{D_{AB}}{V} \tag{5.E.6-17}$$

Note that this estimate for y_s is the same as the characteristic thickness of the 'steady-state exponential layer' cited for flow-field-flow fractionation (see footnote 30). The axial convection must be of the same order as the transverse diffusion near the lower membrane boundary where the particles are concentrated; this provides the scale for the axial concentration gradient as follows:

$$\frac{4\overline{U} y_s^3 c_{xs}}{D_{AB} H c_s} = \frac{4\overline{U} D_{AB}^2 c_{xs}}{V^3 H c_{A0}} = 1 \Rightarrow c_{xs} = \frac{V^3 H c_{A0}}{4\overline{U} D_{AB}^2} \tag{5.E.6-18}$$

Since this is inherently an unsteady-state problem, the time scale is the observation time; that is, $t_s = t_o$.

Our dimensionless describing equations then assume the form

$$\begin{aligned} &\frac{1}{\text{Fo}_m \text{Pe}_V^2} \frac{\partial c_A^*}{\partial t^*} + \left(y^* - \frac{1}{2} \frac{1}{\text{Pe}_V} y^{*2} - \frac{1}{4} \text{Pe}_V \frac{\overline{U}_w}{\overline{U}} \right) \left(\frac{\partial c_A}{\partial \tilde{x}} \right)^* - \frac{\partial c_A^*}{\partial y^*} \\ &= \frac{1}{4} \frac{\text{Pe}_V}{\text{Pe}_m} \frac{H}{L} \frac{\partial}{\partial \tilde{x}^*} \left(\frac{\partial c_A}{\partial \tilde{x}} \right)^* + \frac{\partial^2 c_A^*}{\partial y^{*2}} \end{aligned} \tag{5.E.6-19}$$

$$\left.\begin{array}{lll} c_A^* = 1 & \text{for} & 0 \le \tilde{x}^* \le \dfrac{L_i}{L} \\ c_A^* = 0 & \text{for} & \dfrac{L_i}{L} \le \tilde{x}^* \le 1 \end{array}\right\} \quad \text{at} \quad t^* = 0 \tag{5.E.6-20}$$

$$c_A^* = 0 \qquad \text{at} \quad \tilde{x}^* = -\frac{\overline{U}_w t_o}{L} t^* \tag{5.E.6-21}$$

$$c_A^* = f(y^*) \qquad \text{at} \quad \tilde{x}^* = 1 - \frac{\overline{U}_w t_o}{L} t^* \tag{5.E.6-22}$$

$$\frac{\partial c_A^*}{\partial y^*} + c_A^* = 0 \qquad \text{at} \quad y^* = 0 \tag{5.E.6-23}$$

$$\frac{\partial c_A^*}{\partial y^*} + c_A^* = 0 \qquad \text{at} \quad y^* = 2\text{Pe}_V \tag{5.E.6-24}$$

where $\text{Pe}_m \equiv \overline{U}H/D_{AB}$ and $\text{Pe}_V \equiv VH/D_{AB}$ are the Peclet numbers for mass transfer based on $\overline{U}$ and V, respectively, and $\text{Fo}_m \equiv t_o D_{AB}/H^2$ is the solutal Fourier number.

Now let us examine the scaled describing equations to assess how these might be simplified (step 8). Let us first estimate the thickness of the layer wherein the particles are concentrated near the lower membrane boundary. This thickness is characterized by y_s, whose dimensionless value is inversely proportional to Pe_V; that is,

$$\frac{y_s}{H} = \frac{D_{AB}}{VH} = \frac{1}{\text{Pe}_V} \tag{5.E.6-25}$$

For typical Flow-Field-Flow Fractionation operating conditions, $\text{Pe}_V \cong 10^2$; hence, the particles are confined to a thin layer very close to the lower membrane boundary. Equation (5.E.6-19) indicates that only the linear portion of the velocity profile near the lower membrane need be considered if the following conditions apply:

$$\text{Pe}_V \gg 1 \Rightarrow \text{linear velocity-profile approximation} \tag{5.E.6-26}$$

This velocity profile can be obtained by expanding equation (5.E.6-3) in a Taylor series, which permits determining $\overline{U}_w$, the average velocity within the thin particle layer:

$$u_x \cong 4\overline{U}\frac{y}{H} \Rightarrow \overline{U}_w = 2\overline{U}\frac{y_s}{H} = \frac{2\overline{U}D_{AB}}{VH} = \frac{2\overline{U}}{\text{Pe}_V} \tag{5.E.6-27}$$

When this value for $\overline{U}_w$ is substituted into equation (5.E.6-19), we see that the pseudo-convection term arising from transforming to a convected coordinate system is an O(1) term, as might be expected.

Quasi-steady-state can be assumed if the following condition is satisfied:

$$\text{Fo}_m \cdot \text{Pe}_V^2 \gg 1 \Rightarrow \text{quasi-steady-state} \tag{5.E.6-28}$$

Typical conditions for flow-field-flow fractionation indicate that this condition is satisfied within approximately 20 seconds for an elution process that lasts several hours. Note that dropping the unsteady-state term in equation (5.E.6-19) does not imply that we are assuming that the process is steady-state; indeed, time enters through the spatial coordinate in our convected coordinate system; that is, $\tilde{x} = x - \overline{U}_w t$. The axial diffusion term can be ignored if the following condition is satisfied:

$$\frac{\text{Pe}_V}{\text{Pe}_m}\frac{H}{L} \ll 1 \Rightarrow \text{axial diffusion can be neglected} \tag{5.E.6-29}$$

This is also easily satisfied for flow-field-flow fractionation, for which typical values are $\text{Pe}_m = 10^4$ and $H/L = 10^{-4}$.

In view of these considerations, our describing equations can be simplified to

$$\left(y^* - \frac{1}{2}\right)\left(\frac{\partial c_A}{\partial \tilde{x}}\right)^* - \frac{\partial c_A^*}{\partial y^*} = \frac{\partial^2 c_A^*}{\partial y^{*2}} \tag{5.E.6-30}$$

$$c_A^* = 0 \qquad \text{at} \quad \tilde{x}^* = -2\frac{\text{Pe}_m \text{Fo}_m}{\text{Pe}_V}\frac{H}{L} \tag{5.E.6-31}$$

$$\frac{\partial c_A^*}{\partial y^*} + c_A^* = 0 \qquad \text{at} \quad y^* = 0 \tag{5.E.6-32}$$

$$\frac{\partial c_A^*}{\partial y^*} + c_A^* = 0 \qquad \text{at} \quad y^* = \infty \tag{5.E.6-33}$$

An analytical solution to these simplified describing equations can be obtained for the special case of $\partial c_A/\partial \tilde{x} = 0$, which is given by[32]

$$c_A^* = \beta e^{-y^*} \tag{5.E.6-34}$$

Since the describing equations constitute a linear homogeneous differential equation with homogeneous boundary conditions, the solution can be obtained only to within a multiplicative constant, denoted here by β. This unknown constant can be determined by equating the total eluted mass of particles to the initial mass of particles that is injected. Equation (5.E.6-34) then clearly establishes that flow-field-flow fractionation results in a steady-state exponential layer whose thickness is given by D_{AB}/V, as was stated in the introductory remarks for this example.

5.E.7 Mass Transfer in a Membrane Permeation Cell

Consider a gas-permeable polymer film of thickness L that is placed in a cylindrical permeation cell of circular cross-sectional area S_c, as shown in Figure 5.E.7-1. The permeable polymer film divides the permeation cell into an upper and a lower

[32]This is equivalent to assuming no axial concentration changes in the layer of particles that is convected at the average velocity in the region near the lower membrane boundary.

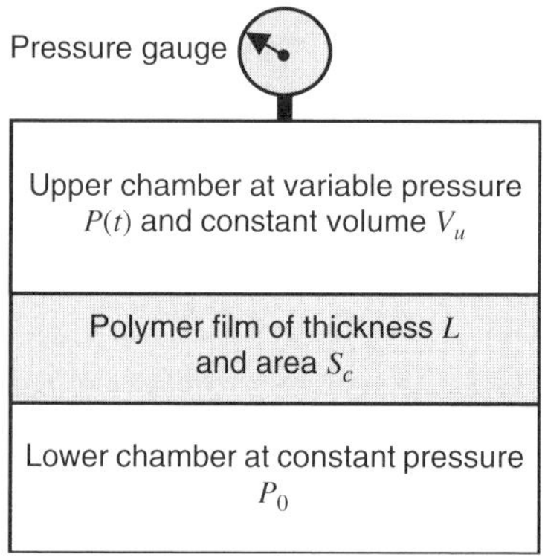

Figure 5.E.7-1 Membrane permeation cell in which a permeable polymer film separates the lower and upper chambers, both of which are evacuated initially; the membrane permeability can be determined by injecting a permeable gas into the lower chamber and then measuring the change in pressure in the upper chamber.

chamber, as shown in the figure; the volume of the upper chamber is denoted by V_u. Initially, both the upper and lower chambers are evacuated such that their initial pressure is $P = 0$. At time $t = 0$ a permeable single-component gas is introduced into the lower chamber at pressure P_0, and maintained at this pressure. This pure gas then begins to permeate through the polymer film into the upper evacuated chamber, causing its pressure to increase gradually. The pressure in the upper chamber at any time is denoted by $P(t)$, implying that the pressure in the upper chamber is a continuously increasing function of time. The permeating component can be assumed to form a dilute solution in the polymer film whose solubility is described by $\rho_A = HP$,where ρ_A is the concentration (mass/volume) and H is the Henry's law constant.

Typical data obtained using this apparatus are shown in Figure 5.E.7-2, in which the pressure in the upper chamber is plotted as a function of time. Note that there is a short period of time during which the pressure in the upper chamber remains at zero. This is followed by another relatively short period of time during which the pressure in the upper chamber increases nonlinearly. Finally, there is a relatively long period of time during which the pressure in the upper chamber increases linearly. We use scaling analysis to explain this interesting behavior and to determine useful properties of the membrane that we can extract from these data.

The appropriate form of the species-balance equations given by equation (G.1-5) in the Appendices and the corresponding initial and boundary conditions are (step 1)

$$\frac{\partial \rho_A}{\partial t} = D_{AB} \frac{\partial^2 \rho_A}{\partial x^2} \tag{5.E.7-1}$$

$$\rho_A = 0, \qquad P = P_0 \qquad \text{at} \quad t = 0 \tag{5.E.7-2}$$

$$\rho_A = HP_0 \qquad \text{at} \quad x = 0 \tag{5.E.7-3}$$

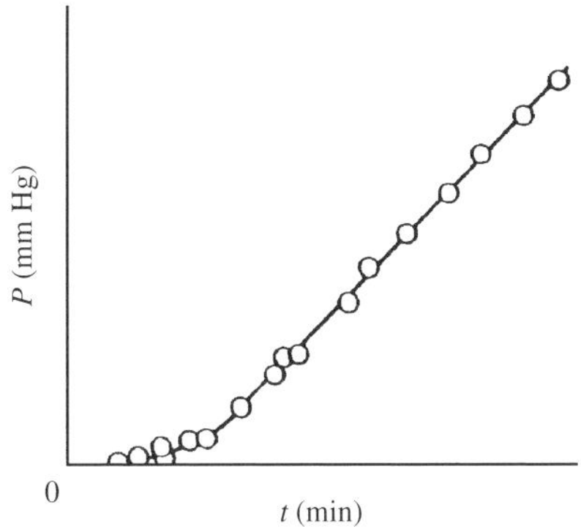

Figure 5.E.7-2 Pressure in the upper chamber versus time from the inception of the permeation process through the membrane separating the upper and lower chambers.

$$\rho_A = HP \qquad \text{at} \quad x = L \tag{5.E.7-4}$$

The boundary condition given by equation (5.E.7-4) is in terms of the unknown instantaneous pressure in the upper chamber. The auxiliary equation needed to determine this pressure can be obtained from an integral mass balance on the upper chamber as follows:

$$\frac{d}{dt}(cV_u) = \frac{V_u}{RT}\frac{dP}{dt} = -\frac{D_{AB}S_c}{M_A}\left.\frac{\partial \rho_A}{\partial x}\right|_{x=L} \tag{5.E.7-5}$$

where c is the molar density of the gas in the upper chamber, R the gas constant, and T the absolute temperature.

Define the following dimensionless variables (steps 2, 3, and 4):

$$\rho_A^* \equiv \frac{\rho_A}{\rho_s}; \qquad P^* \equiv \frac{P}{P_s}; \qquad \left(\frac{dP}{dt}\right)^* \equiv \frac{1}{\dot{P}_s}\frac{dP}{dt}; \qquad x^* \equiv \frac{x}{x_s}; \qquad t^* \equiv \frac{t}{t_s} \tag{5.E.7-6}$$

Note that we have introduced a separate scale, $\dot{P}_s$, for the time derivative of the pressure since there is no reason why this should scale with P_s and t_s.

Introduce these dimensionless variables into the describing equations and divide through by the coefficient of one term in each equation (steps 5 and 6):

$$\frac{x_s^2}{D_{AB}t_s}\frac{\partial \rho_A^*}{\partial t^*} = \frac{\partial^2 \rho_A^*}{\partial x^{*2}} \tag{5.E.7-7}$$

$$\rho_A^* = 0, \qquad P^* = \frac{P_0}{\rho_s} \qquad \text{at} \quad t^* = 0 \tag{5.E.7-8}$$

$$\rho_A^* = \frac{HP_0}{\rho_s} \qquad \text{at} \quad x^* = 0 \tag{5.E.7-9}$$

$$\rho_A^* = \frac{HP_s}{\rho_s} P^* \quad \text{at} \quad x^* = \frac{L}{x_s} \tag{5.E.7-10}$$

$$\left(\frac{dP}{dt}\right)^* = -\frac{D_{AB}S_c RT\rho_s}{M_A V_u x_s \dot{P}_s} \left.\frac{\partial \rho_A^*}{\partial x^*}\right|_{x^*=L/x_s} \tag{5.E.7-11}$$

The scale factors are determined by means of the following considerations (step 7). The dimensionless concentration and pressure can be bounded to be o(1) by setting the dimensionless groups in equations (5.E.7-8), (5.E.7-9), and (5.E.7-10) equal to 1 to obtain $\rho_s = HP_0$ and $P_s = P_0$. Since this is unsteady-state mass transfer, the time scale is the observation time t_o. The length scale is obtained from the appropriate dimensionless group in equation (5.E.7-10) equal to 1 to obtain $x_s = L$. Since the two terms in equation (5.E.7-11) must balance, the dimensionless group in this equation is set equal to 1 to obtain $\dot{P}_s = D_{AB}S_c RTHP_0/M_A V_u L$. These choices for the scale factors then result in the following describing equations:

$$\frac{1}{\text{Fo}_m}\frac{\partial \rho_A^*}{\partial t^*} = \frac{\partial^2 \rho_A^*}{\partial x^{*2}} \tag{5.E.7-12}$$

$$\rho_A^* = 0 \quad \text{at} \quad t^* = 0 \tag{5.E.7-13}$$

$$\rho_A^* = 1 \quad \text{at} \quad x^* = 0 \tag{5.E.7-14}$$

$$\rho_A^* = P^* \quad \text{at} \quad x^* = 1 \tag{5.E.7-15}$$

$$\left(\frac{dP}{dt}\right)^* = -\left.\frac{\partial \rho_A^*}{\partial x^*}\right|_{x^*=1} \tag{5.E.7-16}$$

where $\text{Fo}_m \equiv D_{AB}t_o/L^2$ is the solutal Fourier number.

Now let us consider how our scaled describing equations can be used to interpret the data shown in Figure 5.E.7-2 (step 8). The time required for any pressure increase to occur in the upper chamber can be estimated from the time required for the permeating component to penetrate the membrane. This corresponds to setting the solutal Fourier number equal to 1; that is,

$$\text{Fo}_m = \frac{D_{AB}t_o}{L^2} = 1 \Rightarrow t_o = \frac{L^2}{D_{AB}} \tag{5.E.7-17}$$

Equation (5.E.7-17) then provides an estimate of the dead time for any pressure response to occur in the upper chamber for the data shown in Figure 5.E.7-2. Once the permeating component penetrates through the membrane, a period of unsteady-state mass transfer will occur during which the pressure will increase nonlinearly in time. The duration of the latter period can be estimated from the time required to achieve quasi-steady-state mass-transfer conditions; that is, when

$$\frac{1}{\text{Fo}_m} = \frac{L^2}{D_{AB}t_o} \ll 1 \quad \text{or when} \quad t_o \cong \frac{100L^2}{D_{AB}} \tag{5.E.7-18}$$

This then provides an estimate of the time required from the introduction of the permeating gas to achieve quasi-state mass transfer through the membrane. We will show that the latter condition corresponds to a linear pressure increase in time. For observation times greater than that defined by equation (5.E.7-18), the unsteady-state term in equation (5.E.7-12) can be ignored. If, in addition, $P^* \ll 1$, the concentration driving force across the membrane will be constant, and equation (5.E.7-16) implies that

$$\left(\frac{dP}{dt}\right)^* = 1 \Rightarrow \frac{dP}{dt} = \dot{P}_s = \frac{D_{AB} S_c RTHP_0}{M_A V_u L} \Rightarrow P = \frac{D_{AB} S_c RTHP_0}{M_A V_u L} t \tag{5.E.7-19}$$

That is, the pressure will increase linearly in time, as seen in Figure 5.E.7-2 at longer times. Note that the diffusion coefficient for permeation through the membrane can be obtained from the slope in linear region of the pressure response curve. However, when $P^* \geq 0.1$ the permeation driving force across the membrane will decrease progressively, causing a less than linear increase in the pressure in the upper chamber. Figure 5.E.7-2 does not show this long-time behavior since it obviously does not include data taken for sufficiently long observation times.

For quasi-steady-state conditions, equations (5.E.7-12) and (5.E.7-16) can be solved analytically to obtain the following solution for the pressure in the upper chamber:

$$P = P_0 \left[1 - e^{-(D_{AB} S_c RTH/M_A V_u L)t}\right] \tag{5.E.7-20}$$

Note that for small values of the exponent, equation (5.E.7-20) reduces to the linear response given by equation (5.E.7-19). Hence, in summary, scaling analysis of the describing equations is able to describe all the principal features of the pressure-response curve for this standard membrane characterization test procedure.

5.E.8 Large Damköhler Number Approximation for Laminar Flow with a Heterogeneous Reaction

In Section 5.5 we considered steady-state laminar tube flow containing a solute A that underwent a first-order irreversible reaction at the wall as shown in Figure 5.5-1. We considered a scaling appropriate to a small Damköhler number for which the mass transfer was controlled by the slow rate of heterogeneous reaction. This implied that the concentration gradient across the tube was negligible and the concentration was spatially uniform. Here we apply scaling analysis to the complementary case of a large Damköhler number corresponding to a fast heterogeneous reaction. To supply mass to the tube wall at the same rate that it is consumed by the heterogeneous reaction, the concentration gradient will be very steep and occur over a region of influence or solutal boundary layer whose thickness δ_s is the appropriate radial length scale.

The appropriately simplified species-balance equation and associated boundary conditions are the same as those in Section 5.5 (step 1):

$$2\overline{U}\left(1-\frac{r^2}{R^2}\right)\frac{\partial c_A}{\partial z} = D_{AB}\frac{\partial^2 c_A}{\partial z^2} + D_{AB}\frac{1}{r}\frac{\partial}{\partial r}\left(r\frac{\partial c_A}{\partial r}\right) \tag{5.E.8-1}$$

$$c_A = c_{A0} \quad \text{at} \quad z = 0 \tag{5.E.8-2}$$

$$c_A = f(r) \quad \text{at} \quad z = L \tag{5.E.8-3}$$

$$\frac{\partial c_A}{\partial r} = 0 \quad \text{at} \quad r = 0 \tag{5.E.8-4}$$

$$-D_{AB}\frac{\partial c_A}{\partial r} = k_1^{\bullet} c_A \quad \text{at} \quad r = R \tag{5.E.8-5}$$

It is convenient to reference the coordinate system to the surface of the tube where the solutal boundary layer is located. Hence, we define the following dimensionless variables (steps 2, 3, and 4):

$$c_A^* \equiv \frac{c_A}{c_s}; \qquad r^* \equiv \frac{R-r}{\delta_s}; \qquad z^* \equiv \frac{z}{z_s} \tag{5.E.8-6}$$

Substitute these dimensionless variables into the describing equations and divide through by the coefficient of one term in each equation (steps 4 and 5):

$$\frac{4\overline{U}\delta_s^3}{D_{AB}Rz_s}\left(r^* - \frac{\delta_s}{2R}r^{*2}\right)\frac{\partial c_A^*}{\partial z^*} = \frac{r_s^2}{z_s^2}\frac{\partial^2 c_A^*}{\partial z^{*2}} + \frac{1}{\left(\frac{R}{\delta_s}\right) - r^*}\frac{\partial}{\partial r^*}\left[\left(\frac{R}{\delta_s} - r^*\right)\frac{\partial c_A^*}{\partial r^*}\right] \tag{5.E.8-7}$$

$$c_A^* = \frac{c_{A0}}{c_s} \quad \text{at} \quad z^* = 0 \tag{5.E.8-8}$$

$$c_A^* = f(r^*) \quad \text{at} \quad z^* = \frac{L}{z_s} \tag{5.E.8-9}$$

$$\frac{\partial c_A^*}{\partial r^*} = 0 \quad \text{at} \quad r^* = \frac{R}{\delta_s} \tag{5.E.8-10}$$

$$\frac{\partial c_A^*}{\partial r^*} = \frac{k_1^{\bullet}\delta_s}{D_{AB}}c_A^* \quad \text{at} \quad r^* = 0 \tag{5.E.8-11}$$

The dimensionless groups in equations (5.E.8-8) and (5.E.8-9), when set equal to 1 indicate that $c_s = c_{A0}$ and $z_s = L$ (step 7). Since the convection and radial heat conduction terms must be of the same order, equation (5.E.8-7) provides the following estimate for δ_s:

$$\frac{4\overline{U}\delta_s^3}{D_{AB}RL} = 1 \Rightarrow \delta_s = \left(\frac{D_{AB}RL}{4\overline{U}}\right)^{1/3} = \left(\frac{R^2L}{4\text{Pe}_m}\right)^{1/3} \tag{5.E.8-12}$$

where $\text{Pe}_m \equiv \overline{U}R/D_{AB}$ is the solutal Peclet number. Substitution of these scale factors into equations (5.E.8-7) through (5.E.8-11) then yields the following set of dimensionless describing equations:

$$\left[r^* - \frac{1}{2}\left(\frac{L}{4\text{Pe}_m R}\right)^{1/3} r^{*2}\right]\frac{\partial c_A^*}{\partial z^*} = \left(\frac{R^2}{4\text{Pe}_m L^2}\right)^{2/3}\frac{\partial^2 c_A^*}{\partial z^{*2}} + \frac{1}{\left[(4\text{Pe}_m R/L)^{1/3} - r^*\right]}\frac{\partial}{\partial r^*}\left\{\left[\left(\frac{4\text{Pe}_m R}{L}\right)^{1/3} - r^*\right]\frac{\partial c_A^*}{\partial r^*}\right\} \quad (5.E.8\text{-}13)$$

$$c_A^* = 1 \quad \text{at} \quad z^* = 0 \quad (5.E.8\text{-}14)$$

$$c_A^* = f(r^*) \quad \text{at} \quad z^* = 1 \quad (5.E.8\text{-}15)$$

$$\frac{\partial c_A^*}{\partial r^*} = 0 \quad \text{at} \quad r^* = \left(\frac{4\text{Pe}_m R}{L}\right)^{1/3} \quad (5.E.8\text{-}16)$$

$$\frac{\partial c_A^*}{\partial r^*} = \text{Da}^{\text{II}}\left(\frac{L}{4\text{Pe}_m R}\right)^{1/3} c_A^* \quad \text{at} \quad r^* = 0 \quad (5.E.8\text{-}17)$$

where $\text{Da}^{\text{II}} \equiv k_1^{\bullet} R/D_{AB}$ is the second Damköhler number.

Now let us consider how this set of dimensionless describing equations can be simplified (step 8). If $\left(R^2/4\text{Pe}_m L^2\right)^{2/3} \ll 1$, the axial diffusion term can be ignored in equation (5.E.8-13). If $(L/4\text{Pe}_m R)^{1/3} \ll 1$, the curvature effects and higher-order term in the velocity profile can be ignored in equation (5.E.8-13). Moreover, the boundary condition given by equation (5.E.8-16) can be applied at infinity. A zero mass flux far from the tube wall implies no change in the concentration in the core of the flowing fluid. Hence, equation (5.E.8-16) can be replaced by the condition that $c_A^* = 1$ as $r^* \to \infty$. Equation (5.E.8-17) indicates that as $\text{Da}^{\text{II}} \to \infty$, $c_A^* \to 0$, to ensure that $\partial c_A^*/\partial r^*$ remains bounded of $o(1)$. This implies that the concentration at the tube wall is zero. Hence, for large solutal Peclet numbers and large Damköhler numbers, the describing equations simplify to

$$r^*\frac{\partial c_A^*}{\partial z^*} = \frac{\partial}{\partial r^*}\left(r^*\frac{\partial c_A^*}{\partial r^*}\right) \quad (5.E.8\text{-}18)$$

$$c_A^* = 1 \quad \text{at} \quad z^* = 0 \quad (5.E.8\text{-}19)$$

$$c_A^* = 1 \quad \text{at} \quad r^* \to \infty \quad (5.E.8\text{-}20)$$

$$c_A^* = 0 \quad \text{at} \quad r^* = 0 \quad (5.E.8\text{-}21)$$

This simplified system of describing equations can be solved by standard methods, such as combination of variables.

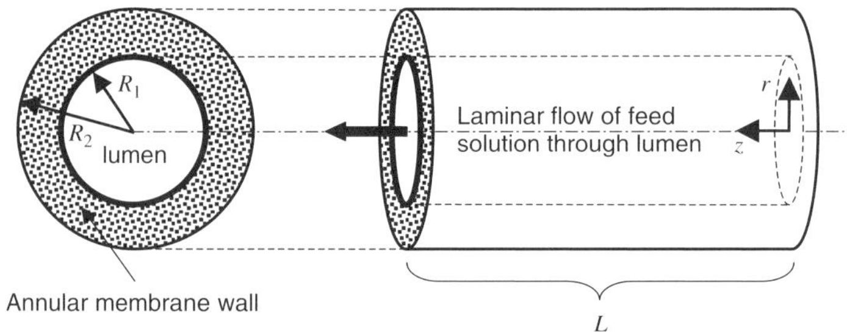

Figure 5.E.9-1 Hollow-fiber membrane of length L, lumen radius R_1, and outer annular membrane wall radius R_2. A feed solution flows axially in fully developed laminar flow. A reacting component in the feed permeates through the inner wall of the hollow fiber and undergoes a first-order homogeneous reaction with an enzyme that is immobilized within the pores of the annular region.

5.E.9 Small Thiele Modulus Approximation for Mass Transfer in a Hollow-Fiber Membrane

The development of hollow-fiber membranes with diameters of at most a few hundred micrometers has made it possible to design membrane separation systems having a very high area-to-volume ratio. This same feature has also permitted designing efficient catalytic membrane bioreactors by immobilizing an enzyme within the porous matrix of the tubular membrane. Figure 5.E.9-1 shows both cross-sectional and profile views of a single hollow-fiber membrane of length L and inner and outer radii R_1 and R_2, respectively. The annular membrane wall is made from a synthetic polymer that is microporous and serves as host to the immobilized enzyme. An ultrathin layer at the inner surface of the microporous annular region is impermeable to the enzyme and thereby confines it, but is highly permeable to low-molecular-weight solute(s). A solution of a low-molecular-weight solute is then pumped through the lumen or hollow core of the hollow fiber in steady-state laminar flow. This solute diffuses through the solution into the annular wall of the hollow fiber, where it reacts with the enzyme via a first-order reaction in the solute concentration. The aspect ratio of hollow fibers is such that axial diffusion can be ignored in most applications. Moreover, the concentration of the solute in both the solution and membrane is sufficiently dilute so that binary diffusion can be assumed; that is, the reaction product(s) does(do) not influence the diffusion of the solute through either the solution or the membrane. We use scaling to determine how the describing equations for the membrane enzyme reactor can be simplified.

The species-balance equation in cylindrical coordinates given by equation (G.2-10) in the Appendices and the corresponding boundary conditions for both the lumen and annular wall regions of the membrane, after appropriate simplification, are given by (step 1)

$$U_0\left[1-\left(\frac{r}{R_1}\right)^2\right]\frac{\partial c_{Al}}{\partial z} = D_l\frac{1}{r}\frac{\partial}{\partial r}\left(r\frac{\partial c_{Al}}{\partial r}\right), \qquad 0 \le r \le R_1 \tag{5.E.9-1}$$

$$0 = D_m \frac{1}{r}\frac{\partial}{\partial r}\left(r\frac{\partial c_{Am}}{\partial r}\right) - k_1 c_{Am}, \qquad R_1 \le r \le R_2 \tag{5.E.9-2}$$

$$c_{Al} = c_{A0} \quad \text{at} \quad z = 0 \tag{5.E.9-3}$$

$$\frac{\partial c_{Al}}{\partial r} = 0 \quad \text{at} \quad r = 0, \quad 0 \le z \le L \tag{5.E.9-4}$$

$$D_l \frac{\partial c_{Al}}{\partial r} = D_m \frac{\partial c_{Am}}{\partial r} \quad \text{at} \quad r = R_1, \quad 0 \le z \le L \tag{5.E.9-5}$$

$$c_{Am} = K_A c_{Al} \quad \text{at} \quad r = R_1, \quad 0 \le z \le L \tag{5.E.9-6}$$

$$\frac{\partial c_{Am}}{\partial r} = 0 \quad \text{at} \quad r = R_2, \quad 0 \le z \le L \tag{5.E.9-7}$$

where c_{Al} and c_{Am} are the molar concentrations of the solute in the lumen solution and annular membrane wall, respectively, U_0 the maximum fluid velocity at the centerline of the lumen, D_l and D_m the effective binary diffusion coefficients of the solute in the lumen solution and annular wall, respectively, c_{A0} the initial concentration in the feed solution to the lumen, and K_A the distribution coefficient for the thermodynamic equilibrium of component A between the lumen solution and the annular membrane wall, respectively.

Define the following dimensionless variables (steps 2, 3, and 4):

$$c_{Al}^* \equiv \frac{c_{Al}}{c_{ls}}; \qquad c_{Am}^* \equiv \frac{c_{Am}}{c_{ms}}; \qquad \left(\frac{\partial c_{Al}}{\partial r}\right)^* \equiv \frac{1}{c_{rls}}\frac{\partial c_{Al}}{\partial r};$$
$$\left(\frac{\partial c_{Al}}{\partial z}\right)^* \equiv \frac{1}{c_{zls}}\frac{\partial c_{Al}}{\partial z}; \qquad z^* \equiv \frac{z}{z_s}; \qquad r_l^* \equiv \frac{r}{r_{ls}}; \qquad r_m^* \equiv \frac{r - r_{lr}}{r_{ms}} \tag{5.E.9-8}$$

Note that we have introduced separate scales for the concentration and radial coordinate within the lumen and the annular wall. We also have allowed separate scales for both the axial and radial concentration gradients within the lumen since these do not necessarily scale with the concentration scale divided by the length scale. If we had naively scaled these derivatives with the concentration scale divided by a length scale, the forgiving nature of scaling would have indicated a contradiction.[33] We have also allowed for a reference length factor for the radial coordinate in order to reference it to zero within the annular membrane wall.

Introduce these dimensionless variables into the describing equations and divide through by the coefficient of one term in each equation (steps 5 and 6):

$$\left[1 - \left(\frac{r_{ls}}{R_1}\right)^2 r_l^{*2}\right]\frac{\partial c_l^*}{\partial z^*} = \frac{D_l c_{rls}}{U_0 c_{zls} r_{ls}} \frac{1}{r_l^*}\frac{\partial}{\partial r_l^*}\left(r_l^* \frac{\partial c_l^*}{\partial r_l^*}\right), \qquad 0 \le r_l^* \le \frac{R_1}{r_{ls}} \tag{5.E.9-9}$$

[33] This is explored further in Practice Problem 5.P.18.

$$0 = \frac{1}{r_m^* + r_{mr}/r_{ms}} \frac{\partial}{\partial r_m^*}\left[\left(r_m^* + \frac{r_{mr}}{r_{ms}}\right)\frac{\partial c_m^*}{\partial r_m^*}\right] - \frac{k_1 r_m^{*2}}{D_m} c_{Am}^*$$

$$\frac{R_1 - r_{mr}}{r_{ms}} \le r_m^* \le \frac{R_2 - r_{mr}}{r_{ms}} \tag{5.E.9-10}$$

$$c_l^* = \frac{c_{A0}}{c_{ls}} \quad \text{at} \quad z^* = 0 \tag{5.E.9-11}$$

$$\frac{\partial c_l^*}{\partial r^*} = 0 \quad \text{at} \quad r_l^* = 0, \quad 0 \le z^* \le \frac{L}{z_s} \tag{5.E.9-12}$$

$$\frac{\partial c_l^*}{\partial r_l^*} = \frac{D_m c_{ms}}{D_l c_{rls} r_{ms}} \frac{\partial c_m^*}{\partial r_m^*} \quad \text{at} \quad r_l^* = \frac{R_1}{r_{ls}}, \quad 0 \le z^* \le \frac{L}{z_s} \tag{5.E.9-13}$$

$$c_m^* = \frac{K_A c_{ls}}{c_{ms}} c_l^* \quad \text{at} \quad r_m^* = \frac{R_1 - r_{lr}}{r_{ls}}, \quad 0 \le z^* \le \frac{L}{z_s} \tag{5.E.9-14}$$

$$\frac{\partial c_m^*}{\partial r_m^*} = 0 \quad \text{at} \quad r_m^* = \frac{R_2 - r_{mr}}{r_{ms}}, \quad 0 \le z^* \le \frac{L}{z_s} \tag{5.E.9-15}$$

The following considerations dictate determining our scale factors (step 7). The concentration in the lumen and annular membrane wall can be bounded to be o(1) by setting the appropriate dimensionless groups in equations (5.E.9-11) and (5.E.9-14) equal to 1 to obtain $c_{ls} = c_{A0}$ and $c_{ms} = K_A c_{A0}$. The axial length scale is obtained from the dimensionless group L/z_s. The radial length scale in the annular membrane wall can be referenced to zero by setting the appropriate dimensionless group equal to zero in equation (5.E.9-14) to obtain $r_{ls} = R_1$. Since the radial diffusion must balance the axial convection in equation (5.E.9-9), the dimensionless group, which is a measure of this ratio, must be set equal to 1; this establishes a relationship between the axial and radial concentration gradient scales:

$$\frac{D_l c_{rls}}{U_0 c_{zls} r_{ls}} = \frac{D_l c_{rls}}{U_0 c_{zls} R_1} = 1 \Rightarrow c_{zls} = \frac{D_l c_{rls}}{U_0 R_1} \tag{5.E.9-16}$$

For steady-state to exist, all the diffusing solute must react within the annular membrane wall. This implies that radial diffusion must balance the homogeneous reaction within the annular membrane wall. Hence, by setting the dimensionless group in equation (5.E.9-10) that is a measure of this ratio equal to 1, we obtain the radial length scale factor:

$$\frac{k_1 r_m^{*2}}{D_m} = 1 \Rightarrow r_m = \sqrt{\frac{D_m}{k_1}} \tag{5.E.9-17}$$

Setting the dimensionless group in equation (5.E.9-13) equal to 1 provides the scale for the radial concentration gradient in the lumen:

$$\frac{D_m c_{ms}}{D_l c_{rls} r_{ms}} = \frac{D_m K_A c_{A0}}{D_l c_{rls}\sqrt{D_m/k_1}} \Rightarrow c_{rls} = \sqrt{\frac{D_m}{D_l}} K_A \mathrm{Th} \frac{c_{A0}}{R_1} \tag{5.E.9-18}$$

where $\text{Th} \equiv \sqrt{k_1 R_1^2/D_l}$ is the Thiele modulus, a dimensionless group that is a measure of the characteristic time for diffusion in the lumen to homogeneous reaction in the annular membrane wall. It is convenient to express the scale for the radial concentration gradient within the lumen in terms of the Thiele modulus, since this provides a means for assessing its magnitude relative to the maximum possible concentration gradient.[34]

These scale factors then result in the following set of dimensionless describing equations:

$$\left(1 - r_l^{*2}\right) \frac{\partial c_l^*}{\partial z^*} = \frac{1}{r_l^*} \frac{\partial}{\partial r_l^*} \left(r_l^* \frac{\partial c_l^*}{\partial r_l^*} \right), \qquad 0 \le r_l^* \le 1 \tag{5.E.9-19}$$

$$0 = \frac{1}{r_m^* + R_1/\sqrt{D_m/k_1}} \frac{\partial}{\partial r_m^*} \left[\left(r_m^* + \frac{R_1}{\sqrt{D_m/k_1}} \right) \frac{\partial c_m^*}{\partial r_m^*} \right] - c_{Am}^*$$

$$0 \le r_m^* \le \frac{R_2 - R_1}{\sqrt{D_m/k_1}} \tag{5.E.9-20}$$

$$c_l^* = 1 \qquad \text{at} \quad z^* = 0 \tag{5.E.9-21}$$

$$\frac{\partial c_l^*}{\partial r^*} = 0 \qquad \text{at} \quad r_l^* = 0, \quad 0 \le z^* \le 1 \tag{5.E.9-22}$$

$$\frac{\partial c_l^*}{\partial r_l^*} = \frac{\partial c_m^*}{\partial r_m^*} \tag{5.E.9-23}$$

$$c_m^* = c_l^* \qquad \text{at} \quad r_m^* = 0, \quad 0 \le z^* \le 1 \tag{5.E.9-24}$$

$$\frac{\partial c_m^*}{\partial r_m^*} = 0 \qquad \text{at} \quad r_m^* = \frac{R_2 - R_1}{\sqrt{D_m/k_1}}, \quad 0 \le z^* \le 1 \tag{5.E.9-25}$$

Now let us consider how these dimensionless describing equations can be simplified (step 8). If the dimensionless group $R_1/\sqrt{D_m/k_1} \gg 1$, implying that the region of influence wherein the homogeneous reaction converts all of the reacting solute to product is much thinner than the radius of the lumen, curvature effects on the mass-transfer process can be ignored. Moreover, if $(R_2 - R_1)/\sqrt{D_m/k_1} \gg 1$, the aforementioned region of influence is much thinner than the thickness of the annular membrane wall, the boundary condition given by equation (5.E.9-25) can be applied at infinity.

The solution to equations (5.E.9-19) and (5.E.9-20) can be simplified for the special case of very small Thiele moduli, that is, for $\text{Th} \ll 1$. For this case, equation (5.E.9-18) indicates that the concentration gradient within the lumen of the hollow fiber is negligibly small, thereby implying that the concentration within

[34]This problem is a mass-transfer analog to the heat-transfer problem considered in Section 4.4; that is, the low Biot number approximation made in the latter is analogous to the small Thiele modulus approximation made here.

the lumen is uniform. Hence, equation (5.E.9-19) can be integrated as follows:

$$\int_0^1 \left(1 - r_l^{*2}\right) \frac{\partial c_l^*}{\partial z^*} 2\pi r_l^* dr_l^* = \int_0^1 \frac{1}{r_l^*} \frac{\partial}{\partial r_l^*} \left(r_l^* \frac{\partial c_l^*}{\partial r_l^*}\right) 2\pi r_l^* dr_l^* \tag{5.E.9-26}$$

$$\int_0^1 \left(1 - r_l^{*2}\right) \frac{\partial c_l^*}{\partial z^*} r_l^* dr_l^* = \int_0^{\partial c_m^*/\partial r_m^*} \partial \left(r_l^* \frac{\partial c_l^*}{\partial r_l^*}\right) \tag{5.E.9-27}$$

Use of Leibnitz's rule given by equation (H.1-2) in the Appendices to integrate the first term and the fact that the concentration is essentially uniform within the lumen for very small Thiele moduli then yields

$$\frac{d}{dz^*} \int_0^1 c_l^* \left(1 - r_l^{*2}\right) r_l^* dr_l^* = \frac{dc_l^*}{dz^*} \int_0^1 \left(1 - r_l^{*2}\right) r_l^* dr_l^* = \frac{1}{4} \frac{dc_l^*}{dz^*} = \frac{\partial c_m^*}{\partial r_m^*} \tag{5.E.9-28}$$

Equation (5.E.9-28) now replaces equation (5.E.9-23) as one of the boundary conditions on the differential equation for the mass transfer with the annular membrane wall. If the conditions $R_1/\sqrt{D_m/k_1} \gg 1$ and $(R_2 - R_1)/\sqrt{D_m/k_1} \gg 1$ apply, the solution of the resulting simplified system of describing equations is straightforward and given by[35]

$$c_l^* = e^{-4z^*}; \qquad c_m^* = e^{-(4z^* + r_m^*)} \tag{5.E.9-29}$$

5.E.10 Dimensional Analysis for Oxygen Diffusion into a Spherical Red Blood Cell

Consider a spherical red blood cell of radius R in the blood stream, as shown in Figure 5.E.10-1. The cell is sufficiently large that it moves more slowly than the blood flow far removed from its surface (where the no-slip condition has to be satisfied). Hence, there is a velocity of the bloodstream U_∞ relative to that of the red blood cell. Oxygen diffuses through the bloodstream to this red blood cell, after which it diffuses into the cell and reacts with the hemoglobin. The oxygen and blood are referred to as components A and B, respectively. We wish to model the convective oxygen mass transfer through the bloodstream to the cell. However, since this is a rather complicated problem to solve, we employ the scaling method for dimensional analysis to obtain a correlation for the dimensionless mass-transfer coefficient, defined by

$$k_L \equiv \frac{N_{Aw}}{c_{Aw} - c_{A\infty}} \tag{5.E.10-1}$$

where N_{Aw} is the molar flux (moles/area·time) at the outer cell wall, c_{Aw} the oxygen concentration (moles/volume) at the outer cell wall, and $c_{A\infty}$ the oxygen

[35] An analytical solution to this problem accounting for curvature effects is given by W. Lewis and S. Middleman, *A.I.Ch.E.J.*, **20**(5), 1012–1014 (1974).

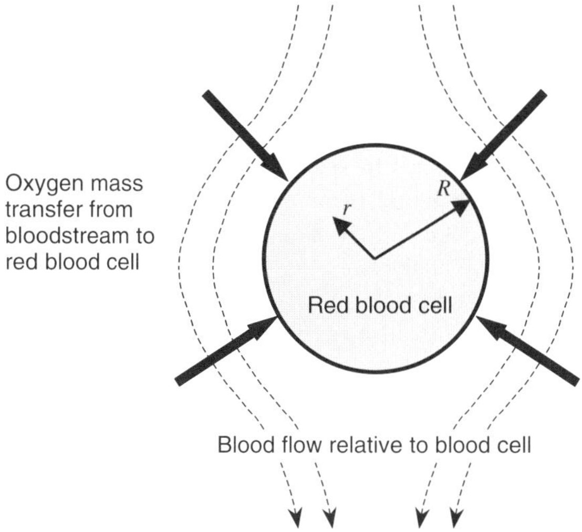

Figure 5.E.10-1 Steady-state oxygen mass transfer from a flowing bloodstream to a spherical red blood cell.

concentration in the bulk of the bloodstream far from the red blood cell. We assume that c_{Aw}, $c_{A\infty}$, U_∞, and the physical properties of the blood are known. Moreover, we assume that the oxygen concentration in the blood is dilute.

Equation (5.E.10-1) requires determination of N_{Aw}, which is given by

$$N_{Aw} = -\frac{D_{AB}}{1 - x_{A\infty}} \left. \frac{\partial c_A}{\partial r} \right|_R \cong -D_{AB} \left. \frac{\partial c_A}{\partial r} \right|_R \tag{5.E.10-2}$$

where D_{AB} is the effective binary diffusion coefficient of oxygen in blood and x_{Aw} is the oxygen mole fraction at the outer wall of the red blood cell. The molar concentration in equation (5.E.10-2) would have to be determined from a solution to the coupled species-balance equation and equations of motion given by equations (B.5-3) and (B.2-3) in the Appendices, respectively, which when appropriately simplified in a spherical coordinate system referenced to the center of the spherical red blood cell are given by

$$\vec{u} \cdot \nabla c_A = D_{AB} \nabla^2 c_A \tag{5.E.10-3}$$

$$\rho \vec{u} \cdot \nabla \vec{u} = \mu \nabla^2 \vec{u} - \nabla P \tag{5.E.10-4}$$

where $\vec{u}$ is the vector velocity in the bloodstream, ρ and μ are the mass density and shear viscosity of the blood, respectively, and P is the pressure. Note that each term in equation (B.5-3) has been divided by the molecular weight of component A in arriving at equation (5.E.10-3). The boundary conditions required to solve equations (5.E.10-3) and (5.E.10-4) are given formally as follows:

$$c_A = c_{Aw} \quad \text{at} \quad r = R \tag{5.E.10-5}$$

$$\vec{n} \cdot \vec{u} = 0 \quad \text{at} \quad r = R \tag{5.E.10-6}$$

$$c_A = c_{A\infty} \qquad \text{as} \quad r \to \infty \tag{5.E.10-7}$$

$$\vec{u} \cdot \vec{\delta}_r = U_\infty \cos\theta, \quad \vec{u} \cdot \vec{\delta}_\theta = U_\infty \sin\theta \qquad \text{as} \quad r \to \infty \tag{5.E.10-8}$$

where $\vec{\delta}_r$ and $\vec{\delta}_\theta$ are unit vectors in the radial and circumferential directions, respectively. It is not necessary to specify any boundary conditions on the pressure if the velocity of the bloodstream far from the red blood cell is specified. Equations (5.E.10-1) through (5.E.10-8) constitute step 1 in the scaling analysis procedure for dimensional analysis.

Define arbitrary scale factors for all the dependent and independent variables and reference factors for those variables not naturally referenced to zero (steps 2 and 3):

$$c^* \equiv \frac{c - c_r}{c_s}; \qquad P^* \equiv \frac{P}{P_s}; \qquad \vec{u}^* \equiv \frac{\vec{u}}{u_s}; \qquad r^* \equiv \frac{r}{r_s} \tag{5.E.10-9}$$

Note that we do not need to scale the angular coordinates θ and ϕ, since they are dimensionless and bounded of o(1).

Introduce these dimensionless variables into the describing equations and divide through by the dimensional coefficient of one term in each equation (steps 4 and 5):

$$\frac{k_L(c_{Aw} - c_{A\infty})r_s}{D_{AB}c_s} = -\frac{\partial c_A^*}{\partial r^*} \qquad \text{at} \quad r^* = \frac{R}{r_s} \tag{5.E.10-10}$$

$$\frac{u_s r_s}{D_{AB}} \vec{u}^* \cdot \nabla^* c_A^* = \nabla^{*2} c_A^* \tag{5.E.10-11}$$

$$\frac{\rho v_s r_s}{\mu} \vec{u}^* \cdot \nabla^* \vec{u}^* = \nabla^{*2} \vec{u}^* - \frac{P_s r_s}{\mu u_s} \nabla^* P^* \tag{5.E.10-12}$$

$$c_A^* = \frac{c_{Aw} - c_r}{c_s} \qquad \text{at} \quad r^* = \frac{R}{r_s} \tag{5.E.10-13}$$

$$\vec{n} \cdot \vec{u}^* = 0 \qquad \text{at} \quad r^* = \frac{R}{r_s} \tag{5.E.10-14}$$

$$c_A^* = \frac{c_{A\infty} - c_r}{c_s} \qquad \text{as} \quad r^* \to \infty \tag{5.E.10-15}$$

$$\vec{u}^* \cdot \vec{\delta}_r = \frac{U_\infty}{u_s} \cos\theta, \quad \vec{u}^* \cdot \vec{\delta}_\theta = \frac{U_\infty}{u_s} \sin\theta \qquad \text{as} \quad r^* \to \infty \tag{5.E.10-16}$$

Step 6 involves setting the various groups equal to 1 or zero to determine the scale and reference factors, respectively. Since there is no attempt to achieve O(1) in dimensional analysis scaling, it makes no difference which groups we choose. Hence, the two dimensionless groups in equation (5.E.10-13), when set equal to zero and 1, respectively, indicate that $c_r = c_{Aw}$ and $r_s = R$. The dimensionless group in equation (5.E.10-15), when set equal to 1, indicates that $c_s = c_{A0} - c_{Aw}$. Setting the dimensionless group in equation (5.E.10-16) equal to 1 indicates that $u_s = U_\infty$. The dimensionless group in equation (5.E.10-12) multiplying the dimensionless pressure gradient, when set equal to 1, indicates that $P_s = \mu U_\infty / R$. These choices then yield the following minimum parametric representation of the describing equations:

$$\text{Sh} = \frac{\partial c_A^*}{\partial r^*} \quad \text{at} \quad r^* = 1 \tag{5.E.10-17}$$

$$\text{Pe}_m \cdot \vec{u}^* \cdot \nabla^* c_A^* = \nabla^{*2} c_A^* \tag{5.E.10-18}$$

$$\text{Re} \cdot \vec{u}^* \cdot \nabla^* \vec{u}^* = \nabla^{*2} \vec{u}^* - \nabla^* P^* \tag{5.E.10-19}$$

$$c_A^* = 0 \quad \text{at} \quad r^* = 1 \tag{5.E.10-20}$$

$$\vec{n} \cdot \vec{u}^* = 0 \quad \text{at} \quad r^* = 1 \tag{5.E.10-21}$$

$$c_A^* = 1 \quad \text{as} \quad r^* \to \infty \tag{5.E.10-22}$$

$$\vec{u}^* \cdot \vec{\delta}_r = \cos\theta, \quad \vec{u}^* \cdot \vec{\delta}_\theta = \sin\theta \quad \text{as} \quad r^* \to \infty \tag{5.E.10-23}$$

where

$$\text{Sh} \equiv \frac{k_L R}{D_{AB}} \quad \text{is the Sherwood number} \tag{5.E.10-24}$$

$$\text{Pe}_m \equiv \frac{U_\infty R}{D_{AB}} \quad \text{is the solutal Peclet number} \tag{5.E.10-25}$$

$$\text{Re} \equiv \frac{\rho U_\infty R}{\mu} \quad \text{is the Reynolds number} \tag{5.E.10-26}$$

Note that Sherwood number provides a measure of the overall mass transfer to that by diffusion alone; as such, it is analogous to the Nusselt number in heat transfer. The solutal Peclet number is a measure of the convective to diffusional transport of species. Equation (5.E.10-17) implies that the Sherwood number is a function of the dimensionless groups involved in determining $\partial c^*/\partial r^*|_{r^*=1}$ and hence will be a function of only the dimensionless groups involved in solving equations (5.E.10-18) and (5.E.10-19), which introduce the Peclet and Reynolds numbers. Hence, we conclude that oxygen transfer to a red blood cell can be correlated in terms of three dimensionless groups, that is,

$$\text{Sh} = f(\text{Re}, \text{Pe}_m) \tag{5.E.10-27}$$

Note that a naive application of the Pi theorem would imply that five dimensionless groups would be required (i.e., $n = 8$ and $n = 3$).

The three dimensionless groups in the correlation given by equation (5.E.10-27) are not unique. It is convenient to isolate the velocity into just one group by applying the formalism indicated in equation (2.4-2); that is,

$$\text{Sc} = \text{Pe}_m \text{Re}^{-1} = \frac{\mu}{\rho D_{AB}} \equiv \text{the Schmidt number} \tag{5.E.10-28}$$

Hence, our modified correlation for the Sherwood number is given by

$$\text{Sh} = f(\text{Re}, \text{Sc}) \tag{5.E.10-29}$$

A frequently used equation for the Sherwood number for mass transfer to a single sphere that has a constant velocity relative to a fluid far removed from it is[36]

$$\mathrm{Sh} = 1 + 0.4536\sqrt{\mathrm{Re}\cdot\mathrm{Sc}} = 1 + 0.4536\sqrt{\mathrm{Pe}_m} \tag{5.E.10-30}$$

Equation (5.E.10-30) indicates that the Sherwood number can be correlated in terms of just one dimensionless group, the solutal Peclet number. Note that this correlation is applicable only in the limit of very small Reynolds numbers. This more restrictive correlation can be obtained from the general correlation given by equation (5.E.10-29) by invoking the formalism suggested by equation (2.4-3); that is, by expanding equation (5.E.10-29) in a Taylor series for small Reynolds number (step 9):

$$\mathrm{Sh} = f(\mathrm{Re}, \mathrm{Pe}_m) = f|_{\mathrm{Re}=0} + \left.\frac{\partial f}{\partial \mathrm{Re}}\right|_{\mathrm{Re}=0} \mathrm{Re} + O(\mathrm{Re}^2) \tag{5.E.10-31}$$

Hence, in the limit of very small Reynolds number, we conclude that

$$\mathrm{Sh} = f(\mathrm{Pe}_m) = f(\mathrm{Re}\cdot\mathrm{Sc}) \tag{5.E.10-32}$$

which is consistent with the form of equation (5.E.10-30).

5.P PRACTICE PROBLEMS

5.P.1 Penetration Theory Approximation for a Specified Equation of State

Consider unsteady-state mass transfer in the liquid film considered in Section 5.3 and shown in Figure 5.2.1. The boundary conditions at $z = 0$ and $z = H$ remain the same. However, rather than specifying the ratio of the mass fluxes, an equation of state is given for the mass density of the form

$$\rho = \omega_A \rho_A^0 + \omega_B \rho_B^0 \tag{5.P.1-1}$$

where ω_i and ρ_i^0 are the mass fraction and pure component mass density of component i. Use scaling analysis to determine when the convective transport arising from the diffusive mass transfer can be neglected.

5.P.2 Error Estimate for Penetration Theory Approximation

Consider unsteady-state mass transfer through the liquid film considered in Section 5.3 and shown in Figure 5.2-1.

[36]Bird et al., *Transport Phenomena*, 2nd ed., p. 678. Note that this reference defines the Sherwood and Reynolds numbers in terms of the sphere diameter rather than the radius.

(a) Estimate the time required for the mass transfer to penetrate the entire thickness of the film.

(b) Consider the special case of unimolecular diffusion of component A in stationary component B for which equation (5.2-6) implies that $\kappa = 0$. Solve equations (5.3-2) through (5.3-6) for the concentration profile $\rho_A^*(z^*, t^*)$, invoking the penetration theory approximation given by equation (5.3-7) while retaining the convective mass-transfer contribution to the mass flux in equation (5.3-3).

(c) Plot $\rho_A^*(z^*, t^*)$ as a function of the dimensionless group Π_1 defined by equation (5.2-26) to show the error encountered when the convective contribution to the total mass-transfer flux defined by equation (5.3-3) is ignored.

5.P.3 Diffusion in a Tapered Pore

In Section 5.E.3 we considered steady-state binary gas-phase diffusion at constant pressure and temperature through the pore shown in Figure 5.E.3-1, which has a nonconstant circular cross-sectional area with a radius $R = R_0 - \beta\sqrt{z}$, where β is a constant. The concentration at the mouth of the pore was held constant at c_{A0} (moles/volume), whereas it was maintained at zero at $z = L$. We scaled this problem to determine a criterion to assess when radial diffusion could be neglected. Determine when this assumption is no longer valid.

5.P.4 Liquid Evaporation for Short Contact Times

In Section 5.7 we considered the unsteady-state evaporation of a pure liquid contained in a cylindrical tube that initially contained none of the evaporating component in the gas phase and for which the evaporating component concentration was maintained at zero, as shown in Figure 5.7-1. We considered a long-time scaling to assess when quasi-steady-state could be assumed. Here we consider this same problem for very short contact times.

(a) Write the appropriately simplified species-balance equation and its initial and boundary conditions.

(b) Consider an integral mass balance in order to derive an auxiliary condition needed to determine the location of the moving liquid–gas interface.

(c) Scale the describing equations appropriate to very short contact times to determine the thickness of the region of influence wherein all the mass transfer is effectively concentrated.

(d) Develop a criterion for assuming that the boundary condition at the entrance of the tube can be applied at infinity; note that if this criterion is satisfied, a pseudo-convection term is not generated when converting the describing equations to a translated coordinate system.

(e) Use your scaling analysis result in part (c) to determine when the region of influence penetrates the entrance of the tube.

(f) Determine the criterion for ignoring the convective contribution to the mass-transfer flux.

5.P.5 Mass Transfer to Film Flow Down a Vertical Plane

In Section 5.6 we applied scaling analysis to the absorption of a soluble gas into falling film flow. We found that if the solutal Peclet number were sufficiently large, the mass transfer was confined to a region of influence in the vicinity of the liquid–gas interface. Moreover, if the solutal Peclet number were sufficiently large, we found that the convective mass transfer was influenced only by the velocity at the liquid–gas interface. However, both of these approximations have limitations that we explore in this problem.

(a) Based on the scaling analysis in Section 5.6, determine when the region of influence or mass-transfer boundary layer penetrates the full distance from the liquid–gas interface to the solid boundary.
(b) Determine when the assumption that the convective mass transfer is influenced only by the surface velocity breaks down.
(c) Determine the thickness of the region of influence near the top of the liquid film wherein axial diffusion cannot be neglected.
(d) Discuss how the scaling analysis done in Section 5.6 can be used to solve the full set of elliptic differential equations that are required to describe mass transfer at the top of the liquid film.

5.P.6 Mass Transfer to Film Flow Down a Vertical Cylinder

Consider the absorption of a sparingly soluble component from an inviscid gas into a liquid film in fully developed laminar flow down a cylindrical wire of radius R_1. The liquid–gas interface is located at R_2, as shown in Figure 5.P.6-1. The liquid film has an initial concentration c_{A0} at $z = 0$ and an interfacial concentration c_{AI} established through equilibrium with the adjacent gas phase. The velocity profile in the liquid film is given by

$$u_z = U_m \frac{(r^2 - R_1^2) - 2R_2^2 \ln(r/R_1)}{(R_2^2 - R_1^2) - 2R_2^2 \ln(R_2/R_1)} \tag{5.P.6-1}$$

where U_m is the maximum liquid velocity: namely, at the liquid–gas interface.

(a) Use scaling analysis to develop a criterion for assuming that the mass transfer is confined to a region of influence or boundary layer near the liquid–gas interface.
(b) Develop a criterion for assuming that the convective mass transfer is influenced only by the surface velocity of the liquid film.
(c) Develop a criterion for ignoring the axial diffusion of species.
(d) Develop a criterion for assuming that the curvature effects can be ignored.

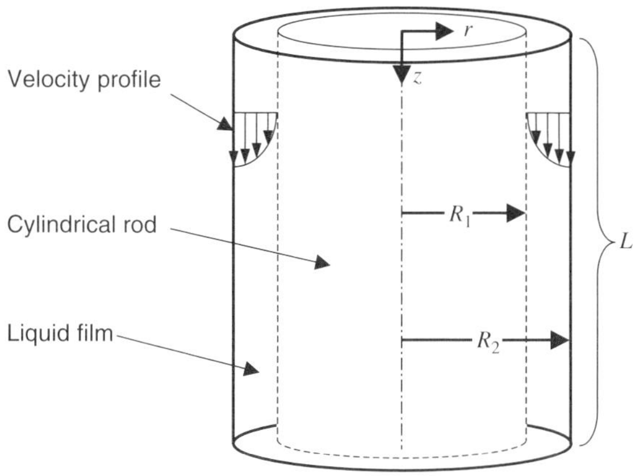

Figure 5.P.6-1 Mass transfer of a sparingly soluble solute from an inviscid gas to a liquid film flowing down a vertical cylindrical rod.

5.P.7 Mass Transfer with Chemical Reaction for Flow Between Semipermeable Membranes

In Section 5.4 we considered fully developed laminar flow between two parallel semipermeable membrane boundaries through which a component was injected that reacted with the liquid feed stream. We considered a low solutal Peclet number scaling for which the convective transport of species could be neglected. However, we considered the implications of both a low and a high Thiele modulus, corresponding to very slow and very fast homogeneous reaction conditions. Consider this same flow geometry, as shown in Figure 5.4-1 for the special case of a high Thiele modulus when convective mass transfer is important. For this scaling use the following dimensionless variables:

$$c_A^* \equiv \frac{c_A}{c_s}; \qquad \left(\frac{\partial c_A}{\partial x}\right)^* \equiv \frac{1}{c_{xs}}\frac{\partial c_A}{\partial x}; \qquad x^* \equiv \frac{x}{x_s}; \qquad y^* \equiv \frac{y_r - y}{y_s} \tag{5.P.7-1}$$

(a) Indicate why a separate scale is needed for the axial derivative of the concentration and why a reference factor is needed for the transverse coordinate.

(b) Determine the scale and reference factors for this convective mass-transfer problem.

(c) Determine the thickness of the region of influence near the membrane boundaries within which all the permeating reactant is consumed.

(d) Determine the criterion for assuming that the velocity profile can be linearized.

(e) Determine the criterion for ignoring the axial diffusion of species.

(f) Indicate when the criterion in part (e) breaks down.

(g) When the criterion in part (e) is not satisfied, the full elliptic problem must be considered. Often an appropriate downstream boundary condition to solve this elliptic problem is not known. Indicate how the results of the scaling analysis in parts (a) through (e) can be used to solve the full elliptic problem in the entry region.

5.P.8 Entrance Effect Limitations for Laminar Tube Flow with a Fast Heterogeneous Reaction

In Example Problem 5.E.8 we considered fully developed laminar flow in a cylindrical tube for which a solute underwent an irreversible heterogeneous reaction at the wall. We considered the special case of a very large Damköhler number for which our scaling analysis indicated that there was a region of influence or solutal boundary layer near the tube wall across which the concentration dropped from its initial value to essentially zero.

(a) Scaling analysis for large Damköhler numbers led to a simplified set of describing equations given by (5.E.8-18) through (5.E.8-21). Indicate when the criteria leading to these simplified equations break down.
(b) Determine the thickness of the region of influence near the entrance of the tube wherein the axial diffusion term cannot be neglected.
(c) When the axial diffusion term cannot be neglected, one is faced with solving an elliptic system of equations for which a downstream boundary condition is required. Indicate how the results of the scaling analysis in Example Problem 5.E.8 can be used to solve the full elliptic problem in the entry region.
(d) In Section 5.4 we considered steady-state convective mass transfer for the case of a homogeneous chemical reaction and found that for sufficiently small Peclet numbers, the convective transport of species could be neglected. Is a small Peclet number approximation ever justified for steady-state mass transfer in the convective mass-transfer problem being considered in Example Problem 5.E.8?

5.P.9 Aeration of Water Containing Aerobic Bacteria

Consider a spherical bubble consisting of pure oxygen with a radius R rising at its terminal velocity U_t through at stationary tank of water of depth L. The water contains aerobic bacteria that consume dissolved oxygen via a zeroth-order reaction whose rate constant is k_0 (moles/volume·time). The water is assumed to have no oxygen at the bottom of the tank where the bubbles enter. The bubbles can be assumed to become saturated with water vapor very quickly relative to the oxygen transfer to the liquid. The corresponding equilibrium concentration of oxygen in water, denoted by c_{A0} (moles/volume), will be assumed to be unaffected by the small increase in the bubble pressure associated with the decrease in bubble size due to oxygen absorption. We assume that the bubbles rise such that the hydrodynamics cause the mass transfer to be confined to a film of liquid having a constant thickness δ_m that surrounds each bubble as shown in Figure 5.P.9-1. However, δ_m is not necessarily thin in comparison to the radius of the bubble. We ignore any effects

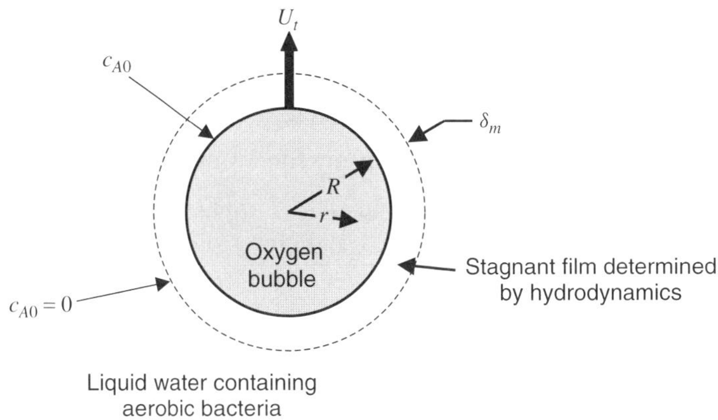

Figure 5.P.9-1 Water-saturated oxygen bubble of radius R rising at its terminal velocity U_t in liquid water containing aerobic bacteria that consume the oxygen via a zeroth-order reaction.

of the change in bubble size and also assume that the molar density of the liquid, c, remains constant. Use the following dimensionless variables containing arbitrary scale and reference factors for scaling this problem:

$$c_A^* \equiv \frac{c_A}{c_s}; \qquad N_A^* \equiv \frac{N_A}{N_s}; \qquad r^* \equiv \frac{r - r_r}{r_s}; \qquad t^* \equiv \frac{t}{t_s} \tag{5.P.9-1}$$

(a) Consider a spherical coordinate system located at the center of a single rising bubble as shown in Figure 5.P.9-1; write the appropriate form of the species-balance equation in spherical coordinates along with the requisite initial and boundary conditions; do not ignore the bulk-flow contribution to the molar flux; note that if the oxygen bubbles are saturated with water vapor, this is unimolecular diffusion.

(b) Introduce the dimensionless variables defined above into your describing equations and determine the relevant scale and reference factors; note that the observation time for this problem is well-defined since it will be the time required for a bubble to rise at its terminal velocity through the entire depth of liquid.

(c) Determine the criterion for ignoring curvature effects.

(d) Determine the criterion for assuming quasi-steady-state.

(e) Determine the criterion for ignoring the effect of the bacterial consumption of oxygen.

(f) Consider the case of a very fast bacterial consumption of oxygen such that the oxygen concentration is essentially reduced to zero within a distance much smaller than the film thickness δ_m; determine the region of influence or boundary-layer thickness wherein all the diffusive mass transfer occurs.

(g) Discuss how the scaling of this problem would change if we allowed for the effects of the decrease in bubble size due to the oxygen absorption.

5.P.10 Dissolution of a Spherical Capsule for a Concentrated Solution

In Section 5.E.4 we considered the dissolution of a spherical capsule of pure solid component A as shown in Figure 5.E.4-1 and used scaling analysis to assess how the describing equations could be simplified. One assumption that we made was that the solution was dilute, which permitted us to ignore the bulk-flow contribution to Fick's law of diffusion. In this problem we no longer assume dilute solutions but assume constant physical properties and that the effect of any reaction products resulting from the dissolution can be ignored.

(a) Rescale this problem appropriate to a slow chemical reaction to assess when the reaction and the bulk-flow contributions can be ignored.
(b) Rescale this problem appropriate to a fast chemical reaction to assess when quasi-steady-state can be assumed and when the bulk-flow contribution can be ignored.

5.P.11 Dissolution of a Spherical Capsule for a Bimolecular Reaction

In Section 5.E.4 we considered the mass transfer from a spherical capsule of pure solid component A as shown in Figure 5.E.4-1 for the case of a first-order dissolution chemical reaction. Assume now that the dissolution kinetics are governed by a bimolecular reaction of component A with the stomach liquid B, for which the reaction rate (moles/volume·time) is given by $R_A = k_2 c_A c_B$, where k_2 is the second-order reaction-rate constant. We do not assume dilute solutions in this problem but assume constant physical properties and that the effect of any reaction products resulting from the dissolution can be ignored.

(a) Write the appropriately simplified species-balance equations and their initial and boundary conditions in spherical coordinates; express the concentrations in terms of mole fractions.
(b) Carry out an integral mass balance to obtain the auxiliary equation required to determine the location of boundary that defines the interface between the spherical capsule and the stomach liquid.
(c) Scale the describing equations for conditions appropriate to a fast homogeneous chemical reaction.
(d) Determine the criterion for quasi-steady-state mass transfer.
(e) Determine the criterion for ignoring the bulk-flow contribution to the mass-transfer flux.
(f) Determine the criterion for ignoring the curvature effects on the mass transfer.

(g) Use your scaling analysis with the simplification considered in part (e) to estimate the time required for the capsule to dissolve.

5.P.12 Slow Dissolution of a Cylindrical Capsule

Consider the dissolution of a medication in the form of a cylindrical capsule of pure solid component A with initial radius and length, R_0 and L_0, respectively, as shown in Figure 5.P.12-1. We assume that this cylindrical solid capsule is swallowed and dissolves in the acidic aqueous liquid environment of the stomach via binary diffusion accompanied by a relatively slow zeroth-order homogeneous chemical reaction for which the rate constant is k_0 (moles/volume·time). The equilibrium solubility mole fraction of component A in the stomach liquid is x_{A0}. The binary solution can be assumed to be dilute and to have constant physical and transport properties. However, mass transfer must be considered from both the circumferential area and the end of the cylindrical capsule. We scale this problem to assess when the mass transfer from the circular ends of the cylindrical capsule can be neglected; this has significant consequences for how you determine your length scales. We also make the assumption that the capsule remains a cylinder throughout the dissolution process. In fact, variations in the mass-transfer fluxes along the length and ends of the cylinder will gradually cause it to become ellipsoidal and eventually spherical. However, during the early stage of the dissolution process, the cylindrical shape will be maintained at least approximately. Note that because of the plane of symmetry in the capsule, one can consider the mass-transfer problem only for one-half of the capsule.

(a) Write the appropriately simplified species-balance equation and its initial and boundary conditions in cylindrical coordinates; express the concentration in terms of mole fraction.

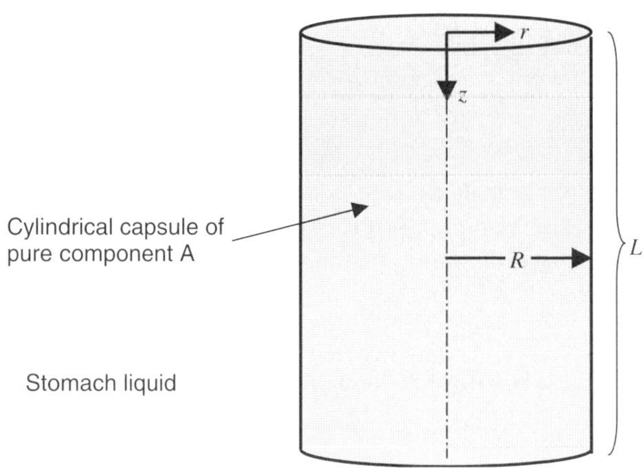

Figure 5.P.12-1 Unsteady-state dissolution of a medication in the form of a cylindrical capsule of pure component A into the stomach liquid, where it is consumed via a zeroth-order homogeneous chemical reaction.

(b) Carry out an integral mass balance to obtain the auxiliary equation required to determine the location of boundary that defines the interface between the cylindrical capsule and the stomach liquid; account for both a changing cylindrical radius and length.

(c) Scale the describing equations for conditions appropriate to a relatively slow homogeneous chemical reaction.

(d) Based on your scaling analysis, discuss whether quasi-steady-state mass transfer is ever possible.

(e) Determine the criterion for ignoring mass transfer from the ends of the cylindrical capsule.

(f) Determine the criterion for ignoring the curvature effects on the mass transfer.

(g) Use your scaling analysis with the simplification considered in part (e) to estimate the time required for the cylindrical capsule to dissolve.

5.P.13 Dissolution of a Cylindrical Capsule in the Fast Reaction Limit

Let us consider the mass-transfer problem defined in Practice Problem 5.P.12 for the special case of a fast reaction.

(a) Write the appropriately simplified species-balance equation and its initial and boundary conditions in cylindrical coordinates; express the concentration in terms of mole fraction.

(b) Carry out an integral mass balance to obtain the auxiliary equation required to determine the location of boundary that defines the interface between the cylindrical capsule and the stomach liquid; account for both a changing cylindrical radius and length.

(c) Scale the describing equations for conditions appropriate to a fast homogeneous chemical reaction.

(d) Determine the criterion for assuming quasi-steady-state mass transfer.

(e) Determine the criterion for ignoring the curvature effects on the mass transfer.

(f) Use your scaling analysis with the simplification considered in part (e) to estimate the time required for the cylindrical capsule to dissolve completely; do not ignore the dissolution from the ends of the capsule in determining your estimate.

5.P.14 Diffusional Growth of a Nucleated Water Droplet

Nucleation of a liquid droplet from a gas phase requires that the latter be sufficiently supersaturated with the nucleating component. This is required because a submicroscopic nucleus has a large surface area relative to its volume, which increases its Gibbs free energy. Supersaturation in turn increases the Gibbs free energy of this component in the gas phase, thereby permitting a decrease in the

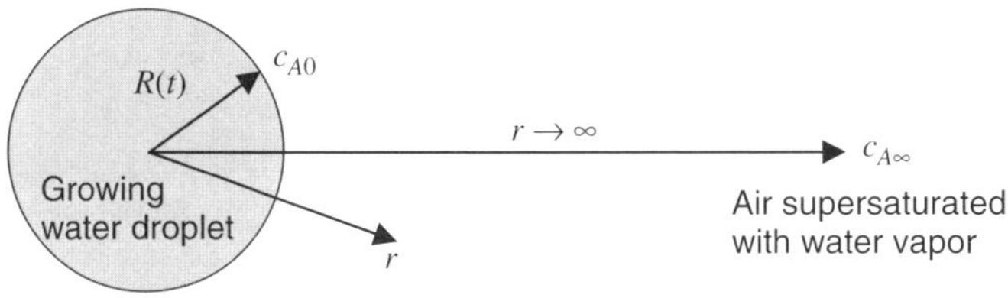

Figure 5.P.14-1 Water droplet of instantaneous radius $R(t)$ growing in supersaturated air at concentration $c_{A\infty}$ due to diffusion and instantaneous heterogeneous nucleation on its surface, at which the saturated gas-phase concentration is c_{A0}.

system Gibbs free energy via nucleation. Once nucleation occurs, the dispersed phase droplets grow via coalescence with neighboring droplets and diffusion of the nucleating component from the supersaturated gas accompanied by heterogeneous nucleation on their surface. Here we explore the diffusive growth of a water droplet in supersaturated air as shown schematically in Figure 5.P.14-1. We assume that the heterogeneous nucleation occurs instantaneously. The droplet has a negligible initial radius and an instantaneous radius $R(t)$. The supersaturation concentration is $c_{A\infty}$, whereas the thermodynamic equilibrium concentration of water in the atmosphere at the prevailing temperature is c_{A0}. Since nuclei are quite small, we assume that this water droplet initiates from a zero radius. We use scaling to explore how this mass-transfer problem can be simplified.

(a) Write the appropriately simplified species-balance equation and its initial and boundary conditions in spherical coordinates; express the concentration in terms of mole fraction and assume dilute solutions with constant physical properties.
(b) Carry out an integral mass balance to obtain the auxiliary equation required to determine the location of boundary that defines the interface between the water droplet and the atmosphere.
(c) Use scaling analysis to estimate the thickness of the region of influence or solutal boundary layer wherein essentially all the mass transfer is occurring.
(d) Determine the criterion for ignoring the curvature effects on the mass transfer.
(e) Can quasi-steady-state conditions ever be achieved in this mass-transfer problem?

5.P.15 Crystallization from a Supersaturated Liquid

Consider the one-dimensional diffusional growth of a planar crystal of pure component A from its binary solution with component B as shown in Figure 5.P.15-1. The binary solution is assumed to be supersaturated with a concentration $\rho_{A\infty}$ far from the growing crystal face, whereas the equilibrium concentration at the boundary between the solid crystal and liquid solution is ρ_{A0}.

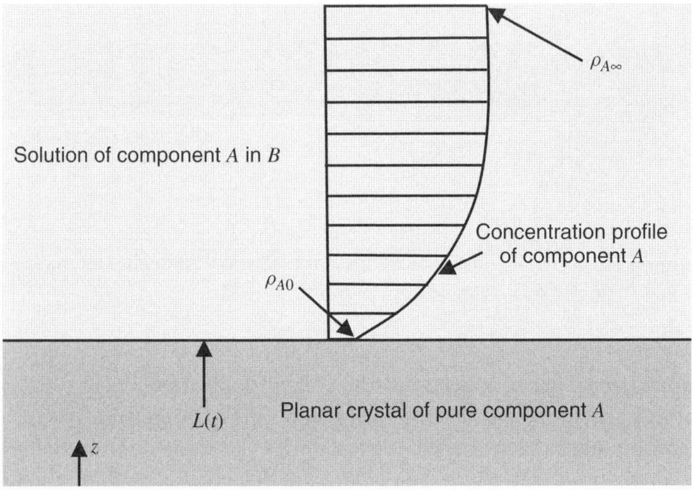

Figure 5.P.12-1 One-dimensional growth of a planar crystal of pure component A from its supersaturated solution with component B.

(a) Write the appropriately simplified species-balance equation and its initial and boundary conditions in rectangular coordinates; since this involves an incompressible liquid phase, express the concentration in terms of mass density and mass fraction; do not ignore the bulk flow contribution to the mass-transfer flux.

(b) Carry out an integral mass balance to obtain the auxiliary equation required to determine the location of boundary that defines the interface between the crystal and the liquid solution.

(c) Use scaling analysis to estimate the thickness of the region of influence wherein all the mass transfer is effectively occurring.

(d) Is quasi-steady-state ever possible for this mass-transfer problem?

(e) Use scaling analysis to determine the criterion for ignoring the bulk-flow contribution to the mass-transfer flux.

5.P.16 Growth of a Liquid Droplet by means of Diffusion and Heterogeneous Nucleation

In Practice Problem 5.P.14 we considered the diffusional growth of a liquid droplet assuming that the heterogeneous nucleation was instantaneous. The latter is a reasonable assumption for heterogeneous nucleation from the gas phase that involves relatively low molecular weight "simple" molecules. However, for heterogeneous nucleation of a liquid phase in another immiscible liquid phase involving more complex molecules (e.g., an amorphous polymer component from its solution in an organic solvent), steric effects can result in noninstantaneous heterogeneous nucleation. In this problem we assume that the heterogeneous nucleation of the pure dispersed phase component from its solution, which constitutes the continuous phase, is characterized by a zeroth-order rate constant $\hat{k}_0$ (moles/area·time). The supersaturation concentration far from the surface of the growing droplet is denoted $\rho_{A\infty}$; the thermodynamic equilibrium concentration of the nucleating component in the continuous phase is denoted by ρ_{A0}. Mass rather than molar concentrations are used here since we are considering a liquid phase. The molecular weight of the crystallizing component is M_A.

(a) Write the appropriately simplified species-balance equation and its initial and boundary conditions in spherical coordinates; express the concentration in terms of mass fraction.
(b) Carry out an integral species balance to obtain the auxiliary equation required to determine the location of boundary that defines the interface between the liquid droplct and thc continuous liquid phase; consider carefully how the heterogeneous nucleation affects this integral balance.
(c) Use scaling analysis to estimate the thickness of the region of influence wherein all the mass transfer effectively is occurring.
(d) Determine the criterion for ignoring the curvature effects on the mass transfer.
(e) Determine the criterion for ignoring the heterogeneous nucleation.
(f) Determine the criterion for assuming quasi-steady-state.

5.P.17 Rusting of a Planar Surface

Consider a flat piece of initially pure iron that is immersed so that its upper surface is exposed continuously to liquid water saturated with dissolved oxygen (O_2), whose concentration is denoted by $c_{A\infty}$. The oxygen in the water will diffuse to the surface of the iron and promote rusting via the formation of iron oxide, Fe_2O_3, based on the reaction

$$\tfrac{4}{3}Fe + O_2 \rightarrow \tfrac{2}{3}Fe_2O_3 \qquad (5.P.17\text{-}1)$$

Note that reaction (5.P.17-1) implies a relationship between the molar fluxes of oxygen and iron as well as the molar rate of growth of the rust layer. The rate of conversion of oxygen to rust is assumed to occur by means of a first-order heterogeneous reaction whose reaction rate is given by

$$R_A = \hat{k}_1 c_A \qquad \text{(moles/area} \cdot \text{time)} \qquad (5.P.17\text{-}2)$$

in which $\hat{k}_1$ (length/time) is the heterogeneous reaction-rate constant and A denotes the molecular oxygen. The rate of growth of the rust layer decreases progressively in time since the oxygen must diffuse through both the water and the increasing thickness of the rust layer in order to reach the surface of the iron. Since the oxygen diffuses through the water-saturated microporous structure of the rust layer, its local concentration in the latter is equal to εc_A, in which ε is the porosity and c_A is the local concentration of oxygen in the pores. Owing to the microporous structure, the effective oxygen diffusivity in the rust, which is denoted by D_{AC}, will be less than that in the water, which is denoted by D_{AB}. The densities of the water, water-saturated rust, and pure iron are denoted by ρ_B, ρ_C, and ρ_D, respectively, in which B, C, and D denote the water, rust, and iron, respectively. The corresponding molecular weights are denoted by M_B, M_C, and M_D, respectively. Note that this problem involves two moving boundaries: the interface between the rust and the water and that between the iron and the rust, as shown in Figure 5.P.17-1. One might anticipate that initially, oxygen diffusion through the water controls the rate of rusting, whereas at longer times, oxygen diffusion through the rust becomes controlling. We use scaling analysis to explore how the describing equations for this mass-transfer process can be simplified. In scaling this problem, use the following dimensionless variables involving unspecified scale and reference factors:

$$c_{AW}^* \equiv \frac{c_A - c_{Wr}}{c_{Ws}}; \qquad c_{Ar}^* \equiv \frac{c_A - c_{Rr}}{c_{Rs}}; \qquad z_W^* \equiv \frac{z - z_{Wr}}{z_{Ws}};$$
$$z_R^* \equiv \frac{z - z_{Rr}}{z_{Rs}}; \qquad t^* \equiv \frac{t}{t_s}; \tag{5.P.17-3}$$
$$L_R^* \equiv \frac{L - L_R}{L_{Rs}}; \qquad L_I^* \equiv \frac{L}{L_{Is}}; \qquad \left(\frac{dL}{dt}\right)_R^* \equiv \frac{1}{\dot{L}_{Rs}}\left(\frac{dL}{dt}\right)_R;$$
$$\left(\frac{dL}{dt}\right)_I^* \equiv \frac{1}{\dot{L}_{Is}}\left(\frac{dL}{dt}\right)_I$$

where the subscripts W, R, and I refer to the water, rust, and iron layers, respectively.

(a) Explain why it is necessary to define separate reference and scale factors for the concentrations and spatial coordinates in the water and rust layers.

(b) Explain why it is necessary to define separate scale factors for the thicknesses of the rust and iron layers as well as the velocities of each layer.

(c) Write the appropriately simplified species-balance equations and their initial and boundary conditions for both the water and the rust layer; use molar concentrations and assume dilute solutions so that the bulk-flow contribution to the mass-transfer flux can be ignored.

(d) Carry out an integral mass balance on the iron and rust layers to obtain the auxiliary equations required to determine the location of the two moving boundaries. Note that the growth rate of the rust layer must satisfy the molar exchange dictated by reaction (5.P.17-1).

(e) Use scaling analysis to estimate the thickness of the region of influence in the water layer.

(f) Determine the criteria for quasi-steady-state to apply.

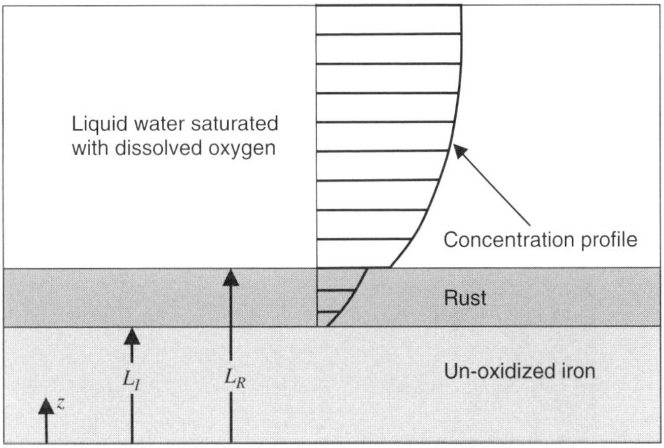

Figure 5.P.17-1 Rusting of a flat piece of pure iron immersed in liquid water that is saturated with dissolved oxygen; this involves two moving boundaries: the interface between the rust layer and the water and that between the iron and the rust layer.

(g) Use scaling analysis to determine when the resistance to mass transfer in the rust layer can be ignored.

(h) Use scaling analysis to determine when the resistance to mass transfer in the water bath can be ignored.

5.P.18 Mass Transfer in a Hollow-Fiber Membrane

In Example Problem 5.E.9 we applied scaling analysis to a hollow-fiber membrane reactor as shown in Figure 5.E.9-1. An enzyme was immobilized in the microporous annular wall of the hollow fiber. A low-molecular-weight solute permeated through the ultrathin inner wall of this annular region and reacted with the confined enzyme. We used scaling analysis to explore how the describing equations in the annular region could be simplified. In doing this we allowed for separate scale factors for the axial and radial concentration gradients. Rather than allowing for separate scales for the axial and radial concentration gradients, assume that these two derivatives scale with the characteristic concentration scale divided by the characteristic length scales in the axial and radial directions, respectively. This will lead to a contradiction in that one or more terms will not be bounded of o(1). This exercise provides another example of the forgiving nature of scaling; that is, scaling tells you if the ordering analysis has been done correctly.

5.P.19 Permeation Accompanied by Membrane Swelling

Consider a mass-transfer process that involves maintaining concentrations ρ_{A0} (mass/unit volume) and zero of component A on opposing sides of a polymeric membrane having thickness L as shown in Figure 5.P.19-1. The permeating

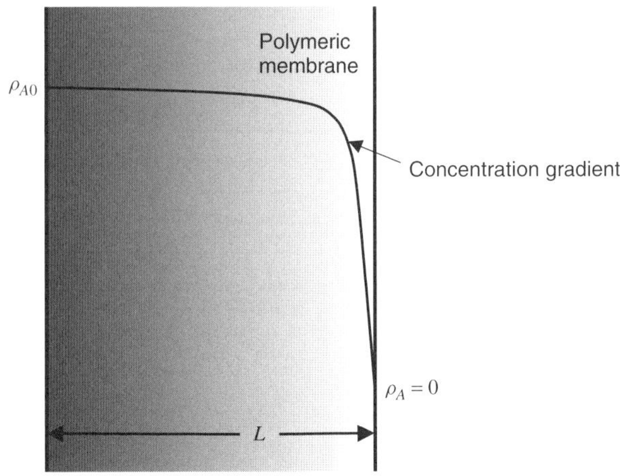

Figure 5.P.19-1 Steady-state diffusion of a solute A that causes membrane swelling, suggested by the progressive shading, which in turn increases the effective diffusion coefficient.

component A swells the membrane, which in turn causes an increase in the effective binary diffusion coefficient that is given by

$$D_{AB} = (1 - \omega_A) D_0 e^{\beta \rho_A} \tag{5.P.19-1}$$

where D_0 is the binary diffusion coefficient in the membrane at infinite dilution of component A, ω_A the mass fraction of component A, and β a positive constant.

(a) Write the appropriately simplified species-balance equation and its boundary conditions; assume dilute solutions so that the bulk-flow contribution to the mass-transfer flux can be ignored.

(b) Scale the describing equations to determine when the concentration dependence of the diffusion coefficient can be neglected.

(c) If the swelling is marked, there will be a region of influence or boundary layer near one side of the membrane within which the resistance to diffusion will be concentrated; use scaling analysis to determine the thickness of this region of influence.

(d) Use the results of your scaling analyses in parts (b) and (c) to develop a criterion for determining when significant swelling is occurring in the membrane. *Hint*: Use scaling to estimate the mass-transfer flux in the presence and absence of swelling.

5.P.20 Evaporative Polymer Film Casting for Short Contact Times

In Example Problem 5.E.1 we considered the evaporative casting of a dense film from a solution of a polymer in a volatile solvent as shown in Figure 5.E.1-1.

We developed a long-contact-time scaling to determine the criterion for assuming quasi-steady-state. In this problem we consider a short-contact-time scaling.

(a) Write the appropriately simplified species-balance equation and its initial and boundary conditions.
(b) Consider an integral mass balance to derive an auxiliary condition needed to determine the location of the moving liquid–gas interface.
(c) Scale the describing equations appropriate to very short contact times to determine the thickness of the region of influence wherein all the mass transfer is effectively concentrated.
(d) Develop a criterion for assuming that the boundary condition at the solid support surface can be applied at infinity; note that if this criterion is satisfied, a pseudo-convection term is not generated when converting the describing equations to a translated coordinate system.
(e) Use your scaling analysis result in part (c) to determine when the region of influence penetrates to the solid support surface.
(f) Determine the criterion for ignoring the convective contribution to the mass-transfer flux.

5.P.21 Mass Transfer with a Homogeneous Chemical Reaction and a Concentration-Dependent Diffusivity

Consider a liquid film consisting of components A and B of thickness L for which the concentration of a diffusing component is maintained at c_{A0} (moles/volume) at $z = 0$ and zero at $z = L$. Component A undergoes a first-order irreversible homogeneous chemical reaction with a rate constant k_1 (time^{-1}) as shown in Figure 5.P.21-1. In addition, the diffusion coefficient D_{AB} decreases linearly with concentration as follows:

$$D_{AB} = D_0 - \beta c_A \tag{5.P.21-1}$$

where D_0 is the diffusion coefficient at infinite dilution and β is a positive constant. The effects of the reaction product(s) and any change in the film thickness can be neglected.

(a) Write the appropriately simplified steady-state species-balance equation and boundary conditions.
(b) Scale these describing equations to determine a criterion for ignoring the homogeneous reaction term.
(c) Scale these describing equations to determine a criterion for ignoring the concentration dependence of the binary diffusion coefficient.
(d) Scale the describing equations for the special case of a very fast reaction that consumes all the diffusing reactant before it can reach the boundary at $z = L$. Use your scaling to determine the thickness of the region of influence wherein all the diffusion takes place.

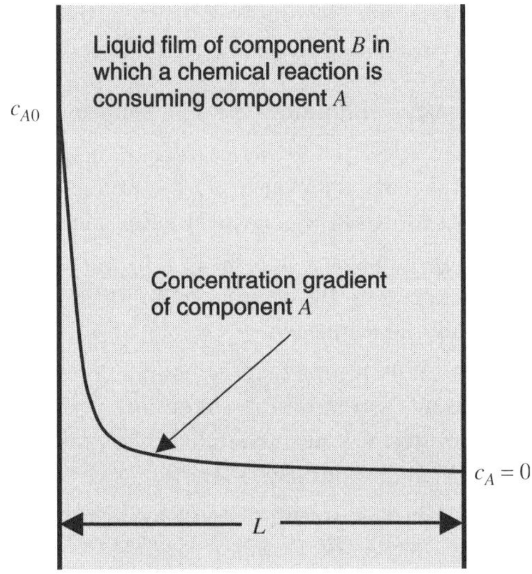

Figure 5.P.21-1 Diffusion of component A in a liquid film of component B within which A is being consumed by a homogeneous first-order chemical reaction and for which the diffusivity is concentration dependent.

5.P.22 Mass Transfer in the Annular Region Between Fixed and Rotating Cylinders

Consider a liquid confined to the annular gap between two concentric cylindrical shells of radii R_1 and R_2, as shown in Figure 5.P.22-1. The inner cylinder rotates at a constant angular velocity of ω (radians/time), while the outer cylinder remains stationary. The cylindrical boundaries at R_1 and R_2 are permeable, so they can be maintained at constant compositions ρ_1 and ρ_2 (mass/volume), respectively. In this case the mass transfer is not influenced by the rotating fluid flow. However, the mass transfer can influence the fluid flow in two ways. First, the shear viscosity μ can be concentration-dependent. Second, the bulk flow velocity established by the radial mass transfer can distort the velocity profile in the circumferential direction. In this problem we use scaling analysis to explore when these effects need to be considered. Assume that the concentration dependence of the viscosity is given by

$$\mu = \mu_0 e^{\beta \rho_A} \tag{5.P.22-1}$$

where β is a constant that can be either positive or negative.

(a) Write the appropriately simplified steady-state continuity and equations of motion in cylindrical coordinates.

(b) Write the requisite boundary conditions on the equations in part (a); note that the radial component of velocity will be equal to the mass flux divided by the mass density at the inner and outer cylindrical boundaries.

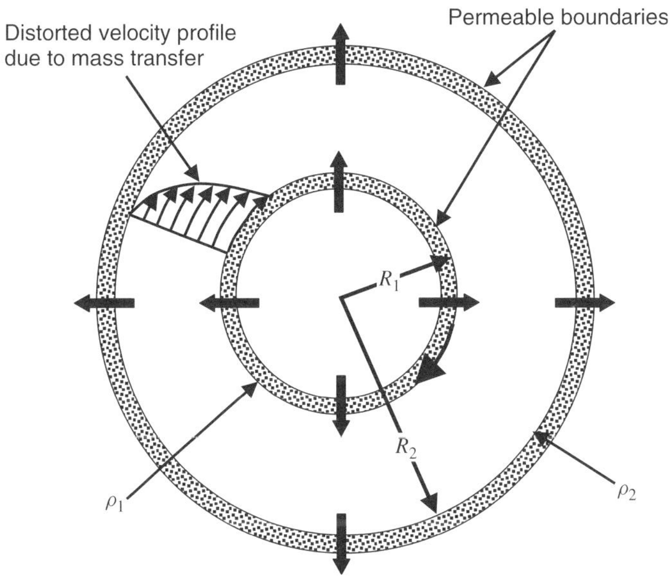

Figure 5.P.22-1 Circumferential flow between an inner permeable boundary at R_1 that is rotating at a constant angular velocity ω (radians/time) and a stationary outer permeable boundary at R_2. Steady-state mass transfer occurs in the radial direction to maintain the mass concentration at the inner and outer boundaries at ρ_1 and ρ_1, respectively. Whereas the circumferential fluid flow does not affect the mass transfer, the radial mass transfer can distort the velocity profile.

(c) Write the appropriately simplified steady-state species-balance equation in cylindrical coordinates in terms of the mass concentration ρ_A and mass flux n_A along with Fick's law of diffusion expressed in terms of the mass flux.

(d) Write the requisite boundary conditions on the equations in part (c).

(e) Introduce the following dimensionless variables and determine the unspecified scale and reference factors:

$$\rho_A^* \equiv \frac{\rho_A - \rho_{Ar}}{\rho_{As}}; \quad n_A^* \equiv \frac{n_A}{n_{As}}; \quad u_r^* \equiv \frac{u_r}{u_{rs}}; \quad u_\theta^* \equiv \frac{u_\theta}{u_{\theta s}};$$
$$\mu^* \equiv \frac{\mu}{\mu_s}; \quad r^* \equiv \frac{r - r_r}{r_s} \tag{5.P.22-2}$$

(f) Determine the criterion for ignoring the curvature effects.

(g) Determine the criterion for ignoring the bulk-flow contribution to Fick's law.

(h) Determine the criterion for ignoring the concentration-dependence of the viscosity.

(i) Determine the criterion for ignoring the effect of the radial velocity on distorting the circumferential velocity profile.

5.P.23 Bulk-Flow Effects for Free-Convection Mass Transfer from a Vertical Cylinder

In Section 5.9 we considered solutally driven free convection caused by the transpiration of water vapor through the inner wall of a vertical annular region. We employed scaling to determine how the describing equations could be simplified for large solutal Grashof numbers. Although we allowed for a radial component of velocity appropriate to a developing free-convection flow, we did not allow for any effect of the water vapor transpiration on the radial velocity. That is, the transpiration of water vapor will cause a nonzero radial velocity component at the inner wall. In this problem we use scaling analysis to determine the criterion for ignoring this blowing or transpiration velocity effect. Assume that the water vapor is transpired into the annular region at a constant radial velocity V_0 so as to maintain a constant concentration c_{A0} at the wall.

(a) Write the appropriately simplified form of the continuity equation, equations of motion, and species-balance equation for this free-convection problem for which the effects of the transpiration velocity must be considered.
(b) Write the requisite boundary conditions for the equations you derived in part (a).
(c) Determine the criterion for ignoring the effect of the transpiration or blowing velocity at the porous inner wall.

5.P.24 Free-Convection Mass Transfer Adjacent to a Transpiring Vertical Flat Wall

Consider an infinitely wide vertical flat porous wall of length L that transpires water vapor into initially quiescent air, as shown in Figure 5.P.24-1. The water

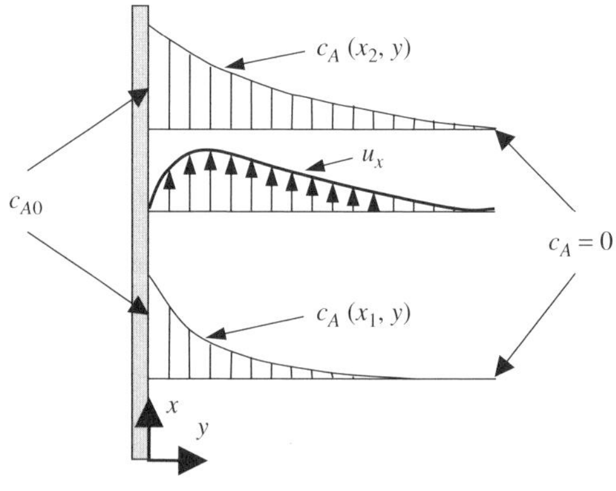

Figure 5.P.24-1 Buoyancy-induced free convection next to a vertical porous wall that transpires water vapor into initially dry air to maintain a concentration c_{A0} at its surface; concentration profiles at positions x_1 and x_2 along the plate, where $x_2 > x_1$, and the velocity profile of the component parallel to the plate are shown.

concentration at the wall is c_{A0}, whereas the quiescent air is assumed to be initially dry. The air next to the porous wall will become less dense than the air farther removed from it. Hence, a hydrostatic pressure imbalance will occur that causes fluid near the wall to rise. We consider this convective flow after the transients have died out when steady-state free convection prevails. We ignore end effects at the top and bottom of the plate and viscous dissipation and assume constant physical properties other than the density in the gravitational body force term in the equations of motion. Note that this is inherently a developing flow due to the progressive water transpiration that occurs as the fluid moves up the wall; therefore, velocity components in both the x- and y-directions must be considered. Since the density is temperature-dependent, we need an appropriate equation of state. We consider small density variations and hence represent the density via a Taylor series expansion about the density ρ_0 of the dry air given by

$$\rho = \rho_0 + \left.\frac{\partial \rho}{\partial c_A}\right|_0 c_A = \rho_0 - \rho_0 \beta_s c_A \tag{5.P.24-1}$$

where β_s is the coefficient of solutal volume expansion. It will be convenient in this problem to split the pressure into dynamic, P, and hydrostatic, P_h, contributions:

$$P = P(x, y) + P_h(x) \tag{5.P.24-2}$$

(a) Write the appropriately simplified steady-state continuity equation, equations of motion, and species-balance equation.

(b) Write the requisite boundary conditions on the equations in part (a).

(c) Introduce the following dimensionless variables and determine the unspecified scale and reference factors:

$$u_x^* \equiv \frac{u_x}{u_{xs}}; \quad u_y^* \equiv \frac{u_y}{u_{ys}}; \quad P^* \equiv \frac{P}{P_s}; \quad c_A^* \equiv \frac{c_A}{c_s};$$
$$x^* \equiv \frac{x}{x_s}; \quad y_m^* \equiv \frac{y}{\delta_m}; \quad y_c^* \equiv \frac{y}{\delta_s} \tag{5.P.24-3}$$

where δ_m and δ_s are the momentum and solutal boundary-layer thicknesses, respectively. Indicate why different y-length scales are used for the equations of motion and species-balance equation.

(d) Determine the scale factors and estimate the thicknesses of both the momentum and solutal boundary layers.

(e) Consider the limit of very large solutal Grashof numbers, where $\text{Gr}_m \equiv L^3 g \beta_s \rho_0^2 \Delta c_A / \mu^2$, in which μ is the shear viscosity, to derive the boundary-layer equations for solutal free convection.

5.P.25 Correlation for Steady-State Mass Transfer from a Sphere

Correlations for mass transfer to standard geometries are usually characterized in terms of a dimensionless mass-transfer coefficient known as the Sherwood number or Nusselt number for mass transfer. The latter is a function of the other dimensionless groups characterizing the mass-transfer process, such as the Reynolds or Peclet number, Schmidt number, and ratios of characteristic lengths that define the geometry. Consider steady-state flow over a sphere having radius R that is transferring a solute A to the surrounding fluid, consisting primarily of a binary solution of components A and B. Assume that the concentrations of solute A in the flowing fluid adjacent to and far from the sphere are c_{A0} and $c_{A\infty}$, respectively. Component B is assumed to be insoluble in the sphere, and the solution is assumed to be dilute. The physical and transport properties of the surrounding fluid stream can be assumed to be constant and constitute the density ρ, viscosity μ, and binary diffusion coefficient D_{AB}. The fluid velocity far from the sphere is U_∞. We seek to develop a correlation for the steady-state mass-transfer coefficient defined by

$$k_x^\bullet \equiv \frac{-cD_{AB}\overline{\dfrac{\partial x_A}{\partial r}}}{x_{A0} - x_{A\infty}} = \frac{\overline{N}_{Ar} - x_{A0}(\overline{N}_{Ar} + \overline{N}_{Br})}{x_{A0} - x_{A\infty}} \cong \frac{\overline{N}_{Ar}}{x_{A0} - x_{A\infty}} \tag{5.P.25-1}$$

where $\overline{N}_{ir}$ is the molar flux of component i in the r-direction averaged over the surface of the sphere and x_A denotes a mole fraction; the simplification in equation (5.P.25-1) follows from applying Fick's law of diffusion given by equation (G.3-7) in the Appendices and the assumption of dilute solutions.

(a) Write the equation that relates the average molar flux at the surface of the sphere to the concentration.
(b) Write the appropriate forms of the continuity equation, equations of motion and species-balance equation and their boundary conditions for this mass-transfer problem.
(c) Use the scaling method for dimensional analysis to obtain the dimensionless groups needed to correlate the mass-transfer coefficient defined by equation (5.P.25-1).
(d) Consider how the correlation that you obtained in part (c) simplifies in the limit of zero flow, that is, $U_\infty = 0$.
(e) Compare the result that you obtained in part (c) to the standard correlation for the Sherwood number given by[37]

$$\mathrm{Sh} = 2 + 0.60\mathrm{Re}^{1/2}\mathrm{Sc}^{1/3} \tag{5.P.25-2}$$

where the Sherwood and Reynolds numbers are based on the sphere diameter.
(f) In Practice Problem 5.P.9, scaling was applied to the aeration of water by the sparging of spherical bubbles; each bubble was assumed to be surrounded by a stagnant film having thickness δ_m; use the correlation you obtained for the mass-transfer coefficient to develop a correlation for the thickness of the region of influence or solutal boundary layer.

[37] Bird et al., *Transport Phenomena*, 2nd ed., p. 681.

5.P.26 Mass-Transfer-Coefficient Correlation for Film Theory

In Section 5.2 we used scaling analysis to develop the classical film theory approximation. Consider the following definition of the mass-transfer coefficient:

$$k_L^\bullet \equiv \frac{-\rho D_{AB}(\partial \omega_A/\partial z)}{\omega_{A1} - \omega_{A0}} = \frac{n_{Az} - \omega_A(n_{Az} + n_{Bz})}{\omega_{A1} - \omega_{A0}} \tag{5.P.26-1}$$

where n_{Az} is the mass-transfer flux (mass/area·time) in the z-direction and ω_{A0} and ω_{A1} are the mass fractions on each side of the film.

(a) Use the scaling approach to dimensional analysis to determine the dimensionless groups required to correlate the dimensionless mass-transfer coefficient (Sherwood number) for film theory. Do not assume that the convective transport is negligible.
(b) Evaluate the Sherwood number using the general solution for the film theory model given by equation (5.2-29), and discuss the physical significance of the result that you obtain; that is, relate the value of the Sherwood number to the nature of the mass transfer through a stationary film.
(c) Determine the ratio of the Sherwood number for the general case of film theory (i.e., when convective transport is included) to the Sherwood number for the special case of negligible convection.
(d) Discuss how the result that you obtained in part (c) could be used to correct a mass-transfer coefficient obtained from some literature correlation for very dilute solutions (low mass transfer rates) to account for the effect of convective transport.
(e) Discuss when using film theory might provide a reliable correction for the effects of convective transport on a mass-transfer coefficient obtained from a correlation applicable only to dilute solutions and low mass-transfer fluxes.

5.P.27 Correlation for Free-Convection Mass Transfer from an Evaporating Liquid

Consider a horizontal film of width W on each side that consists of a pure volatile liquid A exposed on its upper surface to dry air, referred to a component B, as shown in Figure 5.P.27-1. The volatile component is less dense than the surrounding air, so that steady-state free convection develops. The equilibrium concentration of the volatile component in the air adjacent to the liquid–gas interface is c_{A0}, which can be assumed to be dilute. The air is assumed to be insoluble in the liquid film. We seek to develop a correlation for the steady-state mass-transfer coefficient defined by

$$k_x^\bullet \equiv \frac{-cD_{AB}\overline{\dfrac{\partial x_A}{\partial y}}}{x_{A0} - x_{A\infty}} = \frac{\overline{N}_{Ay} - x_{A0}(\overline{N}_{Ay} + \overline{N}_{By})}{x_{A0} - x_{A\infty}} \cong \frac{\overline{N}_{Ay}}{x_{A0} - x_{A\infty}} \tag{5.P.27-1}$$

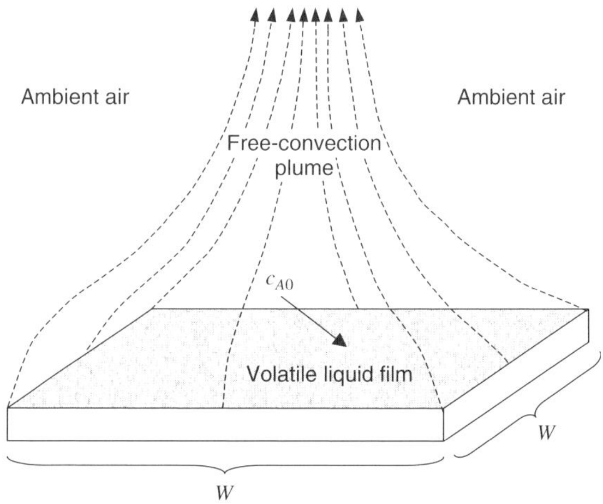

Figure 5.P.27-1 Solutally driven buoyancy-induced free convection due to the evaporation of a pure volatile liquid contained in a trough with sides of width W; the vapor of the evaporating liquid is less dense than that of the ambient air.

where $\overline{N}_{iy}$ is the molar flux of component i in the y-direction (perpendicular to the horizontal plate) averaged over the surface of the film and x_A denotes a mole fraction; the simplification in equation (5.P.27-1) follows from Fick's law of diffusion given by equation (G.1-7) in the Appendices and the assumption of dilute solutions.

(a) Write the equation that relates the average molar flux at the surface of the liquid film to the concentration.
(b) Write the appropriate forms of the continuity equation, equations of motion, and species-balance equation and their boundary conditions for this mass-transfer problem.
(c) Use the scaling method for dimensional analysis to obtain the dimensionless groups needed to correlate the mass-transfer coefficient defined by equation (5.P.27-1).
(d) Compare the result that you obtained in part (c) to the standard correlation for the Sherwood number given by[38]

$$\text{Sh} = 0.816(\text{Gr}_m \text{Sc})^{0.2} \tag{5.P.27-2}$$

where the Sherwood and solutal Grashof numbers are based on the half-width of the film.

5.P.28 Correlation for a Tubular Photocatalytic Reactor

Photocatalytic reactors can be used to employ solar energy for promoting heterogeneous natural reactions. A typical geometry involves a feed stream containing a

[38] S. N. Singh, R. C. Birkebak, and R. M. Drake, *Prog. Heat Mass Transfer*, 2:87 (1969).

solute that diffuses to the transparent walls of the flow channel that are exposed to sunlight. The solute decomposes due to a photocatalytic reaction at the tube wall.[39] Consider steady-state fully developed laminar flow of a Newtonian fluid having constant physical properties in a cylindrical tube of radius R containing a solute A having an initial concentration c_{A0} that undergoes a photocatalytic reaction along length L, such as that shown in Figure 5.5-1. The heterogeneous reaction is assumed to be irreversible and first-order with a reaction-rate constant $\hat{k}_1$ (length/time).

(a) Develop a correlation for the Sherwood number or Nusselt number for mass transfer as a function of the Reynolds number, $Re \equiv R\rho\overline{U}/\mu$, and other relevant dimensionless groups, where the mass-transfer coefficient is defined as

$$k_x^{\bullet} \equiv \frac{-cD_{AB}\left.\overline{\dfrac{\partial x_A}{\partial r}}\right|_{r=R}}{x_{A0} - x_{A\infty}} = \frac{\overline{N}_{Ar}|_{r=R} - x_{A0}(\overline{N}_{Ar} + \overline{N}_{Br})|_{r=R}}{x_{A0} - x_{A\infty}} \quad (5.P.28\text{-}1)$$

where $\overline{N}_{Ar}$ denotes the molar flux in the r-direction averaged over the inner wall of the photocatalytic reactor.

(b) Sketch a plot of n_A^* versus Re based on your dimensional analysis correlation and the results of Section 5.5 and Example Problem 5.E.8.

(c) Based on the results of Example Problem 5.E.8, derive an equation for the Reynolds number beyond which no further increase in the dimensionless mass-transfer flux to the tube wall is possible for a photocatalytic reactor system with fixed physical and chemical properties.

(d) Develop an equation for the maximum possible mass-transfer flux to the tube wall for a photocatalytic reactor system with fixed physical and chemical properties.

[39] An annular photocatalytic reactor has been considered for removing trace amounts of trichloroethane (TCE) from air; TCE is a common solvent and chemical intermediate used in industry; owing to its volatility, it evaporates readily and poses a health hazard if present in air.

6 Applications in Mass Transfer with Chemical Reaction

Taking all the mutually interfering phenomena which constitute a process of mass transfer with chemical reaction into account simultaneously is so difficult a task that, in practice, simplifying hypotheses are always needed. It is more useful in my opinion to recognize these simplifying hypotheses as justified on the grounds of physical intuition, than to perceive a strictly mathematical justification.[1]

6.1 INTRODUCTION

Mass transfer with chemical reaction can be very challenging to model since it not only can be affected by the complex hydrodynamics discussed in Chapter 3 and heat-transfer effects considered in Chapter 4, but necessarily includes mass transfer, discussed in Chapter 5. Indeed, reacting systems involve at least one or more fluid phases and can involve the release or absorption of energy arising from heat-of-reaction effects. Chemical species are no longer conserved in reacting systems but can be depleted or generated. Hence, we would certainly agree with the first part of the quotation at the top of the page; that is, simplifications need to be made to model mass transfer with chemical reaction. In this chapter we address the second part of this quotation; that is, scaling analysis provides a mathematical formalism for justifying these approximations that obviates the need to be gifted with physical intuition.

In Chapter 5 scaling analysis was applied to several examples that involved either homogeneous or heterogeneous chemical reactions. However, these examples involved mass transfer, with chemical reaction occurring on only one length scale. We considered problems for which mass transfer with chemical reaction occurred over the length of the reactor, such as in Section 5.4, or over a small element such as a rising bubble in Practice Problem 5.P.9. It is easy to see how problems can

[1]G. Astarita, *Mass Transfer with Chemical Reaction*, Elsevier, New York, 1967, pp. v–vi.

Scaling Analysis in Modeling Transport and Reaction Processes: A Systematic Approach to Model Building and the Art of Approximation, By William B. Krantz

arise involving transport phenomena on multiple scales. Consider, for example, the bubbling of a gas into a countercurrently flowing liquid in which it is chemisorbed; that is, the gaseous component dissolves in the liquid while reacting with it. The solute from the gas phase can enter the liquid only during the short time that it passes over a gas bubble. If the reaction is relatively slow, none of the transferring solute will react during the short contact time for passage of the liquid over a gas bubble. However, once the solute is in the bulk of the liquid, it can both diffuse and react. Hence, this example involves mass transfer, with chemical reaction occurring on two scales. There is the *microscale*, characterizing the flow of the liquid over a bubble, and the *macroscale*, characterizing the flow through the chemisorption device. In this chapter we employ scaling analysis to assess appropriate simplifications that can be made in both the microscale and macroscale describing equations for mass transfer with chemical reaction.

To provide a coherent focus in this chapter, we restrict our attention to a particular form of mass transfer with a chemical reaction: namely, chemisorption. The latter refers to the use of chemical reaction to enhance the solubility of a solute in a fluid into which it is being transferred. However, the scaling protocols illustrated in the chapter can be applied more generally to any type of process involving mass transfer with chemical reaction.

This chapter has a dual focus in that it applies scaling analysis to mass transfer with chemical reaction, but uses the latter as an example of scaling analysis in microscale–macroscale modeling. In general, any system involving dispersed phases will involve phenomena occurring on multiple scales. For example, phase-transition phenomena such as crystallization, boiling, and condensation will involve transport occurring on the microscale of a dispersed phase particle and on the macroscale of the bulk liquid. A fluidized bed reactor will involve heat and or mass transfer on the microscale of the particles as well as on the macroscale of the reactor. The scaling protocols illustrated in this chapter can be applied to any modeling problem involving transport phenomena and chemical reaction occurring on multiple scales.

The organization of this chapter is somewhat different from that of the preceding chapters. Rather than considering several different examples to illustrate how scaling is used to arrive at the various approximations made in transport processes, we focus here on illustrating how scaling is applied to microscale–macroscale modeling using chemisorption as the example.[2] We begin with a discussion of the *microscale* element in Section 6.2, since it is critical to understand precisely what is meant by this concept. Since scaling on the microscale involves some concepts that are different from those for the macroscale, we treat these two topics in separate sections. In Section 6.3 we focus on applying systematic scaling analysis to the describing equations for the microscale element. Scaling of the complementary describing equations for the *macroscale element* is considered in Section 6.8. Scaling analysis

[2]Microscale–macroscale modeling of mass transfer with chemical reaction has been treated by other authors, although there is no general agreement on the terminology used to describe this. Prior treatments rely on intuition rather than systematic scaling analysis in order to simplify the describing equations at the two scales. See, for example, Astarita, *Mass Transfer with Chemical Reaction*.

of the microscale element leads to the identification of various reaction regimes discussed in Sections 6.4 through 6.7. Scaling analysis of the macroscale element leads to the identification of various reaction domains in the slow reaction regime that are discussed in Sections 6.9 and 6.10. Scaling analysis will not only develop criteria for determining the relative importance of diffusion and reaction on both the micro- and macroscale but will also lead to a set of precepts for selecting the type of reactor that will be most effective for a particular reaction regime or domain, as shown in Section 6.11. Scaling analysis will be used to explain why the performance of a contacting device for mass transfer with chemical reaction depends on the hydrodynamics for the slow and instantaneous reaction regimes but not for the fast reaction regime. In Section 6.12, simple models suggested by scaling analysis will be developed to interrelate the mass-transfer coefficients in the presence and absence of chemical reaction. The power of scaling analysis to provide considerable insight into the design of devices involving mass transfer with chemical reaction without the need to actually solve any model equations analytically or numerically is shown in Section 6.11 and again in Sections 6.13 and 6.14. In Section 6.13 we illustrate how microscale–macroscale scaling can be used to design and interpret performance data for a continuous-stirred tank reactor. Section 6.14 does the same for a packed gas-absorption column. In Section 6.15 we summarize the implications of scaling analysis for microscale–macroscale modeling and, in particular, its application to chemisorption. Unworked practice problems are included at the end of the chapter.

6.2 CONCEPT OF THE MICROSCALE ELEMENT

In developing models for complex mass-transfer processes, in particular those involving dispersed phases, it is convenient to consider the concept of the *microscale element*. The latter is associated with mass transfer at the smallest continuum scale. For example, in a gas-absorption process in which a gas containing a soluble component is bubbled through a countercurrently flowing liquid phase, the microscale element is a gas bubble. In a fluidized bed reactor involving the flow of a gas relative to suspended recirculating solid particles, the microscale element is a solid particle. In phase-transition processes involving growth kinetics, the microscale element is a dispersed phase particle. In turbulent flows the microscale element can be an eddy that transports species to and from an interface. The mass-transfer rate on the microscale is controlled by a region of influence within the fluid adjacent to the microscale element. A pivotal consideration in microscale–macroscale modeling is that the microscale element is a mathematical point on the macroscale; that is, the mass-transfer flux from or to a microscale element becomes in the macroscale balance a homogeneous source or sink term, respectively[3].

[3]Some authors consider the microscale element to be the thin region of fluid adjacent to what we call the microscale element here. For example, these authors would consider the microscale element to be a thin element of liquid flowing adjacent to and around a gas bubble rather than the bubble itself. The definition of the microscale element used in this chapter is preferred since this is what is reduced to a mathematical point when one considers the implications of the microscale mass transfer on the

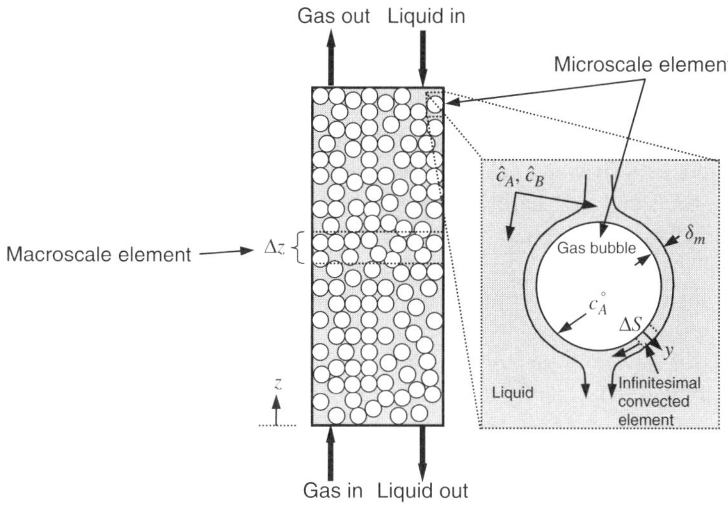

Figure 6.2-1 Bubble column for countercurrent gas absorption; the gas diffuses into the liquid, where it is chemisorbed; the microscale element is a gas bubble; the flux of the soluble solute is controlled on the microscale by a thin region of influence within the surrounding liquid having thickness δ_m, as shown in the enlarged view; the macroscale element is an incremental slice Δz of the entire absorption column.

The microscale element concept can be grasped most easily by means of an example. Consider chemisorption[4] in a bubble column that involves the absorption of a component from gas bubbles moving upward through a downward-flowing liquid, as shown in Figure 6.2-1. Chemisorption is used to increase both the amount that can be transferred and the mass-transfer rates, thereby allowing the use of a smaller and less costly column to accomplish the same absorption. The microscale element in this case is a gas bubble. The mass transfer of the soluble component from this bubble occurs on the microscale within a thin region of influence or boundary layer in the liquid whose thickness is δ_m. This thickness is controlled by the hydrodynamics; that is, the influence of the gas–liquid interface on the flow will create a relatively stagnant region near the gas bubble wherein the mass transfer on the microscale will be confined. The concentration gradients of the diffusing components will occur over the entire thickness δ_m for sufficiently long contact times and slow reaction rates. However, for short contact times or sufficiently fast reaction rates, they will occur over shorter length scales, identified in Section 6.3 using scaling analysis. An upper bound on δ_m is provided by the mass-transfer coefficient $k^{\bullet}_{L0}$ for purely

macroscale equations. However, both conventions for defining the microscale element lead to the same final results for describing the various mass-transfer regimes.

[4]*Chemisorption* refers to a component going into solution by a chemical potential driving force accompanied by chemical reaction. In contrast, *physical absorption* involves only a chemical potential driving force that leads to a thermodynamic equilibrium concentration of the transferring component for a sufficiently long contact time.

diffusive mass transfer in the absence of chemical reaction as shown in Section 6.3. The contact time between the gas and liquid that is available for mass transfer on the microscale can be estimated from the size of a bubble and the relative velocity between the gas and liquid. Depending on the contact time, a simple model can be developed to describe the mass-transfer flux from the gas to the liquid. Whether this simple model needs to include the effects of chemical reaction on the consumption of the reactants and steepening of the concentration profiles depends on the reaction time scale relative to the characteristic diffusion time. In any event, the mass-transfer flux occurring on the microscale of a bubble is considered to occur at a point on the macroscale of the gas-absorption column; that is, one converts the mass-transfer flux per unit area of the microscale element into a species-generation rate per unit volume of contacting device in the species-balance equations for the absorbing component by multiplying the former by the packing area per unit volume of column. One then integrates the species-balance equation for the liquid phase over the length of the absorption column to determine its overall performance.

6.3 SCALING THE MICROSCALE ELEMENT

As an example of scaling the describing equations for mass transfer from a microscale element, we consider the chemisorption of component A from a gas that is bubbled upward through a liquid consisting initially of a nonreacting nonvolatile solvent S and a reacting nonvolatile solute B that is flowing downward in a vertical column as shown in Figure 6.2-1. Although a chemisorption process is considered here, the methodology outlined for the scaling analysis is general and can be applied to other microscale elements involving mass transfer with chemical reaction as well as to other systems, such as phase-transition phenomena that can be described by microscale–macroscale modeling.

In this example we assume irreversible kinetics involving reactants A and B but will not specify the form of the reaction-rate equation, in order to keep the results as general as possible. *Irreversible kinetics* means that the reaction can proceed in only one direction; that is, the products formed by the reaction of A and B cannot decompose to reform A and B. We will assume that component A is the *limiting reactant*; that is, its concentration limits the amount of reaction that can occur. For chemisorption the component being absorbed is usually the limiting reactant; that is, the concentration of component B is maintained at a sufficiently high level to permit the maximum possible absorption of component A in the bulk liquid. However, for an instantaneous reaction it is possible that component B will become the limiting reactant at least within some region on the microscale. This is explored further in Section 6.7.

Solute transfer from the microscale element consisting of the gas bubble will be controlled by mass transfer through a region of influence having thickness δ_m. The system considered for developing the describing equations will be an infinitesimal element of liquid having area ΔS being convected through the region of influence as shown by the dotted lines in the enlarged view in Figure 6.2-1. We describe the mass transfer within this system in a coordinate system convected at the local relative velocity between the liquid and the gas bubble. We ignore both transverse and streamwise convective mass transfer within the convected infinitesimal element; recall from Chapter 5 that these approximations are justified for dilute solutions of the diffusing species and for small solutal Peclet numbers; the latter will be small, owing to the small characteristic length for the thickness of the region of influence. We also ignore both streamwise diffusion and curvature effects on the mass transfer; we saw in Chapter 5 that these approximations are justified if $\delta_m/L \ll 1$, in which L is the characteristic streamwise length, which is the diameter of the gas bubble in this example. These assumptions are not limiting in practice since these neglected effects are accounted for, at least in part, by using mass-transfer coefficient correlations for this contacting device in order to determine the thickness of the region of influence. The dilute solution assumption implies that in the absence of chemical reaction on the microscale, there will be diffusion only of component A since B will not have a concentration gradient. We also assume that the mass transfer is liquid-phase controlled; that is, any resistance to mass transfer due to the diffusion of component A in the gas phase is negligible. However, in Section 6.7 we consider instantaneous reaction conditions for which the mass transfer can become gas-phase controlled.

Mass transfer from the microscale element to the convected infinitesimal element can occur only during this characteristic time that they are in contact. Hence, the appropriate contact time for the unsteady-state mass transfer from the microscale element to the liquid is the time required for the convected infinitesimal element to pass around a gas bubble. Once this convected infinitesimal element passes over the gas bubble, it is assumed to mix intimately with the bulk liquid. The microscale model will provide the continuous mass-transfer flux of solute from the gas to the liquid phase, which then will be converted into a homogeneous source term in the species balance on the macroscale. This mass-transfer flux can influence the local concentration of both reactants on the macroscale of the absorption column, for which both reactants can continue to undergo convective diffusion and reaction.

In view of the considerations above, the resulting describing equations in general will involve unsteady-state one-dimensional diffusion and homogeneous chemical reaction. Hence, equation (G.1-5) in the Appendices simplifies to the following for the two transferring components (step 1):

$$\frac{\partial c_A}{\partial t} + u_x \frac{\partial c_A}{\partial x} = D_{AS} \frac{\partial^2 c_A}{\partial y^2} + R_A \tag{6.3-1}$$

$$\frac{\partial c_B}{\partial t} + u_x \frac{\partial c_B}{\partial x} = D_{BS} \frac{\partial^2 c_B}{\partial y^2} + \kappa R_A \tag{6.3-2}$$

where c_A and c_B denote the molar concentrations of components A and B in the liquid within the microscale element, u_x is the liquid mass-average velocity relative to a coordinate system located on the gas bubble, y is a coordinate normal to the surface of the gas bubble as shown in Figure 6.2-1, D_{AS} and D_{BS} are the effective binary diffusion coefficients of components A and B in the liquid S

that are assumed constant for the dilute solutions being considered here, κ is a stoichiometric coefficient that relates how many moles of component B are consumed per mole of A, and R_A is the rate of production (moles/volume ·time) of component A by the homogeneous chemical reaction; note that for chemisorption, $R_A < 0$. Note that each term in equation (G.1-5) has been divided by the molecular weight of component A or B in arriving at equations (6.3-1) and (6.3-2). It is convenient to transform to a coordinate system that is convected at the liquid velocity relative to the gas bubble, which will be assumed to be constant; that is, equations (6.3-1) and (6.3-2) are transformed to a moving coordinate system for which $\tilde{x} \equiv x - u_x t$, in which u_x is assumed to be the average x-component of the velocity in the region of influence around the gas bubble. Equations (6.3-1) and (6.3-2) assume the following form in this convected coordinate system:

$$\frac{\partial c_A}{\partial t} = D_{AS}\frac{\partial^2 c_A}{\partial y^2} + R_A \tag{6.3-3}$$

$$\frac{\partial c_B}{\partial t} = D_{BS}\frac{\partial^2 c_B}{\partial y^2} + \kappa R_A \tag{6.3-4}$$

in which the time derivatives are now evaluated at a constant value of $\tilde{x}$ rather than at a constant value of x.

The initial and boundary conditions for equations (6.3-3) and (6.3-4) are given by

$$c_A = \hat{c}_A, c_B = \hat{c}_B \quad \text{at} \quad t = 0 \tag{6.3-5}$$

$$c_A = c_A^\circ, \frac{\partial c_B}{\partial y} = 0 \quad \text{at} \quad y = 0 \tag{6.3-6}$$

$$c_A = \hat{c}_A, c_B = \hat{c}_B \quad \text{at} \quad y = \delta_m \tag{6.3-7}$$

where $\hat{c}_A$ and $\hat{c}_B$ are the molar concentrations of components A and B in the bulk liquid on the macroscale and c_A° is the equilibrium molar concentration of component A in the liquid at the interface of the gas bubble. Equation (6.3-5) states that the initial concentrations at the upstream leading edge of the microscale element are those of the local bulk liquid, as shown in Figure 6.2-1; this translates to an initial condition for the convected infinitesimal element being considered here. Although the liquid entering the absorption column is assumed to consist of only the nonreacting solvent S and reacting solute B, $\hat{c}_A$ can be nonzero and $\hat{c}_B$ can be less than its inlet concentration, due to the transfer of component A to the bulk liquid and depletion of component B by chemical reaction that occur due to upstream microscale elements; that is, these are local bulk concentrations that can change with axial position on the macroscale of the absorption column. Equation (6.3-6) states that thermodynamic equilibrium prevails at the gas–liquid interface for component A and that component B is insoluble in the gas phase; that is, it has no mass flux into the gas phase. Equation (6.3-7) states that the undisturbed bulk liquid concentrations prevail at the edge of the region of influence as shown in Figure 6.2-1.

Introduce the following dimensionless variables containing unspecified scale and reference factors (steps 2, 3, and 4):

$$c_A^* \equiv \frac{c_A - c_{Ar}}{c_{As}}; \qquad c_B^* \equiv \frac{c_B - c_{Br}}{c_{Bs}}; \qquad R_A^* \equiv \frac{R_A}{R_{As}}; \qquad y^* \equiv \frac{y}{y_s}; \qquad t^* \equiv \frac{t}{t_s} \tag{6.3-8}$$

We have introduced reference factors for the concentrations of both components since neither is naturally referenced to zero. Substitute these dimensionless variables into the describing equations and divide through by the dimensional coefficient of one term in each equation (steps 5 and 6):

$$\frac{y_s^2}{D_{As}t_s}\frac{\partial c_A^*}{\partial t^*} = \frac{\partial^2 c_A^*}{\partial y^{*2}} + \frac{R_{As}y_s^2}{D_{AS}c_{As}}R_A^* \tag{6.3-9}$$

$$\frac{y_s^2}{D_{Bs}t_s}\frac{\partial c_B^*}{\partial t^*} = \frac{\partial^2 c_B^*}{\partial y^{*2}} + \frac{\kappa R_{As}y_s^2}{D_{BS}c_{Bs}}R_A^* \tag{6.3-10}$$

$$c_A^* = \frac{\hat{c}_A - c_{Ar}}{c_{As}}, \quad c_B^* = \frac{\hat{c}_B - c_{Br}}{c_{Bs}} \quad \text{at} \quad t^* = 0 \tag{6.3-11}$$

$$c_A^* = \frac{c_A^\circ - c_{Ar}}{c_{As}}, \quad \frac{\partial c_B^*}{\partial y_B^*} = 0 \quad \text{at} \quad y^* = 0 \tag{6.3-12}$$

$$c_A^* = \frac{\hat{c}_A - c_{Ar}}{c_{As}}, \quad c_B^* = \frac{\hat{c}_B - c_{Br}}{c_{Bs}} \quad \text{at} \quad y^* = \frac{\delta_m}{y_s} \tag{6.3-13}$$

We divided through by the coefficient of the diffusion term in equations (6.3-9) and (6.3-10) because it provides the only means by which mass can get into the convected infinitesimal element to react or accumulate and hence must be retained.

We now set appropriate dimensionless groups equal to zero or 1 to determine the reference and scale factors, respectively (step 7). If we assume that the reaction and diffusion of both components occur over the entire thickness of the liquid layer on the microscale, the appropriate length scale is obtained by setting $\delta_m/y_s = 1$ to obtain $y_s = \delta_m$. The scale factor for the reaction rate is chosen to be some characteristic maximum value R_A^m. If the form of the reaction-rate equation were specified, R_A^m would be determined from the reaction-rate constant and relevant concentrations. We can reference c_A^* to zero by setting $(\hat{c}_A - c_{Ar})/c_{As} = 0$ to obtain $c_{Ar} = \hat{c}_A$ since $\hat{c}_A$ is its smallest possible value. We can bound c_A^* to be o(1) by setting $(c_A^\circ - c_{Ar})/c_{As} = (c_A^\circ - \hat{c}_A)/c_{As} = 1$ to obtain $c_{As} = c_A^\circ - \hat{c}_A$, which is the driving force for the diffusion of component A. We can bound c_B^* to be o(1) by setting $(\hat{c}_B - c_{Br})/c_{Bs} = 1$ to obtain $c_{Bs} = \hat{c}_B - c_{Br}$. The reference factor c_{Br} is obtained by recognizing that component B will have a concentration gradient only if there is significant chemical reaction on the microscale. This implies that the reaction term in equation (6.3-10) must be of the same order as the diffusion term; hence, we set $\kappa R_{As}y_s^2/D_{BS}c_{Bs} = \kappa R_A^m \delta_m^2/D_{BS}(\hat{c}_B - c_{Br}) = 1$

to obtain $c_{Br} = \hat{c}_B - \kappa R_A^m \delta_m^2 / D_{BS}$, which provides us with an estimate of the minimum concentration of component B. When these scale and reference factors are substituted into the describing equations, we obtain

$$\frac{\delta_m^2}{D_{AS} t_o} \frac{\partial c_A^*}{\partial t^*} = \frac{\partial^2 c_A^*}{\partial y^{*2}} + \frac{R_A^m \delta_m^2}{D_{AS}(c_A^\circ - \hat{c}_A)} R_A^* \tag{6.3-14}$$

$$\frac{\delta_m^2}{D_{BS} t_o} \frac{\partial c_B^*}{\partial t^*} = \frac{\partial^2 c_B^*}{\partial y_B^{*2}} + R_A^* \tag{6.3-15}$$

$$c_A^* = 0, \quad c_B^* = 1 \quad \text{at} \quad t^* = 0 \tag{6.3-16}$$

$$c_A^* = 1, \quad \frac{\partial c_B^*}{\partial y_B^*} = 0 \quad \text{at} \quad y^* = 0 \tag{6.3-17}$$

$$c_A^* = 0, \quad c_B^* = 1 \quad \text{at} \quad y^* = 1 \tag{6.3-18}$$

Let us now use the results of our scaling analysis to explore possible simplifications in the describing equations (step 8). Equation (6.3-14) indicates that there are three characteristic times for the microscale mass-transfer process:

t_o contact time between the liquid and gas on the microscale (6.3-19)

$\dfrac{\delta_m^2}{D_{AS}}$ characteristic time for diffusion of A on the microscale (6.3-20)

$\dfrac{c_A^\circ - \hat{c}_A}{R_A^m}$ characteristic time for reaction on the microscale (6.3-21)

Let us first consider the implications of the contact time relative to the characteristic time for diffusion. Note that the characteristic time for diffusion must be at least of the same order as the contact time if any mass transfer is to occur, since diffusion is the only transport mechanism available to transfer component A from the gas to the liquid phase. Steady-state mass transfer will apply on the microscale if the following criterion is satisfied:

$$\frac{\delta_m^2}{D_{AS} t_o} \ll 1 \Rightarrow \text{steady-state mass transfer on the microscale} \tag{6.3-22}$$

Equation (6.3-22) implies that the characteristic time for diffusion is much shorter than the contact time. When this criterion is satisfied, the diffusion can penetrate the entire thickness of the liquid region δ_m within the time that the liquid flows over the gas bubble. Hence, the mass transfer on the microscale can be described by a film theory model, for the reasons discussed in Section 5.2.

Let us use the results of our scaling analysis for steady-state mass transfer to establish a general relationship between $k_{L0}^\bullet$, the mass-transfer coefficient for purely physical absorption, and δ_m, the effective thickness of the fluid layer for the microscale element. Let us scale the equation that defines the mass-transfer coefficient, which is given by

$$k_L^{\bullet} \equiv \frac{[N_A - x_A(N_A + N_B)]|_{y=0}}{c_A^{\circ} - \hat{c}_A} \cong \frac{-D_{AS}\,(\partial c_A/\partial y)|_{y=0}}{c_A^{\circ} - \hat{c}_A} \tag{6.3-23}$$

This equation for the mass-transfer coefficient has been simplified consistent with the dilute solution assumption, for which the molar density is approximately constant. Introduce the scale factors obtained from our scaling analysis for steady-state mass transfer on the microscale into equation (6.3-23):

$$\frac{k_L^{\bullet}\delta_m}{D_{AS}} = -\left.\frac{\partial c_A^*}{\partial y^*}\right|_{y^*=0} \tag{6.3-24}$$

Since $\partial c_A^*/\partial y_A^*|_{y^*=0} = \bigcirc(1)$, it follows that for steady-state purely physical absorption,

$$k_{L0}^{\bullet} = \frac{D_{AS}}{\delta_m} \qquad \text{steady-state physical absorption} \tag{6.3-25}$$

where $k_{L0}^{\bullet}$ denotes the value of $k_L^{\bullet}$ for purely physical absorption. It is determined solely by the hydrodynamics for purely physical absorption in the absence of any chemical reaction. Equation (6.3-25) then establishes that the mass-transfer coefficient for purely physical absorption can be used to estimate the maximum diffusive penetration distance δ_m, which was stated without proof in Section 6.2.

Consider now the case when the contact time is very short. This causes a contradiction in equation (6.3-14) since the diffusion term must balance the unsteady-state term; that is, component A can accumulate in the liquid solution only by diffusing into it from the gas phase. The forgiving nature of scaling indicates that δ_m is not the proper characteristic length scale for short contact times; that is, there is a region of influence or solutal boundary layer δ_s within which component A experiences a characteristic change in concentration. The thickness of this solutal boundary layer is determined by setting the dimensionless group multiplying the unsteady-state term in equation (6.3-9) equal to 1, thereby obtaining

$$\delta_s = \sqrt{D_{AS}t_o} \tag{6.3-26}$$

Choosing this characteristic length ensures that both the diffusion and unsteady-state terms are retained in equation (6.3-9). If $\delta_s \ll \delta_m$, the boundary condition on component A given in equation (6.3-13) can be applied at infinity and a penetration theory model can be developed to describe the mass transfer with chemical reaction on the microscale, for the reasons discussed in Section 5.3. This penetration boundary-layer approximation can be made when the following criterion is satisfied:

$$1 \ll \frac{\delta_m^2}{D_{AS}t_o} = \frac{1}{\text{Fo}_m} \Rightarrow \text{unsteady-state penetration theory applies} \tag{6.3-27}$$

where $\text{Fo}_m \equiv D_{AS}t_o/\delta_s^2$ is the solutal Fourier number.

Let us use the results of our scaling analysis for unsteady-state mass transfer to establish the relationship between $k_{L0}^{\bullet}$, the mass-transfer coefficient for purely physical absorption, and t_o, the contact time between the liquid and the gas for the microscale element. Introduce the scale factors obtained from our scaling analysis for unsteady-state mass transfer in the microscale element into equation (6.3-23):

$$\frac{k_L^{\bullet}\delta_s}{D_{AS}} = \frac{k_L^{\bullet}\sqrt{D_{AS}t_o}}{D_{AS}} = k_L^{\bullet}\sqrt{\frac{t_o}{D_{AS}}} = -\left.\frac{\partial c_A^*}{\partial y^*}\right|_{y^*=0} \tag{6.3-28}$$

Since $\partial c_A^*/\partial y_A^*|_{y^*=0} = \bigcirc(1)$, it follows that for unsteady-state purely physical absorption

$$k_{L0}^{\bullet} = \sqrt{\frac{D_{AS}}{t_o}} \qquad \text{unsteady-state physical absorption} \tag{6.3-29}$$

where $k_{L0}^{\bullet}$ again denotes the value of $k_L^{\bullet}$ for purely physical absorption, which is determined solely by the hydrodynamics.

In the following sections we consider the implications of the reaction time scale relative to that for diffusion. The mass transfer from the microscale to the macroscale can be markedly different depending on the relative magnitude of these two time scales. Figure 6.3-1 shows the concentration profiles of components A and B for the assumed irreversible chemical reaction for five characteristic reaction regimes that can occur in the microscale element, depending on the magnitude of the reaction time scale relative to that for diffusion for component A. Scaling analysis will now be used to determine the conditions for these various reaction regimes.

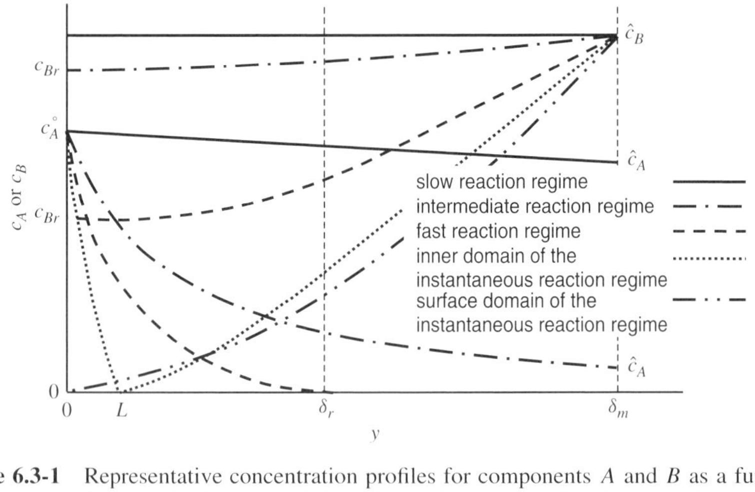

Figure 6.3-1 Representative concentration profiles for components A and B as a function of distance within the liquid layer of the microscale element for chemisorption with an irreversible reaction.

6.4 SLOW REACTION REGIME

The characteristic time for reaction can be much longer than that for diffusion, in which case no chemical reaction will occur on the microscale, thereby justifying ignoring the reaction term in equation (6.3-14). This is referred to as the *slow reaction regime approximation*. Equation (6.3-14) indicates that this approximation is applicable when the following criterion is satisfied:

$$\frac{R_A^m \delta_m^2}{D_{AS}(c_A^\circ - \hat{c}_A)} \ll 1 \Rightarrow \text{slow reaction regime} \qquad (6.4\text{-}1)$$

The criterion above involves the characteristic diffusion time for component A rather than B because A must diffuse from the gas to the liquid phase for any reaction to occur, irrespective of how rapidly B can diffuse from the bulk liquid to the microscale element. Note that the slow reaction regime implies that $c_{Br} = \hat{c}_B - \kappa R_A^m \delta_m^2 / D_{BS} \cong \hat{c}_B$; that is, it implies that there is no change in the concentration of component B on the microscale. Figure 6.3-1 shows that for the slow reaction regime the concentration gradient of component A occurs over the entire thickness δ_m, whereas there is no concentration gradient for component B, owing to the absence of any chemical reaction; the linear concentration profile for component A for the slow reaction regime shown in Figure 6.3-1 is for the special case of steady-state diffusion corresponding to a long contact time. For short contact times, the concentration profile of component A would be nonlinear and extend over a thickness $\delta_s < \delta_m$. Note that in the slow reaction regime, the bulk concentration of component A can range from $0 \le \hat{c}_A \le c_A^\circ$ for an irreversible reaction, depending on how fast the reaction is on the macroscale relative to mass transfer from the microscale element. The slow reaction regime implies that correlations for purely physical absorption in the absence of chemical reaction can be used to describe the mass transfer of component A from the microscale to the macroscale. Note that the slow reaction regime approximation implies only that chemical reaction does not influence the mass transfer on the microscale; it could influence the transport on the macroscale, as we will see in Section 6.8.

6.5 INTERMEDIATE REACTION REGIME

If the characteristic time for reaction is of the same order as that for diffusion, the reaction term must be retained in equation (6.3-14). This will accelerate the absorption of component A from that for purely physical absorption. Moreover, it will cause some depletion and thereby diffusion of component B. We refer to this as the *intermediate reaction regime approximation*[5]; it applies when the following criterion is satisfied:

$$\frac{R_A^m \delta_m^2}{D_{AS}(c_A^\circ - \hat{c}_A)} = O(1) \Rightarrow \text{intermediate reaction regime} \qquad (6.5\text{-}1)$$

[5]The intermediate reaction regime defined here is sometimes referred to as the *transition between the slow to fast reaction regimes*.

The minimum concentration of component B due to the chemical reaction can be estimated from the reference factor that we determined for this component from our scaling analysis; that is,

$$c_{Br} = \hat{c}_B - \frac{\kappa R_A^m \delta_m^2}{D_{BS}} \tag{6.5-2}$$

Figure 6.3-1 shows that in the intermediate reaction regime, components A and B both undergo a characteristic change in concentration over the length δ_m. However, the reaction is not fast enough to reduce the concentration of either component to zero within the microscale element. In the intermediate reaction regime, the bulk concentration of component A can range from $0 \leq \hat{c}_A < c_A^\circ$ for an irreversible reaction, depending on the time scale for reaction on the macroscale relative to the time scale for mass transfer from the microscale element. The mass-transfer flux in the intermediate reaction regime is larger than that for the slow reaction regime, as shown by the increase in the slope of the concentration profile for component A at $y = 0$. It would appear from equation (6.5-2) that the concentration of component B could be reduced to zero for sufficiently fast reaction rates. However, this is not possible in the intermediate reaction regime since the concentration of component A would have to be reduced to zero first because it is the limiting reactant. The latter can occur within the microscale element only in the fast or instantaneous reaction regimes.

6.6 FAST REACTION REGIME

Consider now conditions for which the characteristic time for reaction is much shorter than the characteristic time for diffusion. This is referred to as the fast reaction regime, which applies when the following criterion is satisfied:

$$\frac{R_A^m \delta_m^2}{D_{AS}(c_A^\circ - \hat{c}_A)} \gg 1 \Rightarrow \text{fast reaction regime} \tag{6.6-1}$$

For these conditions the forgiving nature of scaling indicates a contradiction in equation (6.3-14); that is, the reaction term no longer balances the diffusion term that must be retained in order to satisfy the boundary conditions. This implies that the length scale $y_s = \delta_m$ is not appropriate for this reaction regime and suggests that there is a region of influence or reaction boundary layer δ_r within which the irreversible reaction of component A goes to completion at which $\hat{c}_A = 0$. The thickness of this reaction boundary layer is obtained by setting the dimensionless group that multiplies the reaction term in equation (6.3-14) equal to 1 to obtain

$$\delta_r = \left(\frac{D_{AS} c_A^\circ}{R_A^m}\right)^{1/2} \tag{6.6-2}$$

Figure 6.3-1 shows that in the fast reaction regime the concentration of component A is reduced to zero within the distance δ_r. However, since component A is the limiting reactant, component B is not necessarily reduced to zero; moreover, it diffuses over the full thickness of the microscale liquid layer and reacts with component A over the thickness δ_r. The minimum concentration of component B can again be estimated from our scaling analysis and is given by equation (6.5-2). The mass-transfer flux is increased for the fast relative to the slow or intermediate reaction regimes, as shown by the steeper slope of the concentration profile for component A at $y = 0$.

6.7 INSTANTANEOUS REACTION REGIME

As the reaction rate increases in the fast reaction regime, the concentration of component B at the gas–liquid interface continues to decrease. For sufficiently fast irreversible reaction conditions, it is reduced to zero. An estimate of the reaction rate required to achieve this condition can be obtained from our reference factor for component B given by equation (6.5-2):

$$c_{Br} = \hat{c}_B - \frac{\kappa R_A^m \delta_m^2}{D_{BS}} = 0 \Rightarrow R_A^m = \frac{D_{BS}\hat{c}_B}{\kappa \delta_m^2} \tag{6.7-1}$$

When the reaction rate is increased from this value, component B becomes the rate-limiting reactant in the vicinity of the gas–liquid interface; that is, there is a region near the gas–liquid interface wherein there is no reaction, due to the depletion of component B. Hence, the reaction will occur only in the region wherein both components A and B have concentrations greater than zero. Equation (6.6-2) indicates that the thickness of the region wherein the chemical reaction occurs decreases to zero as the reaction rate become infinite. This limiting condition is referred to as the *instantaneous reaction regime*, for which the criterion is

$$\frac{R_A^m \delta_m^2}{D_{AS}(c_A^\circ - \hat{c}_A)} \to \infty \Rightarrow \text{instantaneous reaction regime} \tag{6.7-2}$$

This condition implies that the irreversible reaction is so fast that the two reacting components cannot coexist anywhere in the microscale element. When instantaneous reaction conditions prevail, the resistance to mass transfer in the liquid phase is greatly reduced. Hence, it is possible for the mass transfer on the microscale to become gas-phase controlled. If the mass transfer remains liquid-phase controlled, the interfacial concentration of component A will remain at c_A° and the reaction plane will be in the liquid phase; this is referred to as the *inner-reaction domain* of the instantaneous reaction regime. However, if the mass transfer becomes gas-phase controlled, the interfacial concentration of component A will become zero and the reaction plane will be at the gas–liquid interface; this is referred to as the *surface-reaction domain* of the instantaneous reaction regime. Figure 6.3-1 shows that for the inner-reaction domain of the instantaneous reaction regime the reaction plane separates a region in which component A is diffusing in the absence of any B from a region in which component B is diffusing in the absence of any A. However, for the surface-reaction domain of the instantaneous reaction regime the concentration of component A remains at zero throughout the microscale element, owing to the extremely fast reaction that consumes it at the interface between the two phases. In addition, the concentration of component B becomes zero at the gas–liquid interface. Figure 6.3-1 shows that the mass-transfer flux is increased for the instantaneous relative to the slow, intermediate, and fast reaction regimes, as indicated by the steeper concentration gradient of component A at $y = 0$. However, the surface-reaction domain of the instantaneous reaction regime provides the fastest absorption since the resistance to mass transfer in the liquid becomes negligible in comparison to that in the gas phase.

The inner-reaction domain of the instantaneous reaction regime is no longer described by equations (6.3-3) through (6.3-7) since the homogeneous reaction term does not appear in the describing equations. Moreover, the concentrations of both components are forced to zero at the reaction plane as shown in Figure 6.3-1. Since the instantaneous reaction is confined to a reaction plane, the mass transfer again becomes diffusion-limited; that is, the diffusion of the two components occurs within the liquid layer thickness defined by δ_m. The resulting describing equations then are given by (step 1)

$$\frac{\partial c_A}{\partial t} = D_{AS} \frac{\partial^2 c_A}{\partial y^2} \tag{6.7-3}$$

$$\frac{\partial c_B}{\partial t} = D_{BS} \frac{\partial^2 c_B}{\partial y^2} \tag{6.7-4}$$

$$c_A = 0, \quad c_B = \hat{c}_B \quad \text{at} \quad t = 0 \tag{6.7-5}$$

$$c_A = c_A^\circ \quad \text{at} \quad y = 0 \tag{6.7-6}$$

$$c_A = 0, \quad c_B = 0 \quad \text{at} \quad y = L(t) \tag{6.7-7}$$

$$c_B = \hat{c}_B \quad \text{at} \quad y = \delta_m \tag{6.7-8}$$

The initial condition given by equation (6.7-5) states that for an instantaneous reaction component A, the limiting reactant cannot be present in the bulk liquid, whereas component B has a concentration dictated by the local position in the absorber. Equation (6.7-6) states that the equilibrium concentration of component A is maintained at the interface. Equation (6.7-7) states that the concentrations of both reactants are zero at the instantaneous reaction front. Equation (6.7-8) states that component B has its bulk concentration at the edge of the liquid layer in the microscale element. An auxiliary condition is needed to locate $L(t)$, the instantaneous position of the reaction plane; that is, the instantaneous reaction case involves a moving boundary problem. Consider an integral species balance on component A:

$$\frac{d}{dt} \int_0^L c_A \, dy = N_A|_{y=0} + \frac{1}{\kappa} \frac{d}{dt} \int_L^{\delta_m} c_B \, dy \tag{6.7-9}$$

Equation (6.7-9) states that the change in the concentration of component A in the liquid layer of the microscale element is equal to the rate at which it diffuses in from the gas phase minus the rate at which it is consumed by the irreversible instantaneous chemical reaction; the latter is equal to the change in the concentration of component B (a negative number) divided by the stoichiometric coefficient for the chemical reaction. Equation (6.7-9) can be rearranged using Leibnitz's rule for differentiation of an integral given by equation (H.1-2) in the Appendices and equations (6.7-3) and (6.7-4) to obtain the following auxiliary condition for determining the instantaneous location of the reaction plane[6]:

$$D_{AS}\frac{\partial c_A}{\partial y} = -\frac{D_{BS}}{\kappa}\frac{\partial c_B}{\partial y} \quad \text{at} \quad y = L \tag{6.7-10}$$

Note that velocity of the moving boundary does not appear in equation (6.7-10) because the concentrations of both A and B are zero at the reaction plane. Our describing equations for instantaneous reaction conditions then are given by equations (6.7-3) through (6.7-8) and equation (6.7-10).

Introduce the following dimensionless variables containing unspecified scale and reference factors (steps 2, 3, and 4):

$$c_A^* \equiv \frac{c_A}{c_{As}}; \qquad c_B^* \equiv \frac{c_B}{c_{Bs}}; \qquad y_A^* \equiv \frac{y}{y_{As}}; \qquad y_B^* \equiv \frac{y - y_{Bs}}{y_{Bs}}; \qquad t^* \equiv \frac{t}{t_s} \tag{6.7-11}$$

We do not need to introduce reference factors for c_A and c_B since they are naturally referenced to zero. We have introduced a reference length factor for component B since its diffusion occurs only on one side of the reaction front. Substitute these dimensionless variables into the describing equations and divide through by the dimensional coefficient of one term in each equation (steps 5 and 6):

$$\frac{y_{As}^2}{D_{As}t_s}\frac{\partial c_A^*}{\partial t^*} = \frac{\partial^2 c_A^*}{\partial y_A^{*2}} \tag{6.7 12}$$

$$\frac{y_{Bs}^2}{D_{Bs}t_s}\frac{\partial c_B^*}{\partial t^*} = \frac{\partial^2 c_B^*}{\partial y_B^{*2}} \tag{6.7-13}$$

$$c_A^* = 0, \quad c_B^* = \frac{\hat{c}_B}{c_{Bs}} \quad \text{at} \quad t^* = 0 \tag{6.7-14}$$

[6]In some formulations of the instantaneous reaction problem, an additional condition that is based on the fact that the moving boundary is a plane having constant concentration is used to obtain an equation that relates the velocity of the reaction front dL/dt to the ratio of the partial derivatives of the concentration $\partial c_A/\partial t$ and $\partial c_A/\partial y$. However, this condition is redundant since it is embodied in equation (6.7-10). Indeed, the time dependence of L can be obtained by demanding that the solution to equations (6.7-3) through (6.7-8) satisfy equation (6.7-10). Note, however, that we will see that our scaling analysis gives us the time dependence of L to within a multiplicative constant of $\bigcirc(1)$: that is, $L \propto \sqrt{t}$.

$$c_A^* = \frac{c_A^\circ}{c_{As}} \quad \text{at} \quad y^* = 0 \tag{6.7-15}$$

$$c_A^* = 0 \quad \text{at} \quad y_A^* = \frac{L}{y_{As}} \tag{6.7-16}$$

$$c_B^* = 0 \quad \text{at} \quad y_B^* = \frac{L - y_{Br}}{y_{Bs}} \tag{6.7-17}$$

$$c_B^* = \frac{\hat{c}_B}{c_{Bs}} \quad \text{at} \quad y_B^* = \frac{\delta_m - y_{Br}}{y_{Bs}} \tag{6.7-18}$$

$$\frac{\partial c_A^*}{\partial y_A^*} = -\frac{D_{BS}c_{Bs}y_{As}}{\kappa D_{AS}c_{As}y_{Bs}}\frac{\partial c_B^*}{\partial y_B^*} \quad \text{at} \quad y_A^* = \frac{L}{y_{As}} \tag{6.7-19}$$

We again set appropriate dimensionless groups equal to zero or 1 to determine the reference and scale factors, respectively (step 7). The concentration scales are obtained by setting $c_A^\circ/c_{As} = 1$ and $\hat{c}_B/c_{Bs} = 1$ to obtain $c_{As} = c_A^\circ$ and $c_{Bs} = \hat{c}_B$, respectively. Since this is inherently an unsteady-state problem, the characteristic time is the observation time t_o. Careful consideration must now be given to determining the proper scales for y_{As} and y_{Bs}. Note that for an instantaneous reaction, we expect that $L \ll \delta_m$ since the reaction is limited by depletion of component B, which must diffuse to the reaction front. Hence, we set the dimensionless group in equation (6.7-16) equal to 1 to obtain $y_{As} = L$. Since the two terms in equation (6.7-13) must balance, we set $y_{Bs}^2/D_{BS}t_o = 1$ to obtain $y_{Bs} = \sqrt{D_{BS}t_o}$. Since the diffusion of component B is inherently unsteady-state, the two terms in equation (6.7-19) must balance; thus, we set $D_{BS}c_{Bs}y_{As}/\kappa D_{AS}c_{As}y_{Bs} = D_{BS}\hat{c}_B L/\kappa D_{AS}c_A^\circ\sqrt{D_{BS}t_o} = 1$, thereby obtaining $L = (\kappa D_{AS}c_A^\circ/\hat{c}_B)\sqrt{t_o/D_{BS}}$. When these scale and reference factors are substituted into the describing equations, we obtain

$$\left(\frac{\kappa c_A^\circ}{\hat{c}_B}\right)^2 \frac{D_{AS}}{D_{BS}}\frac{\partial c_A^*}{\partial t^*} = \frac{\partial^2 c_A^*}{\partial y_A^{*2}} \tag{6.7-20}$$

$$\frac{\partial c_B^*}{\partial t^*} = \frac{\partial^2 c_B^*}{\partial y_B^{*2}} \tag{6.7-21}$$

$$c_A^* = 0, \quad c_B^* = 1 \quad \text{at} \quad t^* = 0 \tag{6.7-22}$$

$$c_A^* = 1 \quad \text{at} \quad y^* = 0 \tag{6.7-23}$$

$$c_A^* = 0 \quad \text{at} \quad y_A^* = 1 \tag{6.7-24}$$

$$c_B^* = 0 \quad \text{at} \quad y_B^* = 0 \tag{6.7-25}$$

$$c_B^* = 1 \quad \text{at} \quad y_B^* = \frac{\delta_m - L}{\sqrt{D_{BS}t_o}} \tag{6.7-26}$$

$$\frac{\partial c_A^*}{\partial y_A^*} = -\frac{\partial c_B^*}{\partial y_B^*} \quad \text{at} \quad y_A^* = 1 \tag{6.7-27}$$

Now let us consider how the dimensionless describing equations can be simplified (step 8). The boundary condition given by equation (6.7-26) can be applied at infinity if the following condition is satisfied:

$$\frac{\delta_m - L}{\sqrt{D_{BS} t_o}} \gg 1 \tag{6.7-28}$$

This condition will be satisfied for relatively short contact times. The unsteady-state term in equation (6.7-20) can be dropped if the following condition is satisfied:

$$\left(\frac{\kappa c_A^\circ}{\hat{c}_B}\right)^2 \frac{D_{AS}}{D_{BS}} \ll 1 \tag{6.7-29}$$

This condition will be satisfied if the nonlimiting reactant is present in considerable excess relative to the reactant that is being absorbed, since in most cases one would expect $D_{AS} \cong D_{BS}$. If the condition given by equation (6.7-29) is satisfied, a quasi-steady-state linear concentration profile of component A is obtained. One can then solve equation (6.7-21) via a penetration theory approach if the condition given by equation (6.7-28) is satisfied.

6.8 SCALING THE MACROSCALE ELEMENT

In Section 6.3 we focused on scaling the describing equations for the microscale element. The mass-transfer flux determined for the microscale analysis will become a homogeneous source or sink term in the describing equations for the macroscale that we consider now. The analysis for the macroscale is greatly simplified for the fast and instantaneous reaction regimes, since they reduce the concentration of the absorbing component to its reaction equilibrium value, which is zero for an irreversible reaction. Since the latter concentration will prevail on the macroscale everywhere in the contacting device, no further reaction can occur. However, the concentration of the nonlimiting reactant, component B, is depleted on the macroscale, due to its consumption by the chemical reaction on the microscale. However, as we will see in Section 6.14, the mass-transfer flux of the absorbing component can be obtained without having to solve for the concentration profile of the liquid-phase reactant.

Consider now the slow reaction regime for which no reaction occurs on the microscale and there is no diffusive flux of component B. However, there is a diffusive flux N_A of component A from the gas into the liquid phase. This has the effect of being a point source of component A and a point sink of component B on the macroscale. We assume that conditions are laterally uniform so that the bulk concentrations of components A and B change only in the axial direction in Figure 6.2-1. Consider a molar balance on components A and B over a control volume consisting of a differential slice of the bubble column Δz that is convected at the average axial velocity of the liquid, as shown in Figure 6.2-1 (step 1):

$$N_A a \Delta z - R_A \phi a \,\Delta z = \frac{d}{dt}(\hat{c}_A \phi a \,\Delta z) \tag{6.8-1}$$

$$-\kappa R_A \phi a \,\Delta z = \frac{d}{dt}(\hat{c}_B \phi a \,\Delta z) \tag{6.8-2}$$

where a is the interfacial area per unit volume of the contacting device, ϕ the volume of liquid per unit interfacial area, R_A the rate of production (moles/volume ·time) of component A by the homogeneous chemical reaction, κ the stoichiometric coefficient for the chemical reaction, and d/dt denotes the total derivative.[7] In the slow reaction regime N_A can be obtained directly from $k^{\bullet}_{L0}$, the mass-transfer coefficient for purely diffusive transfer in the liquid phase via the equation

$$N_A = k^{\bullet}_{L0}(c^{\circ}_A - \hat{c}_A) \tag{6.8-3}$$

Hence, equations (6.8-1) and (6.8-2) can be written as

$$k^{\bullet}_{L0}(c^{\circ}_A - \hat{c}_A) - R_A \phi = \phi \frac{\partial \hat{c}_A}{\partial t} \tag{6.8-4}$$

$$-\kappa R_A = \frac{\partial \hat{c}_B}{\partial t} \tag{6.8-5}$$

The initial conditions required to solve equations (6.8-4) and (6.8-5) are given by

$$\hat{c}_A = 0, \quad \hat{c}_B = \hat{c}_{B0} \qquad \text{at} \quad t = 0 \tag{6.8-6}$$

that is, we are assuming that the nonreacting solvent that enters the bubble column contains only liquid component B at an initial concentration $\hat{c}_{B0}$.

Introduce the following dimensionless variables and scale factors (steps 2, 3, and 4):

$$\hat{c}^*_A \equiv \frac{\hat{c}_A}{\hat{c}_{As}}; \qquad \hat{c}^*_B \equiv \frac{\hat{c}_B - \hat{c}_{Br}}{\hat{c}_{Bs}}; \qquad R^*_A \equiv \frac{R_A}{R_s}; \qquad t^* \equiv \frac{t}{t_s} \tag{6.8-7}$$

Introduce these variables into equations (6.8-4) through (6.8-6) and divide through by the coefficient of one term in each equation to obtain the following dimensionless describing equations (steps 5 and 6):

$$\left(\frac{c^{\circ}_A}{\hat{c}_{As}} - \hat{c}^*_A\right) - \frac{R_s \phi}{k^{\bullet}_{L0}\hat{c}_{As}} R^*_A = \frac{\phi}{k^{\bullet}_{L0} t_s} \frac{\partial \hat{c}^*_A}{\partial t^*} \tag{6.8-8}$$

$$-\frac{\kappa R_s t_s}{\hat{c}_{Bs}} R^*_A = \frac{\partial \hat{c}^*_B}{\partial t^*} \tag{6.8-9}$$

[7] In a convected coordinate system, one takes the total derivative of an extensive quantity such as the total number of moles, but the partial derivative of an intensive quantity such as the molar concentration; the partial derivative with respect to the temporal coordinate is evaluated at a fixed value of the spatial coordinate in the convected coordinate system.

$$\hat{c}_A^* = 0, \quad \hat{c}_B^* = \frac{\hat{c}_{B0} - \hat{c}_{Br}}{\hat{c}_{Bs}} \quad \text{at} \quad t^* = 0 \tag{6.8-10}$$

Note that we have divided through by the coefficient of the first term in equation (6.8-4) since this term must be retained if component A is transferred from the gas to the liquid phase. Equation (6.8-8) indicates that the dimensionless concentration can be bounded of O(1) if we set $\overset{\circ}{c}_A/\hat{c}_{As} = 1$ in order to obtain $\hat{c}_{As} = \overset{\circ}{c}_A$. The appropriate time scale is the observation or contact time t_o that is required for the liquid to flow through the bubble column. The dimensionless reaction rate is bounded to be O(1) by again choosing its scale factor to be some characteristic maximum value R_A^m, which could be determined if the reaction rate were specified. Setting the dimensionless group in equation (6.8-10) equal to 1 indicates that $\hat{c}_{Bs} = \hat{c}_{B0} - \hat{c}_{Br}$. When this scale is substituted into the dimensionless group in equation (6.8-9) and the latter is set equal to -1 (because $R_A^m < 0$), we obtain $\hat{c}_{Br} = \hat{c}_{B0} - \kappa R_A^m t_o$. Substituting these scale and reference factors into the describing equations then yields the following set of dimensionless describing equations for the macroscale element (step 7):

$$\underbrace{(1 - \hat{c}_A^*)}_{\substack{\text{mass transfer} \\ \text{from microscale} \\ \text{to macroscale}}} - \underbrace{\frac{R_A^m \phi}{\overset{\bullet}{k}_{L0} \overset{\circ}{c}_A} R_A^*}_{\text{chemical reaction}} = \underbrace{\frac{\phi}{\overset{\bullet}{k}_{L0} t_o} \frac{\partial \hat{c}_A^*}{\partial t^*}}_{\substack{\text{accumulation in} \\ \text{convected system}}} \tag{6.8-11}$$

$$R_A^* = \frac{\partial \hat{c}_B^*}{\partial t^*} \tag{6.8-12}$$

$$\hat{c}_A^* = 0, \quad \hat{c}_B^* = 1 \quad \text{at} \quad t^* = 0 \tag{6.8-13}$$

Now let us consider how the describing equations for the macroscale element can be simplified (step 8). These equations again involve three characteristic time scales:

$$t_o \sim \text{contact or residence time in the bubble column} \tag{6.8-14}$$

$$\frac{\phi}{\overset{\bullet}{k}_{L0}} \sim \text{characteristic time for interphase mass transfer on the macroscale} \tag{6.8-15}$$

$$\frac{\overset{\circ}{c}_A}{R_A^m} \sim \text{characteristic time for reaction on the macroscale} \tag{6.8-16}$$

Consider first the case where the characteristic time for reaction is extremely long compared to that for interphase mass transfer. This corresponds to purely physical absorption in the absence of any chemical reaction for which the criterion is

$$\frac{R_A^m \phi}{\overset{\bullet}{k}_{L0} \overset{\circ}{c}_A} \cong 0 \Rightarrow \text{purely physical absorption} \tag{6.8-17}$$

Now let us consider the more interesting case for which chemical reaction occurs. Note that the accumulation or unsteady-state term in equation (6.8-11) is equal to the difference between the rate of mass transfer from the microscale to the macroscale and the rate at which the transferring component is consumed by the chemical reaction. Although both the mass transfer and reaction terms can be large, their difference can be quite small under some conditions that we will now explore. The accumulation or unsteady-state term will be very small when the following criterion is satisfied:

$$\frac{\phi}{\dot{k}_{L0} t_o} \ll 1 \Rightarrow \text{species accumulation is very small} \tag{6.8-18}$$

Equation (6.8-18) implies that the time scale for interphase mass transfer is much shorter than the residence time in the contacting device. When the criterion given by equation (6.8-18) is satisfied, it means that the mass transfer and reaction terms are nearly equal in magnitude such that their difference is very small. This approximation is referred to as the *quasistationary hypothesis*. The quasistationary hypothesis is rigorous for a continuous stirred tank reactor (CSTR), within which there is a uniform concentration. However, it is an approximation for contactors such as packed columns, for which the concentration changes with axial position.

6.9 KINETIC DOMAIN OF THE SLOW REACTION REGIME

Now let us assume that the quasistationary hypothesis applies, in which case we can write equation (6.8-11) as follows:

$$\underbrace{(1 - \hat{c}_A^*)}_{\substack{\text{mass transfer}\\ \text{from microscale}\\ \text{to macroscale}}} - \underbrace{\frac{R_A^m \phi}{\dot{k}_{L0} c_A^\circ} R_A^*}_{\text{chemical reaction}} = \underbrace{\frac{\phi}{\dot{k}_{L0} t_o} \frac{\partial \hat{c}_A^*}{\partial t^*}}_{\substack{\text{accumulation in}\\ \text{convected system}}} \ll 1 \tag{6.9-1}$$

Let us first consider the case for which the characteristic time for mass transfer from the microscale to macroscale is much shorter than the characteristic time for reaction. This means that component A is transferred from the gas to the liquid phase much faster than it can be consumed by the chemical reaction. This is referred to as the *kinetic domain of the slow reaction regime*.[8] The criterion for its applicability is

$$\frac{R_A^m \phi}{\dot{k}_{L0} c_A^\circ} \ll 1 \Rightarrow \text{kinetic domain of the slow reaction regime} \tag{6.9-2}$$

Since the mass transfer and reaction terms in equation (6.9-1) must nearly balance, the kinetic domain of the slow reaction regime implies that $\hat{c}_A^* \cong 1$ or that $\hat{c}_A \cong c_A^\circ$; that is, the bulk liquid is maintained nearly at the thermodynamic equilibrium concentration for component A. The driving force for the chemical reaction then is proportional to c_A° for an irreversible reaction.

[8]The kinetic domain of the slow reaction regime is often referred to as the kinetic regime of the slow reaction regime; the terminology used here is preferred because it more clearly indicates that the domain is a subcategory of the regime.

6.10 DIFFUSIONAL DOMAIN OF THE SLOW REACTION REGIME

Let us again assume that the quasi-stationary hypothesis applies so that equation (6.9-1) applies. Now consider the case for which the characteristic time for mass transfer from the microscale to macroscale is much longer than the characteristic time for reaction. This means that component A is transferred from the gas to the liquid phase much slower than it is consumed by the chemical reaction. This is referred to as the *diffusional domain of the slow reaction regime*.[9] The criterion for its applicability is the following:

$$\frac{R_A^m \phi}{\dot{k}_{L0} c_A^\circ} \gg 1 \Rightarrow \text{diffusional domain of the slow reaction regime} \qquad (6.10\text{-}1)$$

For the diffusional domain of the slow reaction regime, it would appear that the reaction term is much larger than the mass-transfer term in equation (6.9-1). However, since these two terms must nearly balance, we conclude that $R_A^* \ll 1$. This in turn implies that the driving force for the chemical reaction must be very small (i.e., $\hat{c}_A^* \cong 0$); that is, the chemical reaction is sufficiently fast so that it maintains the bulk concentration at the reaction equilibrium concentration. This in turn implies that the driving force for mass transfer from the microscale to the macroscale is c_A° for an irreversible reaction. It might seem contradictory that the chemical reaction is fast enough in the diffusional domain of the slow reaction regime to reduce the bulk concentration to zero, whereas in the intermediate reaction regime for which the reaction is faster, the bulk concentration is not necessarily reduced to zero. In the intermediate reaction regime the reaction occurs on the microscale and therefore steepens the concentration gradient of component A at the gas–liquid interface and increases the mass-transfer coefficient from that for purely physical absorption. In contrast, the mass-transfer coefficient for the diffusional domain of the slow reaction regime is the same as that for purely physical absorption. The faster mass-transfer rate in the intermediate regime relative to the diffusional domain of the slow reaction regime corresponds to a steeper concentration gradient of component A at $y = 0$ in Figure 6.3-1.

6.11 IMPLICATIONS OF SCALING ANALYSIS FOR REACTOR DESIGN

The scaling analysis that we have applied to the microscale–macroscale modeling of mass transfer with chemical reaction can be used to develop a set of design precepts for selecting the optimal reactor. These will be determined by considering how the process parameters affect the total absorption rate per unit volume $N_A a$. We will proceed from the slowest to the fastest reaction conditions. Hence, let us first consider purely diffusional mass transfer in the absence of any chemical

[9] The diffusional domain of the slow reaction regime is often referred to as the diffusional regime of the slow reaction regime.

reaction for which the mass-transfer rate per unit volume of reactor is given by

$$N_A a = k_{L0}^{\bullet} a(c_A^{\circ} - \hat{c}_A) \qquad \text{for purely physical absorption} \tag{6.11-1}$$

Correlations for $k_{L0}^{\bullet}$, the mass-transfer coefficient for purely physical absorption, in terms of the Sherwood number or Nusselt number for mass transfer are available in the literature for many contacting geometries. If the contact time is long in comparison to the diffusion time, a film theory model applies and the mass-transfer coefficient for purely physical absorption is given by equation (6.3-25); that is,

$$k_{L0}^{\bullet} = \frac{D_{AS}}{\delta_m} \qquad \text{steady-state physical absorption} \tag{6.11-2}$$

If the contact time is short in comparison to the diffusion time, a penetration theory model applies and the mass-transfer coefficient for purely physical absorption is given by equation (6.3-29); that is,

$$k_{L0}^{\bullet} = \sqrt{\frac{D_{AS}}{t_o}} \qquad \text{unsteady-state physical absorption} \tag{6.11-3}$$

Equation (6.11-1) implies that the total absorption rate per unit volume for purely physical absorption is:

- Proportional to the interfacial area
- Independent of the liquid-phase holdup ϕa
- Proportional to the mass-transfer coefficient $k_{L0}^{\bullet}$ that characterizes the transport between the microscale and macroscale elements
- Proportional to the overall driving force for the absorbing component

These considerations indicate that appropriate contactors for purely physical absorption are packed columns that provide large interfacial area (i.e., large values of a), good hydrodynamics (i.e., high Reynolds numbers) to promote large mass-transfer coefficients (i.e., large values of $k_{L0}^{\bullet}$), and maintain large overall driving forces (i.e., large values of $c_A^{\circ} - \hat{c}_A$ are maintained by avoiding global mixing). On the other hand, contactors such as stirred tanks are not appropriate since they have low interfacial area and promote large scale mixing that reduces the overall driving force for mass transfer.

Now let us consider the kinetic domain of the slow reaction regime. For this domain the chemical reaction is very slow on both the micro- and macroscales. This means that the mass-transfer coefficient for transfer from the microscale to the macroscale is the same as for mass transfer in the absence of chemical reaction. On the macroscale the kinetic domain of the slow reaction regime means that the reaction is so slow that the local bulk fluid is nearly at the thermodynamic equilibrium concentration c_A°. Since the mass-transfer and chemical reaction terms in equation (6.9-1) must nearly balance if the quasi-stationary hypothesis is applicable, the absorption rate per unit volume of reactor is given by

$$N_A a \cong R_A \phi a \tag{6.11-4}$$

where the reaction rate is determined using $\hat{c}_A \cong c_A^{\circ}$ and the local value of $\hat{c}_B$. Equation (6.11-4) implies that the total absorption rate per unit volume for the kinetic domain of the slow reaction regime is:

- Independent of the interfacial area (note that the product ϕa is the liquid-phase holdup since the interfacial area cancels in the product)
- Proportional to the liquid-phase holdup ϕa
- Independent of the mass-transfer coefficient $k_{L0}^{\bullet}$ that characterizes the transport between the microscale and macroscale elements
- Proportional to the reaction rate per unit volume R_A
- Influenced by the overall driving force for the limiting reactant c_A° only insofar as it enters through the reaction rate R_A
- Dependent on the concentration of the nonlimiting reactant $\hat{c}_B$ insofar as it enters through the reaction rate R_A

These considerations indicate that appropriate contactors for mass transfer with chemical reaction operating in the kinetic domain of the slow reaction regime are stirred tank reactors that provide large liquid holdups (i.e., large values of ϕa) and for which the large-scale mixing that reduces the driving force has no effect on the total absorption rate per unit volume (i.e., c_A° is a constant not affected by mixing). On the other hand, contactors such as packed columns are not appropriate since they have low liquid holdups and a large interfacial area. These considerations also indicate that changing the hydrodynamics to increase the Reynolds number, which will increase the interfacial area a and the mass-transfer coefficient $k_{L0}^{\bullet}$, will have no effect on the total absorption rate per unit volume in the kinetic domain. The fact that the kinetic domain of the slow reaction regime is independent of the interfacial area is advantageous in the use of laboratory absorbers to determine the kinetics of a reaction; that is, one can employ a packed column whose interfacial area is unknown in order to determine the unknown kinetics of a reaction.

Consider now the diffusional domain of the slow reaction regime. For this domain the chemical reaction is very slow on the microscale but fast on the macroscale. Again the mass-transfer coefficient for transfer from the microscale to the macroscale is the same as for mass transfer in the absence of chemical reaction. On the macroscale the diffusional domain of the slow reaction regime means that the reaction is sufficiently fast to maintain the local bulk concentration of the component A, the limiting reactant, at nearly zero for the assumed irreversible reaction. Hence, equation (6.8-3) implies that the absorption rate per unit volume of reactor is given by

$$N_A a \cong k_{L0}^{\bullet} a c_A^{\circ} \tag{6.11-5}$$

Equation (6.11-5) implies that the total absorption rate per unit volume for the diffusional domain of the slow reaction regime is:

- Proportional to a, the interfacial area per unit volume
- Independent of the liquid-phase holdup ϕa
- Proportional to the mass-transfer coefficient $k^{\bullet}_{L0}$ that characterizes the transport between the microscale and macroscale elements
- Independent of the reaction rate per unit volume R_A
- Proportional to the overall driving force $c^{\circ}_A - \hat{c}_A$
- Independent of the concentration of the nonlimiting reactant component B

These considerations indicate that appropriate contactors for mass transfer with chemical reaction operating in the diffusional domain of the slow reaction regime are packed columns that provide large interfacial area, good hydrodynamics, and maintain large overall driving forces. Contactors such as stirred tanks are not appropriate since they have low interfacial area and promote large-scale mixing that reduces the overall driving force for mass transfer.

Now let us consider the intermediate reaction regime for which both diffusion and reaction on the microscale are of equal importance. The reaction is sufficiently fast to steepen the concentration profile significantly near the gas–liquid interface on the microscale, but not large enough necessarily to reduce the bulk concentration to its reaction equilibrium value of zero for an irreversible reaction. We use the results of our scaling analysis for the intermediate reaction regime to obtain an estimate of $N_A a$, the total absorption rate per unit volume of reactor; that is,

$$N_A \cong -D_{AS}\frac{\partial c_A}{\partial y} \Rightarrow N^*_A \cong -\frac{D_{AS}c_{As}}{N_{As}y_{As}}\frac{\partial c^*_A}{\partial y^*_A}$$
$$= -\frac{D_{AS}(c^{\circ}_A - \hat{c}_A)}{N_{As}\delta_m}\frac{\partial c^*_A}{\partial y^*_A} = -\frac{k^{\bullet}_{L0}(c^{\circ}_A - \hat{c}_A)}{N_{As}}\frac{\partial c^*_A}{\partial y^*_A} \tag{6.11-6}$$

where we have used equation (6.11-2) to introduce $k^{\bullet}_{L0}$, the mass-transfer coefficient for purely physical absorption. To bound N^*_A to be O(1), we set $k^{\bullet}_{L0}(c^{\circ}_A - \hat{c}_A)/N_{As} = 1 \quad \Rightarrow N_{As} = k^{\bullet}_{L0}(c^{\circ}_A - \hat{c}_A)$. Hence, we obtain the following estimate for the mass-transfer rate per unit volume:

$$N_A a \cong k^{\bullet}_{L0}a(c^{\circ}_A - \hat{c}_A) \qquad \text{for the intermediate reaction regime} \tag{6.11-7}$$

Although equation (6.11-7) is identical in form to equation (6.11-1) for purely physical absorption, the driving force in the former is much larger. This follows from the reduction in $\hat{c}_A$ due to the chemical reaction on the microscale that is implied by equation (6.5-2). Equation (6.11-7) implies that the total absorption rate per unit volume for the intermediate reaction regime is:

- Proportional to a, the interfacial area per unit volume

- Independent of the liquid-phase holdup ϕa
- Proportional to the mass-transfer coefficient $k^{\bullet}_{L0}$
- Dependent on the reaction rate per unit volume R_A insofar as it reduces $\hat{c}_A$
- Proportional to the overall driving force $c^{\circ}_A - \hat{c}_A$
- Dependent on the concentration of the nonlimiting reactant component B insofar as it influences the reaction rate per unit volume R_A

These considerations indicate that appropriate contactors for mass transfer with chemical reaction operating in the intermediate reaction regime are packed columns that provide large interfacial area, good hydrodynamics, and maintain a large driving force. Contactors such as stirred tanks are not appropriate since they have small interfacial area and a reduced driving force.

Consider now the fast reaction regime that maintains the bulk liquid at the reaction equilibrium concentration within a distance that is less than the thickness δ_m, which is determined by the hydrodynamics. In this case we use the results of our scaling analysis for the fast reaction regime to obtain an estimate of $N_A a$:

$$N_A \cong -D_{AS}\frac{\partial c_A}{\partial y} \Rightarrow N^*_A \cong -\frac{D_{AS}c_{As}}{N_{As}y_{As}}\frac{\partial c^*_A}{\partial y^*_A} = -\frac{D_{AS}c^{\circ}_A}{N_{As}\delta_r}\frac{\partial c^*_A}{\partial y^*_A} = -\frac{\sqrt{D_{AS}c^{\circ}_A R^m_A}}{N_{As}}\frac{\partial c^*_A}{\partial y^*_A} \tag{6.11-8}$$

To bound N^*_A to be $\text{O}(1)$, we set $\sqrt{D_{AS}c^{\circ}_A R^m_A}/N_{As} = 1$ to obtain $N_{As} = \sqrt{D_{AS}c^{\circ}_A R^m_A}$. Hence, we obtain the following estimate for the mass-transfer rate per unit volume:

$$N_A a = a\sqrt{D_{AS}c^{\circ}_A R^m_A} \tag{6.11-9}$$

Equation (6.11-9) implies that the total absorption rate per unit volume for the fast reaction regime is:

- Proportional to a, the interfacial area per unit volume
- Independent of the liquid-phase holdup ϕa
- Independent of the mass-transfer coefficient $k^{\bullet}_{L0}$
- Proportional to the reaction rate per unit volume R_A
- Proportional to the overall driving force c°_A both explicitly and implicitly through the reaction rate R^m_A
- Dependent on the nonlimiting reactant concentration through the reaction rate R^m_A

The fast reaction regime is distinguished from the other reaction regimes because the hydrodynamics affect the total absorption rate only through a, the interfacial area per unit volume. In general, changing the hydrodynamics (i.e., the Reynolds number) can affect the interfacial area, mass-transfer coefficient, and liquid holdup.

If changing the hydrodynamics does not affect the total absorption rate per unit volume, it indicates that the device is operating in the fast reaction regime. The characteristics of the fast reaction regime imply that laboratory contactors for which the interfacial area per unit volume is known, such as wetted-wall columns, liquid jets, or falling film sphere absorbers, can be used to determine the reaction kinetics.

Consider now the inner domain of the instantaneous reaction regime. For this case we also use the results of our scaling analysis to obtain an estimate of $N_A a$:

$$N_A \cong -D_{AS}\frac{\partial c_A}{\partial y} \Rightarrow N_A^* \cong -\frac{D_{AS}c_{As}}{N_{As}y_{As}}\frac{\partial c_A^*}{\partial y_A^*} = -\frac{D_{AS}c_A^\circ}{N_{As}L}\frac{\partial c_A^*}{\partial y_A^*} = -\frac{k_{L0}^\bullet}{N_{As}}\frac{\hat{c}_B}{\kappa}\sqrt{\frac{D_{BS}}{D_{AS}}}\frac{\partial c_A^*}{\partial y_A^*} \tag{6.11-10}$$

where we have made the substitution $L = (\kappa D_{AS}c_A^\circ/\hat{c}_B)\sqrt{t_o/D_{BS}}$ and used equation (6.3-29) to identify $\sqrt{D_{AS}/t_o}$, with $k_{L0}^\bullet$ the mass-transfer coefficient obtained for unsteady-state physical absorption. To bound N_A^* to be $\circ(1)$, we set $(k_{L0}^\bullet\hat{c}_B/N_{As}\kappa)\sqrt{D_{BS}/D_{As}} = 1$ to obtain $N_{As} = (k_{L0}^\bullet\hat{c}_B/\kappa)\sqrt{D_{BS}/D_{As}}$. Hence, we obtain the following estimate for the mass-transfer rate per unit volume:

$$N_A a = \frac{k_{L0}^\bullet a\hat{c}_B}{\kappa}\sqrt{\frac{D_{BS}}{D_{AS}}} \tag{6.11-11}$$

Equation (6.11-11) implies that the total absorption rate per unit volume for the inner domain of the instantaneous reaction regime is:

- Proportional to a, the interfacial area per unit volume
- Independent of the liquid-phase holdup ϕa
- Proportional to the mass-transfer coefficient $k_{L0}^\bullet$
- Independent of the reaction rate per unit volume R_A
- Independent on the concentration of the absorbing component
- Proportional to $\hat{c}_B$, the driving force of the liquid-phase reactant

Contactors such as packed columns that provide large interfacial area, good hydrodynamics, and maintain a large driving force are appropriate for mass transfer with chemical reaction operating in the inner domain of the instantaneous reaction regime. In contrast, contactors such as stirred tanks are not appropriate.

We did not apply scaling analysis to the surface reaction domain of the instantaneous reaction regime since the mass transfer becomes controlled by the gas phase rather than by the liquid phase. Since the resistance to mass transfer in the liquid phase is negligible and the reaction is instantaneous, the interfacial concentration of the absorbing component is zero. Since the liquid-phase reactant is assumed to be nonvolatile, the gas-phase mass transfer will not involve any chemical reaction.

Hence, the mass-transfer rate per unit volume is determined directly from the gas-phase mass-transfer coefficient $k^{\bullet}_{G0}$ and is given by

$$N_A a = k^{\bullet}_{G0} a \overline{p}_A \tag{6.11-12}$$

in which $\overline{p}_A$ is the partial pressure of the absorbing component A in the gas phase. Equation (6.11-12) implies that the total absorption rate per unit volume for the surface domain of the instantaneous reaction regime is:

- Proportional to a, the interfacial area per unit volume
- Independent of the liquid-phase holdup ϕa
- Proportional to the gas-phase mass-transfer coefficient $k^{\bullet}_{G0}$
- Independent of the reaction rate per unit volume R_A
- Proportional to the gas-phase concentration of the absorbing component
- Independent of the concentration of the liquid-phase reactant

Appropriate contactors are those that provide a large interfacial area such as packed columns.

In this section we have shown that scaling analysis provides valuable insight into selecting the proper contacting equipment and operating conditions for mass transfer with chemical reaction. In particular, scaling analysis permitted us to develop a set of design precepts for each operating regime without the need for solving any of the describing equations.

6.12 MASS-TRANSFER COEFFICIENTS FOR REACTING SYSTEMS

The performance of contacting devices for mass transfer is characterized by correlations for the dimensionless mass-transfer coefficient, that is, the Sherwood number or Nusselt number for mass transfer. In microscale–macroscale modeling this mass-transfer coefficient characterizes the transfer between the micro- and macroscales. We have established that in the slow reaction regime the same mass-transfer coefficient for the particular contacting geometry applies to both mass transfer with and without chemical reaction. However, for the intermediate, fast, and instantaneous reaction regimes the mass-transfer coefficient will be increased from that for purely physical absorption. In this section we demonstrate how the approximations suggested by scaling analysis can be used to determine the mass-transfer coefficient under reaction conditions that influence it.

Since the concentration profiles in the microscale element are influenced by the chemical reaction, the mass-transfer coefficient will be greater than that for purely physical absorption. Unfortunately, it is not possible to obtain a general solution for the mass-transfer coefficient for the intermediate reaction regime. However, we can invoke the steady-state approximation, for which the criterion is given by equation (6.3-22), to obtain a solution for the mass-transfer coefficient for specified reaction kinetics. Let us consider the special case of a zeroth-order irreversible reaction for which $R_A = -k_0$. When the criterion for the applicability of the intermediate reaction regime given by equation (6.5-1) is satisfied, our dimensionless describing equations corresponding to a film theory model for the microscale element reduce to the following for this special case:

$$0 = \frac{d^2 c_A^*}{dy^{*2}} - \frac{k_0 \delta_m^2}{D_{AS} c_A^\circ} = \frac{d^2 c_A^*}{dy^{*2}} - \Psi_r \tag{6.12-1}$$

$$c_A^* = 1 \quad \text{at} \quad y^* = 0 \tag{6.12-2}$$

$$c_A^* = 0 \quad \text{at} \quad y^* = 1 \tag{6.12-3}$$

where $\Psi_r \equiv k_0 \delta_m^2 / D_{AS} c_A^\circ$. The corresponding solution for the mass-transfer flux is given by

$$N_A = k_{L0}^{\bullet} c_A^\circ \left(1 + \frac{\Psi_r}{2}\right) \tag{6.12-4}$$

where $k_{L0}^{\bullet}$ has been introduced via the substitution $D_{AS}/\delta_m = k_{L0}^{\bullet}$. The corresponding mass-transfer coefficient characterizing the intermediate reaction regime then is given by

$$k_L^{\bullet} = k_{L0}^{\bullet} \left(1 + \frac{\Psi_r}{2}\right) \tag{6.12-5}$$

Hence, the mass-transfer coefficient for a zeroth-order reaction in the intermediate reaction regime can be obtained directly from that for purely physical absorption using equation (6.12-5). Note that in the limit of $\Psi_r \to 0$ corresponding to no chemical reaction, $k_L^{\bullet}$ reduces to $k_{L0}^{\bullet}$. Film theory can be used to develop similar relationships between the mass-transfer coefficients in the presence and absence of chemical reaction for other kinetic expressions.

In the case of the fast reaction regime it is possible to obtain a general equation for the mass-transfer coefficient if steady-state can be assumed: that is, if the criterion given by equation (6.3-22) is satisfied. For the fast reaction regime, $y_s = \delta_r$ and $\hat{c}_A = 0$ for an irreversible reaction; this implies that equations (6.3-14) and (6.3-15) assume the form

$$\frac{c_A^\circ}{R_A^m t_o} \frac{\partial c_A^*}{\partial t^*} = \frac{\partial^2 c_A^*}{\partial y^{*2}} + R_A^* \tag{6.12-6}$$

$$\frac{D_{AS} c_A^\circ}{D_{BS} R_A^m t_o} \frac{\partial c_B^*}{\partial t^*} = \frac{\partial^2 c_B^*}{\partial y_B^{*2}} + R_A^* \tag{6.12-7}$$

For all but the shortest contact times, $c_A^\circ / R_A^m t_o \ll 1$ and $D_{AS} c_A^\circ / D_{BS} R_A^m t_o \ll 1$, due to the large value of R_A^m for the fast reaction regime. Hence, the unsteady-state term can be ignored and equations (6.12-6) and (6.12-7) simplify to

$$0 = \frac{d^2 c_A^*}{d y_A^{*2}} + R_A^* \tag{6.12-8}$$

$$0 = \frac{d^2 c_B^*}{d y_B^{*2}} + R_A^* \tag{6.12-9}$$

The boundary conditions on these simplified describing equations for the microscale element are the following:

$$c_A^* = 1 \quad \text{at} \quad y_A^* = 0$$

$$c_B^* = 0 \quad \text{and} \quad \frac{dc_B^*}{dy^*} = 0 \quad \text{at} \quad y_B^* = \sqrt{\frac{\delta_m^2 \kappa R_A^m}{D_{BS}\hat{c}_B}} \tag{6.12-10}$$

$$c_A^* = 0 \quad \text{and} \quad \frac{dc_A^*}{dy^*} = 0 \quad \text{at} \quad y_A^* = \sqrt{\frac{\delta_m^2 R_A^m}{D_{AS} c_A^\circ}} \tag{6.12-11}$$

$$c_B^* = 1 \quad \text{at} \quad y_B^* = 0$$

Note that when the concentration of a reactant is reduced to zero, there is no further diffusion of this reactant, thereby implying that its concentration gradient is zero as well. If the reaction time is very fast, that is, if $\sqrt{\delta_m^2 R_A^m / D_{AS} c_A^\circ} \gg 1$ and $\sqrt{\delta_m^2 \kappa R_A^m / D_{BS}\hat{c}_B} \gg 1$, the boundary conditions containing these dimensionless groups can be applied at infinity. Equation (6.12-8) can be integrated subject to the appropriate boundary conditions given in equations (6.12-10) and (6.12-11) by defining $\varphi \equiv dc_A^*/dy^*$ and substituting $dy^* \equiv dc_A^*/\varphi$ to obtain the following general solution for the mass-transfer flux of component A from the microscale to macroscale in the fast reaction regime:

$$N_A = -\left(2 D_{AS} R_A^m c_A^\circ \int_0^1 R_A^* \, dc_A^*\right)^{1/2} \tag{6.12-12}$$

The corresponding mass-transfer flux of component B can be obtained by integrating equation (6.12-9) in a similar manner or by recognizing that the stoichiometry of the chemical reaction implies

$$N_B = -\kappa N_A \tag{6.12-13}$$

Since we have established that the mass transfer is not affected by the hydrodynamics in the fast reaction regime, N_A is not a function of $k_{L0}^\bullet$. The corresponding mass-transfer coefficient characterizing the fast reaction regime then is given by

$$k_L^\bullet = -\sqrt{\frac{2 D_{AS} R_A^m}{c_A^\circ} \int_0^1 R_A^* dc_A^*} \tag{6.12-14}$$

It is possible to obtain a solution for the mass-transfer coefficient for the inner domain of the instantaneous reaction regime if steady-state can be assumed; that is, if a film theory model is used. In this case the describing equations given by equations (6.7-3) through (6.7-10) simplify to[10]

$$0 = \frac{d^2 c_A}{dy^2} \tag{6.12-15}$$

$$0 = \frac{d^2 c_B}{dy^2} \tag{6.12-16}$$

$$c_A = c_A^{\circ} \quad \text{at} \quad y = 0 \tag{6.12-17}$$

$$c_A = 0, \quad c_B = 0 \quad \text{at} \quad y = L \tag{6.12-18}$$

$$c_B = \hat{c}_B \quad \text{at} \quad y = \delta_m \tag{6.12-19}$$

$$D_{AS}\frac{\partial c_A}{\partial y} = -\frac{D_{BS}}{\kappa}\frac{\partial c_B}{\partial y} \quad \text{at} \quad y = L \tag{6.12-20}$$

The solution to these equations is straightforward and leads to the following equation for the mass-transfer flux:

$$N_A = k_{L0}^{\bullet}\left(c_A^{\circ} + \frac{D_{BS}\hat{c}_B}{D_{AS}\kappa}\right) \tag{6.12-21}$$

The corresponding mass-transfer coefficient characterizing the inner domain of the instantaneous reaction regime is then given by

$$k_L = k_{L0}^{\bullet}\left(1 + \frac{D_{BS}\hat{c}_B}{D_{AS}\kappa c_A^{\circ}}\right) \tag{6.12-22}$$

Hence, the mass-transfer coefficient for the inner domain of the instantaneous reaction regime can be obtained directly from that for purely physical absorption using equation (6.12-22).

The three examples considered in this section are representative of how scaling analysis can be applied to developing models that can be used to obtain mass-transfer coefficients for the design of contacting devices involving mass transfer with chemical reaction. Other models can be developed based on different reaction kinetics and approximations, some of which are explored in the practice problems at the end of the chapter.

6.13 DESIGN OF A CONTINUOUS STIRRED TANK REACTOR

In this section we apply microscale–macroscale scaling analysis to interpret performance data for a continuous stirred tank reactor (CSTR). This CSTR is used to absorb component A from a gas that is bubbled continuously into a liquid that flows into and out of it as shown in Figure 6.13-1. The incoming and outgoing

[10]Scaling analysis can be used to develop the criteria for assuming steady-state for the inner domain of the instantaneous reaction regime; this is considered in Practice Problem 6.P.17.

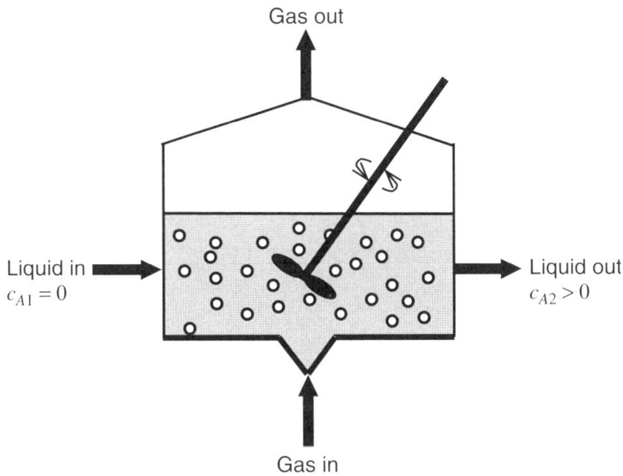

Figure 6.13-1 Continuous stirred tank reactor for chemisorption of a soluble component from a gas that is bubbled through a liquid.

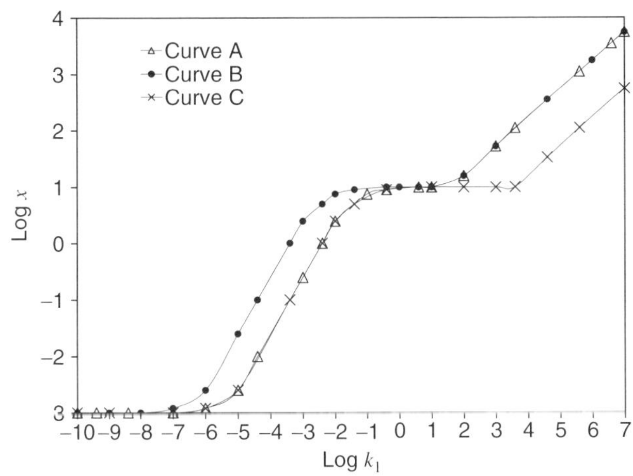

Figure 6.13-2 Log of the ratio of the total absorption rate divided by the maximum physical absorption rate (χ) as a function of the log of the first-order reaction rate constant k_1 for the three sets of system parameter values given in Table 6.13-1.

liquid streams have compositions $c_{A1} = 0$ and $c_{A2} \geq 0$, respectively. The CSTR is assumed to be perfectly mixed so that all the liquid in it has the same composition as the exiting stream. Its total volume is V_T; the volumetric flow rate of the liquid is Q_L; and the interfacial area per unit volume of the CSTR is a. The solute is assumed to be chemisorbed via a first-order irreversible reaction given by $R_A = k_1 c_A$. Figure 6.13-2 shows performance data in the form of a log-log plot of $\chi \equiv N_A a V_T / N_A^m a V_T$, the ratio of the total absorption rate (moles/time) relative to

TABLE 6.13-1 Operating Conditions for the Three CSTR Absorber Performance Curves Shown in Figure 6.13-2

Curve	D_{AS} (cm^2/s)	Mass-Transfer Coefficient for Physical Absorption $\overset{\bullet}{k}_{L0}$ (cm/s)	Contact Time t_o (s)	Interfacial Area per Unit Volume a (cm^2/cm^3)
A	5×10^{-5}	0.04	250	1.0
B	5×10^{-5}	0.04	2500	0.1
C	5×10^{-5}	0.4	250	0.1

its maximum value for purely physical absorption, as a function of the reaction rate constant k_1 for the three sets of operating conditions defined in Table 6.13-1. We use the results of our scaling analysis for the microscale and macroscale elements to explain the shape and characteristics of these curves.

Note that the quasi-stationary hypothesis is rigorous for a CSTR since there is no accumulation for steady-state operation. Moreover, for a CSTR, $\phi a = 1$, since there are no packing elements to occupy volume and the microscale elements (bubbles) are assumed to be mathematical points on the macroscale.

We first need to determine the maximum possible total absorption rate for purely physical absorption. The mass-transfer rate per unit volume, $N_A a$, can be determined from a material balance on the transferring component over the entire volume of the CSTR:

$$Q_L(c_{A2} - c_{A1}) = N_A a V_T \Rightarrow N_A a = \frac{Q_L c_{A2}}{V_T} = \frac{c_{A2}}{t_o} \tag{6.13-1}$$

where t_o is the contact or residence time of the liquid in the CSTR.[11] The maximum possible value of the mass-transfer rate per unit volume for purely physical absorption denoted here by $N_A^m a$ corresponds to the exiting liquid having the thermodynamic equilibrium concentration c_A°; hence,

$$N_A^m a = \frac{Q_L c_A^\circ}{V_T} = \frac{c_A^\circ}{t_o} \tag{6.13-2}$$

Let us first consider the flat portion of all three curves at the lowest values of k_1 for which the mass transfer must be occurring via physical absorption in the absence of any reaction. The relative absorption ratio in the purely physical absorption regime is given by

$$\chi = \frac{N_A a V_T}{N_A^m a V_T} = \frac{c_{A2}}{c_A^\circ} \tag{6.13-3}$$

[11]Note that the control volume here is the total volume of the CSTR rather than the differential slice used in the macroscale balance that led to equations (6.8-1) and (6.8-2). There is an accumulation term (in a convected coordinate system) in the latter that balances the mass transfer from the microscale to the macroscale. However, in the former there is no accumulation term since the mass transfer from the microscale to the macroscale element balances the reaction term.

Figure 6.13-2 indicates that $\chi \cong 1 \times 10^{-3}$ ($\log \chi = -3$) for $k_1 < 1 \times 10^{-8}\ \text{s}^{-1}$ ($\log k_1 = -8$) for curve B and for $k_1 < 1 \times 10^{-7}\ \text{s}^{-1}$ ($\log k_1 = -7$) for curves A and C, implying that the liquid exiting this CSTR is far from thermodynamic equilibrium.

The criterion for purely physical absorption is given by equation (6.8-17), which assumes the following form for first-order kinetics:

$$\frac{R_A^m \phi}{k_{L0}^{\bullet} c_A^{\circ}} = \frac{k_1 c_A^{\circ} \phi}{k_{L0}^{\bullet} c_A^{\circ}} = \frac{k_1}{k_{L0}^{\bullet} a} \cong 0 \qquad (6.13\text{-}4)$$

Table 6.13-1 indicates that the product $k_{L0}^{\bullet} a$ is an order of magnitude smaller for curve B than for curves A and C. Hence, the criterion given by equation (6.13-4) explains why curve B begins deviating from purely physical absorption at a value of $k_1 (\cong 1 \times 10^{-8})$ one order of magnitude smaller than for curves A and C ($\cong 10^{-7}$).

Once the criterion given by equation (6.13-4) is no longer satisfied, χ begins to increase, owing to a transition to the kinetic domain of the slow reaction regime for which the transfer from the microscale to the macroscale is sufficiently fast so that $\hat{c}_A \cong c_A^{\circ}$; hence, equation (6.11-4) implies the following for χ:

$$\chi \equiv \frac{N_A a V_T}{N_A^m a V_T} = \frac{R_A \phi a V_T}{Q_L c_A^{\circ}} = \frac{k_1 V_T}{Q_L} = k_1 t_o \qquad (6.13\text{-}5)$$

where we have used the fact that $\phi a = 1$ for a CSTR. It follows that a log-log plot of χ versus k_1 should be linear with a slope of $+1$ and an intercept of $\log t_o$ when $k_1 = 1 (\log k_1 = 0)$. Figure 6.13-2 indicates that the slopes of all three curves are $+1$ in the linear region (on a log-log plot), but that curve B has an intercept of 3.4 at $k_1 = 1$, corresponding to $t_o = 2500$, whereas curves A and C have an intercept of 2.4 at $k_1 = 1$, corresponding to $t_o = 250$.

As k_1 increases there will be a transition from the kinetic to the diffusional domain of the slow reaction regime for which $\hat{c}_A = c_{Ar} = 0$ for the assumed irreversible kinetics. The criterion for this transition based on our scaling analysis is given by equation (6.10-1) and becomes the following for this example:

$$\frac{R_A^m \phi}{k_{L0}^{\bullet} c_A^{\circ}} = \frac{k_1}{k_{L0}^{\bullet} a} \gg 1 \qquad (6.13\text{-}6)$$

where we have again used the fact that $\phi a = 1$. Table 6.13-1 indicates that the product $k_{L0}^{\bullet} a$ is an order of magnitude smaller for curve B than for curves A and C. Hence, the criterion given by equation (6.13-6) explains why curve B enters the diffusional domain at a value of k_1 ($\cong 1 \times 10^{-1}$) one order of magnitude smaller than for curves A and C ($\cong 1$). In the diffusional domain of the slow reaction regime equation (6.11-5) implies the following for χ:

$$\chi \equiv \frac{N_A a V_T}{N_A^m a V_T} = \frac{k_{L0}^{\bullet} a c_A^{\circ} V_T}{Q_L c_A^{\circ}} = k_{L0}^{\bullet} a t_o \qquad (6.13\text{-}7)$$

Equation (6.13-7) explains why χ is independent of k_1 and equal to 10 ($\log \chi = 1$) for all three curves in the diffusional domain of the slow reaction regime; that is, $\overset{\bullet}{k}_{L0} a t_o = 10$ for all three curves.

Eventually, k_1 will become sufficiently large, so that the chemical reaction can occur on the microscale; that is, there will be a transition to the fast reaction regime for which $\hat{c}_A = c_{Ar} = 0$ for the assumed irreversible kinetics. The criterion for this transition given by equation (6.6-1) becomes the following for this example:

$$\frac{R_A^m \delta_m^2}{D_{AS}(c_A^\circ - \hat{c}_A)} = \frac{k_1 c_A^\circ D_{AS}}{\overset{\bullet}{k}{}_{L0}^2 c_A^\circ} = \frac{k_1 D_{AS}}{\overset{\bullet}{k}{}_{L0}^2} \gg 1 \tag{6.13-8}$$

Table 6.13-1 indicates that the quotient $D_{AB}/\overset{\bullet}{k}{}_{L0}^2$ is two orders of magnitude larger for curves A and B than for curve C. Hence, the criterion given by equation (6.13-8) explains why curves A and B enter the fast reaction regime at a value of $k_1(\cong 10)$ two orders of magnitude smaller than for curve C ($\cong 10^3$). In the fast reaction regime, equation (6.11-9) implies the following for χ:

$$\chi \equiv \frac{N_A a V_T}{N_A^m a V_T} = \frac{a\sqrt{D_{AS} c_A^\circ R_A^m}\, V_T}{Q_L c_A^\circ} = \frac{a t_o \sqrt{D_{AS} c_A^\circ R_A^m}}{c_A^\circ} = a t_o \sqrt{D_{AS} k_1} \tag{6.13-9}$$

It follows that a log-log plot of χ versus k_1 should be linear with a slope of $+\frac{1}{2}$ and an intercept of $a t_o \sqrt{D_{AS}}$ when $k_1 = 1$. Figure 6.13-2 indicates that the slopes of all three curves are $+\frac{1}{2}$ in the linear region (on a log-log plot), but that curve C has an intercept of -0.75 at $k_1 = 1 (\log k_1 = 0)$, corresponding to $a t_o \sqrt{D_{AS}} = 0.177$, whereas curves A and B have an intercept of 0.25 at $k_1 = 1$, corresponding to $a t_o \sqrt{D_{AS}} = 1.77$.

This example demonstrates that systematic scaling analysis can provide considerable insight into interpreting performance data for chemisorption as well as other contacting devices that involve transport phenomena. Indeed, we were able to obtain quantitative estimates of where the various reaction regimes apply as well as the reaction rates in each of these regimes without the need to solve any differential equations.

6.14 DESIGN OF A PACKED COLUMN ABSORBER

In this section we apply the results of our microscale–macroscale scaling analysis in order to design a packed gas-absorption column. This contacting device, shown schematically in Figure 6.14-1, involves the use of small packing elements such as spheres, hollow cylinders, saddle-shaped particles, and other specially designed structures to provide a large contact area between the liquid and gas streams. The liquid flows over the packing elements cocurrently or countercurrently relative to an upward-flowing gas stream from which it absorbs some soluble component. In this example a solute A is to be removed from a nonabsorbing gas by cocurrent absorption in a liquid solvent S. The solvent S contains an active component B

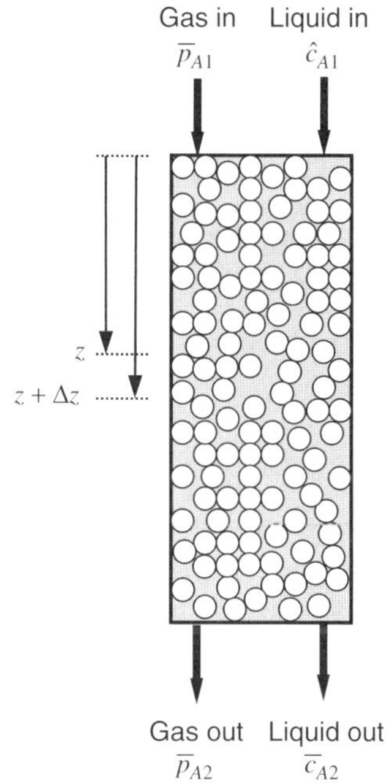

Figure 6.14-1 Packed column for absorbing a solute A from a nonabsorbing gas by cocurrent absorption in a liquid solvent S that contains an active component B which reacts with A by a first-order irreversible reaction.

TABLE 6.14-1 Performance Data for a Packed Gas-Absorption Column

Gas molar flow rate, W_G	0.3 mol/s
Solute concentration in inlet liquid, $\hat{c}_{A1}$	0
Solute partial pressure in inlet gas, $\overline{p}_{A1}$	1.0×10^4 Pa
Solute partial pressure in outlet gas, $\overline{p}_{A2}$	5 Pa
Reaction rate constant, k_1	$1.0 \times 10^3\ s^{-1}$
Liquid holdup, ϕ	$1 \times 10^{-3}\ m^3/m^2$
Interfacial area per unit volume, a	$10\ m^2/m^3$
Liquid-phase mass-transfer coefficient for physical absorption, $k^{\bullet}_{L0}$	1.0×10^{-4} m/s
Effective diffusivity of solute in liquid, D_{AS}	$1.2 \times 10^{-9}\ m^2/s$
Henry's law constant, H	$70\ Pa \cdot m^3/mol$
Column pressure, P	1×10^5 Pa
Column cross-sectional area, S_c	$0.07\ m^2$

that reacts with component A, which is assumed to be the limiting reactant, by a first-order irreversible reaction. Performance data for this absorber are given in Table 6.14-1.

Let us first determine the ratio of the characteristic diffusion time to that for chemical reaction to assess the reaction regime in which this chemisorption process occurs:

$$\frac{R_A^m \delta_m^2}{D_{AS}(c_A^\circ - \hat{c}_A)} = \frac{k_1 c_A^\circ D_{AS}}{k_{L0}^{\bullet 2}(c_A^\circ - 0)} = \frac{(1.0 \times 10^3\ \mathrm{s}^{-1})\left(1.2 \times 10^{-9}\ \mathrm{m}^2/\mathrm{s}\right)}{\left(1.0 \times 10^{-4}\ \mathrm{m/s}\right)^2} = 120 \tag{6.14-1}$$

In evaluating the above, we have estimated the diffusion penetration thickness on the microscale using film theory; that is, $\delta_m = D_{AS}/k_{L0}^{\bullet}$. The criterion given by equation (6.6-1) then indicates that this chemisorption occurs in the fast reaction regime, for which $\hat{c}_A = c_{Ar} = 0$ for the assumed irreversible reaction and the mass-transfer flux is given by equation (6.12-12); that is,

$$N_A = -\left(2D_{AS} R_A^m c_A^\circ \int_0^1 R_A^*\, dc_A^*\right)^{1/2} = -(D_{AS} k_1)^{1/2} c_A^\circ \tag{6.14-2}$$

For the fast reaction regime the mass-transfer coefficient for transfer from the microscale to the macroscale is greater than that for purely physical absorption and can be determined via equation (6.12-14):

$$k_L^\bullet = -\frac{\left(2D_{AS} R_A^m c_A^\circ \int_0^1 R_A^* dc_A^*\right)^{1/2}}{c_A^\circ} = (D_{AS} k_1)^{1/2} \tag{6.14-3}$$

$$= \left[\left(1.2 \times 10^{-9}\ \mathrm{m}^2/\mathrm{s}\right)\left(1.0 \times 10^3\ \mathrm{s}^{-1}\right)\right]^{1/2} = 1.1 \times 10^{-3}\ \mathrm{m/s}$$

We see that the mass-transfer coefficient for chemisorption in the fast reaction regime is an order of magnitude larger than that for purely physical absorption.

Now let us determine the height of the absorption column required to achieve the given change in the gas-phase composition. Note that no data are given for the change in the liquid-phase composition. This is because this reaction occurs in the fast reaction regime, which keeps the bulk concentration of component A on the macroscale equal to the reaction equilibrium concentration, which is zero for the assumed irreversible kinetics. Under fast reaction conditions, a material balance in the liquid phase on the transferring component provides no information whatsoever. One could determine the column height from a knowledge of the increase in concentration of the product of the chemisorption reaction in the liquid phase. However, no information is provided on this component. Hence, to determine the column height, we will carry out a material balance on the transferring component in the gas phase. A species balance on component A in the gas phase for a macroscale element consisting of a differential length Δz in a fixed coordinate system yields

$$N_A a S_c \Delta z = \left.\frac{W_G \overline{p}_A}{P}\right|_z - \left.\frac{W_G \overline{p}_A}{P}\right|_{z+\Delta z}$$

$$k_L^\bullet a S_c c_A^\circ = \frac{k_L^\bullet a S_c \overline{p}_A}{H} = -\frac{W_G}{P}\frac{d\overline{p}_A}{dz} \tag{6.14-4}$$

where W_G is the molar flow rate of the gas. Integrating equation (6.14-4) over the length of the column then yields the following for the absorption column height required:

$$L = -\frac{W_G H}{P k_L^{\bullet} a S_c} \ln \frac{\overline{p}_{A1}}{\overline{p}_{A2}}$$

$$= -\frac{(0.3 \text{ mol/s})\left(70 \text{ Pa} \cdot \text{m}^2/\text{mol}\right)}{(1 \times 10^5 \text{ Pa})\left(1.1 \times 10^{-3} \text{ m/s}\right)\left(10 \text{ m}^2/\text{m}^3\right)(0.07 \text{ m}^2)} \ln \frac{5}{1 \times 10^4} = 2.1 \text{ m} \tag{6.14-5}$$

Note that the liquid-phase concentration does not enter into either equation (6.14-2) for the mass-transfer flux or equation (6.14-5) for the column height. This is because the chemisorption is in the fast reaction regime that maintains the liquid at the reaction equilibrium concentration. One implication of this is that the column height required, determined by equation (6.14-5), would not change if we were to employ countercurrent rather than cocurrent flow. However, under fast reaction conditions one would choose cocurrent flow because of the reduced pressure drop requirement. Moreover, flooding of the packed column owing to the gas holding up the liquid in countercurrent flow is avoided in cocurrent flow. Hence, we see that scaling the describing equations to determine the reaction regime is pivotal in choosing both the contacting device and the flow configuration.

Now let us consider how we can use the results of our scaling analysis to improve the performance of this packed absorption column. In doing this we refer to the design precepts developed in Section 6.11. Consider first an increase in the liquid flow rate. This will improve the performance of this packed column operating in the fast reaction regime only if it increases a, the interfacial area per unit volume; this increase will not be significant. However, a larger increase in a could be achieved by changing the type or size of packing elements. Decreasing the gas flow rate will decrease the required column height since less of the transferring solute will enter the column. This will also cause an increase in the bubble size and therefore a decrease in a; this will cause a small decrease in the mass-transfer rate per unit volume. Increasing P, the operating pressure, or decreasing the temperature in the column will improve the performance by increasing the thermodynamic equilibrium solubility of the transferring solute. Increasing k_1, the reaction-rate constant, will also improve the performance significantly. This could be done by increasing the operating temperature or employing a different reacting component in the liquid. Hence, changing the temperature could either increase or decrease the performance, depending on whether its influence on the equilibrium solubility or the reaction-rate constant dominates.

6.15 SUMMARY

In this chapter we focused on the complications that are encountered in processes involving chemical reaction that often have mass transfer occurring on two markedly different length scales. In particular, we considered chemisorption whereby the solubility of a transferring solute is enhanced by chemical reaction. The microscale–macroscale modeling methodology developed in this chapter for mass transfer with chemical reaction can be applied equally well to other processes involving momentum, heat, or mass transfer occurring on multiple scales, such as phase-transition phenomena, heat transfer in dispersed phase systems, and others.

In Section 6.2 we provided an example illustrating microscale and macroscale elements. An important consideration is that for mass transfer with chemical reactions the microscale element becomes a homogeneous point source or sink term in the macroscale balance.

Scaling analysis was applied to the microscale species-balance equations for mass transfer in Section 6.3. Scaling led to three time scales associated with the diffusion, reaction, and contact times. When the ratio of the diffusion to contact time was very small, steady-state mass transfer can be assumed for the microscale element for which film theory models are applicable. When this ratio is large, the mass transfer is inherently unsteady-state, for which penetration theory models can be used. For purely diffusive mass transfer a film theory model interrelated the mass-transfer coefficient and the microscale fluid layer thickness, whereas a penetration theory model interrelated it to the contact time.

Scaling analysis was used in Section 6.4 to develop a criterion for the slow reaction regime for which the ratio of the diffusion to the reaction time was very small for the microscale element. Since the chemical reaction does not occur in the microscale element, the mass-transfer coefficient for the slow reaction regime is the same as for purely physical absorption.

In Section 6.5, scaling analysis was used to identify a criterion for the intermediate reaction regime for which the diffusion and reaction times are of equal magnitude. For this reaction regime the chemical reaction occurs on the microscale, which implies that the mass-transfer coefficient is greater than that for purely physical absorption.

In Section 6.6, scaling analysis was used to develop the criterion for the fast reaction regime for which the reaction time is sufficiently fast to maintain the bulk liquid at the reaction equilibrium concentration. In the fast reaction regime a reaction boundary layer or region of influence exists within the microscale liquid layer, within which the concentration of the absorbing component undergoes a characteristic change.

In Section 6.7 we considered the criterion for the instantaneous reaction regime for which the reaction is so fast that the reacting components cannot co-exist at any point within the microscale element. This regime implies that the nonvolatile liquid-phase reactant becomes rate-limiting, owing to its depletion in the microscale liquid layer. The inner domain of the instantaneous reaction regime corresponds to a reaction plane within the liquid layer of the microscale element at which the concentrations of the reacting components are reduced to zero. The surface domain of the instantaneous reaction regime corresponds to the reaction plane being at the gas–liquid interface, for which the mass transfer becomes gas-phase controlled.

The describing equations for the macroscale element for which the transfer from the microscale element appears as a homogeneous source term were scaled in Section 6.8. This macroscale element is relevant only for the slow or intermediate reaction regimes, for which the reaction is not sufficiently fast to maintain the bulk liquid at the reaction equilibrium concentration. Scaling led to three time scales associated with the mass transfer from the microscale, reaction, and macroscale contact times.

In Section 6.9 we considered the kinetic domain of the slow reaction regime. In this domain the mass transfer from the microscale is sufficiently fast to maintain the bulk liquid at the thermodynamic equilibrium concentration.

In Section 6.10 we considered the diffusional domain of the slow reaction regime. In this domain the chemical reaction is sufficiently fast on the macroscale to maintain the bulk liquid at the reaction equilibrium concentration.

The results of scaling analysis for the various reaction regimes were used in Section 6.11 to develop a set of precepts for the design of contacting equipment for mass transfer with chemical reaction. The power of scaling analysis was particularly apparent in this section since it provided invaluable information for choosing and operating mass-transfer contacting devices without the need for developing any models or solving any differential equations.

The mass-transfer coefficients correlated in the literature can be used to design contacting devices only for the slow reaction regime. For faster kinetics, microscale models need to be developed to determine the mass-transfer coefficients. In Section 6.12, film theory was used to demonstrate how a model can be developed for the mass-transfer coefficients in the intermediate, fast, and instantaneous reaction regimes.

The results of scaling analysis were used in Section 6.13 to design a mass transfer process involving a first-order irreversible reaction in a CSTR. In particular, it was used to interpret performance data for the increase in rate of chemisorption as a function of the reaction-rate constant. Nearly all features of the complex performance curves for three representative sets of operating conditions were explained using the scaling analysis results without the need to solve any differential equations.

In Section 6.14, the results of scaling analysis were used to design a continuous packed gas-absorption column. Scaling permitted determining how the various process parameters affected the performance of this device. Again, emphasis was placed on how much information could be obtained from scaling analysis without the need for developing and solving complex models.

6.P PRACTICE PROBLEMS

6.P.1 Criterion for Ignoring the Gas-Phase Resistance to Mass Transfer

Throughout this chapter, with the exception of the surface domain of the instantaneous reaction regime, we have assumed that the resistance offered to mass transfer by the gas phase was negligible in comparison to that of the liquid phase. Assume that the gas-phase mass transfer rate per unit volume $N_A a$ is described by

$$N_A a = k_{G0}^{\bullet} a \overline{p}_A \tag{6.P.1-1}$$

in which $k_{G0}^{\bullet}$ is the gas-phase mass-transfer coefficient, a the interfacial area per unit volume of contacting device, and $\overline{p}_A$ the partial pressure of the absorbing component A in the gas phase. The thermodynamic equilibrium liquid-phase composition c_A° is related to the partial pressure $\overline{p}_A$ by Henry's law given by $\overline{p}_A = Hc_A^{\circ}$, where H is the Henry's law constant. Use scaling analysis to develop a criterion for ignoring the gas-phase relative to the liquid-phase mass transfer.

6.P.2 Penetration Theory Model for a Slow Reaction Regime

In Section 6.3 we scaled the definition of the mass-transfer flux for purely physical unsteady-state absorption and obtained equation (6.3-29) that interrelates the mass-transfer coefficient and the contact time. In principle, scaling analysis can provide only an order-of-magnitude estimate of any quantity such as the mass-transfer coefficient.

(a) Develop a penetration theory model to obtain a rigorous solution for the interrelationship between the mass-transfer coefficient and the contact time for unsteady-state purely physical absorption.
(b) Compare the result you obtained in part (a) with that obtained via scaling analysis.

6.P.3 Correlation for a Liquid-Phase Mass-Transfer Coefficient

Sherwood and Holloway provide the following correlation for the liquid-phase mass-transfer coefficient for packed gas-absorption columns $k_{L0}^{\bullet}$ (having dimensions of L^3/t)[12]:

$$\frac{k_{L0}^{\bullet} a}{D_{AS}} = \left(\frac{\varsigma U_0}{\mu}\right)^{1-n} \left(\frac{\mu}{\rho D_{AS}}\right)^{0.5} \tag{6.P.3-1}$$

where a is the interfacial area per volume of contacting device, D_{AS} the effective binary diffusion coefficient, U_0 the superficial velocity of the liquid flow, μ the shear viscosity of the liquid, ρ the mass density of the liquid, and ς and n are dimensional (M/L^2) and dimensionless constants, respectively, that depend on the type and size of the packing elements.

(a) Compare the estimate that we obtained from scaling analysis using the film theory model given by equation (6.3-25) with the correlation of Sherwood and Holloway.

[12] T. K. Sherwood and F. A. L. Holloway, *Trans. A.I.Ch.E.*, **36**, 21, 39, 181 (1940).

(b) Compare the estimate that we obtained from scaling analysis using the penetration theory model given by equation (6.3-29) with the correlation of Sherwood and Holloway.

6.P.4 Slow Reaction Regime for Nondilute Solutions

Throughout this chapter we have assumed dilute solutions of the absorbing component so that the convective contribution to Fick's law could be ignored. Assume now that the solutions are sufficiently concentrated that the convective term in Fick's law cannot be neglected; that is,

$$N_A = -D_{AS}c\frac{\partial x_A}{\partial z} + c_A\hat{u} = N_A = -D_{AS}c\frac{\partial x_A}{\partial z} + x_A(N_A + N_B + N_S) \quad (6.P.4\text{-}1)$$

in which $\hat{u}$ is the molar-average velocity and x_A is the mole fraction of the absorbing component A.

(a) Scale the microscale element to determine the criterion for the applicability of the slow reaction regime; assume that the liquid-phase component S is nonvolatile, which implies unimolecular diffusion of component A in the microscale element.
(b) Determine the criteria for ignoring the convective term in Fick's law.

6.P.5 Microscale Element Scaling for a Reversible Unimolecular Reaction

In Section 6.3 we scaled the describing equations for the microscale element for an irreversible bimolecular first-order reaction. Now let us consider the situation where solute A is chemisorbed in the liquid S via a reversible unimolecular reaction given by $R_A = k_1(c_A - c^{\circ}_{Ar})$, where c°_{Ar} is the reaction equilibrium concentration; that is, the concentration that is achieved when the reaction goes to completion.

(a) Scale the describing equations for the microscale element to determine the relevant time scales.
(b) Determine the criterion for the slow reaction regime.
(c) Determine the criterion for the intermediate reaction regime.
(d) Determine the criterion for the fast reaction regime.
(e) Determine the criterion for the instantaneous reaction regime.
(f) Determine the criterion for steady-state mass transfer on the microscale.
(g) Estimate the thickness of the region of influence for very short contact times.

6.P.6 Microscale Element Scaling for an Irreversible *n*th-Order Reaction

In Section 6.3 we scaled the describing equations for the microscale element for an irreversible bimolecular first-order reaction. Now let us consider the situation where solute A is chemisorbed in the liquid S via an irreversible nth-order reaction given by $R_A = k_n c_A^n$.

(a) Scale the describing equations for the microscale element to determine the relevant time scales.
(b) Determine the criterion for the slow reaction regime.
(c) Determine the criterion for the intermediate reaction regime.
(d) Determine the criterion for the fast reaction regime.
(e) Determine the criterion for the instantaneous reaction regime.
(f) Determine the criterion for quasi-steady-state.
(g) Estimate the thickness of the region of influence for very short contact times.

6.P.7 Applicability of the Quasi-stationary Hypothesis for Physical Absorption

In Section 6.8 we considered the quasi-stationary hypothesis whereby the accumulation term in the macroscopic balance is assumed to be very small. We then invoked this assumption to explore the implications of a very slow and very fast reaction time relative to the time scale for interphase mass transfer. This led to identifying the kinetic and diffusional domains of the slow reaction regime. Consider the applicability of the quasi-stationary hypothesis for the macroscale balance given by equation (6.8-11) for purely physical absorption for which there is no chemical reaction. Discuss any anomalies that result from the application of the quasi-stationary hypothesis and their implications for the applicability of the quasi-stationary hypothesis.

6.P.8 Effect of Axial Dispersion in the Macroscale Balance for a Packed Absorption Column

In carrying out the macroscale balance in Section 6.8, we did not include the term associated with axial dispersion of species. In this problem we include this term in the macroscale balance for a packed gas-absorption column to develop a criterion for when it can be neglected. We assume that the axial dispersion is characterized by a constant dispersion coefficient D_L that accounts for both axial diffusion due to the concentration gradients and dispersion due to the presence of the packing elements.

(a) Use scaling analysis to develop a criterion for ignoring the axial dispersion term. Scaling the macroscale balance equation for the packed absorption column is facilitated by recasting it in a fixed coordinate rather than a convected coordinate system.
(b) In Chapter 6 we found that the axial diffusion term could not be ignored within a region of influence whose thickness we estimated using scaling analysis. Use scaling analysis to estimate the thickness of the region of influence wherein the axial dispersion term cannot be ignored.

6.P.9 Intermediate Domain of a Slow Reaction Regime

In the slow reaction regime we considered two limiting cases, corresponding to the kinetic and diffusional domains that applied when the reaction rate was very slow and very fast, respectively, in comparison to the mass transfer from the microscale to the macroscale. There must also be an intermediate domain between these two limiting cases for which the characteristic times for reaction and mass transfer from the microscale to the macroscale are comparable.

(a) Determine the criterion for applicability of this intermediate domain of the slow reaction regime.
(b) Discuss the implications of this intermediate domain of the slow reaction regime for reactor design; that is, indicate how the various process parameters will affect the performance in this domain.
(c) Contrast the equation that you obtained for the mass-transfer rate per unit volume for the intermediate reaction domain with those that were obtained for the kinetic and diffusional domains of the slow reaction regime.

6.P.10 Macroscale Element Scaling for a Reversible Unimolecular Reaction

Consider the describing equations for the macroscale element for the reversible unimolecular reaction described in Practice Problem 6.P.5 for the special case of the slow reaction regime.

(a) Determine the criterion for the applicability of the quasi-stationary hypothesis.
(b) Determine the criterion for the kinetic domain.
(c) Determine the criterion for the intermediate domain. *Note*: See Practice Problem 6.P.9 regarding the intermediate domain of the slow reaction regime.
(d) Determine the criterion for the diffusional domain.

6.P.11 Macroscale Element Scaling for an Irreversible *n*th-Order Reaction

Consider the describing equations for the macroscale element for the irreversible *n*th-order reaction unimolecular reaction described in Practice Problem 6.P.6 for the special case of the slow reaction regime.

(a) Determine the criterion for the applicability of the quasi-stationary hypothesis.
(b) Determine the criterion for the kinetic domain.
(c) Determine the criterion for the intermediate domain. *Note*: See Practice Problem 6.P.9 regarding the intermediate domain of the slow reaction regime.
(d) Determine the criterion for the diffusional domain.

6.P.12 Implications of the Intermediate Reaction Regime for a Microscale Element on the Macroscale Balance

The slow reaction regime differs from the fast reaction regime in that the mass-transfer coefficient is the same as that for physical absorption in the former. Moreover, in the fast reaction regime the reaction is fast enough to maintain the bulk fluid at the reaction equilibrium concentration. Consider now the intermediate reaction regime for the microscale element. The reaction is sufficiently fast so that the mass-transfer coefficient is greater than that for purely physical absorption. However, it is not necessarily fast enough to maintain the bulk fluid at the reaction equilibrium concentration. In this problem we consider the implications of the intermediate reaction regime for the macroscale element describing equations.

(a) Scale the describing equations for the macroscale element for which the mass-transfer term corresponds to the intermediate reaction regime; in particular, identify the three time scales appropriate to this scaling.
(b) Consider the applicability of the kinetic domain; if this domain is possible, develop a criterion for its occurrence and contrast the criteria for this domain for the slow and intermediate reaction regimes.
(c) Consider the applicability of the diffusional domain; if this domain is possible, develop a criterion for its occurrence and contrast the criteria for this domain for the slow and intermediate reaction regimes.
(d) Based on your results in parts (b) and (c), discuss critically how the slow and intermediate reaction regimes differ on both the micro- and macroscales.

6.P.13 Comparison of the Fast Reaction Regime and the Diffusional Domain of the Intermediate Reaction Regime

Consider the intermediate reaction regime for which the reaction is not necessarily fast enough to maintain the bulk fluid at the reaction equilibrium concentration. In addition, the mass-transfer coefficient is not equal to that for purely physical absorption. When the macroscale balance is considered for the intermediate reaction regime, it is possible that it will admit both a kinetic and a diffusional domain for conditions similar to those determined by scaling for the slow reaction regime. If the intermediate reaction regime admits a diffusional domain, it would imply that the bulk fluid concentration is equal to the reaction equilibrium concentration. Hence, for both the fast reaction regime and the diffusional domain of the intermediate reaction regime, the mass-transfer coefficient is not that for purely physical absorption and the bulk-fluid concentration is the reaction equilibrium concentration. Superficially, these two reaction regimes appear to be the same when, in fact, they are not. Discuss critically how the diffusional domain of the intermediate reaction regime differs from the fast reaction regime.

6.P.14 Discriminating Between the Fast Reaction Regime and the Kinetic Domain of the Slow Reaction Regime

In Section 6.11 we found for both the fast reaction regime and the kinetic domain of the slow reaction regime that the mass-transfer rate per unit volume of reactor was independent of the mass-transfer coefficient for physical absorption $k^{\bullet}_{L0}$, which is strongly dependent on the hydrodynamics. Moreover, in Section 6.11 we stated that if one observes that the contactor performance is independent of the hydrodynamics, it indicates that the chemisorption is occurring in the fast reaction regime. Critically assess the validity of the latter statement based on the design precepts for the two reaction regimes discussed in Section 6.11; that is, it would appear that if neither of these two reaction regimes depends on $k^{\bullet}_{L0}$, which depends on the hydrodynamics, how in fact can we make the claim that if the contactor performance is independent of the hydrodynamics, it is operating in the fast reaction regime?

6.P.15 Transition Between the Inner and Surface Domains in the Instantaneous Reaction Regime for Gas Absorption

The instantaneous reaction regime for gas absorption with a bimolecular reaction implies that the chemical reaction is so fast that the two reactants cannot coexist. Hence, the instantaneous reaction is confined to a plane that separates two regions, in each of which only one of the reactants is diffusing. In the inner reaction domain of the instantaneous reaction regime, this reaction plane is in the liquid phase, whereas in the surface domain it is at the interface between the gas and liquid phases. In contactors such as packed columns, the reaction regime can change along the length of the column due to changes in the concentrations that affect the reaction rate. In this problem we explore both the criterion and its implications for chemisorption in a packed gas absorption column. We assume that the gas-phase partial pressure of the absorbing component is $\overline{p}_A$ and that the absorption equilibrium at the gas–liquid interface is defined by $\overline{p}_A = Hc^{\circ}_A$, where H is a constant.

(a) Use the results of our scaling analyses for the inner and surface domains of the instantaneous reaction regime to develop a criterion for the transition between these two domains. Note that for the inner reaction domain the mass-transfer rate per unit volume in the gas phase must be equal to that in the liquid phase.

(b) Assume that the gas and liquid flow cocurrently down a packed gas absorption column operation in the instantaneous reaction regime. If a transition between the inner and surface domains occurs in this column, indicate where each domain will occur relative to the top of the column.

6.P.16 Fast Reaction Regime for an *n*th-Order Reaction

Consider the fast reaction regime for the special case of an nth-order reversible reaction given by $R_A = k_n(c_A - c^{\circ}_{Ar})^n$, in which c°_{Ar} is the reaction equilibrium concentration of the absorbing component.

(a) Integrate equation (6.12-12) for this nth-order reaction to determine the concentration profile for the microscale element.

(b) Use the result of part (a) to derive an equation for the thickness of the region required for the concentration to decrease from its maximum value of c_A° to its minimum value for the fast reaction regime of c_{Ar}°.

(c) For $n < 1$ the thickness you calculated in part (b) is finite, whereas for $n > 1$, it is infinite; reconcile this with the result of scaling analysis for the thickness of the region of influence for the fast reaction regime given by equation (6.6-2), which predicts a finite thickness for any value of n. *Hint*: Use your result from part (a) in a film theory model to estimate the effective thickness.

6.P.17 Steady-State Approximation for the Inner Domain of the Instantaneous Reaction Regime

Use scaling analysis to develop criteria for assuming steady-state for the inner domain of the instantaneous reaction regime; that is, develop criteria for using a film theory model to describe this reaction regime.

6.P.18 Improved Model for the Inner Domain of the Instantaneous Reaction Regime

The steady-state approximation will apply for the inner domain of the instantaneous reaction regime only for very long contact times for which the diffusive penetration of the liquid-phase reactant extends from the reaction plane to the outer edge of the microscale element. An improved model for this reaction regime can be developed by recognizing that the reaction plane will be very close to the gas–liquid interface if the contact time is not too long. Hence, quasi-steady-state conditions can be assumed for the region in which the absorbing component is diffusing, whereas the unsteady-state term is retained in the region in which the liquid-phase reactant is diffusing.

(a) Use scaling analysis to develop the criterion for assuming quasi-steady-state conditions in the region wherein the absorbing component is diffusing.

(b) Develop an analytical solution for the mass-transfer flux and mass-transfer coefficient, assuming quasi-steady-state in the region where the absorbing component is diffusing but unsteady-state in the region where the liquid-phase reactant is diffusing. Note that the differential equation in the unsteady-state region can be solved using either the method of combination of variables or Laplace transforms.

(c) Compare the result you obtained in part (b) for the mass-transfer coefficient to that obtained for the steady-state film theory model developed in Section 6.12.

6.P.19 Model for the Mass-Transfer Coefficient for a First-Order Reversible Reaction in the Intermediate Reaction Regime

In Section 6.12 we developed a film theory model to obtain an equation for the mass-transfer coefficient for a zeroth-order reaction. Develop a film theory model for the mass-transfer coefficient appropriate to a first-order reversible reaction given by $R_A = k_n(c_A - c_{Ar}^{\circ})$, in which c_{Ar}° is the reaction equilibrium concentration of the absorbing component.

6.P.20 Model for the Mass-Transfer Coefficient for a First-Order Reversible Reaction in the Fast Reaction Regime

In Section 6.12 we developed a film theory model to obtain an integral equation for the mass-transfer coefficient for the fast reaction regime with unspecified reaction kinetics.

(a) Develop a film theory model for the mass-transfer coefficient appropriate to a first-order reversible reaction for which $R_A = k_n(c_A - c_{Ar}^{\circ})$, in which c_{Ar}° is the reaction equilibrium concentration of the absorbing component.

(b) Compare the result you obtained in part (a) with that given by equation (6.12-14); in particular, discuss any anomalies that are observed when you make this comparison.

6.P.21 Macroscale Element Scaling for a Zeroth-Order Reversible Reaction

In Section 6.12 we developed an equation for the mass-transfer coefficient for an irreversible zeroth-order reaction in the intermediate reaction regime. In this problem we will consider the slow reaction regime for a zeroth-order reversible reaction for which the reaction rate per unit volume R_A is given by

$$R_A = k_0\xi \quad \begin{cases} \xi = 0 & \text{for } \hat{c}_A \leq c_{Ar}^{\circ} \\ \xi - 1 & \text{for } \hat{c}_A > c_{Ar}^{\circ} \end{cases} \tag{6.P.21-1}$$

(a) Scale the macroscale element for the slow reaction regime for this reversible zeroth-order reaction and determine the criterion for the kinetic domain.

(b) Derive an equation for the mass-transfer rate per unit volume $N_A a$ for the kinetic domain.

(c) Compare the design precepts for the kinetic domain for the special case of the reversible zeroth-order reaction to those implied by equation (6.11-4) for unspecified irreversible reaction kinetics.

(d) Scale the macroscale element for the slow reaction regime for this reversible zeroth-order reaction and determine the criterion for the diffusional domain.

(e) Derive an equation for the mass-transfer rate per unit volume $N_A a$ for the diffusional domain.

(f) Compare the design precepts for the diffusional domain for the special case of the reversible zeroth-order reaction to those implied by equation (6.11-5) for unspecified irreversible reaction kinetics.

(g) Discuss whether an intermediate domain of the slow reaction regime is possible for a zeroth-order reversible reaction.

(h) Discuss the nature of the transition between the kinetic and diffusional domains of the slow reaction regime for a gas-absorption column employing chemisorption with a zeroth-order reversible reaction.

6.P.22 Design of a CSTR for an *n*th-Order Reversible Reaction Operating in the Slow Reaction Regime

In Section 6.13 we considered the design of a CSTR for a first-order irreversible reaction. In this problem we consider the design of a CSTR such as that shown in Figure 6.13-1 for the special case of an *n*th-order reversible reaction operating in the slow reaction regime for which the reaction rate per unit volume is given by $R_A = k_n(c_A - c^{\circ}_{Ar})^n$.

(a) Derive an equation for the total absorption rate $N_A a V_T$ if the chemisorption occurs in the kinetic regime of the slow reaction regime.

(b) Derive an equation for the total absorption rate $N_A a V_T$ if the chemisorption occurs in the diffusional regime of the slow reaction regime.

(c) Discuss the relative merits of using a CSTR if the chemisorption occurs in the slow reaction regime.

6.P.23 Determining Kinetic Parameters for a CSTR

In Section 6.13 we considered how the results of our scaling analyses could be used to interpret performance data for chemisorption through a first-order irreversible reaction in a CSTR. In this problem we explore how performance data for chemisorption in a CSTR can be used to determine the kinetic parameters. We assume that the chemisorption occurs via a first-order irreversible reaction for which the reaction rate constant is k_1. Moreover, we assume that the CSTR liquid volume V_T, diffusion coefficient D_{AS}, thermodynamic equilibrium concentration c°_A, and inlet liquid concentration c_{A1} are known. Let us assume that we can vary both the contact time t_o by changing the volumetric flow rate Q_L, and the mixing speed; the latter will change both $k^{\bullet}_{L0}$ and a. Of course, a change in either the contact time or mixing speed will change the outlet liquid concentration c_{A2}, which we assume can be measured. We wish to design a series of experiments to determine the reaction regime and the following design parameters for the CSTR: k_1, the reaction rate constant; $k^{\bullet}_{L0}$, the mass-transfer coefficient for physical absorption; and a, the interfacial area per unit volume of contacting device.

(a) Assume that we increase the contact time t_o and find that $\chi = N_A/N^m_A > 1$ and increases linearly with t_o. This can occur for the kinetic and diffusional domains of the slow reaction regime as well as for the fast reaction regime. If we hold the contact time constant, what additional experiment(s) can be done by changing the process parameters under our control to determine if the chemisorption is occurring in the kinetic domain?

(b) Assume in part (a) that we establish that the chemisorption is in the kinetic domain. What information associated with the kinetic parameters can we determine from a measurement of the outlet liquid concentration c_{A2}?

(c) What is the particular advantage of operating in the kinetic domain for characterizing the reaction kinetics?

6.P.24 Determining the Interfacial Area per Unit Volume for a CSTR

In designing a CSTR for chemisorption, one would like to know $k^{\bullet}_{L0}$, the mass-transfer coefficient for purely physical absorption, and a, the interfacial area per unit volume of contacting device, as functions of the mixing speed or equivalently, the Reynolds number based on the impeller diameter. In this problem we consider how experiments might be designed for a CSTR that will permit determining both of these parameters. We assume that the CSTR liquid volume V_T, diffusion coefficient D_{AS}, thermodynamic equilibrium concentration c°_{A}, and inlet liquid concentration c_{A1} are known and that the liquid volumetric flow rate Q_L and outlet liquid concentration c_{A2} can be measured. We assume that we have the option of operating in either purely physical absorption or chemisorption by adding an excess of an appropriate liquid-phase reactant for which the first-order irreversible reaction rate constant is k_1.

(a) Assume that we have done a series of experiments for physical absorption in which we have measured the outlet liquid concentration c_{A2} as we progressively increased the mixing speed. What can these data tell us about the parameters $k^{\bullet}_{L0}$ and a?

(b) Assume now that we have the parameters obtained in part (a) and that we add the liquid-phase reactant so that chemisorption occurs. What types of experiments could we carry out to determine definitively in which reaction regime the chemisorption occurs?

(c) Assume that we determine that the chemisorption is occurring in the fast reaction regime. What can data from a series of experiments in which we measure the outlet liquid concentration c_{A2} as a function of mixing speed tell us about the parameters $k^{\bullet}_{L0}$ and a?

6.P.25 Design of a CSTR for a Zeroth-Order Irreversible Reaction

Consider a perfectly mixed CSTR that is used to absorb a pure gas by means of a zeroth-order irreversible reaction operating in the batch mode. By this we mean that the gas consisting of pure component A flows into and out of the CSTR, which contains a fixed amount of liquid that is contained in it; there is no inflow or outflow of the liquid. We assume that the liquid is initially pure component B.

TABLE 6.P.25-1 Design Parameters and Physical Properties for a Batch CSTR Chemisorption Process Employing a Zeroth-Order Irreversible Reaction

Zeroth-order reaction-rate constant, k_0	3.0×10^{-6} gmol/cm$^3 \cdot$ s
Diffusion coefficient, D_{AB}	3.0×10^{-6} gmol/cm$^3 \cdot$ s
Molar density of bulk liquid, c	0.06 gmol/cm^3
Henry's law constant relating gas and liquid mole fractions, y_A and x_A, respectively	$y_A = 100x_A$
Mass-transfer coefficient for physical absorption, $k^{\bullet}_{L0}$	6.0×10^{-4} gmol/cm$^2 \cdot$ s
Liquid-phase holdup, ϕ	2 cm^3/cm^2
Interfacial area per unit volume, a	0.5 cm^2/cm^3
Gas bubble radius, R	0.15 cm

The concentration of component A in the liquid will increase in time, owing to the chemisorption. Table 6.P.25-1 summarizes the design parameters and physical properties for this batch chemisorption process. Assume constant physical properties and that the gas bubbles do not change significantly in size, due to the chemisorption process. Note in working this problem that it is possible to answer all parts using the results of the scaling analyses done in Chapter 6.

(a) Estimate the thickness δ_m of the liquid film surrounding each gas bubble.

(b) Use an appropriate criterion emanating from scaling analysis to determine if it is reasonable to ignore curvature effects when modeling the mass transfer from the microscale element.

(c) Determine the maximum possible concentration in the liquid.

(d) Determine if this chemisorption process takes place in the slow, fast, or instantaneous reaction regime; if it occurs in the slow reaction regime, determine if it is in the kinetic, intermediate (see Practice Problem 6.P.9), or diffusional reaction domain of the slow reaction regime.

(e) Determine the mass-transfer rate per unit volume $N_A a$ at the initiation of this chemisorption process.

(f) Determine the mass-transfer rate per unit volume $N_A a$ applicable at very long contact times.

(g) Suggest how the performance of this chemisorption process could be improved; that is, how can the mass-transfer rate per unit volume be increased?

6.P.26 Use of a Packed Column Absorber to Determine Reaction Order

A packed absorption column can be used to determine the reaction order for a chemical reaction for which the kinetics are unknown under some conditions. Note that a, the interfacial area per unit volume, is usually unknown for packed columns. Carefully consider the properties of each reaction regime and recommend the regime most appropriate for determining the reaction order from the usual quantities measured for packed absorption column performance (inlet and outlet concentrations and flow rates).

6.P.27 Cocurrent Versus Countercurrent Flow in a Packed Gas-Absorption Column

A packed gas-absorption column can operate with both streams flowing either in parallel (cocurrent) or antiparallel (countercurrent) directions. In this problem we consider the merits of each flow configuration relative to the reaction regime.

(a) In which reaction regimes will the amount of the soluble gas-phase component that is absorbed into the liquid be independent of the flow configuration?
(b) Which flow configuration would be preferred for the reaction regimes you identified in part (a)? Indicate the reason(s) for your answer.
(c) Which flow configuration would be preferred for those reaction regimes that do depend on the contacting scheme? Indicate the reason(s) for your answer.

6.P.28 Design of a Packed Column Absorber for an *n*th-Order Irreversible Reaction

In Section 6.14 we considered the design of a packed column absorber for a first-order irreversible reaction. In this problem we consider the more general case of an nth-order irreversible reaction for which $R_A = k_1 c_A^n$. Assume that the data given in Table 6.14-1 apply for the nth-order reaction as well.

(a) Determine the reaction regime in which this chemisorption process takes place.
(b) Derive an equation for the mass-transfer flux at any point within the microscale element.
(c) Determine the concentration distribution of component A at any point within the microscale element.
(d) Consider the implications for your results in parts (b) and (c) for the special case of $n = 1$; explain any anomalies you observe for this special case.
(e) Derive an equation for the mass-transfer coefficient for an nth-order irreversible reaction.
(f) Derive an equation for the column height for an nth-order irreversible reaction.
(g) Discuss any differences in the effects that the various parameters will have on the mass-transfer rate per unit volume for $n > 1$ versus $n = 1$.

6.P.29 Design of a Packed Gas-Absorption Column for a First-Order Irreversible Reaction

Consider a packed gas-absorption column such as that shown in Figure 6.2-1 in which an initially pure liquid stream consisting of component S that enters at the top

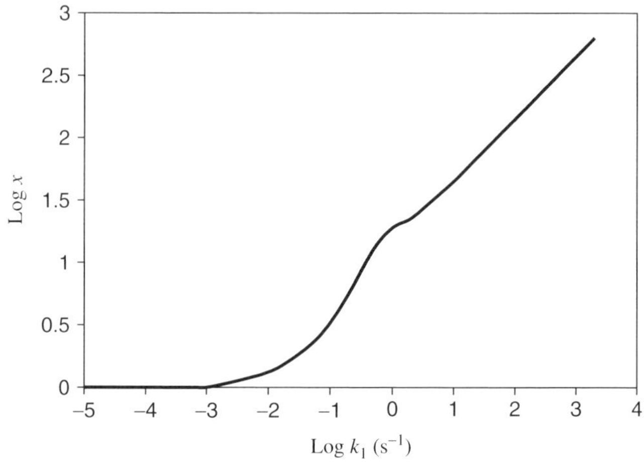

Figure 6.P.29-1 Performance data for a packed gas-absorption column plotted as log χ versus log k_1, where χ is the ratio of the total absorption rate to the maximum possible rate for physical absorption and k_1 is the reaction-rate constant for the first-order irreversible chemisorption reaction.

is used to absorb a component A from a gas stream that enters at the bottom. Component A can be chemisorbed in liquid S by means of a first-order irreversible reaction whose reaction rate per unit volume R_A is given by $R_A = k_1 c_A$. The following information is available for this chemisorption process: mass-transfer coefficient $k^{\bullet}_{L0} = 0.01$ cm/s; effective binary diffusion coefficient $D_{AS} = 5.5 \times 10^{-5}$ cm^2/s; and liquid-phase (macroscale) contact time $t_o = 500$ s. Figure 6.P.29-1 shows performance data for this chemisorption process as a plot of χ, the ratio of the total absorption rate to the maximum possible rate for physical absorption, as a function of k_1, the first-order reaction rate constant. The total absorption rate is defined as

$$W = \int_0^{V_T} N_A a \, dV \tag{6.P.29-1}$$

where V and V_T denote the volume and total volume of the column, respectively.

(a) Develop a general equation for χ in terms of the mass-transfer rate per unit volume $N_A a$, the contact time t_o, and any relevant concentration(s).

(b) Explain the shape of the line for the performance data in Figure 6.P.29-1 in terms of the various reaction regimes and domains; that is, identify which regime or domain applies to each part of this line. When invoking the scaling analysis criteria for the various regimes and domains, use one order of magnitude in your ordering arguments. *Note*: Do not forget to consider the intermediate regime and the intermediate domain of the slow reaction regime (see Practice Problem 6.P.9).

(c) Determine a, the interfacial area per unit volume of column, from the performance data given in Figure 6.P.29-1.

(d) In which ranges of the reaction rate constant would it advantageous to employ cocurrent (parallel) rather than countercurrent (antiparallel) flow?

7 Applications in Process Design

The optimist looked at his glass and said to the bartender that it was half full.
The pessimist looked at his glass and said to the bartender that it was half empty.
The engineer looked at both glasses and said to the bartender that he would O(1).[1]

7.1 INTRODUCTION

The focus and organization of this chapter are a dramatic departure from those of the preceding chapters. In Chapters 3 through 6 we focused primarily on using scaling to justify classical approximations made in fluid dynamics, heat transfer, mass transfer, and mass transfer with chemical reaction. As such, the thrust of the preceding chapters was primarily pedagogical rather than application-oriented. In contrast, in this chapter we show how scaling analysis has been applied to the design of four practical engineering processes. Scaling analysis played a pivotal role in advancing the technologies involved in these four examples. It not only helped in simplifying the describing equations so that a tractable solution could be developed but was also used to design the process; that is, to determine process parameters such as the contact time, oscillation amplitude and frequency, vessel length, adsorbent particle size, and wall temperature that insured effective and efficient operation. We also demonstrate the utility of using o(1) scaling analysis to obtain optimal dimensionless groups for correlating experimental or numerical data.

In Section 7.2 we revisit the design of a membrane–lung oxygenator. In Section 5.10 the scaling approach to dimensional analysis was used to determine the dimensionless groups required to correlate data for the performance enhancement that could be achieved by applying axial vibrations of the hollow-fiber membranes in this oxygenator. In this chapter, o(1) scaling analysis is used to achieve the minimum parametric representation. Subsequently, we will see that o(1) scaling analysis provides considerably more information than just the scaling analysis approach to dimensional analysis or the Pi theorem. Section 7.3 applies scaling analysis to

[1] Anonymous—attributed to a very practical engineer.

Scaling Analysis in Modeling Transport and Reaction Processes: A Systematic Approach to Model Building and the Art of Approximation, By William B. Krantz

the design of the pulsed pressure swing adsorption (PSA) process. This process is of particular interest for use in portable oxygenators for people suffering from severe respiratory problems or chronic obstructive pulmonary disease. This example involves coupled flow through porous media and mass transfer. In Section 7.4 we consider the thermally induced phase-separation (TIPS) process for fabricating semipermeable polymeric membranes. The TIPS process is used to make high flux membranes for water desalination and reclamation as well as for solvent recovery. This example involves coupled heat and mass transfer. In Section 7.5 we use scaling analysis to design the fluid–wall aerosol flow reactor for converting methane into hydrogen. The fluid–wall aerosol flow reactor is being developed to use solar energy to drive the thermal decomposition of methane into hydrogen that can be used as a clean-burning fuel to address the worldwide concern regarding global warming. This example involves coupled fluid dynamics, heat transfer, mass transfer, and chemical reactions in a two-phase system. The examples in Sections 7.3, 7.4, and 7.5 use the microscale–macroscale modeling methodology to handle heterogeneities such as particles or a dispersed phase. In Section 7.6 we summarize the scaling principles employed in each of these examples and the role that scaling analysis played in advancing each of these technologies. Unworked practice problems related to each of the four examples are included at the end of the chapter.

7.2 DESIGN OF A MEMBRANE–LUNG OXYGENATOR

In Section 5.10 we applied the scaling approach to dimensional analysis to develop a correlation for the performance enhancement achieved by applying axial oscillations to the hollow fibers in a membrane–lung oxygenator. Recall that the scaling approach to dimensional analysis outlined in Section 2.4 does not provide as much information as o(1) scaling analysis for achieving the minimum parametric representation. In particular, it does not lead to groups whose magnitude can be used to assess the relative importance of the various terms in the describing equations. Moreover, it does not provide any insight into the fundamental mechanisms involved in the process. It also does not identify regions of influence or boundary layers, which in some cases can reduce the number of dimensionless groups. Here we apply o(1) scaling analysis to design a membrane–lung oxygenator. This example illustrates the value of using o(1) scaling analysis to achieve the minimum parametric representation in contrast to the simple scaling analysis or Pi theorem approaches to dimensional analysis.

The membrane–lung oxygenator involves a bundle of permeable cylindrical hollow-fiber membranes encased in a tubular housing. In this application the hollow-fiber membranes do not cause any separation but rather, serve as a gas-permeable barrier between the blood and the source of oxygen. The mass transfer of oxygen is controlled on the blood side of the membrane because the hemoglobin "particles" that scavenge the oxygen are excluded from the wall region due to the fluid dynamics. Hence, efforts to improve oxygenator performance have focused on various means to reduce the resistance to mass transfer on the blood side of the membrane. A very effective way to enhance the mass transfer is to oscillate the hollow fibers relative to the blood flow to increase the concentration gradients adjacent to the membrane.[2] In this section we use o(1) scaling analysis to design an oxygenator that employs periodic oscillations of the hollow fibers to enhance the mass transfer. It is sufficient to consider the effect of the oscillations on the oxygen mass transfer to the blood flow in a single hollow-fiber membrane having radius R and length L, as shown in Figure 7.2-1.

The initial steps involved in applying o(1) scaling analysis to this problem are similar to those used in the scaling analysis approach to dimensional analysis discussed in Section 5.10. One begins by writing the appropriate describing equations for the oxygen mass transfer to the blood, which is assumed to be in fully developed periodically pulsed laminar flow (step 1). The effect of the oscillating wall on the oxygen mass transfer is assessed in terms of the mass-transfer coefficient $k_L^\bullet$ defined in terms of $\overline{N}_{Aw}$, the time-average molar flux of oxygen at the inner membrane wall, as follows:

$$k_L^\bullet \equiv \frac{\overline{N}_{Aw}}{\Delta c_{\text{lm}}} \tag{7.2-1}$$

where Δc_{lm} is the log-mean concentration driving force, defined as

$$\Delta c_{\text{lm}} \equiv \frac{(c_{Aw} - c_{AL}) - (c_{Aw} - c_{A0})}{\ln[(c_{Aw} - c_{AL})/(c_{Aw} - c_{A0})]} \tag{7.2-2}$$

in which c_{Aw} is the oxygen concentration at the blood side of the membrane that is dictated by thermodynamic equilibrium considerations, c_{A0} and c_{AL} are the average concentrations in the blood at $z = 0$ and at $z = L$, respectively, and $\overline{N}_{Aw}$ is defined as

$$\overline{N}_{Aw} = \frac{\int_0^{2\pi/\omega} N_{Aw}\, dt}{\int_0^{2\pi/\omega} dt} = \frac{\omega}{2\pi} \int_0^{2\pi/\omega} N_{Aw}\, dt \tag{7.2-3}$$

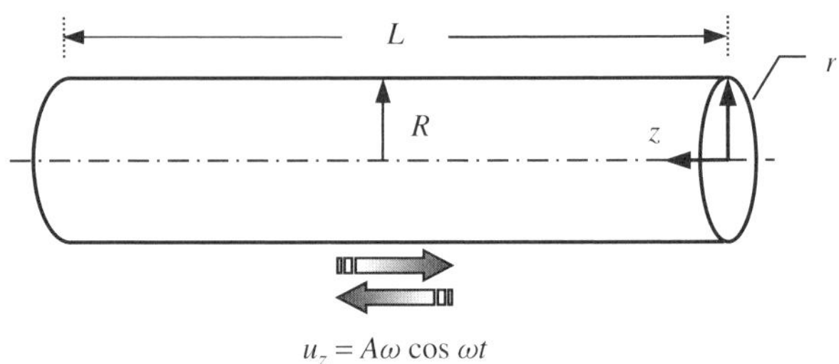

Figure 7.2-1 Single hollow fiber of radius R and length L in a membrane–lung oxygenator; axial oscillations of amplitude A and angular frequency ω are used to increase the concentration gradients at the interior wall where the resistance to mass transfer is concentrated.

[2] R. R. Bilodeau, R. J. Elgas, W. B. Krantz, and M. E. Voorhees, U.S. patent 5,626,759, issued May 6, 1997.

where ω is the angular frequency of the wall oscillations. The molar flux N_{Aw} in equation (7.2-3) is the local molar flux averaged over the length of the hollow-fiber membrane and is defined as

$$N_{Aw} = \frac{\int_0^L D_{AB}\,(\partial c_A/\partial r)|_{r=R}\,dz}{\int_0^L dz} = \frac{1}{L}\int_0^L D_{AB}\left.\frac{\partial c_A}{\partial r}\right|_{r=R} dz \tag{7.2-4}$$

where D_{AB} is the binary diffusion coefficient for oxygen in blood, which is assumed to be constant. The bulk flow contribution to the molar flux has been ignored because the solutions are dilute. When equations (7.2-3) and (7.2-4) are substituted into equation (7.2-1), we obtain

$$k_L^{\bullet} = \frac{\omega D_{AB}}{2\pi L\,\Delta c_{\text{lm}}}\int_0^{2\pi/\omega}\int_0^L \left.\frac{\partial c_A}{\partial r}\right|_{r=R} dz\,dt \tag{7.2-5}$$

Both c_{AL} appearing in Δc_{lm} and c_A in equation (7.2-5) are obtained from a solution to the axisymmetric form of the species-balance equation in cylindrical coordinates given by equation (G.2-5) in the Appendices, which when simplified appropriately for this unsteady-state convective diffusion problem assumes the form

$$\frac{\partial c_A}{\partial t} + u_z\frac{\partial c_A}{\partial z} = \frac{D_{AB}}{r}\frac{\partial}{\partial r}\left(r\frac{\partial c_A}{\partial r}\right) \tag{7.2-6}$$

When simplifying equation (G.2-5) to arrive at equation (7.2-6), note that each term has been divided by the molecular weight, which can be assumed to be constant for the dilute solutions. The axial diffusion term has also been neglected in equation (7.2-6), owing to the small aspect ratio.[3] The mass-average velocity u_z is obtained from a solution to the equations of motion given by equations (D.2-10) through (D.2-12) in the Appendices. If entrance and exit effects can be ignored, this will be an unsteady-state fully developed flow for which the equations of motion reduce to

$$\rho\frac{\partial u_z}{\partial t} = \frac{\Delta P}{L} + \mu\frac{1}{r}\frac{\partial}{\partial r}\left(r\frac{\partial u_z}{\partial r}\right) \tag{7.2-7}$$

The boundary and periodic solution conditions are given by[4]

$$\frac{\partial c_A}{\partial r} = 0 \quad \text{at} \quad r = 0, \quad 0 \le z \le L \tag{7.2-8}$$

$$\frac{\partial u_z}{\partial r} = 0 \quad \text{at} \quad r = 0, \quad 0 \le z \le L \tag{7.2-9}$$

$$c_A = c_{Aw} \quad \text{at} \quad r = R, \quad 0 \le z \le L \tag{7.2-10}$$

$$u_z = A\omega\cos\omega t \quad \text{at} \quad r = R, \quad 0 \le z \le L \tag{7.2-11}$$

$$c_A = c_{A0} \quad \text{at} \quad z = 0 \tag{7.2-12}$$

[3]The criterion for ignoring this term is considered in Practice Problem 7.P.1.

[4]The criterion for ignoring transient flow effects is considered in Practice Problem 7.P.2.

$$c_A|_t = c_A|_{t+2\pi/\omega} \tag{7.2-13}$$

$$u_z|_t = u_z|_{t+2\pi/\omega} \tag{7.2-14}$$

where A is the amplitude of the wall oscillations.

In carrying out an o(1) scaling analysis for this problem, one needs to recognize that even in the absence of any axial oscillations, the mass-transfer problem would involve a region of influence or boundary layer; that is, a solutal boundary layer develops at the wall from the upstream edge of the hollow-fiber membrane. The wall oscillations will influence the thickness of this solutal boundary layer by altering the velocity profile near the wall. This suggests that our scaling analysis should include a reference factor so that the coordinate system can be relocated on the wall of the hollow-fiber membrane. A second consideration is that the axial oscillations will define a hydrodynamic region of influence or momentum boundary layer.[5] The momentum and solutal boundary-layer thicknesses need not be the same; indeed, the high Schmidt numbers that characterize liquids imply that the region of influence for the mass transfer will be considerably thinner than that for the fluid flow. This then suggests that the radial coordinate in the equations of motion should be scaled differently than those for the species-balance equation. Another consideration is the scaling of the time derivative of the concentration. Whereas the membrane oscillations clearly establish the amplitude of the velocity at the wall, it is not clear what the amplitude of the concentration oscillations will be. In particular, there is no reason to expect that this derivative will scale with the characteristic concentration change divided by the characteristic time. Hence, it is appropriate to introduce a separate scale for this time derivative. In view of these considerations, the following dimensionless variables involving unspecified scale and reference factors are defined (steps 2, 3, and 4):

$$c_A^* \equiv \frac{c_A - c_{Ar}}{c_{As}}; \qquad \left(\frac{\partial c_A}{\partial t}\right)^* \equiv \frac{1}{\dot{c}_{As}}\frac{\partial c_A}{\partial t}; \qquad u^* \equiv \frac{u_z}{u_s}; \qquad y_c^* \equiv \frac{r - r_r}{\delta_s};$$

$$y_m^* \equiv \frac{r - r_r}{\delta_m}; \qquad z^* \equiv \frac{z}{z_s}; \qquad t^* \equiv \frac{t}{t_s} \tag{7.2-15}$$

where δ_s and δ_m are the solutal and momentum boundary-layer thicknesses, respectively. Substitute these dimensionless variables into the describing equations and divide through by the dimensional coefficient of one term in each equation (steps 5 and 6):

$$\frac{2\pi k_L^\bullet L\delta_s}{\omega D_{AB} z_s t_s} = -\frac{1}{\Delta c_{\text{lm}}^*}\int_0^1\int_0^1 \left.\frac{\partial c_A^*}{\partial y_c^*}\right|_{y_c^*=(r_r-R)/\delta_s} dz^*\, dt^* \tag{7.2-16}$$

$$\Delta c_{\text{lm}}^* \equiv \frac{-c_{AL}^* + (c_{A0} - c_{Ar})/c_{As}}{\ln\left(\dfrac{c_{Aw} - c_{Ar}}{c_{Aw} - c_{A0}} - \dfrac{c_{As}}{c_{Aw} - c_{A0}}c_{AL}^*\right)} \tag{7.2-17}$$

[5]The region of influence associated with an oscillating boundary was discussed in Section 3.5.

$$\frac{\delta_s^2 \dot{c}_{As}}{D_{AB} c_{As}} \left(\frac{\partial c_A}{\partial t} \right)^* + \frac{u_{zs} \delta_s^2}{D_{AB} z_s} u_z^* \frac{\partial c_A^*}{\partial z^*} = \frac{1}{(r_r/\delta_s) - y_c^*} \frac{\partial}{\partial y_c^*} \left[\left(\frac{r_r}{\delta_s} - y_c^* \right) \frac{\partial c_A^*}{\partial y_c^*} \right] \tag{7.2-18}$$

$$\frac{\delta_m^2}{\nu t_s} \frac{\partial u_z^*}{\partial t^*} = \frac{\delta_m^2}{\mu u_{zs}} \frac{\Delta P}{L} + \frac{1}{(r_r/\delta_m) - y_m^*} \frac{\partial}{\partial y_m^*} \left[\left(\frac{r_r}{\delta_m} - y_m^* \right) \frac{\partial u_z^*}{\partial y_m^*} \right] \tag{7.2-19}$$

$$\frac{\partial c_A^*}{\partial y_c^*} = 0 \quad \text{at} \quad y_c^* = \frac{y_r}{\delta_s}, \quad 0 \le z^* \le \frac{L}{z_s} \tag{7.2-20}$$

$$\frac{\partial u_z^*}{\partial y_m^*} = 0 \quad \text{at} \quad y_m^* = \frac{r_r}{\delta_m}, \quad 0 \le z^* \le \frac{L}{z_s} \tag{7.2-21}$$

$$c_A^* = \frac{c_{Aw} - c_{Ar}}{c_{As}} \quad \text{at} \quad y_c^* = \frac{y_r - R}{\delta_s}, \quad 0 \le z^* \le \frac{L}{z_s} \tag{7.2-22}$$

$$u_z^* = \frac{A\omega}{u_{zs}} \cos \omega t_s t^* \quad \text{at} \quad y_m^* = \frac{r_r - R}{\delta_m} \tag{7.2-23}$$

$$c_A^* = \frac{c_{A0} - c_r}{c_{As}} \quad \text{at} \quad z^* = 0 \tag{7.2-24}$$

$$c_A^*|_{t^*} = c_A^*|_{t^* + 2\pi/\omega t_s} \tag{7.2-25}$$

$$u_z^*|_{t^*} = u_z^*|_{t^* + 2\pi/\omega t_s} \tag{7.2-26}$$

The scale and reference factors are determined by bounding the dependent variables, their derivatives, and the independent variables to be $\circ(1)$ (step 7). The reference and scale factors for the concentration are determined by setting the appropriate dimensionless groups in equations (7.2-22) and (7.2-24) equal to 1 and zero, respectively, to obtain

$$\frac{c_{A0} - c_r}{c_{As}} = 0 \Rightarrow c_r = c_{A0}; \qquad \frac{c_{Aw} - c_{Ar}}{c_{As}} = 1 \Rightarrow c_{As} = c_{Aw} - c_{A0} \tag{7.2-27}$$

The radial reference factor and axial length scale are determined by setting the appropriate dimensionless groups in equation (7.2-22) equal to zero and 1, respectively, to obtain

$$\frac{y_r - R}{\delta_s} = 0 \Rightarrow y_r = R; \qquad \frac{L}{z_s} = 1 \Rightarrow z_s = L \tag{7.2-28}$$

The time scale is obtained by setting the dimensionless group appearing in equations (7.2-25) and (7.2-26) equal to 1 to obtain

$$\frac{\omega t_s}{2\pi} = 1 \Rightarrow t_s = \frac{2\pi}{\omega} \tag{7.2-29}$$

The radial length scale for the momentum equation is equal to the penetration depth of the axial oscillations, which is determined by the viscous diffusion of the wall motion into the fluid; hence, we set the following dimensionless group in equation (7.2-19) equal to 1:

$$\frac{\delta_m^2}{\nu t_s} = 1 \Rightarrow \delta_m = \sqrt{\frac{2\pi\nu}{\omega}} \tag{7.2-30}$$

The velocity scale is obtained by balancing the pressure term with the viscous term in equation (7.2-19); hence, the following dimensionless group is set equal to 1:

$$\frac{\delta_m^2}{\mu u_{zs}}\frac{\Delta P}{L} = \frac{2\pi\nu}{\omega\mu u_{zs}}\frac{\Delta P}{L} = 1 \Rightarrow u_{zs} = \frac{2\pi}{\omega\rho}\frac{\Delta P}{L} = \frac{16\pi\nu\overline{U}}{\omega R^2} \tag{7.2-31}$$

where $\overline{U}$ is the average velocity for steady-state fully developed flow through a circular tube; expressing u_{zs} in terms of $\overline{U}$ is done so that the dimensionless groups emanating form this o(1) scaling analysis can be recast in terms of those obtained via the simpler scaling analysis approach to dimensional analysis considered in Section 5.10. The radial length scale for the species-balance equation is the solutal boundary-layer thickness, which is determined by balancing the axial convection and radial diffusion terms; hence, we set the following dimensionless group in equation (7.2-18) equal to 1:

$$\frac{u_{zs}\delta_s^2}{D_{AB}z_s} = \frac{16\pi\nu\overline{U}\delta_s^2}{\omega R^2 D_{AB}L} = 1 \Rightarrow \delta_s = \sqrt{\frac{\omega R^2 D_{AB}L}{16\pi\nu\overline{U}}} \tag{7.2-32}$$

Determining the proper scale for the time derivative of the concentration is problematic. An upper bound on this derivative can be obtained by balancing the unsteady-state term with the radial diffusion term; hence, we set the following dimensionless group in equation (7.2-18) equal to 1:

$$\frac{\delta_s^2\dot{c}_{As}}{D_{AB}c_{As}} = \frac{\omega R^2 L\dot{c}_{As}}{16\pi\nu\overline{U}(c_{Aw} - c_{A0})} = 1 \Rightarrow \dot{c}_{As} = \frac{16\pi\nu\overline{U}(c_{Aw} - c_{A0})}{\omega R^2 L} \tag{7.2-33}$$

Substitute the scale and reference factors defined by equations (7.2-27) through (7.2-33) into the describing equations given by (7.2-16) through (7.2-26) to obtain the following:

$$\mathrm{Sh}_v = -\sqrt{32\frac{\nu}{\omega R^2}\mathrm{Gz}}\frac{1}{(\Delta c_{\mathrm{lm}}^*)_v}\int_0^1\int_0^1 \left.\frac{\partial c_A^*}{\partial y_c^*}\right|_{y_c^*=0} dz^*\,dt^* \tag{7.2-34}$$

$$(\Delta c_{\mathrm{lm}}^*)_v \equiv \frac{-c_{AL}^*}{\ln(1 - c_{AL}^*)} \tag{7.2-35}$$

$$64\pi \frac{\text{Gz}}{\text{Sc}} \frac{\nu^2}{\omega^2 R^4} \frac{\partial c_A^*}{\partial t^*} + u_z^* \frac{\partial c_A^*}{\partial z^*} = \frac{1}{\sqrt{32(\nu/\omega R^2)\text{Gz}} - y_c^*} \times \frac{\partial}{\partial y_c^*}\left[\left(\sqrt{32 \frac{\nu}{\omega R^2}\text{Gz}} - y_c^*\right) \frac{\partial c_A^*}{\partial y_c^*}\right] \quad (7.2\text{-}36)$$

$$\frac{\partial u_z^*}{\partial t^*} = 1 + \frac{1}{\sqrt{(1/2\pi)(\omega R^2/\nu)} - y_m^*} \times \frac{\partial}{\partial y_m^*}\left[\left(\sqrt{\frac{1}{2\pi}\frac{\omega R^2}{\nu}} - y_m^*\right) \frac{\partial u_z^*}{\partial y_m^*}\right] \quad (7.2\text{-}37)$$

$$\frac{\partial c_A^*}{\partial y_c^*} = 0 \quad \text{at} \quad y_c^* = \sqrt{32 \frac{\nu}{\omega R^2}\text{Gz}}, \quad 0 \le z^* \le 1 \quad (7.2\text{-}38)$$

$$\frac{\partial u_z^*}{\partial y_m^*} = 0 \quad \text{at} \quad y_m^* = \sqrt{\frac{1}{2\pi}\frac{\omega R^2}{\nu}}, \quad 0 \le z^* \le 1 \quad (7.2\text{-}39)$$

$$c_A^* = 1 \quad \text{at} \quad y_c^* = 0, \quad 0 \le z^* \le 1 \quad (7.2\text{-}40)$$

$$u_z^* = \frac{1}{16\pi} \frac{A\omega}{\overline{U}} \frac{\omega R^2}{\nu} \cos 2\pi t^* \quad \text{at} \quad y_m^* = 0 \quad (7.2\text{-}41)$$

$$c_A^* = 0 \quad \text{at} \quad z^* = 0 \quad (7.2\text{-}42)$$

$$c_A^*|_{t^*} = c_A^*|_{t^*+1} \quad (7.2\text{-}43)$$

$$u_z^*|_{t^*} = u_z^*|_{t^*+1} \quad (7.2\text{-}44)$$

We have recast the describing equations in terms of the dimensionless groups identified in Section 5.10, where

$$\text{Sh} \equiv \frac{k_L^\bullet R}{D_{AB}} \quad \text{is the Sherwood number} \quad (7.2\text{-}45)$$

$$\text{Gz} \equiv \frac{\pi \overline{U} R^2}{2 D_{AB} L} \quad \text{is the Graetz number} \quad (7.2\text{-}46)$$

and Sh_v and $(\Delta c_{\text{lm}}^*)_v$ denote the Sherwood number and the dimensionless log-mean concentration driving force, respectively, in the presence of axial oscillations. Step 8 in the procedure for o(1) scaling analysis would involve assessing whether any simplifications can be made in the dimensionless form of the describing equations. Depending on the oscillation frequency, it might be possible to ignore curvature effects in equations (7.2-36) and (7.2-37) and to apply the boundary conditions given by equations (7.2-38) and (7.2-39) at infinity. However, owing to the range of oscillation frequencies that must be considered, it is not possible in general to make these simplifications. Moreover, our interest here is not in simplifying these equations, but rather, in developing a dimensional analysis correlation for the performance enhancement.

Equations (7.2-34) through (7.2-44) indicate that the Sherwood number for mass transfer with applied axial oscillations will be a function of four dimensionless groups; that is,

$$\mathrm{Sh}_v = -\sqrt{32\frac{\nu}{\omega R^2}\mathrm{Gz}}\frac{1}{(\Delta c^*_{\mathrm{lm}})_v}\int_0^1\int_0^1 \left.\frac{\partial c^*_A}{\partial y^*_c}\right|_{y^*_c=0} dz^*\,dt^* = f\left(\frac{\omega R^2}{\nu}, \frac{A\omega}{\overline{U}}, \mathrm{Gz}, \mathrm{Sc}\right) \tag{7.2-47}$$

since the integral and $(\Delta c^*_{\mathrm{lm}})_v$ introduce only those dimensionless groups appearing in equations (7.2-36) through (7.2-44). The scaling approach to dimensional analysis developed in Section 5.10 also indicated that the Sherwood number was a function of four dimensionless groups. However, the o(1) scaling analysis result carried out here provides additional information in identifying the momentum and solutal boundary layers; that is, the mass transfer is confined to a region-of-influence near the wall that is dictated by the effect of the oscillating boundary on the hydrodynamics in the wall region. The oscillating boundary changes the velocity profile near the wall that in turn increases the concentration profiles in this region. Note that the Schmidt number appeared explicitly only when we recast the temporal concentration derivative $(\partial c_A/\partial t)^*$ in terms of $\partial c^*_A/\partial t^*$; that is, to carry out the time integration of equation (7.2-36) the concentration and time must be nondimensionalized as they are in the other terms in the describing equations.

An effective way to assess the performance enhancement for this novel oxygenator design is to consider the ratio of the Sherwood number in the presence and absence of axial oscillations. An o(1) scaling analysis for a conventional oxygenator that does not employ axial oscillations indicates that

$$\mathrm{Sh} = \sqrt{\frac{4}{\pi}\mathrm{Gz}}\frac{1}{\Delta c^*_{\mathrm{lm}}}\int_0^1 \left.\frac{\partial c^*_A}{\partial y^*_c}\right|_{y^*_c=0} dz^* = f(\mathrm{Gz}) \tag{7.2-48}$$

That is, the Sherwood number is a function of only the Graetz number.[6] Hence, we conclude that a correlation for the ratio of the Sherwood number in the presence and absence of axial oscillations is of the form

$$\frac{\mathrm{Sh}_v}{\mathrm{Sh}} = \sqrt{8\frac{\pi\nu}{\omega R^2}}\frac{\left[\frac{1}{(\Delta c^*_{\mathrm{lm}})_v}\int_0^1\int_0^1 \left.\frac{\partial c^*_A}{\partial y^*_c}\right|_{y^*_c=0} dz^*\,dt^*\right]_{\text{with oscillations}}}{\left[\frac{1}{\Delta c^*_{\mathrm{lm}}}\int_0^1 \left.\frac{\partial c^*_A}{\partial y^*_c}\right|_{y^*_c=0} dz^*\right]_{\text{without oscillations}}}$$

$$= f\left(\frac{\omega R^2}{\nu}, \frac{A\omega}{\overline{U}}, \mathrm{Gz}, \mathrm{Sc}\right) \tag{7.2-49}$$

[6] An O(1) scaling analysis to obtain this result is considered in Practice Problem 7.P.4.

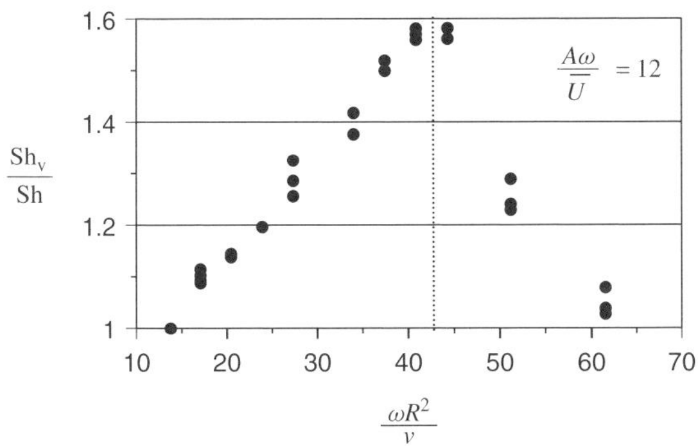

Figure 7.2-2 Effect of the dimensionless group $\omega R^2 / \nu$ on the ratio of the Sherwood number with and without axial vibrations for $A\omega / \overline{U} = 12$ and Gz = 25; maximum enhancement occurs at $\omega R^2 / \nu \cong 43$. Data are for oxygen mass transfer to water. [From W. B. Krantz, R. R. Bilodeau, M. E. Voorhees, and R. J. Elgas, *J. Membrane Sci.*, **124**, 283–299 (1997).]

A plot of Sh_υ / Sh as a function of $\omega R^2 / \nu$ for $A\omega / \overline{U} = 12$ and Gz = 25 is shown in Figure 7.2-2. One observes the dramatic "tuned" response that was anticipated in Section 5.10; a maximum enhancement of approximately 1.6 is observed. Our o(1) scaling analysis can provide considerable insight into the reason for this tuned response and the maximum that occurs at $\omega R^2/\nu \cong 43$. Equation (7.2-30) indicates that the momentum boundary-layer thickness decreases with an increase in the oscillation frequency, whereas equation (7.2-32) indicates that the solutal boundary-layer thickness increases with an increase in the oscillation frequency; for this reason, the performance enhancement displays a maximum with respect to the oscillation frequency. The fact that this maximum occurs at $\omega R^2/\nu \cong 43$ can be explained by considering the ratio of the momentum to solutal boundary-layer thickness for the conditions in Figure 7.2-2:

$$\frac{\delta_m}{\delta_s} = \frac{\sqrt{2\pi\nu/\omega}}{\sqrt{\omega R^2 D_{AB} L / 16\pi\nu\overline{U}}} = 8\frac{\nu}{\omega R^2}\sqrt{\pi \mathrm{Gz}} = \frac{8}{43}\sqrt{25\pi} = 1.65 \qquad (7.2\text{-}50)$$

That is, the maximum enhancement in the performance occurs at a frequency at which the momentum and solutal boundary layers have the same thickness to within the O(1) accuracy of scaling analysis.

This example indicates that an o(1) scaling analysis provides considerably more information on a process than does simple dimensional analysis through either the scaling approach or the Pi theorem. Indeed, the o(1) scaling analysis indicated that the Sherwood number ratio could be correlated in terms of the same four dimensionless groups that we obtained from the simple scaling analysis approach to dimensional analysis that was carried out in Section 5.10. Moreover, o(1) scaling analysis provided valuable insight into why the performance enhancement displays a maximum with respect to the applied frequency. It also provided an explanation for why this maximum occurred at $\omega R^2 / \nu \cong 43$ for the process conditions in Figure 7.2-2. Indeed, equation (7.2-50) can serve as a guide to the design of a membrane–lung oxygenator; that is, it indicates how the frequency required to achieve optimum performance enhancement changes as a function of the Graetz number, kinematic viscosity, and radius of the hollow-fiber membrane.

7.3 PULSED SINGLE-BED PRESSURE-SWING ADSORPTION FOR PRODUCTION OF OXYGEN-ENRICHED AIR

Pressure swing adsorption (PSA) is a separations technology that is used to selectively adsorb one or more components from a gas mixture in order to produce a product stream enriched in the less strongly adsorbed component(s). In particular, it is used to produce oxygen for both industrial as well as medical applications; indeed, at the dawn of the twentieth-first century, 20% of the world's oxygen production was achieved by PSA. To facilitate selective adsorption and regeneration, respectively, conventional PSA uses two parallel packed adsorbent beds that are alternately subjected to pressurization and depressurization. However, conventional PSA has several disadvantages: low separation efficiency per unit mass of adsorbent material; large capital investment for replacement of adsorbent material due to attrition; and complexity of the apparatus. Moreover, since PSA is used for critical life support for patients suffering from chronic obstructive pulmonary disease (COPD) and other respiratory problems, there is considerable motivation to develop a compact portable oxygenator that would greatly improve their quality of life. Hence, research in PSA has been directed toward improving the separation efficiency by decreasing the size of the adsorbent particles, reducing the complexity of the apparatus by using a single adsorption bed, and minimizing the impact of the pressure drop by making the adsorption bed shorter. In 1981 the pulsed packed single-bed PSA process was developed, which employed much smaller adsorbent particles (20 to 120 mesh).[7] A more recently proposed innovation is the use of a thin monolithic single-bed adsorbent composed of integrally bound micrometer-scale crystals which avoids particle attrition problems and provides exceptionally large surface area and high gas throughputs.[8] This novel technology offers the promise of very efficient oxygen production through use of a device that is significantly smaller than current oxygenators. A major problem in developing any new process is to determine appropriate design parameters to carry out bench- or pilot-scale testing. Scaling analysis was used in developing the pulsed single-bed PSA process for a thin monolithic adsorbent in order to determine how to model it as well as to specify principal design parameters such as the applied pressure and pressurization time. In this section we illustrate how scaling analysis was applied to design this process and to show how it differs from a pulsed packed single-bed PSA that employs particulate adsorbents. We will also use the scaling approach to dimensional analysis outlined in Chapter 2 to obtain the optimal dimensionless groups for correlating experimental or numerical data for the PSA process.

[7]R. L. Jones and G. E. Keller, *Sep. Process Technol.*, **2**, 17 (1981).

[8]E. M. Kopaygorodsky, V. V. Guliants, and W. B. Krantz, *A.I.Ch.E. Jl.*, **50**(5), 953 (2004).

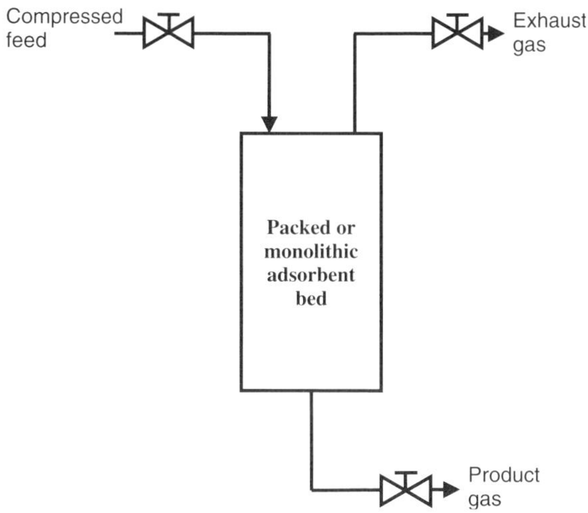

Figure 7.3-1 Pulsed single-bed pressure swing adsorption (PSA) process, which employs cyclic pressurization and depressurization steps to selectively adsorb one or more components from a gas mixture to produce a product stream enriched in the less strongly adsorbed component(s).

A schematic of the pulsed single-bed PSA process is shown in Figure 7.3-1. This process involves using either a packed bed or a monolith adsorbent with a PSA cycle consisting of pressurization and depressurization steps. During the pressurization step for air separation, a high pressure is applied to the surface of the adsorbent bed for a short time. Since the product side of the bed is maintained at atmospheric pressure, a pressure gradient is created to cause airflow through the adsorbent bed. Nitrogen is preferentially retained in the adsorbent bed, and oxygen-enriched gas is collected as a product. During the depressurization step, the product flow outlet at the end of the bed is closed and the adsorbent bed is exposed to a lower pressure (typically, atmospheric) at the feed end for a short time in order to regenerate the adsorbent.

In writing the describing equations for the PSA process, we assume a binary gas mixture consisting of components A and B, corresponding to nitrogen and oxygen, respectively; constant physical properties; and plug flow through the porous adsorbent bed. Since the pressurization and depressurization steps involve different boundary and initial conditions, they are scaled differently. We focus on the pressurization step here because it determines whether the desired enrichment of the product gas can be achieved.[9]

Species balances for each component over a differential volume of the adsorbent bed are given in terms of the molar concentrations c_A and c_B by

$$\varepsilon\frac{\partial c_A}{\partial t} = -\varepsilon\frac{\partial N_A}{\partial z} - (1-\varepsilon)\dot{q}_A \text{ where } N_A = -D_L\frac{\partial c_A}{\partial z} + c_A\hat{u}_z \cong -D_L\frac{\partial c_A}{\partial z} + c_A u_z \tag{7.3-1}$$

[9]Scaling the depressurization step is considered in Practice Problem 7.P.9.

$$\varepsilon \frac{\partial c_B}{\partial t} = -\varepsilon \frac{\partial N_B}{\partial z} - (1-\varepsilon)\dot{q}_B \text{ where } N_B = -D_L \frac{\partial c_B}{\partial z} + c_B \hat{u}_z \cong -D_L \frac{\partial c_B}{\partial z} + c_B u_z \tag{7.3-2}$$

where ε is the porosity of the packed adsorbent column, D_L is the effective axial dispersion coefficient,[10] $\hat{u}_z$ and u_z are the molar- and mass-average velocities, and U is the superficial velocity through the adsorbent bed described by Darcy's law:

$$\varepsilon u_z = U = -\frac{k_p}{\mu}\frac{\partial P}{\partial z} \tag{7.3-3}$$

where k_p is the permeability of the porous adsorbent bed and μ is the shear viscosity of the gas. The molar adsorption rate of component i per unit volume of adsorbent, $\dot{q}_i$, is given by

$$\dot{q}_i = k_i(q_i^e - q_i) \tag{7.3-4}$$

where k_i is the mass-transfer coefficient for component i, q_i the adsorbent concentration (moles per unit volume of adsorbent) of component i, and q_i^e the equilibrium adsorbent concentration of component i given by

$$q_i^e = \frac{q_i^\infty l_i \overline{p}_i}{1 + l_A \overline{p}_A + l_B \overline{p}_B} \tag{7.3-5}$$

where q_i^∞ is the equilibrium adsorbent concentration of component i as $\overline{p}_i \to \infty$ and l_i is the equilibrium adsorption distribution coefficient between the gas and solid phases for component i. Note that the microscale–macroscale methodology introduced in Chapter 6 to model mass transfer with chemical reaction is being employed here; that is, the discrete nature of the adsorbent particles on the microscale is incorporated into a homogeneous sink term on the macroscale. The convective mass transfer with adsorption is described by a mass-transfer coefficient analogous to the manner in which the microscale was handled for mass transfer with chemical reaction. The appropriate form of the overall continuity equation is obtained by adding equations (7.3-1) and (7.3-2) to obtain

$$\varepsilon \frac{\partial c}{\partial t} = -\frac{\partial (Uc)}{\partial z} - (1-\varepsilon)(\dot{q}_A + \dot{q}_B) \tag{7.3-6}$$

where c is the molar density given by $c = c_A + c_B$. Note that in arriving at equation (7.3-6) we have used the definition of the molar-average velocity, $\hat{u}_z = (N_A + N_B)/c$, and have made the assumption that $u = \varepsilon u_z \cong \varepsilon \hat{u}_z$, which is reasonable for gases such as air for which the two components have nearly the same molecular weight.

The operating pressure is usually sufficiently low to permit assuming that the molar density and concentration can be described by the ideal gas law, given by

$$c = \frac{P}{RT}, \qquad c_i = \frac{\overline{p}_i}{RT} \tag{7.3-7}$$

[10]The effective dispersion coefficient incorporates the effects of both molecular diffusion within the void volume and the dispersion due to the hydrodynamics of the flow through the porous media.

where P is the total pressure, $\overline{p}_i$ the partial pressure of component i, and R the gas constant. Hence, equations (7.3-1), (7.3-2), and (7.3-6), only two of which are independent, can be expressed in terms of the partial and total pressures as follows:

$$\varepsilon\frac{\partial \overline{p}_A}{\partial t}+\frac{\partial (U\overline{p}_A)}{\partial z}=\varepsilon D_L\frac{\partial^2 \overline{p}_A}{\partial z^2}-(1-\varepsilon)RT\dot{q}_A \tag{7.3-8}$$

$$\varepsilon\frac{\partial \overline{p}_B}{\partial t}+\frac{\partial (U\overline{p}_B)}{\partial z}=\varepsilon D_L\frac{\partial^2 \overline{p}_B}{\partial z^2}-(1-\varepsilon)RT\dot{q}_B \tag{7.3-9}$$

$$\varepsilon\frac{\partial P}{\partial t}+\frac{\partial (UP)}{\partial z}-=(1-\varepsilon)RT(\dot{q}_A+\dot{q}_B) \tag{7.3-10}$$

Two spatial boundary conditions are required for both P and $\overline{p}_A$. During pressurization one can operate PSA by controlling the upstream and downstream total pressures, P_0 and P_L, respectively, or by controlling the flow rate U and one of the pressures. We assume the former mode of operation. This implies that during pressurization the partial pressures of the components in the feed, $\overline{p}_{A0}$ and $\overline{p}_{B0}$, will be known at the entrance to the adsorption column. However, if axial dispersion cannot be neglected, there necessarily has to be a jump in the species concentration at $z=0$. Jump species balances for components A and B at $z=0$ are given by

$$N_A|_{z=0^-}=\varepsilon N_A|_{z=0^+}\Rightarrow \overline{p}_AU|_{z=0^-}=\left(-\varepsilon D_L\frac{\partial \overline{p}_A}{\partial z}+\overline{p}_AU\right)\Bigg|_{z=0^+} \tag{7.3-11}$$

$$N_B|_{z=0^-}=\varepsilon N_B|_{z=0^+}\Rightarrow \overline{p}_BU|_{z=0^-}=\left(-\varepsilon D_L\frac{\partial \overline{p}_B}{\partial z}+\overline{p}_BU\right)\Bigg|_{z=0^+} \tag{7.3-12}$$

where the notation 0^- and 0^+ denotes evaluation on the upstream and downstream sides, respectively, of the plane at $z=0$. The addition of equations (7.3-11) and (7.3-12) gives the appropriate upstream boundary condition for the total pressure:

$$(N_A+N_B)|_{z=0^-}=\varepsilon(N_A+N_B)|_{z=0^+}\Rightarrow UP|_{z=0^-}=UP|_{z=0^+}\Rightarrow P=P_0 \tag{7.3-13}$$

Adding up similar jump balances on components A and B at the downstream end of the column yields the following downstream boundary condition for the total pressure:

$$\varepsilon(N_A+N_B)|_{z=L^-}=(N_A+N_B)|_{z=L^+}\Rightarrow UP|_{z=L^-}=UP|_{z=L^+}\Rightarrow P=P_L \tag{7.3-14}$$

where the notation L^- and L^+ denotes evaluation on the upstream and downstream sides, respectively, of the plane at $z=L$. Owing to the presence of the axial dispersion term in equation (7.3-8) or (7.3-9), a second boundary condition is required for the partial pressure or its derivative. Frequently, it is assumed without justification that the spatial derivative of the partial pressure is zero at the downstream end of the adsorption bed. However, this implies that there is no further adsorption at the end of the bed. This is not possible if there is an overall pressure drop since the amount of species adsorbed will change in response to the total pressure even if local adsorption equilibrium is achieved. Hence, we merely indicate this downstream boundary condition formally via an unspecified function of time. However, we use the results of our scaling analysis later in this development to suggest a downstream boundary condition that allows for a change in adsorption at the downstream end of the bed. Hence, for the pressurization step the boundary conditions for the describing equations are given by

$$U(y_{A0}P_0 - \overline{p}_A) = -\varepsilon D_L \frac{\partial \overline{p}_A}{\partial z} \quad \text{at} \quad z = 0, \quad 0 \le t \le t_p \tag{7.3-15}$$

$$P = P_0 \quad \text{at} \quad z = 0 \quad 0 \le t \le t_p \tag{7.3-16}$$

$$\overline{p}_A = f(t) \quad \text{at} \quad z = L, \quad 0 \le t \le t_p \tag{7.3-17}$$

$$P = P_L \quad \text{at} \quad z = L, \quad 0 \le t \le t_p \tag{7.3-18}$$

where y_{A0} is the mole fraction of component A in the feed. The initial conditions are assumed to be the following:

$$P = P_L \quad \text{at} \quad t = 0, \quad 0 \le z \le L \tag{7.3-19}$$

$$\overline{p}_A = y_{A0}P_L \quad \text{at} \quad t = 0, \quad 0 \le z \le L \tag{7.3-20}$$

These conditions imply that the depressurization step is done using the same gas as the feed at a pressure equal to that maintained at the end of the bed during the pressurization step. These are reasonable initial conditions for the use of PSA to produce an oxygen-enriched product from air for oxygenator applications. Equations (7.3-3) through (7.3-5), (7.3-8), and (7.3-10), along with the boundary and initial conditions given by equations (7.3-15) through (7.3-20), constitute the describing equations for the pressurization step in pulsed PSA (step 1).

Define the following dimensionless variables in terms of unspecified scale and reference factors (steps 2, 3, and 4):

$$P^* \equiv \frac{P - P_r}{P_s}; \quad \overline{p}_A^* \equiv \frac{\overline{p}_A - \overline{p}_{Ar}}{\overline{p}_{As}}; \quad U^* \equiv \frac{U}{U_s}; \quad q_A^* \equiv \frac{q_A}{q_{As}};$$
$$q_B^* \equiv \frac{q_B}{q_{Bs}}; \quad \dot{q}_A^* \equiv \frac{\dot{q}_A}{\dot{q}_{As}}; \quad \dot{q}_B^* \equiv \frac{\dot{q}_B}{\dot{q}_{Bs}}; \quad t^* \equiv \frac{t}{t_s}; \quad z^* \equiv \frac{z}{z_s} \tag{7.3-21}$$

We have introduced reference factors for the dimensionless total pressure and partial pressure since they are not naturally referenced to zero. Substitute these dimensionless variables into the describing equations and divide through by the dimensional coefficient of one term in each equation to obtain the following dimensionless describing equations (steps 5 and 6):

$$\frac{\varepsilon z_s}{U_s t_s}\frac{\partial P^*}{\partial t^*} + \frac{\partial\left[U^*(P^* + P_r/P_s)\right]}{\partial z^*} = -\frac{(1-\varepsilon)RT\dot{q}_{As}z_s}{U_s P_s}\left(\dot{q}_A^* + \frac{\dot{q}_{Bs}}{\dot{q}_{As}}\dot{q}_B^*\right) \tag{7.3-22}$$

$$\frac{\varepsilon z_s}{U_s t_s}\frac{\partial \overline{p}_A^*}{\partial t^*} + \frac{\partial[U^*(\overline{p}_A^* + \overline{p}_{Ar}/\overline{p}_{As})]}{\partial z^*} = \frac{\varepsilon D_L}{U_s z_s}\frac{\partial^2 \overline{p}_A^*}{\partial z^{*2}} - \frac{(1-\varepsilon)RT\dot{q}_{As}z_s}{U_s\overline{p}_{As}}\dot{q}_A^* \tag{7.3-23}$$

$$U^* = -\frac{k_p P_s}{\mu U_s z_s}\frac{\partial P^*}{\partial z^*} \tag{7.3-24}$$

$$\dot{q}_A^* = \frac{k_A q_{As}}{\dot{q}_{As}}(q_A^{e*} - q_A^*) \tag{7.3-25}$$

$$\dot{q}_B^* = \frac{k_B q_{Bs}}{\dot{q}_{Bs}}(q_B^{e*} - q_B^*) \tag{7.3-26}$$

$$q_A^{e*} = \frac{q_A^{\infty} l_A \overline{p}_{As}}{q_{As}} \times \frac{\overline{p}_A^* + \overline{p}_{Ar}/\overline{p}_{As}}{1 + l_A\overline{p}_{As}\left(\overline{p}_A^* + \overline{p}_{Ar}/\overline{p}_{As}\right) + l_B\overline{p}_{As}\left[(P_s/\overline{p}_{As})P^* - \left(\overline{p}_A^* + \overline{p}_{Ar}/\overline{p}_{As}\right)\right]} \tag{7.3-27}$$

$$q_B^{e*} = \frac{q_B^{\infty} l_B \overline{p}_{As}}{q_{Bs}} \times \frac{(P_s/\overline{p}_{As})P^* - \left(\overline{p}_A^* + (\overline{p}_{Ar}/\overline{p}_{As}\right)}{1 + l_A\overline{p}_{As}\left(\overline{p}_A^* + \overline{p}_{Ar}/\overline{p}_{As}\right) + l_B\overline{p}_{As}\left[(P_s/\overline{p}_{As})P^* - \overline{p}_A^* - \overline{p}_{Ar}/\overline{p}_{As}\right]} \tag{7.3-28}$$

$$P^* = \frac{P_0 - P_r}{P_s} \quad \text{at} \quad z^* = 0, \quad 0 \le t^* \le \frac{t_p}{t_s} \tag{7.3-29}$$

$$P^* = \frac{P_L - P_r}{P_s} \quad \text{at} \quad z^* = \frac{L}{z_s}, \quad 0 \le t^* \le \frac{t_p}{t_s} \tag{7.3-30}$$

$$U^*\left(\frac{y_{A0}P_0 - \overline{p}_{Ar}}{\overline{p}_{As}} - \overline{p}_A^*\right) = -\frac{\varepsilon D_L}{U_s z_s}\frac{\partial \overline{p}_A^*}{\partial z^*} \quad \text{at} \quad z^* = 0, \quad 0 \le t^* \le \frac{t_p}{t_s} \tag{7.3-31}$$

$$\overline{p}_A^* = f(t^*) \quad \text{at} \quad z^* = \frac{L}{z_s}, \quad 0 \le t^* \le \frac{t_p}{t_s} \tag{7.3-32}$$

$$P^* = \frac{P_L - P_r}{P_s} \quad \text{at} \quad t^* = 0, \quad 0 \le z^* \le \frac{L}{z_s} \tag{7.3-33}$$

$$\overline{p}_A^* = \frac{1}{\overline{p}_{As}}(y_{A0}P_L - \overline{p}_{Ar}) \quad \text{at} \quad t^* = 0, \quad 0 \le z^* \le \frac{L}{z_s} \tag{7.3-34}$$

We now set appropriate dimensionless groups equal to 1 or zero to ensure that the dimensionless dependent and independent variables are bounded of O(1) and thereby to determine the scale and reference factors (step 7). The scale and reference pressure come from setting the appropriate dimensionless groups in equations (7.3-29) and (7.3-30) equal to 1 and zero, respectively, to obtain

$$\frac{P_r - P_L}{P_s} = 0 \Rightarrow P_r = P_L; \qquad \frac{P_0 - P_r}{P_s} = 1 \Rightarrow P_s = P_0 - P_L \equiv \Delta P \quad (7.3\text{-}35)$$

The absence of an explicit downstream boundary condition for $\overline{p}_A$ requires special consideration in determining the reference partial pressure $\overline{p}_{Ar}$. Clearly, the minimum value of $\overline{p}_A$ occurs at the downstream end of the bed and is such that $0 \leq \overline{p}_A|_{z=L} \leq y_{A0}P_L$; that is, the minimum value of the partial pressure of the preferentially adsorbed component is bounded between the limits of total and negligible adsorption. Since we seek to bound $\overline{p}_A^*$ to be o(1), let us choose the smallest possible value; that is, $\overline{p}_{Ar} = 0$. The partial pressure scale is then determined by setting the appropriate dimensionless group in equation (7.3-31) equal to 1. Hence, we obtain

$$\overline{p}_{Ar} = 0; \qquad \overline{p}_{As} = y_{A0}P_0 \tag{7.3-36}$$

The scale factor for the time is obtained by setting the dimensionless group appearing in equations (7.3-29) through (7.3-32) equal to 1:

$$\frac{t_p}{t_s} = 1 \Rightarrow t_s = t_p \tag{7.3-37}$$

The scale factor for the spatial coordinate can be obtained by setting the dimensionless group appearing in equations (7.3-30) and (7.3-32) through (7.3-34) equal to 1:

$$\frac{L}{z_s} = 1 \Rightarrow z_s = L \tag{7.3-38}$$

The scale for the velocity is obtained by setting the dimensionless group in equation (7.3-24) equal to 1:

$$\frac{k_p P_s}{\mu U_s z_s} = \frac{k_p \Delta P}{\mu U_s L} = 1 \Rightarrow U_s = \frac{k_p \Delta P}{\mu L} \tag{7.3-39}$$

The scales for the equilibrium adsorption concentrations are obtained by setting the dimensionless groups for the leading-order behavior in equations (7.3-27) and (7.3-28) equal to 1:

$$\begin{aligned} \frac{q_A^\infty l_A \overline{p}_{As}}{q_{As}} &= 1 \Rightarrow q_{As} = q_A^\infty l_A y_{A0} P_0 \\ \frac{q_B^\infty l_B \overline{p}_{As}}{q_{Bs}} &= 1 \Rightarrow q_{Bs} = q_B^\infty l_B y_{A0} P_0 \end{aligned} \tag{7.3-40}$$

The scales for the adsorption rates are obtained by setting the dimensionless groups in equations (7.3-25) and (7.3-26) equal to 1:

$$\frac{k_A q_{As}}{\dot{q}_{As}} = 1 \Rightarrow \dot{q}_{As} = k_A q_A^{\infty} l_A y_{A0} P_0$$
$$\frac{k_B q_{Bs}}{\dot{q}_{Bs}} = 1 \Rightarrow \dot{q}_{Bs} = k_B q_B^{\infty} l_B y_{A0} P_0 \qquad (7.3\text{-}41)$$

This scaling identifies four relevant time scales for the pulsed single-bed PSA process, which are defined as

$$t_c = \frac{L}{U_s} = \frac{\mu L^2}{k_p \Delta P} = \text{characteristic contact time} \qquad (7.3\text{-}42)$$

$$t_p = \text{pressurization time} = \text{total time available for adsorption} \qquad (7.3\text{-}43)$$

$$t_d \equiv \frac{L^2}{D_L} \sim \text{characteristic axial dispersion time} \qquad (7.3\text{-}44)$$

$$t_{ad} = \frac{1}{(1-\varepsilon) RT k_i q_i^{\infty} l_i} = \text{characteristic adsorption time for species } i \qquad (7.3\text{-}45)$$

Substituting the scale and reference factors into equations (7.3-22) through (7.3-34) results in the following dimensionless describing equations:

$$\Pi_1 \frac{\partial P^*}{\partial t^*} + \frac{\partial [U^*(P^* + \Pi_6)}{\partial z^*} = -\Pi_3 \Pi_4 [(q_A^{e*} - q_A^*) + \Pi_5 (q_B^{e*} - q_B^*)] \qquad (7.3\text{-}46)$$

$$\Pi_1 \frac{\partial \overline{p}_A^*}{\partial t^*} + \frac{\partial (U^* \overline{p}_A^*)}{\partial z^*} = \Pi_2 \frac{\partial^2 \overline{p}_A^*}{\partial z^{*2}} - \Pi_3 (q_A^{e*} - q_A^*) \qquad (7.3\text{-}47)$$

$$U^* = -\frac{\partial P^*}{\partial z^*} \qquad (7.3\text{-}48)$$

$$q_A^{e*} = \frac{\overline{p}_A^*}{1 + \Pi_4 \Pi_7 \overline{p}_A^* + \Pi_8 (P^* - \Pi_4 \overline{p}_A^*)} \qquad (7.3\text{-}49)$$

$$q_B^{e*} = \frac{(1/\Pi_4) P^* - \overline{p}_A^*}{1 + \Pi_4 \Pi_7 \overline{p}_A^* + \Pi_8 (P^* - \Pi_4 \overline{p}_A^*)} \qquad (7.3\text{-}50)$$

$$P^* = 1 \quad \text{at} \quad z^* = 0, \quad 0 \le t^* \le 1 \qquad (7.3\text{-}51)$$

$$P^* = 0 \quad \text{at} \quad z^* = 1, \quad 0 \le t^* \le 1 \qquad (7.3\text{-}52)$$

$$U^*(1 - \overline{p}_A^*) = -\Pi_2 \frac{\partial \overline{p}_A^*}{\partial z^*} \quad \text{at} \quad z^* = 0, \quad 0 \le t^* \le 1 \qquad (7.3\text{-}53)$$

$$p_A^* = f(t^*) \quad \text{at} \quad z^* = 1, \quad 0 \le t^* \le 1 \qquad (7.3\text{-}54)$$

$$P^* = 0 \qquad \text{at} \quad t^* = 0, \quad 0 \leq z^* \leq 1 \tag{7.3-55}$$

$$\overline{p}_A^* = \frac{\Pi_6}{1 + \Pi_6} \qquad \text{at} \quad t^* = 0, \quad 0 \leq z^* \leq 1 \tag{7.3-56}$$

where we have substituted equations (7.3-25) and (7.3-26) for the adsorption rates into the overall molar and species-balances given by equations (7.3-22) and (7.3-23). The dimensionless groups introduced into equations (7.3-46) through (7.3-56) are defined as

$$\Pi_1 \equiv \frac{\varepsilon \mu L^2}{k_p \, \Delta P t_p} = \varepsilon \frac{\mu L^2 / k_p \, \Delta P}{t_p} = \varepsilon \frac{t_c}{t_p} \propto \frac{\text{contact time}}{\text{pressurization time}} \tag{7.3-57}$$

$$\Pi_2 \equiv \frac{\varepsilon \mu D_L}{k_p \, \Delta P} = \varepsilon \frac{\mu L^2 / k_p \, \Delta P}{L^2 / D_L} = \varepsilon \frac{t_c}{t_d} \propto \frac{\text{contact time}}{\text{axial dispersion time}} \tag{7.3-58}$$

$$\Pi_3 \equiv \frac{(1 - \varepsilon) R T k_A q_A^{\infty} l_A \mu L^2}{k_p \, \Delta P} = \frac{\mu L^2 / k_p \, \Delta P}{1/(1 - \varepsilon) R T k_A q_A^{\infty} l_A}$$

$$= \frac{t_c}{t_{ad}} \propto \frac{\text{contact time}}{\text{adsorption time}} \tag{7.3-59}$$

$$\Pi_4 \equiv \frac{y_{A0} P_0}{\Delta P} = \text{dimensionless partial pressure in feed} \tag{7.3-60}$$

$$\Pi_5 \equiv \frac{k_B q_B^{\infty} l_B}{k_A q_A^{\infty} l_A} = \frac{1/(1 - \varepsilon) R T k_A q_A^{\infty} l_A}{1/(1 - \varepsilon) R T k_B q_B^{\infty} l_B} \sim \frac{\text{adsorption time for species} A}{\text{adsorption time for species} B} \tag{7.3-61}$$

$$\Pi_6 \equiv \frac{P_L}{\Delta P} \sim \text{dimensionless reference pressure} \tag{7.3-62}$$

$$\Pi_7 \equiv l_A \, \Delta P \sim \text{nonlinear adsorption effects for} A \tag{7.3-63}$$

$$\Pi_8 \equiv l_B \, \Delta P \sim \text{nonlinear adsorption effects for} B \tag{7.3-64}$$

Now let us use our scaling analysis to design the pulsed PSA process for both the packed and monolithic single-bed configurations (step 8). Table 7.3-1 summarizes typical design parameters for the use of pulsed single-bed PSA employing both a packed adsorption column and a monolithic adsorbent (zeolite 5 Å) to provide an oxygen-enriched product from an air feed stream. The monolithic adsorbent addresses the problems associated with using a packed adsorption column: namely, the ablation of the particles and low product flow rates associated with a relatively long adsorbent bed. Whereas the adsorbents used in pulsed packed bed PSA have a radius (R_P) on the order of hundreds of micrometers, the microcrystals constituting the adsorbent monolith have a characteristic pore size on the order of 1 μm. Hence, the mass-transfer coefficient is much larger, whereas the permeability is much smaller for the monolithic adsorbent. Note that the pressurization time is unspecified for both the packed and monolithic bed processes in Table 7.3-1 since this is a design parameter that we seek to determine from the scaling analysis.

TABLE 7.3-1 System Parameters for the Pulsed Single-Bed PSA Process

Parameter	Packed Adsorption Column	Monolithic Adsorbent
R_p (m)	5×10^{-4}	1×10^{-6}
k_A, k_B (s^{-1})	6.0×10^{-3}	1.50×10^{3}
k_p (m^2)	6.76×10^{-10}	2.71×10^{-15}
ε	0.35	0.35
D_L (m^2/s)	1.0×10^{-5}	1.0×10^{-6}
q_A^∞ (mol/m^2)	2.8×10^{3}	2.8×10^{3}
q_B^∞ (mol/m^2)	2.2×10^{3}	2.2×10^{3}
l_A (Pa^{-1})	5.0×10^{-7}	5.0×10^{-7}
l_B (Pa^{-1})	1.8×10^{-7}	1.8×10^{-7}
μ (Pa · s)	1.8×10^{-5}	1.8×10^{-5}
P_0 (Pa)	1.3×10^{5}	1.5×10^{5}
P_L (Pa)	1.0×10^{5}	1.0×10^{5}
T (K)	293	293
L (m)	1.0	2.0×10^{-3}
t_p (s)	—	—

Source: E. M. Kopaygorodsky, V. V. Guliants, and W. B. Krantz, *A.I.Ch.E. J.*, **50**(5) 953 (2004).

TABLE 7.3-2 Dimensionless Groups Characterizing Pulsed Packed and Monolithic Single-Bed PSA for the Process Parameters Given in Table 7.3-1 for Producing an Oxygen-Enriched Product from Air

Dimensionless Group	Packed Bed	Monolithic Bed
Π_1	$0.311/t_p$	$0.186/t_p$
Π_2	3.11×10^{-6}	0.0465
Π_3	0.0118	1768
Π_4	2.63	1.58
Π_5	0.283	0.283
Π_6	3.33	2.00
Π_7	0.015	0.025
Π_8	0.00540	0.00900

If the pressurization time is too short, breakthrough of the adsorption front will not have occurred, and no enrichment will be achieved. On the other hand, if the pressurization time is too long, the bed will become completely saturated at the prevailing local pressure and no enrichment will be achieved in the product gas. The optimal pressurization time achieves the maximum possible enrichment of the product gas. The process parameters in Table 7.3-1 lead to the values of the dimensionless groups that characterize the pulsed packed and monolithic single-bed PSA processes summarized in Table 7.3-2.

To have a workable oxygenator, it is essential that the dimensionless group $\Pi_5 < 1$; that is, the adsorbent must preferentially adsorb nitrogen to produce an oxygen-enriched product. We see from Table 7.3-2 that $\Pi_5 = 0.283$, which indicates that this adsorbent is satisfactory. The value of $\Pi_2 \ll 1$ for pulsed packed single-bed PSA, which implies that axial dispersion can be ignored. Although Π_2 is much larger for pulsed monolithic single-bed PSA, axial dispersion can still be neglected in modeling this process without incurring significant error. The small values of Π_7 and Π_8 imply that the adsorption isotherms for both nitrogen and oxygen can be linearized in the partial pressure without incurring significant error.

A dramatic difference between pulsed packed and monolithic single-bed PSA is apparent from the values of Π_3. Whereas $\Pi_3 \ll 1$ for pulsed packed single-bed PSA, $\Pi_3 \gg 1$ for pulsed monolithic single-bed PSA. Since a proper scaling implies that all the terms in the describing equations be bounded of $\circ(1)$, the product of Π_3 with $q_A^{e*} - q_A^*$ must be of $\circ(1)$ in equation (7.3-47); this implies that $q_A^* \cong q_A^{e*}$ for monolithic bed PSA. A similar argument for equation (7.3-46) implies that $q_B^* \cong q_B^{e*}$. Hence, the adsorption time is so fast in comparison to the contact time (i.e., $t_{ad} \ll t_c$) that we conclude that local thermodynamic adsorption equilibrium prevails for pulsed monolithic single-bed PSA. In contrast, the contact time is considerably faster than the adsorption time for pulsed packed single-bed PSA (i.e., $t_c \ll t_{ad}$), thus implying that adsorption in this process is mass-transfer controlled. This marked difference between the adsorption behavior for pulsed packed and monolithic single-bed PSA arises because of the very small particle size associated with the latter. The small particle size implies a much smaller permeability and therefore a much slower contact time relative to the adsorption time; moreover, in contrast to packed bed PSA, the interparticle mass-transfer resistance is negligible for the monolithic bed process.

The marked difference between the characteristic adsorption time and the contact time for packed relative to monolithic bed PSA implies that a different design strategy must be used for each process. Since the adsorption time is much longer than the contact time for pulsed packed single-bed PSA, the pressurization time must be approximately equal to the adsorption time. Scaling analysis then suggests the following criterion for determining the pressurization time for conventional pulsed packed single-bed PSA:

$$t_p = t_{ad} = \frac{1}{(1-\varepsilon)RTk_A q_A^{\infty} l_A} \tag{7.3-65}$$

For the design parameters given in Table 7.3-1, this criterion suggests a pressurization time of 75.2 seconds, which agrees well with typical values used in practice for packed pulsed single-bed PSA. On the other hand, since the pulsed monolithic single-bed PSA process occurs at local adsorption equilibrium, the pressurization time must be approximately equal to the contact time required for the adsorption "wave" to pass through the column. Scaling analysis then suggests the following criterion for determining the pressurization time for monolithic single-bed PSA:

$$t_p = t_c = \frac{\mu L^2}{k_p \, \Delta P} \tag{7.3-66}$$

For the design parameters given in Table 7.3-1, this criterion suggests a pressurization time of 0.532 seconds. These estimates for the optimal pressurization times permit determining the values of Π_1, the dimensionless group that provides a measure of the importance of the unsteady-state term in equations (7.3-46) and (7.3-47). For packed single-bed PSA, $\Pi_1 = 0.00414$, which implies that the pressurization step can be considered to be a steady-state process. However, for monolithic bed PSA, $\Pi_1 = 0.350$, which implies that the pressurization step is inherently unsteady-state.

The fact that $\Pi_3 \gg 1$ for monolithic single-bed PSA implies that the length scale for the pressure and partial pressure gradients for this process is not equal to the length of the adsorption column as was assumed in our scaling analysis. Indeed, this large value of Π_3 suggests that monolithic bed PSA involves a region of influence or boundary layer across which the pressure drop occurs. Hence, the describing equations must be rescaled for conditions such that the appropriate length scale for the pressure gradients is the boundary-layer thickness δ_m. The thickness of the region of influence or boundary layer wherein the total and partial pressures undergo their characteristic change is obtained by balancing the convection term with the accumulation term in equation (7.3-22) or (7.3-23) since the monolithic bed PSA process is inherently unsteady state; this results in the following equation for the boundary-layer thickness:

$$z_s \equiv \delta_m = \sqrt{\frac{k_p \, \Delta P t_o}{\varepsilon \mu}} \tag{7.3-67}$$

where t_o is the observation time. For the design parameters given in Table 7.3-1, the boundary-layer thickness is given by

$$\delta_m = 4.64 \times 10^{-3} \sqrt{t_o} \tag{7.3-68}$$

Clearly, $0 \leq \delta_m \leq L$; that is, once the boundary layer has penetrated to the end of the monolithic bed, breakthrough of the adsorption wave occurs. One implication of this region of influence is that any numerical technique employed to solve the describing equations must allow for adequate resolution of the steep total and partial pressure gradients at very short contact times.

Whereas the process parameters in Table 7.3-1 imply that neglecting axial dispersion is a very good approximation for pulsed packed single-bed PSA, they indicate that it could be important for pulsed monolithic single-bed PSA at least under some operating conditions. When axial dispersion cannot be neglected, it is necessary to apply a downstream boundary condition on the partial pressure. Due to the very rapid adsorption time scale for the monolithic bed PSA process, the total and partial pressure profiles propagate as sharp fronts through the adsorbent bed. Hence, until breakthrough of the adsorption wave occurs, a downstream boundary condition demanding that the spatial derivative of the partial pressure be zero is appropriate. Note that the downstream boundary condition given in dimensional form by equation (7.3-18) can be applied for the total pressure throughout the entire adsorption step. However, once breakthrough has occured, demanding that the derivative of the partial pressure be zero is not justified. This assumption can introduce some error even when equilibrium adsorption applies since the latter changes with the local total pressure. That is, if there is a pressure drop owing to the resistance offered by the adsorption bed to gas flow, the partial pressure must change with axial distance as well. Our scaling analysis suggests an alternative downstream boundary condition for the partial pressure that is far less restrictive than demanding that the partial pressure gradient be zero. Since both Π_1 and Π_2 are small, it is a reasonable assumption after breakthrough of the adsorption wave has occurred to ignore the accumulation and axial dispersion terms in equation (7.3-47) at the downstream end of the adsorption bed. Hence, a more realistic downstream boundary condition for the partial pressure is given by

$$\frac{\partial(U^*\overline{p}_A^*)}{\partial z^*} = \Pi_3(q_A^{e*} - q_A^*) \qquad \text{at} \quad z^* = 1 \tag{7.3-69}$$

This boundary condition states that the change in the net convection of the adsorbed component is due to its adsorption. Although this condition neglects any accumulation or axial dispersion of the adsorbed component at the downstream end of the bed, it does allow for a nonzero partial pressure gradient.

The pulsed single-bed PSA process provides a good example to illustrate the use of advanced dimensional analysis concepts for correlating experimental or numerical data. The quantities of particular interest in characterizing PSA performance are the product purity y_{BL} and recovery ψ_r. The latter is equal to the ratio of the volumetric flow of the product (oxygen) to that of the feed and is defined by the following equation (step 1 in the scaling procedure for dimensional analysis):

$$\psi_r = \frac{y_{BL}U|_{x=L}}{y_{B0}U|_{x=0}} = \frac{y_{BL}(\partial P/\partial x)|_{x=L}}{y_{B0}(\partial P/\partial x)|_{x=0}} \tag{7.3-70}$$

The values of y_{BL} and $\partial P / \partial x$ at the feed and product ends of the adsorption bed would have to be obtained by solving equations (7.3-46) through (7.3-56). Steps 2 through 7 in the scaling approach to dimensional analysis have already been done when we scaled the pulsed PSA process that resulted in the eight dimensionless groups defined by equations (7.3-57) through (7.3-64). Moreover, Table 7.3-2 indicates that for both the packed and monolithic pulsed single-bed PSA process, the groups Π_2, Π_7, and Π_8 are sufficiently small so that axial dispersion and nonlinear adsorption effects can be ignored without incurring significant error; that is, these dimensionless groups can be ignored in developing our dimensional analysis correlations for the product purity and recovery. However, the remaining five dimensionless groups are not optimal for correlating numerical or experimental data for y_{BL} and ψ_r. For an air feed delivering an enriched oxygen product at a fixed low pressure P_L (usually atmospheric) and a specified adsorbent, bed length, and temperature, the performance of pulsed single-bed PSA depends only on the pressurization time t_p and the applied pressure P_0. Ideally, one would like to isolate these two quantities to simplify correlating either experimental or numerical data. This can be done by first applying step 10 in the procedure for dimensional analysis that is indicated formally by equation (2.4-4); that is, since P_0 and P_L

appear independently in the dimensionless groups Π_4 and Π_6, respectively, the quantity ΔP can be replaced by P_L in the dimensionless groups Π_1 and Π_3. This same principle permits replacing ΔP by P_0 in the dimensionless groups Π_4 and Π_6. These operations then result in the following new dimensionless groups that replace Π_1, Π_3, Π_4, and Π_6:

$$\Pi_9 \equiv \frac{k_p P_L t_p}{\varepsilon \mu L^2} \tag{7.3-71}$$

$$\Pi_{10} \equiv \frac{(1-\varepsilon) R T k_A q_A^{\infty} l_A \mu L^2}{k_p P_L} \tag{7.3-72}$$

$$\Pi_{11} \equiv y_{A0} \tag{7.3-73}$$

$$\Pi_{12} \equiv \frac{P_0}{P_L} \tag{7.3-74}$$

where Π_9 and Π_{12} have been defined so that they are directly proportional to t_p and P_0, respectively. Note that we have isolated the pressurization time t_p in Π_9 and the applied pressure P_0 in Π_{12}. Hence, dimensional analysis correlations for the product purity and recovery will be of the form

$$\begin{aligned} y_{BL} &= f_1(\Pi_5, \Pi_9, \Pi_{10}, \Pi_{11}, \Pi_{12}) \\ \psi_r &= f_2(\Pi_5, \Pi_9, \Pi_{10}, \Pi_{11}, \Pi_{12}) \end{aligned} \tag{7.3-75}$$

However, for an air feed delivering an enriched oxygen product at a fixed low pressure P_L and a specified adsorbent, bed length, and temperature, the dimensionless groups Π_5, Π_{10}, and Π_{11} are constant. Hence, experimental or numerical data for the product purity and recovery can be correlated in terms of just two dimensionless groups:

$$\begin{aligned} y_{BL} &= f_1(\Pi_9, \Pi_{12}) = f_1\left(\frac{k_p P_L t_p}{\varepsilon \mu L^2}, \frac{P_0}{P_L}\right) \\ \psi_r &= f_2(\Pi_9, \Pi_{12}) = f_2\left(\frac{k_p P_L t_p}{\varepsilon \mu L^2}, \frac{P_0}{P_L}\right) \end{aligned} \tag{7.3-76}$$

One would anticipate that both the product purity and recovery would be monotonically increasing functions of Π_{12}, due to the increased adsorption of nitrogen at higher operating pressures. However, both the product purity and recovery display a maximum as a function of Π_9, due to the fact that longer pressurization times imply progressively more loading of the adsorbent.

This design problem again illustrates the advantages of employing O(1) scaling to determine the dimensionless groups required to correlate experimental or numerical data. A naive application of the Pi theorem indicates that 11 dimensionless groups would be required to correlate the product purity and product recovery; that is, $n = 16$ (the 15 quantities in Table 7.3-1 and the gas constant R) and $m = 5$ (M, mol, L, t, and T). Scaling analysis is able to reduce the number of dimensionless groups to five by systematically identifying combinations of the quantities (i.e., RT, $k_A q_A^\infty$, and $k_B q_B^\infty$) and assessing reasonable simplifying assumptions (that the groups Π_2, Π_7, and Π_8 can be neglected); that is, the number of quantities is reduced to 13 in terms of five units, and three of the dimensionless groups are sufficiently small to permit neglecting them. The latter approximations involve ignoring axial dispersion and nonlinear adsorption effects. However, by recognizing that several process parameters are fixed once the adsorbent is specified for producing an oxygen-enriched product from air, it was possible to correlate either the product purity or recovery in terms of just two variable dimensionless groups.

7.4 THERMALLY INDUCED PHASE-SEPARATION PROCESS FOR POLYMERIC MEMBRANE FABRICATION

This application of scaling analysis focuses on the formation of microporous membranes through a process known as *thermally induced phase separation* (TIPS). The goal of scaling in this example is to explore the conditions for which the complex describing equations can be simplified to permit a tractable numerical solution. Before considering the describing equations, a brief background on the TIPS process for polymeric membrane formation will be given.

A membrane is a semipermeable medium that permits the passage of some molecules, colloidal aggregates, or particles relative to others. Most microporous membranes are made from polymeric materials through a technique known as *phase inversion* because it involves converting a single-phase solution of polymer in a solvent, referred to as the *casting solution*, into a two-phase dispersion in which the polymer ultimately becomes a continuous microporous solid-phase matrix. The TIPS process involves forming a single-phase solution of the polymer in a solvent, referred to as the *diluent*, at an elevated temperature. However, the diluent becomes a nonsolvent below some lower temperature. Hence, phase separation in the TIPS process occurs by casting a thin film of the hot single-phase polymer solution onto a cold surface. The resulting front that separates the phase-separated region from the single-phase solution then propagates in time away from the cold boundary. For solutions of amorphous (i.e., having no crystalline regions) or nonconcentrated semicrystalline (i.e., having both amorphous and crystalline regions) polymers, TIPS involves the formation of two liquid phases. However, for a more concentrated semicrystalline polymer solution, which is the focus of this example, TIPS involves the formation of solid polymer and a polymer-lean liquid phase. This solid polymer phase will evolve through a nucleation and growth process that can be described by Avrami theory.[11] The dispersed-phase polymer particles eventually fuse together to create a continuous polymer network that provides the structural matrix of the microporous membrane. The dispersed polymer-lean phase then is extracted with an appropriate solvent to create a microporous network. Subsequent post-casting treatment can involve annealing the membrane at a temperature slightly above that

[11] M. Avrami, *J. Chem. Phys.*, **7**, 1103–1112 (1939); **8**, 212–224 (1940); **9**, 177–184 (1941).

of the *glass-transition point* of the polymer (i.e., the temperature above which significant segmental and polymer chain motion becomes possible) to decrease the pore size and increase the selectively of the membrane. In this example, which is based on the work of Li et al.,[12] we are concerned with developing a model to describe the evolution of the microstructure during the phase-separation process. Prior to the work described here, there was no model available to describe structural evolution during the solid–liquid TIPS process.

The physical considerations that need to be incorporated into a model for structural evolution during the TIPS process are summarized here. Commercial TIPS casting is usually done continuously; hence, it is reasonable to assume one-dimensional transport. The TIPS process will involve heat transfer from the hot polymer casting solution to the cold boundary, and possibly to the ambient gas phase as well. This cooling will eventually cause phase separation of pure polymer dispersed in a continuous polymer-lean phase. This phase separation in turn can cause diffusive mass transfer due to the concentration gradients that are created by the nucleation and growth of the solid phase. This constitutes a moving boundary problem since the front separating the single- and two-phase regions will propagate away from the cold boundary. This moving boundary is defined by the nonequilibrium thermodynamic condition, which relates the temperature at which nucleation of the polymer phase begins to the instantaneous cooling rate; this condition can be determined experimentally via differential scanning calorimetry. The temperature for the inception of nucleation increases in time since the cooling rate decreases as the phase-separation front moves away from the cold surface. One complication is that the physical and transport properties of the polymer-lean and solid polymer in the phase-separated region can be different. A model must somehow account for the presence of the two phases in this phase-separated region. Note also that the volume of the dispersed phase will increase in time at any plane within the phase-separated region, owing to the progressive precipitation of the polymer. In the following, an appropriate set of describing equations is developed that accounts for the principal features of the TIPS casting process. This problem involves a microscale element, (i.e., a dispersed polymer phase particle) and a macroscale element, which is a differential thickness of the casting solution.

A schematic of TIPS casting is shown in Figure 7.4-1. The origin of the coordinate system is located at the cold boundary. Allowance is made for possible heat loss to the ambient gas phase at the upper boundary of the casting solution. The moving boundary between the single- and two-phase regions is located at $z = L(t)$, where z is the spatial coordinate, L is the location of the moving boundary, and t is time. Representative temperature and diluent-concentration profiles are shown in the figure. Note that both the temperature and diluent-concentration profiles as well as their slopes are continuous at the moving boundary, owing to the fact that no phase separation has occurred at this plane. That is, this boundary is defined to be the plane at which nucleation becomes possible at the instantaneous cooling rate. However, no nuclei have formed at this boundary, because there has not been

[12] D. Li, A. R. Greenberg, W. B. Krantz, and R. L. Sani, *J. Membrane Sci.*, **279**, 50 (2006).

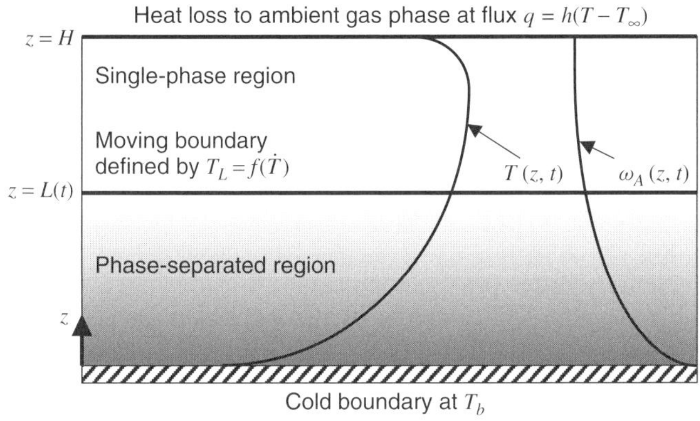

Figure 7.4-1 Thermally induced phase-separation (TIPS) process showing representative temperature and diluent-concentration profiles; phase separation in the initially hot polymer solution propagates from the lower cold boundary; the moving boundary separating the homogeneous and phase-separated regions is denoted by $L(t)$; the solid volume fraction in the phase-separated region increases in time, as shown schematically by the progressively darker shading toward the cold boundary; heat loss at the interface between the casting solution and ambient gas causes a temperature decrease near this upper boundary.

enough time for them to initiate and grow. Nuclei will form and grow at this fixed plane when the moving boundary moves upward due to the progressive cooling. Hence, this moving boundary problem is not a classical Stefan problem defined by a latent heat effect associated with phase change at the moving boundary for which the temperature gradients (and both the concentration and its gradient for a multicomponent phase-change process) are discontinuous.

The appropriately simplified energy equation given by equation (F.1-2) in the Appendices is given by (step 1)

$$\rho C_p \frac{\partial T}{\partial t} = \frac{\partial}{\partial z}\left(k \frac{\partial T}{\partial z}\right) - \rho_p \Delta H_f \frac{\partial \psi}{\partial t} \tag{7.4-1}$$

in which T is the temperature and ψ is the liquid volume fraction in the phase-separated region. The properties ρ, C_p, and k are the mass density, heat capacity, and thermal conductivity, respectively. The heat-generation term introduces ΔH_f, the latent heat of fusion of the polymer, and ρ_p, the density of the pure polymer. The species-balance given by equation (G.1-5) in the Appendices needs to be modified to account for the fact that diffusion can occur only in the continuous liquid phase. For this reason the liquid volume fraction ψ appears in the species-balance, which is given by

$$\frac{\partial(\omega_A \psi)}{\partial t} = \omega_A \frac{\partial \psi}{\partial t} + \psi \frac{\partial \omega_A}{\partial t} = \frac{\partial}{\partial z}\left(D_{AB} \psi \frac{\partial \omega_A}{\partial z}\right) \tag{7.4-2}$$

where ω_A is the mass fraction of the diluent and D_{AB} is the binary diffusion coefficient. Equation (7.4-2) also applies to the single-phase region in which $\psi = 1$.

The physical properties of the polymer solution depend on the concentration. The physical properties in the phase-separated region depend on the liquid volume fraction and the concentration. It is reasonable to assume that the relevant solution properties are weight-fraction-weighted sums of the pure component properties, and that the relevant properties in the two-phase region are volume-fraction-weighted sums of the properties of the solution and pure polymer phases. Hence, the following equations are used for the relevant physical properties:

$$\rho C_p = (1 - \psi)(\rho C_p)_P + \psi[\omega_A(\rho C_p)_D + (1 - \omega_A)(\rho C_p)_P] \tag{7.4-3}$$

$$k = (1 - \psi)k_P + \psi[\omega_A k_D + (1 - \omega_A)k_P] \tag{7.4-4}$$

where the subscripts P and D denote polymer and diluent properties, respectively. Equations (7.4-3) and (7.4-4) apply in both the single- and two-phase regions since in the single-phase region the liquid volume fraction $\psi = 1$.

Equations (7.4-1) and (7.4-2) constitute two equations in three dependent variables, T, ω, and ψ. An additional equation to determine the liquid volume fraction is provided by Avrami theory for the nucleation and growth process[13]:

$$\psi = \psi_\infty + (1 - \psi_\infty)\exp(-Kt_g^n) \tag{7.4-5}$$

where ψ_∞ is the equilibrium liquid volume fraction at the prevailing temperature, K the crystallization rate constant, and n is referred to as the Avrami exponent. Note that t_g is the time for growth of the solid phase; t_g is initialized at each location of the plane that separates the phase-separated and single-phase regions; hence, t_g will become progressively shorter at planes farther removed from the cold boundary, at which phase separation begins.

The initial and boundary conditions required to solve equations (7.4-1) and (7.4-2) are given by

$$T = T_0, \quad \omega_A = \omega_{A0}, \quad \psi = 1 \quad \text{at} \quad t = 0 \tag{7.4-6}$$

$$T = T_b, \quad \frac{\partial \omega_A}{\partial z} = 0 \quad \text{at} \quad z = 0 \tag{7.4-7}$$

$$\psi = 1 \quad \text{for} \quad z \geq L \tag{7.4-8}$$

$$-k\frac{\partial T}{\partial z} = h(T - T_\infty), \quad \frac{\partial \omega_A}{\partial z} = 0 \quad \text{at} \quad z = H \tag{7.4-9}$$

Equation (7.4-7) specifies the temperature at the impermeable cold boundary. Equation (7.4-8) specifies that there is no phase separation at or above the moving boundary at L. Equation (7.4-9) specifies that the heat conducted to the upper

[13] Avrami, *J. Chem. Phys.*, **7**, 1103–1112 (1939); **8**, 212–224 (1940); **9**, 177–184 (1941).

interface is equal to the convective heat transfer into the ambient gas phase, whose temperature is T_∞, and that the diluent is nonvolatile.

An auxiliary equation is needed to locate the position of the moving boundary that separates the single- and two-phase regions. This is obtained from measurements of the nonequilibrium crystallization temperature T_c as a function of cooling rate $\dot{T} \equiv \partial T / \partial t$ using differential scanning calorimetry; that is,

$$T_c = f\left(\frac{\partial T}{\partial t}\right) = f(\dot{T}) \qquad \text{at} \quad z = L \tag{7.4-10}$$

When $\dot{T} \to 0$, $T_c \to$ a constant determined only by the composition of the solution.

To use this model to predict the size of the dispersed-phase particles, it is necessary to have a relationship between the number of nuclei generated as a function of the degree of subcooling at the moving boundary between the phase-separated and single-phase regions. However, this relationship is not needed to scale the describing equations and hence is not given here. This relationship implies that the number of nuclei decreases with a reduced degree of subcooling, which in turn is proportional to the cooling rate. Hence, for a constant cold boundary temperature, the dispersed phase particle size will progressively increase away from this boundary owing to a gradual decrease in the cooling rate. This is the underlying reason why the TIPS process can create an asymmetric membrane structure whereby the permselective layer created near the cold boundary is supported by a more open highly permeable substructure.

The describing equations given (7.4-1) through (7.4-10) are nontrivial to solve numerically, for several reasons. First, this involves a set of coupled parabolic and hyperbolic partial differential equations. These equations are stiff; that is, the time derivatives can be quite large initially. Moreover, this is a moving boundary problem for which it is necessary to obtain the instantaneous cooling rate to determine the extent of the phase-separated region. Another complication is that the origin of the time scale for the time dependence of ψ is different for each plane in the phase-separated region since nucleation begins progressively later as the moving boundary progresses into the single-phase region. Hence, it is necessary to determine both the rate of cooling $\dot{T}$ and the time at which equation (7.4-10) is satisfied for each plane in the phase-separated region. Coping with these complexities certainly contributed to the delay in developing a model for TIPS membrane formation via a nucleation-and-growth process. Hence, we seek to scale these describing equations to determine the conditions for which a tractable numerical solution can be obtained.

We begin by introducing appropriate reference and scale factors via the following dimensionless variables (steps 2, 3, and 4):

$$T^* \equiv \frac{T - T_r}{T_s}; \qquad z^* \equiv \frac{z}{z_s}; \qquad t^* \equiv \frac{t}{t_s}; \qquad \dot{\psi}^* \equiv \frac{1}{\dot{\psi}_s}\frac{\partial \psi}{\partial t}; \qquad k^* \equiv \frac{k}{k_s};$$
$$(\rho C_p)^* \equiv \frac{(\rho C_p)_D}{(\rho C_p)_s}; \qquad D_{AB}^* \equiv \frac{D_{AB}}{D_s} \tag{7.4-11}$$

Note that we did not scale the diluent mass fraction ω_A and the liquid volume fraction ψ since they are dimensionless and bounded of o(1) for typical TIPS processing conditions. However, we have introduced a separate scale for the time rate of change of the liquid volume fraction since this derivative does not necessarily scale with the reciprocal of the characteristic time. These variables result in the following set of dimensionless describing equations (steps 5 and 6):

$$\frac{(\rho C_p)_s z_s^2}{k_s t_s}(\rho C_p)^* \frac{\partial T^*}{\partial t^*} = \frac{\partial}{\partial z^*}\left(k^* \frac{\partial T^*}{\partial z^*}\right) - \frac{\rho_P\, \Delta H_f\, \dot{\psi}_s z_s^2}{k_s T_s}\dot{\psi}^* \tag{7.4-12}$$

$$\dot{\psi}_s t_s \omega_A \dot{\psi}^* + \psi \frac{\partial \omega_A}{\partial t^*} = \frac{D_s t_s}{z_s^2}\frac{\partial}{\partial z^*}\left(D_{AB}^* \psi \frac{\partial \omega_A}{\partial z^*}\right) \tag{7.4-13}$$

$$(\rho C_p)^* = (1-\psi)\frac{(\rho C_p)_P}{(\rho C_p)_s} + \psi\left[\omega_A \frac{(\rho C_p)_D}{(\rho C_p)_s} + (1-\omega_A)\frac{(\rho C_p)_P}{(\rho C_p)_s}\right] \tag{7.4-14}$$

$$k^* = (1-\psi)\frac{k_P}{k_s} + \psi\left[\omega_A \frac{k_D}{k_s} + (1-\omega_A)\frac{k_P}{k_s}\right] \tag{7.4-15}$$

$$T^* = \frac{T_0 - T_r}{T_s}, \quad \omega_A = \omega_{A0}, \quad \psi = 1 \quad \text{at} \quad t^* = 0 \tag{7.4-16}$$

$$T^* = \frac{T_b - T_r}{T_s}, \quad \frac{\partial \omega_A}{\partial z^*} = 0 \quad \text{at} \quad z^* = 0 \tag{7.4-17}$$

$$\psi = 1 \quad \text{for} \quad z^* \geq \frac{L}{z_s} \tag{7.4-18}$$

$$-k^* \frac{\partial T^*}{\partial z^*} = \frac{h z_s}{k_s}\left(T^* + \frac{T_r - T_\infty}{T_s}\right), \quad \frac{\partial \omega_A}{\partial z^*} = 0 \quad \text{at} \quad z^* = \frac{H}{z_s} \tag{7.4-19}$$

$$T_c^* = f(\dot{T}^*) \quad \text{at} \quad z^* = \frac{L}{z_s} \tag{7.4-20}$$

The temperature can be bounded of o(1) by setting the appropriate dimensionless groups in equations (7.4-16) and (7.4-17) equal to 0 and 1, respectively (step 7). Since this is inherently an unsteady-state process, the temporal scale factor is the observation time t_o, that is, the time from the instant of contact between the hot casting solution and the cold boundary. The dimensionless groups in equations (7.4-14) and (7.4-15) suggest that ρC_p and k scale with $(\rho C_p)_P$ and k_P, their corresponding values for the pure polymer. The diffusion coefficient can be bounded of O(1) by scaling it with D_0, its value for the initial casting solution, which is its largest possible value. Since we are interested in the initial stage of TIPS process during which the phase-separated region just penetrates the thickness of the casting solution, the length scale is obtained by balancing the accumulation and conduction terms in equation (7.4-12), which yields $z_s = \sqrt{\alpha_P t_o}$, where α_P is the thermal diffusivity of the pure polymer. It is during this penetration time that the distribution of nuclei is established across the casting solution, which ultimately determines the pore-size distribution in the microporous membrane. The appropriate length scale for longer time scales would be the thickness of the casting solution. The appropriate scale factor for $\dot{\psi}_s$ is obtained by differentiating equation (7.4-5) and then determining the maximum magnitude of $\dot{\psi}$ as follows:

$$\dot{\psi} \equiv \frac{\partial \psi}{\partial t} = -(1-\psi_\infty)Knt_a^{n-1}\exp(-Kt_a^n) \tag{7.4-21}$$

Equation (7.4-21) has an extremum (i.e., a minimum) at

$$t_a = \left(\frac{n-1}{nK}\right)^{1/n} \tag{7.4-22}$$

The corresponding maximum magnitude of $\dot{\psi}$ (note that $\dot{\psi} < 0$) is given by

$$\dot{\psi}_s = (1-\psi_\infty)\sqrt{2K}\exp\left(-\frac{1}{2}\right) \cong (1-\psi_\infty)\sqrt{K} \tag{7.4-23}$$

These considerations for the reference and scale factors then result in the following dimensionless variables:

$$T^* \equiv \frac{T-T_b}{T_0-T_b}; \qquad z^* \equiv \left(\frac{kt_o}{\rho C_p}\right)_P^{1/2} = \sqrt{\alpha_P t_o}; \qquad t^* \equiv \frac{t}{t_o};$$

$$\dot{\psi}^* \equiv \frac{1}{(1-\psi_\infty)\sqrt{K}}\frac{\partial \psi}{\partial t}; \qquad k^* \equiv \frac{k}{k_P}; \qquad (\rho C_p)^* \equiv \frac{(\rho C_p)}{(\rho C_p)_P}; \tag{7.4-24}$$

$$D_{AB}^* \equiv \frac{D_{AB}}{D_0}$$

where the subscripts P and 0 denote the properties for the pure polymer and initial casting solution, respectively.

The dimensionless variables defined by equations (7.4-24) result in the following scaled minimum parametric representation of the describing equations:

$$(\rho C_p)^* \frac{\partial T^*}{\partial t^*} = \frac{\partial}{\partial z^*}\left(k^* \frac{\partial T^*}{\partial z^*}\right) - \frac{\Delta H_f (1-\psi_\infty)\sqrt{K}t_o}{(C_p)_P(T_0-T_b)}\dot{\psi}^* \tag{7.4-25}$$

$$(1-\psi_\infty)\sqrt{K}\,t_o\omega_A\dot{\psi}^* + \psi\frac{\partial \omega_A}{\partial t^*} = \frac{1}{\text{Le}}\frac{\partial}{\partial z^*}\left(D_{AB}^*\psi\frac{\partial \omega_A}{\partial z^*}\right) \tag{7.4-26}$$

$$(\rho C_p)^* = (1-\psi) + \psi\left[\omega_A\frac{(\rho C_p)_D}{(\rho C_p)_P} + (1-\omega_A)\right] \tag{7.4-27}$$

$$k^* = (1-\psi) + \psi\left[\omega_A \frac{k_D}{k_P} + (1-\omega_A)\right] \tag{7.4-28}$$

$$T^* = 1, \quad \omega_A = \omega_{A0}, \quad \psi = 1 \quad \text{at} \quad t^* = 0 \tag{7.4-29}$$

$$T^* = 0, \quad \frac{\partial \omega_A}{\partial z^*} = 0 \quad \text{at} \quad z^* = 0 \tag{7.4-30}$$

$$\psi = 1 \quad \text{for} \quad z^* \geq \frac{L}{\sqrt{\alpha_P t_o}} \tag{7.4-31}$$

$$-k^* \frac{\partial T^*}{\partial z^*} = \frac{h\sqrt{\alpha_P t_o}}{k_P}\left(T^* + \frac{T_b - T_\infty}{T_0 - T_b}\right), \quad \frac{\partial \omega_A}{\partial z^*} = 0 \quad \text{at} \quad z^* = \frac{H}{\sqrt{\alpha_P t_o}} \tag{7.4-32}$$

$$T_c^* = f(\dot{T}^*) \quad \text{at} \quad z^* = \frac{L}{\sqrt{\alpha_P t_o}} \tag{7.4-33}$$

where $\text{Le} \equiv \alpha_P / D_P$ is the Lewis number, which is a measure of the ratio of the heat conduction to species diffusion. Note that our time scaling is appropriate to relatively short observation times such that $0 \leq t_o \leq H^2 / \alpha_P$ since it is on this scale that the phase separation propagates from the cold boundary through the entire casting solution. This time scale dictates the distribution of nuclei across the entire casting solution, which determines the ultimate structure of the microporous membrane.

Now let us consider what approximations can be made to simplify the describing equations (step 8). The heat-generation term in the energy equation can be ignored if the following condition is satisfied:

$$\frac{\Delta H_f (1-\psi_\infty)\sqrt{K t_o}}{C_p (T_0 - T_b)} \ll 1 \tag{7.4-34}$$

The dimensionless group in equation (7.4-34) is a measure of the ratio of the characteristic time for heat conduction to that for heat generation due to polymer solidification. This condition implies that the solution to the energy equation is no longer strongly coupled to the solution for the liquid volume fraction. The species-conservation equation can be greatly simplified if the following condition is satisfied:

$$\frac{1}{\text{Le}} = \frac{D_0}{\alpha_P} \ll 1 \tag{7.4-35}$$

This condition implies that the diffusion term can be neglected in the species-conservation equation. The implication of this for the dimensional form of this equation is that

$$\frac{\partial(\omega_A \psi)}{\partial t} \cong 0 \Rightarrow \omega_A \psi = \text{a constant} = \omega_{A\infty}\psi_\infty \tag{7.4-36}$$

where $\omega_{A\infty}$ and ψ_∞ are the equilibrium diluent concentration and liquid volume fraction as the time approaches infinity and the entire solution is at the cold boundary temperature T_b. When the inequalities in equations (7.4-34) and (7.4-35) are satisfied, the solution for the solid–liquid TIPS process is greatly simplified. The solution to the unsteady-state one-dimensional energy equation permits obtaining the instantaneous cooling rate $\dot{T}$ from which the temperature T_c and location L of the moving boundary can be determined. Equation (7.4-5) then permits determining the liquid volume fraction ψ at any plane in the phase-separated region as a function of the time available for nucleation and growth t_g. Equation (7.4-36) then permits determining the diluent mass fraction ω_A at any plane as a function of time. Knowing $\psi(z, t)$ and $\omega(z, t)$ then permits determining the composition- and liquid-volume-fraction-dependent properties via equations (7.4-3) and (7.4-4).

Two other simplifications are possible for very short observation times. The boundary condition given by equation (7.4-32) is applied at a dimensionless distance that is equal to the ratio of the casting solution thickness divided by the instantaneous thickness of the phase-separated region. When this dimensionless group is very large, this boundary condition can be applied at infinity; that is,

$$\frac{H}{\sqrt{\alpha_P t_o}} \gg 1 \Rightarrow \text{boundary condition can be applied at } \infty \tag{7.4-37}$$

The second dimensionless group involved in this boundary condition is a measure of the ratio of the heat transferred to the ambient gas phase to that transferred to the cold boundary. When this group is very small, heat loss to the ambient gas phase can be neglected; that is,

$$\frac{h\sqrt{\alpha_P t_o}}{k_P} \ll 1 \Rightarrow \text{heat loss to ambient gas phase can be ignored} \tag{7.4-38}$$

We see that the criteria defined by equations (7.4-37) and (7.4-38) are satisfied at sufficiently short observation times. When both criteria are satisfied, equation (7.4-32) reduces to the simplified form

$$\frac{\partial T^*}{\partial z^*} = 0, \quad \frac{\partial \omega_A}{\partial z^*} = 0 \quad \text{as} \quad z^* \to \infty \tag{7.4-39}$$

It is of interest to assess the general applicability of these approximations by evaluating the criteria for a representative TIPS-casting system. This has been done for the isotactic polypropylene–dotriacontane system whose relevant properties and typical processing conditions are summarized in Table 7.4-1. These data and processing conditions result in the following values for the dimensionless groups in equations (7.4-34) and (7.4-35):

$$\frac{\Delta H_f(1-\psi_\infty)\sqrt{K}t_o}{C_p(T_0-T_b)} \le \frac{\Delta H_f(1-\psi_\infty)\sqrt{K}t_a}{C_p(T_0-T_b)} = \frac{\Delta H_f(1-\psi_\infty)\sqrt{K}}{C_p(T_0-T_b)}\left(\frac{n-1}{nK}\right)^{1/n} = 5.5 \times 10^{-2} \tag{7.4-40}$$

TABLE 7.4-1 Properties of Isotactic Polypropylene and Process Parameters for the TIPS Casting of the Isotactic Polypropylene–Dotriacontane, Polymer–Diluent System

Property or Process Parameter	Value
ρ (g/cm^3)	9.03×10^{-1}
k (J/s · cm · °C)	1.00×10^{-3}
C_p (J/g · °C)	1.93
ΔH_f (J/g)	70.6
T_c (°C)	69.5
D_{AB} (cm^2/s)	$\sim 1 \times 10^{-6}$
n	2.0
K (s^{-n})	0.0323
h (J/s · cm^2· s)	1.10×10^{-4}
ω_{A0}	0.60
ψ_0	0.63
T_0 (°C)	200
T_b (°C)	25
T_∞ (°C)	25
L_0 (μm)	200

Source: Physical and transport property data from W. K. Lee and C. L. Choy, *J. Polym. Sci. B Polym. Phys.*, **13**, 619–635 (1975); C. A. Sperati, J. Brandrup, and E. H. Immergut, eds., *Polymer Handbook*, 3rd ed., Wiley-Interscience, New York, 1989; D. R. Lide, ed., *Handbook of Chemistry and Physics*. 77th ed., CRC Press, Boca Raton, FL, 1996; N. B. Vargaftik, L. P. Filippov, A. A. Tarzimanov, and E. E. Totskii, *Handbook of Thermal Conductivity of Liquids and Gases*, CRC Press, Boca Raton, FL, 1994. Avrami coefficient data from G. B. A. Lim, Effects of nucleating agent on thermally induced phase separation membrane formation, Ph.D. dissertation, University of Texas, Austin, TX, 1990.

$$\frac{1}{\text{Le}} = \frac{D_0}{\alpha_P} = 1.74 \times 10^{-3} \tag{7.4-41}$$

In evaluating equation (7.4-40), ψ_∞ was set equal to the initial value of the diluent volume fraction; that is, since the diluent cannot diffuse and the continuous phase is polymer lean, the equilibrium liquid volume fraction at infinite time for growth of the solid phase will become nearly equal to the initial volume fraction of diluent. This underscores one of the advantages of the TIPS process for manufacturing polymeric membranes: that very high porosities can be achieved since the diluent goes from being a good solvent at a high temperature to a pore former at room temperature. Note that the observation time used in evaluating equation (7.4-40) was the time given by equation (7.4-22) at which $\dot{\psi}$ reaches an extremum; this was done to provide a conservative estimate of the magnitude of this dimensionless group since it is proportional to $\dot{\psi}_s$. The condition given by equation (7.4-41) is generally satisfied by nearly all polymers; that is, heat and mass transfer in polymeric systems is characterized by large Lewis numbers. It is not reasonable to assume that the casting solution is infinitely thick since the criterion for this assumption given by equation (7.4-37) requires that $t_o \ll 0.7$. However, the criterion given by equation (7.4-38) for ignoring the heat loss to the ambient is satisfied on the time scale for the phase-separation front to penetrate through the entire casting solution.

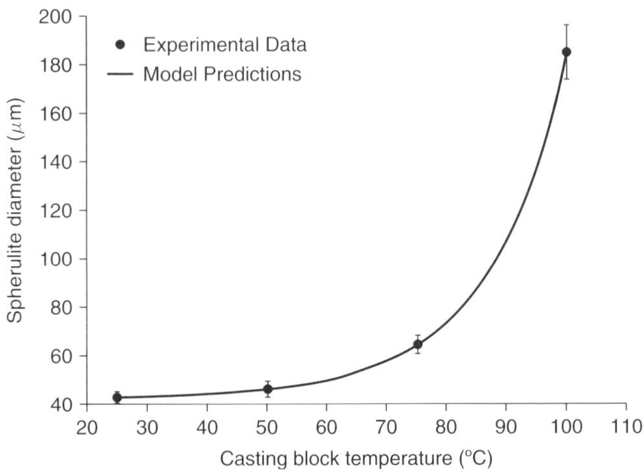

Figure 7.4-2 Corroboration of TIPS model predictions for casting solutions consisting of 40 wt% isotactic polypropylene (iPP) polymer in dotriacontane diluent, showing the final polymer spherulite diameter as a function of the casting-block temperature.

Equations (7.4-40) and (7.4-41) indicate that the describing equations for the solid–liquid TIPS process can be greatly simplified. These simplified describing equations were solved numerically for the isotactic polypropylene–dotriacontane system by Li et al.[14] The resulting solution for ψ in combination with an appropriate equation for the initial nucleation density at L, which was assumed to depend only on the instantaneous degree of subcooling and concentration at the boundary between the phase-separated and single-phase regions, predicted the observed final spherulite diameter distribution across the thickness of the microporous membrane as a function of the TIPS processing conditions quite well. A comparison between observations and the predictions of this TIPS model for the final polymer spherulite diameter as a function of the cold boundary temperature is shown in Figure 7.4-2 for casting solutions having an initial concentration of 40 wt% isotactic polypropylene in dotriacontane diluent. The model is seen to predict the observed spherulite diameter quantitatively for casting-block temperatures above 25°C.

7.5 FLUID-WALL AEROSOL FLOW REACTOR FOR HYDROGEN PRODUCTION FROM METHANE

A hydrogen-fuel-based economy offers considerable promise for providing a clean-burning fuel that will reduce greenhouse gas production at least at the point of use. However, the production of hydrogen by water splitting, that is, by decomposing water into hydrogen and oxygen, requires more energy than is available in the resulting fuel. If the generation of the energy required for water splitting involves

[14] D. Li, A. R. Greenberg, W. B. Krantz, and R. L. Sani, *J. Membrane Sci.*, **279**, 50 (2006).

hydrocarbon-based fossil fuels, carbon dioxide will be produced at the point of production. However, it is possible to thermally dissociate methane at high temperatures directly into hydrogen and carbon, thereby avoiding carbon dioxide generation. Moreover, the energy required to create these high temperatures can be obtained from a solar-heated reactor. Hence, it is possible to use solar energy to drive the thermal decomposition of methane, which is available both naturally and as a product from the decomposition of organic waste, into a hydrogen fuel and carbon particles, a useful by-product. The electrically or solar-heated fluid-wall aerosol flow reactor is being considered for implementing this hydrogen production technology. The manner in which scaling analysis was used to develop a tractable model for the complex describing equations for the fluid-wall aerosol reactor is described in this example.[15]

A schematic of the fluid-wall aerosol flow reactor is shown in Figure 7.5-1. The feed stream consists of methane and carbon particles that facilitate heating the gas. Hydrogen is injected through the inner porous wall of the tubular reactor to prevent deposition of carbon particles; that is, since hydrogen is a product of the reversible methane decomposition, its introduction prevents any reaction at the wall. To achieve significant methane conversion to hydrogen, the inner wall of the reactor is maintained at a very high temperature, typically greater than 1500°C, by either solar or electrical heating. The hot inner wall radiates heat to the carbon particles, which in turn heat the gas. At a sufficiently high temperature the methane decomposes into hydrogen and carbon particles through the reaction

$$CH_4 \rightleftharpoons C + 2H_2 \tag{7.5-1}$$

The effluent product contains hydrogen, carbon particles, and unreacted methane if the conversion is not complete. The carbon particles produced via the thermal decomposition absorb heat via radiation from the reactor walls and thereby enhance heating of the flowing gas.

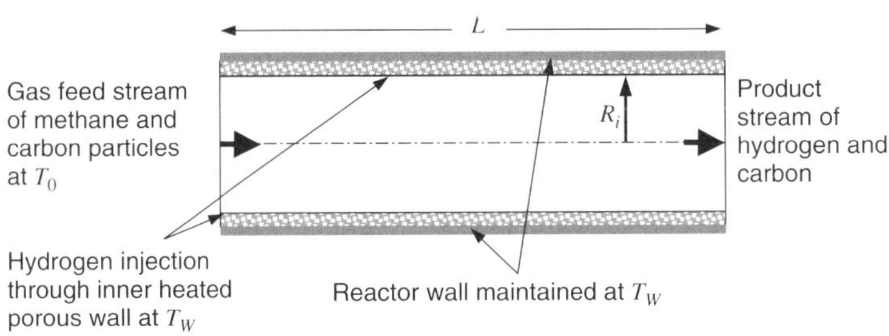

Figure 7.5-1 Fluid-wall aerosol flow reactor for the thermal conversion of methane gas to hydrogen and carbon particles; the reactor walls are maintained at a high temperature via either solar or electrical heating; hydrogen is injected through the inner porous wall of the reactor to prevent carbon formation on the wall.

[15] J. K. Dahl, A. W. Weimer, and W. B. Krantz, *Int. J. Hydrogen Energy*, **29**, 57 (2004).

In developing the describing equations for the fluid-wall aerosol flow reactor, we use the microscale–macroscale modeling approach discussed in Chapter 6; that is, the carbon particles on the microscale will become a volumetric heat-generation term on the macroscale of the reactor. The presence of the carbon particles causes intense mixing in the gas phase, which makes it reasonable to assume that the velocity profile is uniform or pluglike, which obviates the need to solve the equations of motion. The reaction kinetics for the thermal decomposition of methane are expressed in terms of the fractional dissociation of methane, X_c, via the equation

$$\frac{dX_c}{dz} = \frac{k_n}{U_G}(1 - X_c)^n \tag{7.5-2}$$

where z is the axial distance from the inlet of the reactor, U_G is the linear velocity of the gas stream, which increases with the hydrogen production, and k_n is a temperature-dependent kinetic parameter for the pseudo nth-order thermal decomposition reaction, which is given by

$$k_n = k_{n0}e^{-\Delta E/RT_G} \tag{7.5-3}$$

where k_{n0} is a kinetic constant, ΔE the activation energy for reaction, R the gas constant, and T_G the local temperature of the gas phase. Note that equation (7.5-2) follows directly from the species-balance equation for the methane for which the reaction term is assumed to have nth-order kinetics and for which axial diffusion is neglected. By means of the species-balance equation, the concentrations and molar flow rates of both the hydrogen and carbon can be expressed in terms of the fractional conversion. Hence, completing the specification of the describing equations reduces to writing the appropriate forms of the thermal energy balance for the cloud of carbon particles and the gas stream along with their boundary conditions.

A thermal energy balance on the cloud of carbon particles yields

$$\frac{d(W_C H_C)}{dz} = -\frac{h_P a_C W_C M_C}{U_G \rho_C}(T_C - T_G) + \frac{\varepsilon \sigma a_C W_C M_C}{U_G \rho_C}(T_W^4 - T_C^4) \tag{7.5-4}$$

where W_C and H_C are the molar velocity (moles/time) and molar enthalpy (energy/mole) of the cloud of carbon particles, ε is the radiative emissivity (equal to the absorbtivity for a blackbody) of the carbon particles, σ is the Stefan–Boltzmann constant, T_C and T_W are the temperatures of the carbon particles and reactor wall, respectively, h_P is the heat-transfer coefficient between the carbon particles and the gas, a_C is the interfacial area per unit volume of carbon particles, and M_C and ρ_C are the molecular weight and mass density of carbon, respectively. Note that in arriving at this equation the interrelationship between the spatial coordinate z (in a stationary coordinate system) and time t (in a convected coordinate system) is $dz = U_G\,dt$; this is the reason that U_G appears in the denominator of the second and third terms. The first term in equation (7.5-4) is the change in the sensible heat of the carbon particles as they are convected downstream; the second term is the heat transfer from the carbon particles to the gas stream, and the third term is the radiative heat transfer from the reactor wall to the carbon particles. Clearly, the first and third terms must be retained in our scaling analysis.

The corresponding thermal energy balance on the gas phase is given by

$$\frac{d(W_M H_M)}{dz} + \frac{d(W_H H_H)}{dz} = \hat{W}_H(H_{HW} - H_{H0}) + h_W 2\pi R_i (T_W - T_G) + \frac{h_P a_C W_C M_C}{U_G \rho_C}(T_C - T_G) \qquad (7.5\text{-}5)$$

where W_i is the molar velocity (moles/time) of species i; H_i is the molar enthalpy of species i for which the added subscripts 0 and W denote evaluation at the feed and wall temperatures, T_0 and T_W, respectively; $\hat{W}_H$ is the molar injection rate of hydrogen per unit length of reactor tube wall (moles/length·time); and h_W is the heat-transfer coefficient between the heated wall and the gas flow. The first and second terms in equation (7.5-5) are the change in sensible heat of the methane and hydrogen as they are convected downstream, respectively; the third term is the sensible heat added by the injection of hydrogen gas at the wall temperature; the fourth term is the heat transfer from the heated wall to the gas stream; and the fifth term is the heat transfer from the carbon particles to the gas stream. Note that the last term interrelates the heat transfer on the microscale of the carbon particles to the macroscale of the gas flow through the reactor. The heat-transfer coefficient h_P has a role analogous to that of the mass-transfer coefficient in the microscale–macroscale examples involving mass transfer with chemical reaction discussed in Chapter 6. Clearly, the first term must be retained in our scaling analysis since the methane must be heated for thermal decomposition to occur. However, it is not clear which term provides the principal source of heat in this equation.

The molar flow rates of methane, hydrogen, and carbon can be related to that of the methane feed stream through the stoichiometry given by the dissociation reaction 7.5-1:

$$W_M = (1 - X_c) W_M^0; \qquad W_H = 2X_c W_M^0 + \hat{W}_H z; \qquad W_C = X_c W_M^0 + W_C^0 \qquad (7.5\text{-}6)$$

where W_M^0 and W_C^0 are the molar flow rates of the methane and carbon particles in the feed stream to the reactor. The molar flow rate of the hydrogen includes contributions from both the dissociation of the methane and the injection through the porous inner wall of the reactor. The molar flow rate of the cloud of carbon particles includes contributions from both the dissociation of the methane and those contained in the feed stream. The enthalpies of the methane, hydrogen, and carbon are given by

$$H_M = \int_{T_0}^{T_G} C_{pM}\, dT; \qquad H_H = \int_{T_0}^{T_G} C_{pH}\, dT; \qquad H_C = \int_{T_0}^{T_C} C_{pC}\, dT \qquad (7.5\text{-}7)$$

where C_{pi} is the temperature-dependent heat capacity at constant pressure of component i. If there is no relative velocity between the carbon particles and the gas stream, the heat-transfer coefficient can be assumed to be that for pure conduction, for which

$$h_P = \frac{k_G}{R_P} \tag{7.5-8}$$

in which k_G is the thermal conductivity of the gas phase and R_P is the average radius of the carbon particles. To complete the specification of the describing equations, appropriate equations-of-state need to be specified for any temperature-dependent properties, such as C_{pi}, k_G, and h_W. However, for the scaling analysis, we merely need characteristic values of these temperature-dependent quantities.

Equations (7.5-4) and (7.5-5) can be rearranged using equations (7.5-2), (7.5-6), and (7.5-7) into the forms

$$W_C C_{pC} \frac{dT_C}{dz} + \frac{W_M^0 H_C k_n}{U_G}(1 - X_c)^n$$
$$= -\frac{h_P a_C W_C M_C}{U_G \rho_C}(T_C - T_G) + \frac{\varepsilon \sigma a_C W_C M_C}{U_G \rho_C}(T_W^4 - T_C^4) \tag{7.5-9}$$

$$(W_M C_{pM} + W_H C_{pH})\frac{dT_G}{dz} - \frac{W_M^0 H_M k_n}{U_G}(1 - X_c)^n + \frac{2W_M^0 H_H k_n}{U_G}(1 - X_c)^n$$
$$= \hat{W}_H(H_{HW} - H_H) + h_W 2\pi R_i (T_W - T_G) + \frac{h_P a_C W_C M_C}{U_G \rho_C}(T_C - T_G) \tag{7.5-10}$$

The boundary conditions on equations (7.5-2), (7.5-9), and (7.5-10) are given by

$$X_c = 0, \quad T_G = T_0, \quad T_C = T_0 \quad \text{at} \quad z = 0 \tag{7.5-11}$$

Equations (7.5-2), (7.5-3), and (7.5-6) through (7.5-11) constitute the describing equations for the fluid-wall aerosol flow reactor (step 1).

Define the following dimensionless variables containing unspecified scale and reference factors (steps 2, 3, and 4):

$$T_G^* \equiv \frac{T_G - T_{Gr}}{T_{Gs}}; \quad T_C^* \equiv \frac{T_C - T_{Cr}}{T_{Cs}}; \quad \left(\frac{dT_G}{dz}\right)^* \equiv \frac{1}{T_{Gzs}}\frac{dT_G}{dz};$$
$$\left(\frac{dT_C}{dz}\right)^* \equiv \frac{1}{T_{Czs}}\frac{dT_C}{dz}; \quad \left(\frac{dX_c}{dz}\right)^* \equiv \frac{1}{X_{czs}}\frac{dX_c}{dz}; \quad H_M^* \equiv \frac{H_M}{H_{Ms}};$$
$$H_H^* \equiv \frac{H_H}{H_{Hs}}; \quad H_C^* \equiv \frac{H_C}{H_{Cs}}; \quad W_M^* \equiv \frac{W_M}{W_{Ms}}; \quad W_H^* \equiv \frac{W_H}{W_{Hs}};$$
$$W_C^* \equiv \frac{W_C}{W_{Cs}}; \quad C_{pM}^* \equiv \frac{C_{pM}}{C_{pMs}}; \quad C_{pH}^* \equiv \frac{C_{pH}}{C_{pHs}}; \quad C_{pC}^* \equiv \frac{C_{pC}}{C_{pCs}};$$
$$h_P^* \equiv \frac{h_P}{h_{Ps}}; \quad h_W^* \equiv \frac{h_W}{h_{Ws}}; \quad U_G^* \equiv \frac{U_G}{U_{Gs}}; \quad k_n^* \equiv \frac{k_n}{k_{ns}}; \quad k_G^* \equiv \frac{k_G}{k_s} \tag{7.5-12}$$

We have introduced reference factors for the temperature of both the gas and carbon particles since neither is naturally referenced to zero. We have also introduced scale factors for the spatial gradients of the fractional conversion and the temperature in both the gas and cloud of carbon particles since there is no reason to believe that any of these will scale with the length of the reactor.[16] We have not introduced a scale for the axial spatial coordinate since this dependent variable does not enter explicitly into the describing equations. Scales for the enthalpies, molar flow rates, heat capacities, heat-transfer coefficients, gas velocity, kinetic parameter, and thermal conductivity of the gas have also been introduced since these quantities are not necessarily constant for the fluid-wall aerosol flow reactor. We have not scaled the fractional conversion since it is dimensionless and bounded of $\circ(1)$.

Substitute these dimensionless variables into the describing equations and divide each equation through by the dimensional coefficient of one term (steps 5 and 6):

$$\frac{C_{pCs}T_{Czs}U_{Gs}\rho_C}{\varepsilon\sigma a_C M_C T_{Cs}^4} W_C^* C_{pC}^* \left(\frac{dT_C}{dz}\right)^* + \frac{W_M^0 H_{Cs} k_{ns}\rho_C}{\varepsilon\sigma a_C W_C^0 M_C T_{Cs}^4} \frac{k_n^* H_C^*}{U_G^*}(1 - X_c)^n$$
$$= -\frac{h_{Ps}T_{Gs}}{\varepsilon\sigma T_{Cs}^4}\frac{h_P^* W_C^*}{U_G^*}\left[\frac{T_{Cs}}{T_{Gs}}\left(T_C^* + \frac{T_{Cr}}{T_{Cs}}\right) - \left(T_G^* + \frac{T_{Gr}}{T_{Gs}}\right)\right] \quad (7.5\text{-}13)$$
$$+ \frac{W_C^*}{U_G^*}\left[\frac{T_W^4}{T_{Cs}^4} - \left(T_C^* + \frac{T_{Cr}}{T_{Cs}}\right)^4\right]$$

$$\left(\frac{W_{Ms}C_{pMs}}{W_{Hs}C_{pHs}} W_M^* C_{pM}^* + W_H^* C_{pH}^*\right)\left(\frac{dT_G}{dz}\right)^* - \frac{W_M^0 H_{Ms} k_{ns}}{U_{Gs} W_{Hs} C_{pHs} T_{Gzs}} \frac{H_M^* k_n^*}{U_G^*}(1 - X_c)^n$$
$$+ \frac{2W_M^0 H_{Hs} k_{ns}}{W_{Hs}C_{pHs}T_{Gzs}U_{Gs}}\frac{H_H^* k_n^*}{U_G^*}(1 - X_c)^n = \frac{\hat{W}_H H_{Hs}}{W_{Hs}C_{pHs}T_{Gzs}}\left(\frac{H_{HW}}{H_{Hs}} - H_H^*\right)$$
$$+ \frac{2\pi R_i h_{Ws} T_{Gs}}{W_{Hs}C_{pHs}T_{Gzs}} h_W^* \left[\frac{T_W - T_{Gr}}{T_{Gs}} - T_G^*\right] + \frac{h_{Ps} a_C M_C W_{Cs} T_{Gs}}{U_{Gs}\rho_C W_{Hs} C_{pHs} T_{Gzs}} \frac{h_P^* W_C^*}{U_G^*}$$
$$\times \left[\frac{T_{Cs}}{T_{Gs}}\left(T_C^* + \frac{T_{Cr}}{T_{Cs}}\right) - \left(T_G^* + \frac{T_{Gr}}{T_{Gs}}\right)\right] \quad (7.5\text{-}14)$$

$$\left(\frac{dX_c}{dz}\right)^* = \frac{k_{ns}}{X_{czs}U_{Gs}}\frac{k_n^*}{U_G^*}(1 - X_c)^n \quad (7.5\text{-}15)$$

$$k_n^* = \frac{k_{n0}}{k_{ns}} \exp\left[-\frac{\Delta E}{RT_s(T_G^* + T_0/T_s})\right] = \frac{k_{n0}}{k_{ns}} \exp\left[-\frac{\Delta E}{R(T_W - T_0)}\right]$$
$$\times \exp\left[\frac{\Delta E}{R(T_W - T_0)}\frac{(T_G^* + T_0/T_s - 1)}{(T_G^* + T_0/T_s)}\right] \quad (7.5\text{-}16)$$

[16]The scaling done here differs from the development by Dahl et al. (footnote 15) in that a separate scale is introduced here for the gradient of the fractional conversion, whereas Dahl et al. assume that the length scale for this gradient is the length of the reactor; the latter would apply only at very high wall temperatures for which complete thermal dissociation occurs over the full length of the reactor.

$$W_M^* = (1 - X_c)\frac{W_M^0}{W_{Ms}}; \quad W_H^* = \frac{2X_c W_M^0 + \hat{W}_H z_s z^*}{W_{Hs}}; \quad W_C^* = \frac{X_c W_M^0 + W_C^0}{W_{Cs}} \tag{7.5-17}$$

$$H_M^* = \frac{C_{pMs} T_{Gs}}{H_{Ms}} \int_{(T_0 - T_{Gr})/T_{Gs}}^{T_G^*} C_{pM}^* \, dT^*; \quad H_H^* = \frac{C_{pHs} T_{Gs}}{H_{Hs}} \int_{(T_0 - T_{Gr})/T_{Gs}}^{T_G^*} C_{pH}^* dT^*;$$

$$H_C^* = \frac{C_{pCs} T_{Cs}}{H_{Cs}} \int_{(T_0 - T_{Cr})/T_{Cs}}^{T_C^*} C_{pC}^* \, dT^* \tag{7.5-18}$$

$$h_P^* = \frac{k_{Gs}}{h_{Ps} R_P} k_G^* \tag{7.5-19}$$

$$T_G^* = \frac{T_0 - T_{Gr}}{T_{Gs}}, \quad T_C^* = \frac{T_0 - T_{Cr}}{T_{Cs}}, \quad X_c = 0 \quad \text{at} \quad z^* = 0 \tag{7.5-20}$$

We now set appropriate dimensionless groups equal to zero or 1 to determine the reference and scale factors that will ensure that all the dimensionless variables are o(1) (step 7). Setting the two dimensionless groups equal to zero in the boundary conditions given by equation (7.5-20) references both the gas and carbon particle temperatures to zero:

$$\frac{T_0 - T_{Gr}}{T_{Gs}} = 0 \Rightarrow T_{Gr} = T_0; \qquad \frac{T_0 - T_{Cr}}{T_{Cs}} = 0 \Rightarrow T_{Cr} = T_0 \tag{7.5-21}$$

The dimensionless gas and carbon particle temperatures can be bounded to be o(1) by setting the appropriate dimensionless groups in equation (7.5-14) equal to 1[17]:

$$\frac{T_W - T_{Gr}}{T_{Gs}} = 1 \Rightarrow T_{Gs} = T_W - T_{Gr} = T_W - T_0;$$

$$\frac{T_{Cs}}{T_{Gs}} = 1 \Rightarrow T_{Cs} = T_{Gs} = T_W - T_0 \tag{7.5-22}$$

The molar flow rates can be bounded to be o(1) by setting the appropriate dimensionless groups in equations (7.5-17) equal to 1:

$$\frac{W_M^0}{W_{Ms}} = 1 \Rightarrow W_{Ms} = W_M^0$$

$$\frac{2W_M^0 + \hat{W}_H L}{W_{Hs}} = 1 \Rightarrow W_{Hs} = 2W_M^0 + \hat{W}_H L \tag{7.5-23}$$

$$\frac{W_M^0 + W_C^0}{W_{Cs}} = 1 \Rightarrow W_{Cs} = W_M^0 + W_C^0$$

[17]The scaling here differs from that in the development of Dahl et al. (footnote 15) in that the temperature reference scale here is $T_W - T_0$, whereas Dahl et al. use T_W; the former is the appropriate scale for the temperature gradient since the temperatures range between T_0 and T_W.

These scale factors for the molar flow rates are the maximum values, which include both the injected flows and those produced by the thermal decomposition.[18] The scale for the reaction parameter is obtained from the dimensionless group in equation (7.5-16):

$$\frac{k_{n0}}{k_{ns}} e^{-\Delta E/R(T_W - T_0)} = 1 \Rightarrow k_{ns} = k_{n0} e^{-\Delta E/R(T_W - T_0)} \tag{7.5-24}$$

The scale for the heat-transfer coefficient from the carbon particles to the gas is obtained from the dimensionless group in equation (7.5-19):

$$\frac{k_{Gs}}{h_{Ps} R_P} = 1 \Rightarrow h_{Ps} = \frac{k_{Gs}}{R_P} \tag{7.5-25}$$

The dimensionless physical properties and h_W are bounded to be O(1) by choosing the scale factors to be characteristic maximum values.[19] Hence, the heat capacities and thermal conductivities are scaled with their values at T_0, whereas the heat-transfer coefficient h_W is scaled with its value at T_W. The scale factors for the enthalpies are obtained by setting the appropriate dimensionless groups in equations (7.5-18) equal to 1:

$$\begin{aligned}
\frac{C_{pMs} T_{Gs}}{H_{Ms}} &= 1 \Rightarrow H_{Ms} = C_{pMs} T_{Gs} = C_{pMs}(T_W - T_0) \\
\frac{C_{pHs} T_{Gs}}{H_{Hs}} &= 1 \Rightarrow H_{Hs} = C_{pHs} T_{Gs} = C_{pHs}(T_W - T_0) \\
\frac{C_{pCs} T_{Cs}}{H_{Cs}} &= 1 \Rightarrow H_{Cs} = C_{pCs} T_{Cs} = C_{pCs}(T_W - T_0)
\end{aligned} \tag{7.5-26}$$

where C_{pMs}, C_{pHs}, and C_{pCs} are the characteristic maximum values of the heat capacities for the methane, hydrogen, and carbon, respectively. The dimensionless gas velocity is also bounded to be O(1) by basing it on the maximum hydrogen flow rate assuming complete thermal conversion of methane[20]:

$$U_{Gs} = \frac{(2W_M^0 + \hat{W}_H L) M_H}{\rho_H S_c} \tag{7.5-27}$$

where ρ_H and M_H are the mass density and molecular weight of hydrogen, and S_c is the cross-sectional area of the reactor. The scale factor for the temperature

[18]The scaling here differs from that in the development of Dahl et al. (footnote 15) in that the scale for the flow rate of the cloud of carbon particles here is $W_M^0 + W_C^0$, whereas Dahl et al. use W_M^0; the former is the appropriate scale since it is the maximum possible value.

[19]Note that by choosing maximum values for the physical properties, they are bounded to be O(1), in contrast to the scaling of the dependent and independent variables that necessarily are bounded to be O(1).

[20]The scaling here differs from that in the development of Dahl et al. (footnote 15) in that the scale for the gas velocity here includes both the hydrogen produced by thermal decomposition as well as injected, whereas Dahl et al. do not include the hydrogen injected; the former is the appropriate scale since it is the maximum possible value.

gradient in the cloud of carbon particles is obtained by recognizing that it is heated by radiation from the hot reactor wall. Hence, we set the dimensionless group multiplying the first term in equation (7.5-13) equal to one:

$$\frac{C_{pCs}T_{Czs}U_{Gs}\rho_C}{\varepsilon\sigma a_C M_C T_{Cs}^4} = 1 \Rightarrow T_{Czs} = \frac{\varepsilon\sigma a_C M_C \rho_H S_c (T_W - T_0)^4}{(2W_M^0 + \hat{W}_H L) M_H C_{pCs}\rho_C} \tag{7.5-28}$$

To determine the temperature gradient in the gas phase, we need to assess which term in equation (7.5-14) is the principal source of heat. We assume here that the heat is provided mainly by the hydrogen gas that is injected through the porous inner reactor wall. This is a reasonable assumption since considerable hydrogen must be injected to prevent reaction at the wall; moreover, it is injected at the maximum possible temperature, T_W. Hence,

$$\frac{\hat{W}_H H_{Hs}}{W_{Hs}C_{pHs}T_{Gzs}} = 1 \Rightarrow T_{Gzs} = \frac{\hat{W}_H (T_W - T_0)}{2W_M^0 + \hat{W}_H L} \tag{7.5-29}$$

The gas-phase temperature gradient would scale differently if one of the other terms were the principal heat source.[21] Nonetheless, the forgiving nature of scaling will tell us if we have balanced the wrong terms in any of the describing equations; that is, an incorrect scaling will be manifest by a dimensionless group that is not bounded of o(1). Since the two terms in equation (7.5-15) must balance, we set the dimensionless group in this equation equal to 1 to obtain the scale factor for the gradient of the fractional conversion:

$$\frac{k_{ns}}{X_{czs}U_{Gs}} = 1 \Rightarrow X_{czs} = \frac{k_{ns}}{U_{Gs}} = \frac{k_{ns}\rho_H S_c}{(2W_M^0 + \hat{W}_H L) M_H} \tag{7.5-30}$$

The scale and reference factors defined by equations (7.5-21) through 7.5-30 result in the following minimum parametric representation of the dimensionless describing equations:

$$W_C^* C_{pC}^* \left(\frac{dT_C}{dz}\right)^* + \Pi_1 \frac{k_n^* H_C^*}{U_G^*}(1 - X_c)^n = -\Pi_2 \frac{h_P^* W_C^*}{U_G^*}(T_C^* - T_G^*)$$
$$+ \frac{W_C^*}{U_G^*}\left[\left(\frac{1}{1 - \Pi_3}\right)^4 - \left(T_C^* + \frac{\Pi_3}{1 - \Pi_3}\right)^4\right] \tag{7.5-31}$$

$$(\Pi_4 W_M^* C_{pM}^* + W_H^* C_{pH}^*)\left(\frac{dT_G}{dz}\right)^* - \Pi_5 \frac{H_M^* k_n^*}{U_G^*}(1 - X_c)^n + \Pi_6 \frac{H_H^* k_n^*}{U_G^*}(1 - X_c)^n$$
$$= (1 - H_H^*) + \Pi_7 h_W^* (1 - T_G^*) + \Pi_8 \frac{h_P^* W_C^*}{U_G^*}(T_C^* - T_G^*) \tag{7.5-32}$$

[21] Determining the scale for the temperature gradient in the gas phase based on other assumptions for the principal heat source is considered in Practice Problems 7.P.19 and 7.P.20.

$$\left(\frac{dX_c}{dz}\right)^* = \frac{k_n^*}{U_G^*}(1 - X_c)^n \tag{7.5-33}$$

$$k_n^* = e^{\Pi_9[(T_G^*+\Pi_3-1)/(T_G^*+\Pi_3)]} \tag{7.5-34}$$

$$h_P^* = k_G^* \tag{7.5-35}$$

$$T_G^* = 0, T_C^* = 0 \qquad \text{at} \quad z^* = 0 \tag{7.5-36}$$

The dimensionless groups appearing in the minimum parametric representation and their physical significance are defined below:

$$\Pi_1 \equiv \frac{W_M^0 C_{pCs}\rho_C k_{ns}}{\varepsilon\sigma a_C W_C^0 M_C (T_W - T_0)^3} \sim \frac{\text{heat consumed for carbon production}}{\text{radiative heat transfer to particles}} \tag{7.5-37}$$

$$\Pi_2 \equiv \frac{k_{Gs}}{R_P\varepsilon\sigma(T_W - T_0)^3} \sim \frac{\text{heat lost to gas}}{\text{radiative heat transfer to particles}} \tag{7.5-38}$$

$$\Pi_3 \equiv \frac{T_0}{T_W} \sim \frac{\text{initial temperature of feed}}{\text{temperature of reactor wall}} \tag{7.5-39}$$

$$\Pi_4 \equiv \frac{W_M^0 C_{pMs}}{(2W_M^0 + \hat{W}_H L)C_{pHs}} \sim \frac{\text{sensible heat consumed by methane}}{\text{heat supplied by injected hydrogen}} \tag{7.5-40}$$

$$\Pi_5 = \frac{W_M^0 C_{pMs} k_{ns}\rho_H S_c}{(2W_M^0 + \hat{W}_H L)\hat{W}_H C_{pHs} M_H} \sim \frac{\text{heat released by reaction of methane}}{\text{heat supplied by injected hydrogen}} \tag{7.5-41}$$

$$\Pi_6 = \frac{2W_M^0 k_{ns}\rho_H S_c}{(2W_M^0 + \hat{W}_H L)\hat{W}_H M_H} \sim \frac{\text{heat consumed by production of hydrogen}}{\text{heat supplied by injected hydrogen}} \tag{7.5-42}$$

$$\Pi_7 \equiv \frac{2\pi R_i h_{Ws}}{\hat{W}_H C_{pHs}} \sim \frac{\text{heat transferred from reactor wall to gas}}{\text{heat supplied by injected hydrogen}} \tag{7.5-43}$$

$$\Pi_8 \equiv \frac{a_C S_c M_C (W_M^0 + W_C^0)\rho_H k_{Gs}}{(2W_M^0 + \hat{W}_H L)\hat{W}_H M_H \rho_C C_{pHs} R_P} \sim \frac{\text{heat transferred from particles to gas}}{\text{heat supplied by injected hydrogen}} \tag{7.5-44}$$

$$\Pi_9 \equiv \frac{\Delta E}{R(T_W - T_0)} \sim \text{dimensionless activation energy} \tag{7.5-45}$$

Now let us consider how these dimensionless describing equations can be simplified (step 8). Table 7.5-1 summarizes typical system parameters for the fluid-wall aerosol flow reactor. These system parameters lead to the values of the dimensionless groups that characterize the fluid-wall aerosol flow reactor summarized in Table 7.5-2. The additional groups Π_{10} through Π_{17} appearing in this table will be defined when we consider correlating the fractional conversion in terms of

TABLE 7.5-1 System Parameters for the Fluid-Wall Aerosol Flow Reactor

Parameter	Value	Parameter	Value
$a_C(\text{m}^{-1})$	6.0×10^4	$R_P(\text{m})$	5.0×10^{-5}
$C_{pCs}(\text{J / mol} \cdot \text{K})$	8.8	n	4.4
$C_{pHs}(\text{J / mol} \cdot \text{K})$	29.0	$S_c(\text{m}^2)$	4.6×10^{-3}
$C_{pMs}(\text{J / mol} \cdot \text{K})$	34.0	$T_0(\text{K})$	298
$h_{Ps}(\text{W / m}^2 \cdot \text{K})$	3.6×10^3	$T_W(\text{K})$	1200
$h_{Ws}(\text{W / m}^2 \cdot \text{K})$	18.8	$W_C^0(\text{mol / s})$	0.12
$k_{Gs}(\text{W / m} \cdot \text{K})$	0.18	$\hat{W}_H(\text{mol / m} \cdot \text{s})$	0.11
$k_{ns}(\text{s}^{-1})$	0.54	$W_M^0(\text{mol / s})$	0.057
$k_{n0}(\text{s}^{-1})$	6.0×10^{11}	$\Delta E(\text{J / mol})$	2.08×10^5
$L(\text{m})$	0.91	ε	1.0
$M_C(\text{kg / mol})$	0.012	$\sigma(\text{W / m}^2 \cdot \text{K}^4)$	5.67×10^{-8}
$M_H(\text{kg / mol})$	0.002	$\rho_C(\text{kg / m}^3)$	2270
$R(\text{J / mol} \cdot \text{K})$	8.314	$\rho_H(\text{kg / m}^3)$	0.080
$R_i(\text{m})$	0.038	$\rho_M(\text{kg / m}^3)$	0.65

TABLE 7.5-2 Dimensionless Groups Characterizing the Fluid-Wall Aerosol Flow Reactor for Producing Hydrogen from a Methane Feed Stream and the Process Parameters in Table 7.5-1

Dimensionless Group	Value	Dimensionless Group	Value
Π_1	0.17	Π_{10}	180
Π_2	87	Π_{11}	2.4×10^3
Π_3	0.25	Π_{12}	1.17
Π_4	0.31	Π_{13}	1.0×10^{12}
Π_5	0.28	Π_{14}	0.57
Π_6	0.48	Π_{15}	84
Π_7	1.41	Π_{16}	3.8×10^{16}
Π_8	54	Π_{17}	3.8×10^{-6}
Π_9	28		

dimensionless groups that isolate particular dimensional quantities of interest. The magnitudes of the groups Π_{10} through Π_{17} have no relevance to our scaling analysis since these groups do not represent any ratios of the terms appearing in our describing equations.

It might appear that we have scaled equation (7.5-31) incorrectly since $\Pi_2 \gg 1$, yet all the terms in this equation should be bounded of $\circ(1)$. Recall that we balanced the two principal terms in equation (7.5-13) that had to be retained: the term involving the heat required to increase the temperature of the cloud of carbon particles and the heat supplied by radiation from the reactor wall. The fact that $\Pi_2 \gg 1$ means that the dependent variable combination that it multiplies must be very small; that is, $T_C^* - T_G^* \ll 1$, so that the product $\Pi_2(T_C^* - T_G^*)$is $\circ(1)$. This means that the gas is essentially in local thermal equilibrium with the cloud of carbon particles; that is,

$$\Pi_2 \frac{h_P^* W_C^*}{U_G^*}(T_C^* - T_G^*) \cong 87(T_C^* - T_G^*) \cong 1 \Rightarrow T_C^* - T_G^* \cong 0.011 \Rightarrow T_G^* \cong 0.99 T_C^* \tag{7.5-46}$$

Hence, the fluid-wall aerosol reactor can be modeled assuming that the gas is in local thermal equilibrium with the carbon particles; that is, it is not necessary to solve the thermal energy equation for the gas. However, note that this approximation was based on the system parameters given in Table 7.5-1. For sufficiently large particles and/or high reactor wall temperatures, the dimensionless group Π_2 will become less than 1, implying that the gas temperature is much less than that of the carbon. The conclusion that the gas and carbon particles are in thermal equilibrium can also be obtained from the scaled thermal energy balance for the gas phase since $\Pi_8 \gg 1$, which again implies that $T_G^* \cong T_C^*$. The fact that $\Pi_2 \gg 1$ and $\Pi_8 \gg 1$, both of which imply that $T_G^* \cong T_C^*$, confirms that we have chosen our scales correctly for the process parameters given in Table 7.5-1. If one is not aware that rapid thermal equilibrium is achieved between the gas and carbon particles, serious problems can be encountered in solving the describing equations numerically. For sufficiently large values of Π_2 there is a thermal boundary layer at the entrance of the reactor wherein the gas and carbon particles come to thermal equilibrium whose thickness is much less than the length of the reactor. Hence, numerical integration employing a step size based on the reactor length will not resolve the transport processes occurring in this boundary layer.

Let us now consider the temperature gradient in the reactor. This can be estimated from equation (7.5-28). For the parameter values given in Table 7.5-1, we obtain the following estimate:

$$T_{Gzs} \cong T_{Czs} = \frac{\varepsilon \sigma a_C M_C \rho_H S_c (T_W - T_0)^4}{(2W_M^0 + \hat{W}_H L) M_H C_{pCs} \rho_C} = 1163 \text{ K/m} \tag{7.5-47}$$

This estimate for T_{Gzs} implies that the cloud of carbon particles and gas reach the wall temperature before they exit from the downstream end of the reactor; that is,

$$z = \frac{1200 \text{ K} - 298 \text{ K}}{1163 \text{ K/m}} = 0.78 \text{ m} < L = 0.91 \text{ m} \tag{7.5-48}$$

Note that the temperature gradient increases markedly with increasing T_W, which implies that the gas and carbon particles will reach T_W progressively farther upstream as T_W increases.

The dimensionless groups Π_1, Π_5, and Π_6 are of O(1) and thus do not permit us to conclude that the term they multiply, $(1 - X_c)^n$, is small; that is, we cannot conclude from the magnitude of these groups that the conversion is nearly complete. However, our scaling analysis allows us to estimate whether the reactor is sufficiently long to achieve complete conversion of the methane. An estimate of this can be obtained from the scale factor for the spatial gradient of the fractional conversion given by equation (7.5-30). For the parameter values given in Table 7.5-1 we obtain the following estimate:

$$X_{czs} = \frac{k_{ns}}{U_{Gs}} = \frac{k_{ns}\rho_H S_c}{(2W_M^0 + \hat{W}_H L)M_H} = 0.46 \text{ m}^{-1} \tag{7.5-49}$$

This estimate for X_{czs} implies that complete conversion of the methane is not attained for the reactor length given in Table 7.5-1; that is,

$$z = \frac{1-0}{0.46 \text{ m}^{-1}} = 2.2 \text{ m} > L(0.91 \text{ m}) \tag{7.5-50}$$

One way to increase the fractional conversion is to increase the reactor wall temperature. The temperature dependence of X_{czs} enters through the scale factor for the reaction parameter k_{ns}, which is defined by equation (7.5-24). An estimate of the fractional conversion can be obtained from equation (7.5-33), which can be rearranged as follows:

$$\left(\frac{dX_c}{dz}\right)^* = \frac{1}{X_{czs}}\frac{dX_c}{dz} = \frac{k_n^*}{U_G^*}(1-X_c)^n \Rightarrow \frac{dX_c}{dz} = X_{czs}\frac{k_n^*}{U_G^*}(1-X_c)^n \tag{7.5-51}$$

Since both k_n^* and U_G^* are bounded of $\mathrm{O}(1)$, they will be assumed to be constant in order to integrate equation (7.5-46) to obtain an estimate of X_c, which is given by

$$X_c = 1 - [1 + (n-1)X_{czs}L)]^{\frac{1}{(1-n)}} \tag{7.5-52}$$

The effect of the wall temperature on the fractional conversion can be estimated by combining equations (7.5-24), (7.5-30), and (7.5-52) to obtain

$$X_c = 1 - \left\{1 + \left[(n-1)\frac{\rho_H S_c L k_0 e^{-\Delta E/R(T_W - T_0)}}{(2W_M^0 + \hat{W}_H L)M_H}\right]\right\}^{\frac{1}{(1-n)}} \tag{7.5-53}$$

Equation (7.5-53) will provide a reasonable estimate of the conversion only if the dimensionless group Π_2 is sufficiently large to ensure that the gas phase is in thermal equilibrium with the cloud of carbon particles. The magnitude of Π_2 is directly proportional to h_{Ps}, the characteristic heat-transfer coefficient between the carbon particles and the gas, which in turn is inversely proportional to the radius of the carbon particles. Hence, as the radius of the carbon particles increases, Π_2 will become sufficiently small so that the gas is no longer in thermal equilibrium with the cloud of carbon particles. For the parameter values in Table 7.5-1, equation (7.5-53) indicates that $X_c \cong 0.23$ for $T_W = 1200$ K. To achieve a fractional conversion of $X_c = 0.90$ for the parameter values in Table 7.5-1, a reactor wall temperature of 1530 K would be required.

Let us now use the advanced dimensional analysis concepts for presenting numerical or experimental data that were developed in Chapter 2. In particular, let us develop a dimensional analysis correlation for the fractional conversion achieved in the gas exiting the fluid-wall aerosol flow reactor. This is obtained in principle by solving the coupled describing equations for X_c evaluated at $z = L$. This introduces L into our dimensional analysis as an independent quantity, whereas in our scaling analysis it appeared as the product $\hat{W}_H L$. Hence, an additional dimensionless group is required that we will arbitrarily define as

$$\Pi_{10} \equiv \frac{L^2}{S_c} \tag{7.5-54}$$

Therefore, X_c will be a function of the nine dimensionless groups defined by equations (7.5-37) through (7.5-45) and equation (7.5-54). However, if $\Pi_2 \gg 1$, the gas will be in thermal equilibrium with the cloud of carbon particles, in which case the thermal energy equation for the gas need not be considered. This eliminates groups Π_2, Π_4, Π_5, Π_6, Π_7, and Π_8 from the dimensional analysis correlation for X_c; that is, for sufficiently large values of Π_2, the fractional conversion X_c will be a function of only the dimensionless groups Π_1, Π_3, Π_9, and Π_{10}.

Now let us return to the general case for which Π_2 is not necessarily very large. Groups Π_1 through Π_{10} are not particularly convenient for correlating X_c and other quantities of interest in assessing the performance of the fluid-wall aerosol flow reactor. In particular, we would like to isolate the process parameters that can easily be changed, such as the wall temperature T_W, the flow rate of carbon particles injected with the feed W_C^0, and the radius of the carbon particles R_P. In isolating the latter quantity we must be aware that a_C, the surface area per unit volume of particles, depends on the particle size. For spherical particles, $a_C = 3/R_P$, which is a reasonable approximation for the small carbon particles involved in the fluid-wall aerosol flow reactor. Let us first invoke the formalism of steps 9 and 10 in the scaling approach to dimensional analysis that was outlined in Chapter 2. These steps allow us to remove redundant quantities that appear in sums or differences in dimensionless groups. Hence, seven of the nine dimensionless groups that resulted from our scaling analysis can be replaced by the following redefined groups:

$$\Pi_{10} = \Pi_1' \equiv \frac{W_M^0 C_{pCs} \rho_C k_0 R_P}{e\sigma W_C^0 M_C T_0^3} \tag{7.5-55}$$

$$\Pi_{11} = \Pi_2' \equiv \frac{k_{Gs}}{R_P e\sigma T_0^3} \tag{7.5-56}$$

$$\Pi_{12} = \Pi_4' \equiv \frac{C_{pMs}}{C_{pHs}} \tag{7.5-57}$$

$$\Pi_{13} = \Pi_5' \equiv \frac{k_0 \rho_H S_c}{\hat{W}_H M_H} \tag{7.5-58}$$

$$\Pi_{14} = \Pi_6' = \frac{W_M^0}{\hat{W}_H L} \tag{7.5-59}$$

$$\Pi_8' \equiv \frac{S_c M_C \rho_H k_{Gs}}{\hat{W}_H M_H \rho_C C_{pHs} R_P^2} \tag{7.5-60}$$

$$\Pi_{15} = \Pi_9' \equiv \frac{\Delta E}{RT_0} \tag{7.5-61}$$

where the Π_i' denotes the result of redefining Π_i using the formalism of steps 9 and 10. Note that it was not necessary to redefine groups Π_3, Π_7, and Π_{10}. In arriving at equations (7.5-55) through (7.5-61), we have assumed spherical carbon particles and therefore have made the substitution $a_C = 3/R_P$. We have also dropped numerical constants to simplify the dimensionless groups. Although these numerical constants need to be retained in scaling analysis, they serve no purpose in dimensional analysis. Applying steps 9 and 10 has isolated T_W into Π_3 and W_C^0 into Π_1'. However, R_P is still contained in groups Π_1', Π_2', and Π_8'. To isolate R_P into just one dimensionless group, we apply the formalism of step 7 in the scaling analysis approach to dimensional analysis.

$$\Pi_{16} = \Pi_1' \cdot \Pi_2' = \frac{W_M^0 C_{pCs} \rho_C k_0 R_P}{e\sigma W_C^0 M_C T_0^3} \frac{k_{Gs}}{R_P e\sigma T_0^3} = \frac{W_M^0 C_{pCs} \rho_C k_0 k_{Gs}}{(e\sigma)^2 W_C^0 M_C T_0^6} \tag{7.5-62}$$

$$\Pi_{17} = \Pi_8' \frac{1}{(\Pi_2')^2} = \frac{S_c M_C \rho_H k_{Gs}}{\hat{W}_H M_H \rho_C C_{pHs} R_P^2} \left(\frac{R_P e\sigma T_0^3}{k_{Gs}} \right)^2 = \frac{(e\sigma)^2 S_c M_C \rho_H T_0^6}{\hat{W}_H M_H \rho_C C_{pHs} k_{Gs}} \tag{7.5-63}$$

Hence, X_c can now be correlated in terms of the ten dimensionless groups $\Pi_3, \Pi_7, \Pi_{10}, \Pi_{11}, \Pi_{12}, \Pi_{13}, \Pi_{14}, \Pi_{15}, \Pi_{16}$, and Π_{17}, in which T_W, R_P, and W_C^0 have been isolated into groups Π_3, Π_{11}, and Π_{16}, respectively; that is,

$$X_c = f(\Pi_3, \Pi_7, \Pi_{10}, \Pi_{11}, \Pi_{12}, \Pi_{13}, \Pi_{14}, \Pi_{15}, \Pi_{16}, \Pi_{17}) \tag{7.5-64}$$

Note that all the dimensional quantities contained in the original 10 dimensionless groups are also included in the redefined set of dimensionless groups.

The utility of scaling analysis for simplifying the describing equations and for correlating the output from a numerical solution is demonstrated in Figure 7.5-2, which shows a plot of the fractional conversion X_c as a function of the dimensionless group Π_3^{-1} with the group Π_{11} as a parameter; the other eight dimensionless groups are held constant. The solid and dashed lines are the predictions of a numerical solution to the full set of describing equations.[22] The dotted line is the estimate of the fractional conversion obtained from scaling analysis for the limiting case of $\Pi_2 \gg 1$ given by equation (7.5-53). The scaling analysis estimate agrees within the expected O(1)accuracy with the predictions of the numerical solution. One can immediately appreciate the utility of isolating the quantities T_W and R_P into just one dimensionless group; that is, Figure 7.5-2 indicates how increasing the wall temperature or the particle size affects the fractional conversion. Larger values of Π_3^{-1}

[22] J. K. Dahl, A. W. Weimer, and W. B. Krantz, *Int. J. Hydrogen Energy*, **29**, 57 (2004).

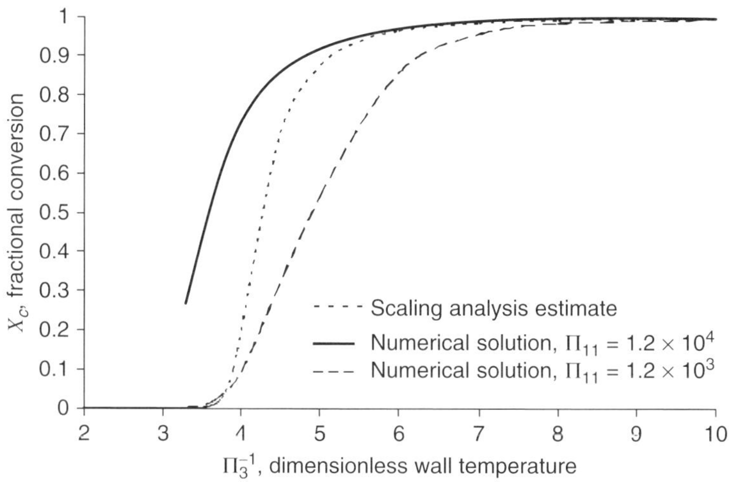

Figure 7.5-2 Fractional conversion as a function of the dimensionless wall temperature estimated by scaling analysis and predicted by a numerical solution to the full set of describing equations for two values of the dimensionless heat-transfer coefficient, Π_{11}.

corresponding to higher reactor wall temperatures result in increased conversion, whereas smaller values of Π_{11} corresponding to larger carbon particles result in decreased conversion.

The redefined dimensionless groups given by equations (7.5-55) through (7.5-61) are useful for isolating various dimensional parameters so that their effect on the fractional conversion and other quantities of interest can be studied. However, to understand the reasons for the trends observed, the original dimensionless groups determined by scaling analysis are required since these groups are proportional to the relative magnitudes of the various terms in the describing equations. The increase in X_c with increasing T_W is a direct effect of increasing the reaction-rate parameter via Π_9 in equation (7.5-34). The decrease in X_c with an increase in particle size can be attributed to a decrease in the heat-transfer coefficient only if the gas and carbon particles are not essentially in thermal equilibrium; that is, if $\Pi_2 = \text{o}(1)$. Note that $\Pi_2 = \Pi_{11} T_0^3 / (T_W - T_0)^3$. Hence, we find that $3.5 \leq \Pi_2 \leq 43$ for the temperature range $2400\text{K} \geq T_W \geqslant 1200$ K and $\Pi_{11} = 1.2 \times 10^3$, whereas $35 \leq \Pi_2 \leq 430$ for $\Pi_{11} = 1.2 \times 10^4$; that is, Π_2 is large at all but the highest temperatures for the larger particle size. Hence, the inverse relationship between R_P and h_{Ps} cannot explain the decrease in X_c with decreasing values of Π_{11}. However, the particle size also affects conversion through the dimensionless group Π_1, which is directly proportional to R_P through the parameter $a_C = 3/R_P$; this group is a measure of the sensible heat required for the carbon particles. Hence, decreasing Π_{11} by an order of magnitude also increases Π_1 proportionately. The larger effect of a decrease in Π_{11} on X_c observed at lower values of Π_3^{-1} is because Π_1 is proportional to $(T_W - T_0)^{-3}$; that is, as the wall temperature increases, Π_1 decreases and the sensible heat requirement for the carbon particles becomes less important.

In summary, one sees in this example the utility of first scaling the describing equations to obtain the minimum parametric representation. The magnitude of the resulting dimensionless groups provides a direct assessment of the relative importance of the various terms in the describing equations. In particular, these dimensionless groups allow one to assess what approximations, if any, might be applicable. However, the dimensionless groups that result from scaling analysis often are not optimal for correlating experimental or numerical data. Hence, it is convenient to redefine the dimensionless groups using the formalisms introduced in Chapter 2 in order to isolate particular dimensional quantities of interest into just one dimensionless group. However, to interpret trends in the observed performance of the particular device or process of interest, one must revert back to the original dimensionless groups obtained from scaling analysis, since the magnitude of the latter is a direct measure of the physicochemical phenomena involved in the describing equations.

7.6 SUMMARY

In contrast to Chapters 3 through 6, in which scaling analysis was used essentially as a pedagogical tool, in this chapter we illustrated how it can be used in the design of improved and new processes. This was demonstrated via four examples that focused on emerging technologies, all of which involved coupled transport. Three of these examples also used the microscale–macroscale modeling concept to address the heterogeneous nature of the systems. Scaling the describing equations for complex technologies is a trial-and-error process; that is, one generally is not certain which terms need to be balanced to determine the proper scale factors. However, the forgiving nature of scaling will always indicate when terms have not been balanced properly. This is indicated when one or more terms are not bounded of o(1) when the relevant physical properties and process parameters are substituted into the dimensionless describing equations. The scaling analysis in each of these examples was the final result of this trial-and-error procedure for specified sets of properties and parameters. However, in the Practice Problems at the end of the chapter we explore other possible scalings.

In Section 7.2 we applied o(1) scaling analysis to the design of a membrane–lung oxygenator. The same problem was analyzed in Section 5.10 using the simple scaling analysis approach to dimensional analysis. Whereas both methods resulted in a correlation for the Sherwood number in terms of the same four dimensionless groups, the o(1) scaling analysis approach provided considerably more information on the performance of this device. Additional insight into the design of the membrane-lung oxygenator was possible because the o(1) scaling analysis identified both a momentum boundary layer within which the effect of the wall oscillations on the velocity profile was confined and a solutal boundary layer across within which the mass transfer occurred. The o(1) scaling analysis provided an explanation for why the enhancement in mass transfer involved a "tuned" response with respect to the oscillation frequency. It also led to a design equation for predicting the effect of the process parameters on the frequency required to achieve optimal enhancement in the mass transfer. This design problem effectively illustrated the advantages of using o(1) scaling analysis to achieve the minimum parametric representation relative to the simple scaling analysis or Pi theorem approaches to dimensional analysis.

In Section 7.3 we considered pulsed single-bed pressure swing adsorption (PSA) for the production of oxygen-enriched air. This example demonstrated the value of an o(1) scaling analysis to design a new process. Whereas the design of conventional pulsed packed single-bed PSA is well established, no performance data are available for the novel pulsed monolithic single-bed PSA process. This example involved coupled fluid dynamics and mass transfer with adsorption and employed microscale–macroscale modeling to address the adsorption from the bulk gas flow onto the discrete adsorbent particles. We found that proper scaling was indicated by all the terms in the describing equations being bounded of o(1). This usually means that all the dimensionless groups are bounded of o(1), although strictly speaking, it means that the product of any dimensionless group and the variables it multiplies must be bounded of o(1). Scaling analysis identified several marked differences between packed single-bed PSA technology and the monolithic bed process. Scaling analysis indicated that local adsorption equilibrium was achieved nearly instantaneously in the monolithic bed PSA process; in contrast, the adsorption in the packed bed PSA process was mass-transfer controlled. This difference between monolithic and packed bed adsorption is analogous to that between the fast and slow reaction regimes in the microscale–macroscale analysis of mass transfer with chemical reaction that was considered in Chapter 6. Scaling analysis indicated that for the monolithic bed process the optimal pressurization time was the contact time required for the adsorption wave to pass through the bed; in contrast, for the packed bed process it was the time required to achieve equilibrium in the adsorption bed. This example also illustrated the advantages of using o(1) scaling analysis for dimensional analysis. Whereas a naïve application of the Pi theorem indicated that 11 dimensionless groups were required, o(1) scaling analysis reduced the number of dimensionless groups required to correlate the product purity and product recovery to just two for the production of oxygen-enriched air using a specified adsorbent.

In Section 7.4 we applied scaling analysis to the thermally induced phase-separation (TIPS) process for the fabrication of polymeric membranes. The TIPS process involves coupled heat and mass transfer as well as nucleation and growth of the dispersed solid phase in the casting solution. As such, this example employed the microscale–macroscale concept to address the growth of the nucleated particles on the microscale that were considered to be a homogeneous source of diluent mass on the macroscale of the casting solution. The dispersed phase particles were assumed to be in both thermal and solutal equilibrium with the bulk of the surrounding casting solution. This assumption is a thermal–solutal analog of the fast reaction regime approximation considered in Chapter 6. This moving boundary problem was atypical because the location of the interface between the phase-separated region and homogeneous solution was not dictated by a classical Stefan condition involving latent heat release but by a dynamic condition that depended on the nucleation process. Scaling analysis was used in this example to determine when the mass transport, latent heat effects, and heat loss to the ambient gas phase could be ignored in the describing equations. Although it was not necessary to scale the liquid volume fraction in this example, since it was dimensionless and bounded of o(1), a separate scale was introduced for its time rate of change. Since we were interested in scaling the TIPS process at short times during which the pore-size distribution of the resulting membrane is created, we determined the length scale by balancing the accumulation and conduction terms. This scaling introduced the Lewis number, which is the ratio of heat conduction to species diffusion. A large Lewis number, which is typical for polymer solutions, implied that mass-transfer effects could be ignored in the describing equations. When the dimensionless group that characterized the ratio of the characteristic time for heat conduction to that for latent heat generation was small, the latter effect could be ignored. For sufficiently short penetration depths relative to the casting solution thickness, heat loss to the ambient gas phase could be ignored. The predictions for the polymer spherulite diameter of a numerical solution to the simplified describing equations compared well with scanning electron microscopy measurements for a typical TIPS membrane-casting system.

In Section 7.5 we applied scaling analysis to the design of the fluid-wall aerosol reactor process for the direct conversion of methane to hydrogen that produces a clean-burning fuel without the production of any greenhouse gases. This example involved coupled heat and mass transfer with chemical reaction. It also involved the concept of microscale–macroscale modeling. In this case the microscale element, a carbon particle, was not necessarily assumed to be in local thermal equilibrium with the gas phase on the macroscale of the reactor. As such, this example was a thermal analog of the intermediate reaction regime considered in Chapter 6; that is, some heat transfer from the carbon particles to the gas phase was assumed to be necessary to promote the decomposition reaction (i.e., the thermal analog of the slow reaction regime would not be practical); however, the gas phase did not need be in thermal equilibrium with the carbon particles (i.e., the fast reaction regime analog). These considerations required including separate energy conservation equations for the dispersed carbon particle and continuous gas phases. Scaling analysis was employed to determine when thermal equilibrium between the carbon particles and gas phase could be assumed and to estimate the local temperature and degree of conversion in the gas phase. Independent scales were introduced for the spatial derivatives of the conversion and temperatures of both the gas phase and carbon particles. One challenge in this scaling analysis was to determine the proper terms to balance in the energy equation for the gas phase; that is, it was not clear whether the principal heat source was sensible heat introduced via the injected hydrogen, transfer from the carbon particles, transfer from the carbon particles, or convective transfer from the heated wall. However, the forgiving nature of scaling ensured that the proper terms were balanced for specific design conditions. This example illustrated the interesting situation where a dimensionless group was very large in a term that had to be bounded of o(1). In particular, a large dimensionless group multiplied a term containing the difference between the dimensionless carbon particle and gas-phase temperatures. This condition then implied that these two temperatures were nearly equal; that is, the carbon particles and gas phase were in local thermal equilibrium (i.e., the thermal analog of the fast reaction regime). The temperature dependence of the *n*th-order decomposition reaction rate kinetics was pivotal in determining the conversion achieved in this reaction. Hence, in this example scaling analysis was used to obtain estimates of the local gas-phase temperature and degree of conversion. The scaling approach to dimensional analysis was then used to isolate into separate dimensionless groups key design parameters such as the reactor wall temperature, carbon particle radius, and flow rate of carbon particles introduced in the feed. Predictions of a numerical solution to the full set of describing equations were then used to illustrate that scaling analysis provides reliable estimates for the degree of conversion. Systematic deviations from the scaling analysis estimates based on the local thermal equilibrium assumption between the carbon particles and the gas phase were then explained in terms of the relevant dimensionless groups. In summary, this example provides a very effective illustration of how scaling analysis can be used to design and estimate the performance of an entirely new process for which no prior operating data are available.

7.P PRACTICE PROBLEMS

7.P.1 Axial Diffusion Effects in an Oscillated Membrane–Lung Oxygenator

In Section 7.2 we scaled the coupled fluid dynamics and mass-transfer problem associated with applying axial oscillations to improve the performance of a hollow-fiber membrane oxygenator. We assumed that axial diffusion effects were negligible in our scaling analysis. Retain the axial diffusion term in the describing equations and use scaling analysis to develop a criterion for assessing when axial diffusion can be ignored.

7.P.2 Transient Flow Effects in an Oscillated Membrane–Lung Oxygenator

In Section 7.2 we scaled the coupled fluid dynamics and mass-transfer problem associated with applying axial oscillations in order to improve the performance of a hollow-fiber membrane oxygenator. We assumed that transient flow effects could be neglected: that is, the unsteady-state flow effects associated with the initiation of the axial oscillations.

(a) Indicate how the describing equations for this problem need to be modified to apply scaling analysis to assess when the transients can be neglected.
(b) Scale the modified describing equations to develop appropriate criteria for neglecting the transient effects.
(c) Discuss how changes in the relevant process parameters influence whether transient effects can be ignored.

7.P.3 Effect of Process Parameters on the Performance of an Oscillated Membrane–Lung Oxygenator

In Section 7.2 we scaled the coupled fluid dynamics and mass-transfer problem associated with applying axial oscillations to improve the performance of a hollow-fiber membrane oxygenator. Our scaling analysis indicated that the dimensionless mass-transfer coefficient or Sherwood number was a function of three dimensionless groups, as indicated in equation (7.2-47). Discuss how changes in the relevant physical properties and process parameters affect the performance of an oscillated membrane–lung oxygenator.

7.P.4 Correlation for the Sherwood Number for a Membrane–Lung Oxygenator Without Axial Oscillations

In Section 7.2 we stated without proof that the Sherwood number for a hollow-fiber membrane oxygenator was correlated in terms of just one dimensionless group, as indicated by equation (7.2-48). In this problem we prove this result.

(a) Consider the mass-transfer geometry shown in Figure 7.2-1; however, assume that no oscillations are applied and that the liquid within the hollow fiber is in fully developed laminar flow. Write the appropriate describing equations and associated boundary conditions.

(b) Scale the describing equations to determine the relevant scale and reference factors. Ignore axial diffusion. In carrying out this scaling analysis, recall that in the absence of axial oscillations the velocity profile is that for fully developed laminar flow in a circular tube. Hence, it is necessary to scale only the species-balance equation and its associated boundary conditions.

(c) Use the results of your scaling analysis to determine the dimensionless group(s) required to correlate the Sherwood number.

7.P.5 Wall Effects in Pulsed PSA

In scaling the pulsed PSA process, we assumed plug flow through the packed adsorbent bed. However, due to the no-slip condition at the wall of the vessel that holds the adsorbent, there is a region of influence wherein a steep radial velocity gradient exists.

(a) Use scaling analysis to estimate the thickness of the region of influence wherein the velocity profile is not uniform; that is, the gas is not in plug flow.

(b) Develop a criterion for ignoring the velocity gradient near the wall for pulsed PSA.

7.P.6 Alternative Boundary Conditions for Pulsed PSA: Specified Superficial Velocity and Downstream Pressure

The pulsed single-bed PSA process can be operated by specifying the feed composition and any two process parameters, consisting of the upstream total pressure, downstream total pressure, and upstream superficial velocity U_0. In Section 7.3 we considered operating the PSA process by controlling the upstream and downstream pressures. In this problem we consider an alternative mode of operation.

(a) Consider operation of the PSA process whereby the feed composition and upstream superficial velocity as well as the downstream total pressure are specified. Scale the modified describing equations and determine the relevant scale factors. Note that in this case it is necessary to introduce a scale factor for the axial pressure gradient.
(b) Determine the characteristic times in terms of the relevant physical properties and design parameters for these modified boundary conditions.
(c) Determine the criterion for ignoring axial dispersion.
(d) Determine the criterion for assuming a linear adsorption isotherm for both components.
(e) Determine the criterion for assuming quasi-steady-state.
(f) Determine the criterion for assuming local adsorption equilibrium.

7.P.7 Alternative Boundary Conditions for Pulsed PSA: Specified Superficial Velocity and Upstream Pressure

The pulsed single-bed PSA process can be operated by specifying the feed composition and any two process parameters, consisting of the upstream total pressure, downstream total pressure, and upstream superficial velocity U_0. In Section 7.3 we considered operating the PSA process by controlling the upstream and downstream pressures. In this problem we consider an alternative mode of operation.

(a) Consider operation of the PSA process whereby the feed composition and upstream superficial velocity pressure are specified. Scale the modified describing equations and determine the relevant scale factors. Note that in this case it is necessary to introduce a scale factor for the axial pressure gradient.
(b) Determine the characteristic times in terms of the relevant physical properties and design parameters for these modified boundary conditions.
(c) Determine the criterion for ignoring axial dispersion.
(d) Determine the criterion for assuming a linear adsorption isotherm for both components.
(e) Determine the criterion for assuming quasi-steady-state.
(f) Determine the criterion for assuming local adsorption equilibrium.

7.P.8 Pressure and Velocity Dependence of the Axial Dispersion Coefficient for Pulsed PSA

In the scaling analysis for the pulsed PSA process that was carried out in Section 7.3, the axial dispersion coefficient was assumed to be constant. In fact, the axial dispersion coefficient depends on both the total pressure and the superficial velocity through the porous media via the equation

$$D_L = 0.7 D_{AB} + \frac{U}{\varepsilon} R_p \tag{7.P.8-1}$$

where R_p is the radius of the adsorbent particles, and D_{AB}, the binary diffusion for components A (nitrogen) and B (oxygen), is given by the following[23]

$$D_{AB} = \frac{0.0018583\sqrt{T^3 \left(1/M_A + 1/M_B\right)}}{P\sigma_{AB}^2 \Omega} \tag{7.P.8-2}$$

in which M_i is the molecular weight of component i, σ_{AB} the collision diameter determined from the Lennard-Jones intermolecular potential function, and Ω a dimensionless function of the temperature and intermolecular potential.

(a) Use scaling analysis to develop a criterion for determining when the effect of the superficial velocity on the axial dispersion coefficient can be ignored.
(b) Use scaling analysis to develop a criterion for determining when the effect of the total pressure on the axial dispersion coefficient can be ignored.

7.P.9 Scaling the Depressurization Step for Pulsed PSA

In Section 7.3 we scaled the pressurization step for a pulsed PSA oxygenator, during which product enrichment in oxygen is effected by the selective adsorption of nitrogen from an air feed stream. The second step in the pulsed PSA process involves depressurization of the packed adsorption bed to cause desorption and regeneration of the adsorbent. Assume that depressurization involves instantaneously reducing the pressure at the feed end of the adsorbent bed to atmospheric, while simultaneously closing a valve at the product end of the bed so that no outward flow can occur. Hence, all the desorbed gas exits through the feed end of the adsorbent bed. The initial condition for this assumed depressurization step constitutes the axial distributions of the pressure and concentration that were established at the end of the pressurization step; these can be indicated formally as $P(0, z)$ and $\overline{p}_A(0, z)$. The spatial distribution of the pressure and concentration necessarily change in time as the adsorbent is regenerated. In particular, the pressure and concentration will display maxima that progressively move through the adsorbent bed. Ideally, one chooses a depressurization time that is sufficiently long to essentially restore the adsorbent bed to its loading in equilibrium with air at atmospheric pressure. In this problem we consider the scaling of this depressurization step.

(a) Write the appropriate describing equations along with the required initial and boundary conditions for the depressurization step in pulsed PSA.

[23] R. B. Bird, W. E. Stewart, and E. N. Lightfoot, *Transport Phenomena*, 2nd ed., Wiley, Hoboken, NJ, 2002, p. 526.

(b) Determine the scale and reference factors as well as the relevant dimensionless groups.

(c) Determine the relevant characteristic times in terms of the physical properties and design parameters.

(d) Estimate the time required to essentially regenerate the adsorbent bed to a loading corresponding to equilibrium with air at atmospheric pressure for the physical properties and design parameters given in Table 7.3-1.

(e) In contrast to the pressurization step, the depressurization step will involve total and partial pressure distributions that display maxima along the adsorbent bed length. Use scaling analysis to estimate the instantaneous location of the maxima in the pressure. Note that this involves estimating the instantaneous thickness of the region of influence wherein the pressure gradient is confined.

(f) Use scaling analysis to estimate the time required for the maxima in pressure to propagate through the entire thickness of the adsorbent bed for the physical properties and design parameters given in Table 7.3-1.

7.P.10 Dimensional Analysis Correlation for a Bed-Size Factor for Pulsed PSA

In Section 7.3 we used the scaling analysis approach for dimensional analysis to develop a correlation for product recovery. In particular, we isolated the applied pressure and pressurization time into separate dimensionless groups. Use the scaling analysis approach for dimensional analysis to develop a correlation for the bed-size factor, ψ_{bf}, which is defined by

$$\psi_{bf} \equiv \frac{m_{\mathrm{ad}}}{m_{\mathrm{O}_2}} \qquad (7.P.10\text{-}1)$$

where m_{ad} is the total mass of adsorbent in the bed and m_{O_2} is the total mass of oxygen produced per day. Isolate the applied pressure and pressurization time into separate groups.

7.P.11 Estimation of the Time Required for TIPS Membrane Casting

The TIPS membrane-casting process was scaled in Section 7.4 to determine the conditions for which the describing equations could be simplified. However, more can be done with the results of this scaling analysis.

(a) Consider conditions for which heat loss to the ambient gas phase can be ignored and estimate the time required for the boundary between the phase-separated and single-phase regions to penetrate through the entire casting solution.

(b) Why would this estimate be inaccurate when significant heat transfer occurs at the upper boundary to the ambient gas phase?

7.P.12 Scaling of the TIPS Process for Concentrated Casting Solutions

The scaling in Section 7.4 assumed that the diluent mass fraction ω was o(1). However, it is possible to have casting solutions concentrated in polymer for which the diluent mass fraction is considerably less than 1. Rescale the describing equations for the TIPS process for a casting solution that is very concentrated in polymer. In this case it will be necessary to scale the diluent mass fraction even though it is dimensionless.

7.P.13 TIPS Membrane Casting with Both Convective and Radiative Heat Transfer to the Ambient Gas Phase

In Section 7.4 we allowed for heat loss from the casting solution to the ambient gas phase only by convective heat transfer, which was described by a lumped-parameter boundary condition given by equation (7.4-9). Consider now heat loss by both radiation and convection, for which the radiative heat flux is given by

$$q_z = \varepsilon\sigma(T^4 - T_\infty^4) \tag{7.P.13-1}$$

where ε and T are the emissivity and absolute temperature of the upper surface of the casting solution, σ is the Stefan–Boltzmann constant, and T_∞ is the temperature in the bulk of the ambient gas phase. In working this problem, assume that the energy and species-balance equations can be decoupled and that latent heat effects can be ignored.

(a) Rescale the TIPS-casting process considered in Section 7.4 for this modified boundary condition.
(b) Determine the criterion for ignoring heat loss to the ambient gas phase at the upper boundary.
(c) Determine the criterion for ignoring the radiative relative to convective heat transfer at the upper boundary.

7.P.14 TIPS Casting on a Cold Boundary with a Constant Heat Flux

The TIPS membrane-casting process considered in Section 7.4 assumed that the cold boundary was maintained at a fixed temperature. Here we consider maintaining a constant heat flux at this boundary, which implies that its temperature must decrease in time as the phase-separation front penetrates farther into the casting solution.

(a) Rescale the describing equations assuming that the heat flux is specified at this cold boundary; that is, assume a boundary condition at the cold surface given by

$$q_z = -k\frac{\partial T}{\partial z} = -q_0, \qquad \text{where } q_0 \text{ is a positive constant} \tag{7.P.14-1}$$

In this scaling it is appropriate to allow a separate scale for the spatial derivative of the temperature. Consider carefully how this scale should be determined to achieve o(1) scaling.

(b) For this case you should obtain a length scale different from that given in equation (7.4-24). This in turn implies that the thickness of the thermal boundary layer or region of influence will have a different time dependence. Compare this time dependence with that obtained in Section 7.4. In particular, discuss the physical reason for this difference in the time dependence.

(c) Determine the criterion for ignoring the latent heat term in the energy equation for the constant heat-flux boundary condition.

(d) Determine the criterion for ignoring the diffusion term in the species-balance equation.

(e) Determine the criterion for applying the boundary condition at the upper liquid–gas interface at infinity.

(f) Determine the criterion for ignoring heat loss to the ambient gas phase.

7.P.15 Effect of Heat Loss in the Ambient Gas Phase in TIPS Casting

In Section 7.4 we scaled the TIPS process for the region of influence near the cold boundary; that is, our temperature scale was based on that of the cold boundary and our length scale was dictated by balancing the accumulation and heat conduction terms in the energy equation. This scaling was appropriate for determining the criteria for neglecting the coupling between the energy and species-balance equations and for ignoring the latent heat effects. However, this scaling provides no information on the thickness of the region of influence near the upper boundary at which heat loss occurs to the ambient gas phase. In working this problem, assume that the energy and species-balance equations can be decoupled and that latent heat effects can be ignored.

(a) Carry a scaling analysis of the describing equations for conditions appropriate to the upper boundary region for short contact times for which the phase separation caused by the cold boundary is far removed. Introduce a separate scale factor for the spatial derivative of the temperature since the latter is unknown at the upper boundary.

(b) Estimate the thickness of the region of influence or thermal boundary layer at the upper surface wherein the conductive heat transfer is essentially confined.

(c) Use the results of your scaling analysis in this problem and those in Section 7.4 to estimate when heat loss at the upper boundary begins to affect the heat transfer to the cold boundary.

7.P.16 Upward- and Downward-Propagating Phase-Separation Fronts in TIPS Casting

In Practice Problem 7.P.15 we considered a scaling analysis appropriate to the region near the upper boundary at which heat transfer to the ambient gas phase occurs. If this heat transfer is sufficiently fast, it is possible that a phase-separation front will propagate downward from the upper boundary as well as upward from the cold lower boundary. In working this problem, assume that the energy and species-balance equations can be decoupled and that latent heat effects can be ignored.

(a) Use the results of the scaling analysis in Practice Problem 7.P.15 to determine the criteria for whether phase separation will also occur in the upper boundary region. Assume that the relationship between the crystallization temperature and cooling rate is of the form given by equation (7.4-10). *Hint*: One needs to use the scale factor for the temperature gradient to estimate the instantaneous temperature at the upper boundary. However, the temperature required for phase separation also depends on the cooling rate, which can also be estimated from your scaling analysis. Note that the temperature of the upper boundary must be greater than or equal to that of the ambient gas phase, which introduces yet another limitation on whether phase separation can occur.

(b) Consider the special case of a very large heat-transfer coefficient for the boundary condition at the upper surface. How does this change the scaling analysis in part (a)?

(c) Determine the criteria for whether phase separation will occur in the upper boundary region for the special case of a very large heat-transfer coefficient.

7.P.17 Low Biot Number Approximation for TIPS Membrane Casting

If the heat flux rather than the temperature is specified at the cold boundary, it is possible under some conditions to achieve a uniform temperature throughout the casting solution during TIPS casting. This might be a useful process for producing a membrane that has a homogeneous pore structure throughout its cross-section. This condition is analogous to that considered in Chapter 4 for low Biot number heat transfer. In working this problem, assume that the energy and species-balance equations can be decoupled and that latent heat effects can be ignored.

(a) Scale the describing equations for a constant heat-flux condition at the cold boundary. Determine the criteria for assuming that the entire casting solution is instantaneously at a uniform temperature that changes only in time. Note that the conditions at both the upper and lower boundary must be considered in developing these criteria.

(b) Discuss why a uniform temperature throughout the casting solution might result in a membrane with a homogeneous structure; that is, a structure in which all the pores have essentially the same size.

7.P.18 Effect of Convective Heat Transfer Due to Densification During TIPS Casting

The scaling analysis carried out in Section 7.4 assumed that the thickness of the casting solution remained constant during the TIPS process. In fact, the thickness of the casting solution can change even when there is no mass loss to the ambient if densification occurs due to the phase separation. Since polymer molecules "unfold" in solution, whereas they are more compact in the solid phase, some degree of densification will occur during TIPS casting. The effect of convective transport due to densification was considered in Example Problem 5.E.1. Assume that the mass density of the binary solution of polymer and diluent is given by

$$\rho = \omega_A \rho_A^0 + \omega_B \rho_B^0 \Rightarrow \rho = \rho_B^0 + \Delta\rho_{AB}^0 \omega_A \tag{7.P.18-1}$$

where ρ_i^0 is the pure component mass density of component i and $\Delta\rho_{AB}^0 \equiv \rho_A^0 - \rho_B^0$. The relationship between the overall mass density and the mass-average velocity u_z is given by the continuity equation

$$\frac{\partial \rho}{\partial t} = -\frac{\partial}{\partial z}(\rho u_z) \Rightarrow \Delta\rho_{AB}^0 \frac{\partial \omega_A}{\partial t} = -(\rho_B^0 + \Delta\rho_{AB}^0 \omega_A)\frac{\partial u_z}{\partial z} - u_z \, \Delta\rho_{AB}^0 \frac{\partial \omega_A}{\partial z} \tag{7.P.18-2}$$

If we assume that the energy and species-balance equations can be decoupled, the diluent mass fraction ω_A is simply related to the volume fraction of the liquid phase ψ by equation (7.4-36), where ψ is determined from equation (7.4-5). Hence, there is no need to consider the species-balance equation to obtain a scale for the velocity arising from densification, as was done in Example Problem 5.E.1.

(a) Use scaling analysis to obtain an estimate of the velocity that arises from densification during the TIPS process.
(b) Determine the criterion for ignoring convective heat transfer during TIPS casting.
(c) Determine the criterion for assuming a constant overall mass density during TIPS casting.

7.P.19 Scaling a Fluid-Wall Aerosol Flow Reactor by Balancing Convection and Heat Transfer from the Wall in the Gas Phase

In Section 7.5 we scaled a fluid-wall aerosol reactor for which we considered energy balances for the gas phase and the carbon cloud. Balancing the principal terms in the carbon-cloud energy equation was straightforward since the carbon particles were heated only by radiation from the reactor walls. However, it was not obvious which terms should be balanced in the energy equation for the gas phase since there were several heat sources. We chose to balance the convection term with the sensible heat supplied by the hydrogen injected. In this problem we consider an alternative scaling for the energy balance in the gas phase.

(a) Scale the energy balance in the gas phase assuming that the convection term is balanced by the convective heat transfer from the reactor wall.
(b) Evaluate the dimensionless groups that emanate from this new scaling using the characteristic values given in Table 7.5-1.

(c) Is this a reasonable scaling based on the values of the dimensionless groups that you determined in part (b)?

(d) How can the values of the dimensionless groups determined for the original scaling in Section 7.5 indicate whether this is a reasonable scaling?

7.P.20 Scaling a Fluid-Wall Aerosol Flow Reactor by Balancing Convection and Heat Transfer from the Carbon Particles in the Gas Phase

In Section 7.5 we scaled a fluid-wall aerosol reactor for which we considered energy balances for the gas phase and the carbon cloud. Balancing the principal terms in the carbon-cloud energy equation was straightforward since the carbon particles were heated only by radiation from the reactor walls. However, it was not obvious which terms should be balanced in the energy equation for the gas phase. We chose to balance the convection term with the sensible heat supplied by the hydrogen injected. We subsequently found that the heat transfer from the carbon particles to the gas was so large that the gas came to thermal equilibrium with the carbon particles within a very short distance down the reactor. This suggests that we should have balanced the convection term with the convective heat transfer from the carbon particles.

(a) Rescale the fluid-wall aerosol reactor assuming that the principal terms in the energy balance for the gas balance are the convection term and the convective heat transfer from the carbon particles.

(b) Evaluate the dimensionless groups that emanate from this new scaling using the characteristic values given in Table 7.5-1.

(c) What do the values of the dimensionless groups that you determined in part (b) indicate insofar as the apparent importance of the heat added to the gas by the hydrogen injected and convective heat transfer from the walls?

(d) Why is it not correct to assume that the heat added to the gas by the hydrogen injected and through convective heat transfer from the walls can be neglected based on this scaling?

(e) What do the values of the dimensionless groups determined in the original scaling in Section 7.5 indicate as to the relative importance of the heat added to the gas by the hydrogen injected and by convective heat transfer from the walls?

7.P.21 Estimation of the Thermal Boundary-Layer Thickness at the Upstream End of a Fluid-Wall Aerosol Flow Reactor

In the scaling analysis carried out in Section 7.5 we found that it was reasonable to assume that the gas and carbon particles were in thermal equilibrium. However, as indicated in Section 7.5, this approximation breaks down near the upstream end of the fluid-wall aerosol flow reactor. Use scaling analysis to estimate the thickness of the region of influence or thermal boundary layer within which the gas and carbon particles are not in thermal equilibrium.

7.P.22 Effect of the Carbon-Particle Radius on the Thermal Equilibrium Approximation for a Fluid-Wall Aerosol Flow Reactor

In Section 7.5 we found that the gas was in local thermal equilibrium with the cloud of carbon particles. This was indicated by the large values of the dimensionless groups Π_2 and Π_8 for the characteristic values given in Table 7.5-1.

(a) Determine the dependence of Π_2 and Π_8 on the radius of the carbon particles.

(b) Although both Π_2 and Π_8 decrease with an increase in the carbon-particle radius, their dependence on this parameter differs. How, then, does one determine how large the carbon particle radius has to be for the local thermal equilibrium approximation to break down; that is, which if either of these dimensionless groups becomes limiting?

(c) For a larger carbon-particle size and the other characteristic values given in Table 7.5-1, what can one conclude if $\Pi_2 = O(1)$ but $\Pi_8 \ll 1$?

(d) How large would the carbon particles have to be for the local thermal equilibrium approximation to break down?

7.P.23 Effect of the Wall Temperature on the Thermal Equilibrium Approximation for a Fluid-Wall Aerosol Flow Reactor

In Section 7.5 we found that the gas was in local thermal equilibrium with the cloud of carbon particles. This was indicated by the large values of the dimensionless groups Π_2 and Π_8 for the characteristic values given in Table 7.5-1. The dimensionless group Π_2 depends markedly on the wall temperature of the reactor, whereas Π_8 is independent of it.

(a) For a higher reactor wall temperature and the other characteristic values given in Table 7.5-1, what can one conclude if $\Pi_2 \ll 1$ but $\Pi_8 \gg O(1)$?

(b) For a higher reactor wall temperature and the other characteristic values given in Table 7.5-1, what can one conclude if $\Pi_2 \ll 1$ but $\Pi_8 = O(1)$?

(c) For a higher reactor wall temperature and the other characteristic values given in Table 7.5-1, what can one conclude if both $\Pi_2 \ll 1$ and $\Pi_8 \ll 1$?

(d) How high would the reactor wall temperature have to be for the local thermal equilibrium approximation to break down?

7.P.24 Implications of Rapid Thermal Decomposition Kinetics for a Fluid-Wall Aerosol Flow Reactor

The dimensionless groups Π_1, Π_5, and Π_6 multiply terms that are associated with heat either released or consumed in the thermal decomposition reaction for methane. These three dimensionless groups are proportional to the parameter k_{ns}, which is the scale for the reaction-rate parameter. In principle it seems possible for k_{ns} to be sufficiently large that one or more of the dimensionless groups Π_1, Π_5, and Π_6 becomes much greater than 1.

(a) Discuss the implications of the condition $\Pi_1 \gg 1$; in particular, is this possible?

(b) Discuss the implications of the condition $\Pi_5 \gg 1$; in particular, is this possible?

(c) Discuss the implications of the condition $\Pi_6 \gg 1$; in particular, is this possible?

7.P.25 Temperature Required to Achieve Complete Fractional Conversion in a Fluid-Wall Aerosol Flow Reactor

Estimate the minimum wall temperature required to achieve essentially complete fractional conversion of methane to hydrogen and carbon in a fluid-wall aerosol reactor for the other characteristic parameters given in Table 7.5-1.

7.P.26 Transformation of the Dimensional Groups Containing the Temperature Difference for a Fluid-Wall Aerosol Flow Reactor

In Section 7.5 we used step 9 in the scaling approach to dimensional analysis outlined in Section 2.4 to transform the dimensionless groups that we obtained by scaling analysis to groups more convenient for correlating numerical data. In particular, we used this step to transform the dimensionless groups containing $T_W - T_0$ into groups that contained either T_W or T_0 but not both. However, we did this without proof. Show how the dimensionless groups given by equations (7.5-37), (7.5-38), and (7.5-45) can be transformed into those given by equations (7.5-55), (7.5-56), and (7.5-61) by invoking step 10 in the scaling approach to dimensional analysis outlined in Section 2.4. The latter step states that if $\Pi_p = f(\Pi_1, \Pi_2, \ldots, \Pi_k)$ and Π_2 through Π_k are redundant (i.e., they appear separately), Π_p can be replaced by Π_1.

7.P.27 Transformation of the Dimensional Groups Containing the Sum of Molar Velocities for a Fluid-Wall Aerosol Flow Reactor

In Section 7.5 we used step 9 in the scaling approach to dimensional analysis outlined in Section 2.4 to transform the dimensionless groups that we obtained via scaling analysis to groups more convenient for correlating numerical data. In particular, we used this step to transform the dimensionless groups containing W_C^0 and $W_M^0 + W_C^0$ into groups that contained either W_M^0 or W_C^0 but not both. However, we did this without proof. Show how the dimensionless groups given by equations (7.5-37) and (7.5-44) can be transformed into those given by equations (7.5-55) and (7.5-60) by invoking step 10 in the scaling approach to dimensional analysis outlined in Section 2.4. The latter step states that if $\Pi_p = f(\Pi_1, \Pi_2, \ldots, \Pi_k)$ and Π_2 through Π_k are redundant (i.e., they appear separately), Π_p can be replaced by Π_1.

7.P.28 Reconciling the Estimate of Fractional Conversion with Dimensional Analysis for a Fluid-Wall Aerosol Flow Reactor

We obtained an estimate of the fractional conversion X_c in Section 7.5 given by equation (7.5-52). This estimate would appear to be inconsistent with the implications of the dimensional analysis given by equation (7.5-64); that is, equation (7.5-52) does not include all the parameters contained in the dimensionless groups in equation (7.5-64).

(a) Discuss why there is no contradiction between equations (7.5-52) and (7.5-64).
(b) Show how equation (7.5-52) can be expressed in terms of the reduced set of dimensionless groups appropriate to the approximations made in obtaining this estimate.
(c) Use the estimate provided by equation (7.5-52) to suggest ways to increase the fractional conversion.

7.P.29 Correlation for the Fractional Conversion When Local Thermal Equilibrium Applies in a Fluid-Wall Aerosol Flow Reactor

In Section 7.5 we used the scaling approach to dimensional analysis to develop a correlation for the fractional conversion X_c for the general case for which local thermal equilibrium between the gas and the carbon particles is not assumed. Develop a dimensional analysis correlation for the fractional conversion for the special case of $\Pi_2 \gg 1$, which implies that the gas can be assumed to be in local thermal equilibrium with the carbon particles.

7.P.30 Isolating the Gas-Phase Mass-Transfer Coefficient in Dimensional Analysis for a Fluid-Wall Aerosol Flow Reactor

In Section 7.5 we used steps 7, 9, and 10 in the scaling approach to dimensional analysis to isolate R_0, W_C^0, and T_W into separate groups. Use this approach to develop a correlation for the fractional conversion for which the gas-phase mass-transfer coefficient, k_{Gs}, is isolated into just one dimensional group.

7.P.31 Isolating the Molar Velocity of Hydrogen in Dimensional Analysis for a Fluid-Wall Aerosol Flow Reactor

In Section 7.5 we used steps 7, 9, and 10 in the scaling approach to dimensional analysis to isolate R_0, W_C^0, and T_W into separate groups. Use this approach to develop a correlation for the fractional conversion for which the molar velocity of hydrogen, $\hat{W}_H$, is isolated into just one dimensional group.

APPENDIX A
Sign Convention for the Force on a Fluid Particle

The equations of motion are a statement of Newton's law of motion applied to a system that consists of a fluid particle. The latter can be thought of as a constituent part of a fluid continuum that moves with the local fluid velocity and has a constant infinitesimal mass; however, it need not have constant volume and density since these two quantities can vary in time while maintaining constant fluid particle mass. The fluid particle is a useful system since its mass is so small that in the limit of vanishingly small volume, its properties become those of a point in the fluid continuum that moves with the local fluid velocity.

To apply Newton's law of motion to a fluid particle, we need to consider the force $\vec{F}_p$ exerted on a fluid particle by the fluid surrounding it. The total stress tensor $\underline{\underline{\sigma}}$ by definition operates on the local unit normal vector $\vec{n}$ to a surface to cause a force. The convention employed here defines $\vec{F}_p$ to be the force exerted on the surface of the particle by the fluid on the side into which the normal vector is pointing; this is shown in Figure A.1-1 and is given by

$$\vec{F}_p = -\int_S \vec{n}\underline{\underline{\sigma}}\, dS \tag{A.1-1}$$

It follows that the force $\vec{F}_f$ that the fluid particle exerts on the surrounding fluid is given by

$$\vec{F}_f = +\int_S \vec{n}\underline{\underline{\sigma}}\, dS \tag{A.1-2}$$

An entirely equivalent definition is the following: σ_{ij} is the total stress exerted in the $+j$-direction on a fluid surface defined by a plane of constant i by the fluid in the region of lesser i. For example, σ_{yx} is the total stress exerted in the $+x$-direction on a fluid surface defined by a plane of constant y by the fluid located beneath this surface.

Scaling Analysis in Modeling Transport and Reaction Processes: A Systematic Approach to Model Building and the Art of Approximation, By William B. Krantz

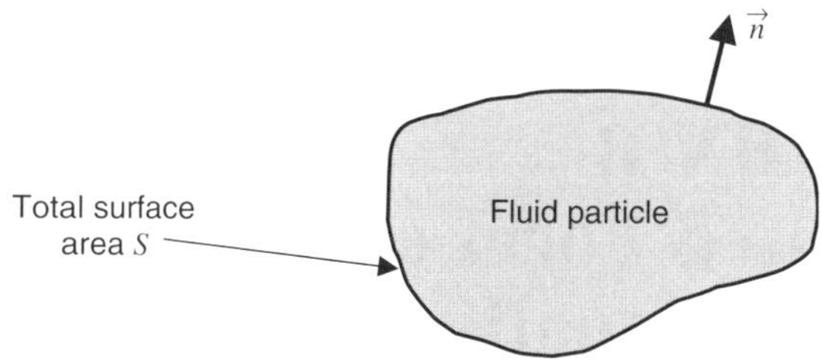

Figure A.1-1 Fluid particle of surface area S showing the unit normal vector to surface $\vec{n}$.

It is important to note that there is no agreement in the technical literature regarding the sign convention for the force acting on a fluid particle. Popular engineering textbooks such as *Transport Phenomena* by Bird ct al.[1] cmploy the sign convention implied by equations (A.1-1) and (A.1-2) whereas other books employ a convention in which the signs are reversed in these two equations. One can use either convention, of course. However, in doing so, one must be certain to use compatible signs in the constitutive equations that relate the stress to the rate of strain; that is, to ensure that positive forces are exerted in the positive coordinate direction, the sign convention given by equations (A.1-1) and (A.1-2) implies a certain set of signs in the constitutive equations that relate the stress to the rate of strain. Hence, the signs given for Newton's constitutive equation in Appendixes B and D are compatible with the sign convention for the force on a fluid particle given by equations (A.1-1) and (A.1-2) When using constitutive equations taken from the technical literature, one has to be careful to employ a form of the equations of motion that has a compatible sign convention.

[1]R. B. Bird, W. E. Stewart, and E. N. Lightfoot, *Transport Phenomena*, 2nd ed., Wiley, Hoboken, NJ, 2002, p. 13.

APPENDIX B Generalized Form of the Transport Equations

B.1 CONTINUITY EQUATION

The following form of the continuity or total mass-balance equation in generalized vector notation is expressed in terms of the mass density ρ, which can be nonconstant, and mass-average velocity vector $\vec{u}$:

$$\frac{\partial \rho}{\partial t} = -\nabla \cdot \rho \vec{u} \tag{B.1-1}$$

The following form of the continuity or total molar-balance equation in generalized vector notation is expressed in terms of the molar density c, which can be nonconstant, and molar-average velocity vector $\vec{\hat{u}}$:

$$\frac{\partial c}{\partial t} = -\nabla \cdot c\vec{\hat{u}} + \hat{G} \tag{B.1-2}$$

where $\hat{G}$ is the total molar generation rate per unit volume.

B.2 EQUATIONS OF MOTION

The following form of the equations of motion in generalized vector–tensor notation allows for nonconstant physical properties, a body force due to a gravitational field, and an unspecified viscous stress tensor $\underline{\underline{\tau}}$:

$$\rho \frac{\partial \vec{u}}{\partial t} + \rho \vec{u} \cdot \nabla \vec{u} = -\nabla P - \nabla \cdot \underline{\underline{\tau}} + \rho \vec{g} \tag{B.2-1}$$

Equation (B.2-1) can be applied to non-Newtonian fluids if the appropriate constitutive equation relating the viscous stress to the rate of strain is known. The viscous stress tensor $\underline{\underline{\tau}}$ follows the sign convection for the force on a fluid particle discussed in Appendix A and for a Newtonian fluid is given by

$$\underline{\underline{\tau}} = -\left(\kappa - \frac{2}{3}\mu\right)(\nabla \cdot \vec{u})\underline{\underline{\delta}} - \mu[\nabla\vec{u} + (\nabla\vec{u})^{\dagger}] \tag{B.2-2}$$

where μ is the shear viscosity, κ the bulk viscosity, $\underline{\underline{\delta}}$ the second-order identity tensor, and $\dagger$ denotes the transpose of a second-order tensor. For the special case of an incompressible Newtonian fluid with constant viscosity, equation (B.2.1) when combined with equation (B.2-2) simplifies to

$$\rho\frac{\partial\vec{u}}{\partial t} + \rho\vec{u} \cdot \nabla\vec{u} = -\nabla P + \mu\nabla^2\vec{u} + \rho\vec{g} \tag{B.2-3}$$

B.3 EQUATIONS OF MOTION FOR POROUS MEDIA

The following form of the equations of motion in generalized vector–tensor notation for flow through porous media is based on Brinkman's empirical modification of Darcy's law and assumes a body force due to a gravitational field, and an incompressible fluid having constant viscosity μ and permeability k_p; $\vec{\hat{u}}$ denotes the superficial velocity based on considering the porous media to be homogeneous[1]:

$$0 = -\nabla P - \frac{\mu}{k_p}\vec{\hat{u}} + \mu\nabla^2\vec{\hat{u}} + \rho\vec{g} \tag{B.3-1}$$

B.4 THERMAL ENERGY EQUATION

The following form of the thermal energy equation in generalized vector–tensor notation allows for nonconstant physical properties, energy generation, and conversion of mechanical to internal energy by means of viscous dissipation, which is expressed in terms of an unspecified viscous stress tensor $\underline{\underline{\tau}}$:

$$\rho C_v\frac{\partial T}{\partial t} + \rho C_v\vec{u} \cdot \nabla T = \nabla \cdot (k\nabla T) - T\left.\frac{\partial P}{\partial T}\right|_{\rho}(\nabla \cdot \vec{u}) - (\underline{\underline{\tau}} : \nabla\vec{u}) + G_e \tag{B.4-1}$$

where C_v is the heat capacity at constant volume, k the thermal conductivity, and G_e the energy generation rate per unit volume. Equation (B.4-1) can be applied to non-Newtonian fluids if the appropriate constitutive equation relating the viscous stress to the rate of strain is known. For the special case of an incompressible Newtonian fluid with constant thermal conductivity k and for which the viscous stress tensor is given by equation (B.2-2), equation (B.4-1) simplifies to

$$\rho C_p\frac{\partial T}{\partial t} + \rho C_p\vec{u} \cdot \nabla T = k\nabla^2 T + \mu[\nabla\vec{u} + (\nabla\vec{u})^{\dagger}] : \nabla\vec{u} + G_e \tag{B.4-2}$$

where C_p is the heat capacity at constant pressure.

[1]H. C. Brinkman, *Appl. Sci. Res.*, **A1**, 27–34, 81–86 (1947).

B.5 EQUATION OF CONTINUITY FOR A BINARY MIXTURE

The following form of the equation of continuity or species-balance equation in generalized vector–tensor notation for component A in a binary system allows for nonconstant physical properties and is expressed in terms of the mass concentration ρ_A and the mass flux vector $\vec{n}_A$:

$$\frac{\partial \rho_A}{\partial t} + \nabla \cdot \vec{n}_A = G_A \tag{B.5-1}$$

where G_A is the mass rate of generation of component A per unit volume. The mass flux vector is given for a binary system by Fick's law of diffusion in the form

$$\vec{n}_A = \omega_A(\vec{n}_A + \vec{n}_B) - \rho D_{AB} \nabla \omega_A = \rho_A \vec{u} - \rho D_{AB} \nabla \omega_A \tag{B.5-2}$$

in which ω_A is the mass fraction of component A, D_{AB} the binary diffusion coefficient, and $\vec{u}$ the mass-average velocity. The form of the species-balance equation given by equation (B.5-1) is particularly useful for describing mass transfer in incompressible liquid and solid systems for which the mass density ρ is constant. For the special case of an incompressible Newtonian fluid with a constant binary diffusion coefficient D_{AB}, equation (B.5-1) when combined with equation (B.5-2) simplifies to

$$\frac{\partial \rho_A}{\partial t} + \vec{u} \cdot \nabla \rho_A = D_{AB} \nabla^2 \rho_A + G_A \tag{B.5-3}$$

The following form of the species-balance equation in generalized vector–tensor notation for component A in a binary system allows for nonconstant physical properties and is expressed in terms of the molar concentration c_A and the molar flux vector $\vec{N}_A$:

$$\frac{\partial c_A}{\partial t} + \nabla \cdot \vec{N}_A = \hat{G}_A \tag{B.5-4}$$

where $\hat{G}_A$ is the molar generation rate of component A per unit volume. The molar flux vector is given for a binary system by Fick's law of diffusion in the form

$$\vec{N}_A = x_A(\vec{N}_A + \vec{N}_B) - c D_{AB} \nabla x_A = c_A \vec{\hat{u}} - c D_{AB} \nabla x_A \tag{B.5-5}$$

in which x_A is the mole fraction of component A and $\vec{\hat{u}}$ is the molar-average velocity. The form of the species-balance equation given by equation (B.5-4) is particularly useful for describing mass transfer in gas systems for which the molar density c is constant at a fixed temperature and pressure. This equation is also used to describe reacting systems for which the rate of generation of species is dictated by the reaction stoichiometry in terms of molar concentrations. For the special case of an incompressible Newtonian fluid with a constant binary diffusion coefficient D_{AB}, equation (B.5-4) when combined with equation (B.5-5) simplifies to

$$\frac{\partial c_A}{\partial t} + \vec{\hat{u}} \cdot \nabla c_A = D_{AB} \nabla^2 c_A + \hat{G}_A \tag{B.5-6}$$

APPENDIX C
Continuity Equation

C.1 RECTANGULAR COORDINATES

The following form of the continuity or total mass-balance equation in rectangular coordinates is expressed in terms of the mass density ρ, which can be nonconstant, and mass-average velocity components u_i:

$$\frac{\partial \rho}{\partial t} + \frac{\partial}{\partial x}(\rho u_x) + \frac{\partial}{\partial y}(\rho u_y) + \frac{\partial}{\partial z}(\rho u_z) = 0 \qquad \text{(C.1-1)}$$

For the special case of an incompressible fluid or fluid having a constant mass density, equation (C.1-1) simplifies to

$$\frac{\partial u_x}{\partial x} + \frac{\partial u_y}{\partial y} + \frac{\partial u_z}{\partial z} = 0 \qquad \text{(C.1-2)}$$

The following form of the continuity or total mass-balance equation in rectangular coordinates is expressed in terms of the molar density c, which can be nonconstant, and molar-average velocity components $\hat{u}_i$:

$$\frac{\partial c}{\partial t} + \frac{\partial}{\partial x}(c\hat{u}_x) + \frac{\partial}{\partial y}(c\hat{u}_y) + \frac{\partial}{\partial z}(c\hat{u}_z) = \hat{G} \qquad \text{(C.1-3)}$$

where $\hat{G}$ is the total molar generation rate per unit volume. For the special case of a fluid having a constant molar density, equation (C.1-3) simplifies to

$$\frac{\partial \hat{u}_x}{\partial x} + \frac{\partial \hat{u}_y}{\partial y} + \frac{\partial \hat{u}_z}{\partial z} = \hat{G} \qquad \text{(C.1-4)}$$

Scaling Analysis in Modeling Transport and Reaction Processes: A Systematic Approach to Model Building and the Art of Approximation, By William B. Krantz

C.2 CYLINDRICAL COORDINATES

The following form of the continuity or total mass-balance equation in cylindrical coordinates is expressed in terms of the mass density ρ, which can be nonconstant, and mass-average velocity components u_i:

$$\frac{\partial \rho}{\partial t} + \frac{1}{r}\frac{\partial}{\partial r}(\rho r u_r) + \frac{1}{r}\frac{\partial}{\partial \theta}(\rho u_\theta) + \frac{\partial}{\partial z}(\rho u_z) = 0 \tag{C.2-1}$$

For the special case of an incompressible fluid or fluid having a constant mass density, equation (C.2-1) simplifies to the following:

$$\frac{1}{r}\frac{\partial}{\partial r}(r u_r) + \frac{1}{r}\frac{\partial u_\theta}{\partial \theta} + \frac{\partial u_z}{\partial z} = 0 \tag{C.2-2}$$

The following form of the continuity or total mass-balance equation in cylindrical coordinates is expressed in terms of the molar density c, which can be nonconstant, and molar-average velocity components $\hat{u}_i$:

$$\frac{\partial c}{\partial t} + \frac{1}{r}\frac{\partial}{\partial r}(c r \hat{u}_r) + \frac{1}{r}\frac{\partial}{\partial \theta}(c \hat{u}_\theta) + \frac{\partial}{\partial z}(c \hat{u}_z) = \hat{G} \tag{C.2-3}$$

where $\hat{G}$ is the total molar generation rate per unit volume. For the special case of a fluid having a constant molar density, equation (C.2-3) simplifies to

$$\frac{1}{r}\frac{\partial}{\partial r}(r \hat{u}_r) + \frac{1}{r}\frac{\partial \hat{u}_\theta}{\partial \theta} + \frac{\partial \hat{u}_z}{\partial z} = \hat{G} \tag{C.2-4}$$

C.3 SPHERICAL COORDINATES

The following form of the continuity or total mass-balance equation in spherical coordinates is expressed in terms of the mass density ρ, which can be nonconstant, and mass-average velocity components u_i:

$$\frac{\partial \rho}{\partial t} + \frac{1}{r^2}\frac{\partial}{\partial r}(\rho r^2 u_r) + \frac{1}{r \sin\theta}\frac{\partial}{\partial \theta}(\rho u_\theta \sin\theta) + \frac{1}{r \sin\theta}\frac{\partial}{\partial \phi}(\rho u_\phi) = 0 \tag{C.3-1}$$

For the special case of an incompressible fluid or fluid having a constant mass density, equation (C.3-1) simplifies to

$$\frac{1}{r^2}\frac{\partial}{\partial r}(r^2 u_r) + \frac{1}{r \sin\theta}\frac{\partial}{\partial \theta}(u_\theta \sin\theta) + \frac{1}{r \sin\theta}\frac{\partial u_\phi}{\partial \phi} = 0 \tag{C.3-2}$$

The following form of the continuity or total mass-balance equation in spherical coordinates is expressed in terms of the molar density c, which can be nonconstant, and molar-average velocity components $\hat{u}_i$:

$$\frac{\partial c}{\partial t} + \frac{1}{r^2}\frac{\partial}{\partial r}(cr^2\hat{u}_r) + \frac{1}{r\sin\theta}\frac{\partial}{\partial\theta}(c\hat{u}_\theta\sin\theta) + \frac{1}{r\sin\theta}\frac{\partial}{\partial\phi}(c\hat{u}_\varphi) = \hat{G} \tag{C.3-3}$$

where $\hat{G}$ is the total molar generation rate per unit volume. For the special case of a fluid having a constant molar density, equation (C.3-3) simplifies to

$$\frac{1}{r^2}\frac{\partial}{\partial r}(r^2\hat{u}_r) + \frac{1}{r\sin\theta}\frac{\partial}{\partial\theta}(\hat{u}_\theta\sin\theta) + \frac{1}{r\sin\theta}\frac{\partial\hat{u}_\phi}{\partial\phi} = \hat{G} \tag{C.3-4}$$

APPENDIX D
Equations of Motion

D.1 RECTANGULAR COORDINATES

The following forms of the x-, y-, and z-components of the equations of motion in rectangular coordinates allow for nonconstant physical properties, a body force due to a gravitational field, and unspecified viscous stress-tensor components τ_{ij}:

$$\rho\frac{\partial u_x}{\partial t} + \rho u_x\frac{\partial u_x}{\partial x} + \rho u_y\frac{\partial u_x}{\partial y} + \rho u_z\frac{\partial u_x}{\partial z} = -\frac{\partial P}{\partial x} - \frac{\partial \tau_{xx}}{\partial x} - \frac{\partial \tau_{yx}}{\partial y} - \frac{\partial \tau_{zx}}{\partial z} + \rho g_x \tag{D.1-1}$$

$$\rho\frac{\partial u_y}{\partial t} + \rho u_x\frac{\partial u_y}{\partial x} + \rho u_y\frac{\partial u_y}{\partial y} + \rho u_z\frac{\partial u_y}{\partial z} = -\frac{\partial P}{\partial y} - \frac{\partial \tau_{xy}}{\partial x} - \frac{\partial \tau_{yy}}{\partial y} - \frac{\partial \tau_{zy}}{\partial z} + \rho g_y \tag{D.1-2}$$

$$\rho\frac{\partial u_z}{\partial t} + \rho u_x\frac{\partial u_z}{\partial x} + \rho u_y\frac{\partial u_z}{\partial y} + \rho u_z\frac{\partial u_z}{\partial z} = -\frac{\partial P}{\partial z} - \frac{\partial \tau_{xz}}{\partial x} - \frac{\partial \tau_{yz}}{\partial y} - \frac{\partial \tau_{zz}}{\partial z} + \rho g_z \tag{D.1-3}$$

Equations (D.1-1) through (D.1-3) can be applied to non-Newtonian fluids if the appropriate constitutive equation relating the viscous stress to the rate of strain is known. The components of the viscous stress tensor τ_{ij} follow the sign convection for the force on a fluid particle discussed in Appendix A and for a Newtonian fluid are given by

$$\tau_{xx} = -\mu\left[2\frac{\partial u_x}{\partial x} - \frac{2}{3}\left(\frac{\partial u_x}{\partial x} + \frac{\partial u_y}{\partial y} + \frac{\partial u_z}{\partial z}\right)\right] \tag{D.1-4}$$

$$\tau_{yy} = -\mu\left[2\frac{\partial u_y}{\partial y} - \frac{2}{3}\left(\frac{\partial u_x}{\partial x} + \frac{\partial u_y}{\partial y} + \frac{\partial u_z}{\partial z}\right)\right] \tag{D.1-5}$$

$$\tau_{zz} = -\mu\left[2\frac{\partial u_z}{\partial z} - \frac{2}{3}\left(\frac{\partial u_x}{\partial x} + \frac{\partial u_y}{\partial y} + \frac{\partial u_z}{\partial z}\right)\right] \tag{D.1-6}$$

$$\tau_{xy} = \tau_{yx} = -\mu\left(\frac{\partial u_x}{\partial y} + \frac{\partial u_y}{\partial x}\right) \tag{D.1-7}$$

Scaling Analysis in Modeling Transport and Reaction Processes: A Systematic Approach to Model Building and the Art of Approximation, By William B. Krantz

$$\tau_{yz} = \tau_{zy} = -\mu\left(\frac{\partial u_y}{\partial z} + \frac{\partial u_z}{\partial y}\right) \tag{D.1-8}$$

$$\tau_{zx} = \tau_{xz} = -\mu\left(\frac{\partial u_z}{\partial x} + \frac{\partial u_x}{\partial z}\right) \tag{D.1-9}$$

For the special case of an incompressible Newtonian fluid with constant viscosity, equations (D.1.1) through (D.1-3) combined with equations (D.1-4) through (D.1-9) simplify to

$$\rho\frac{\partial u_x}{\partial t} + \rho u_x\frac{\partial u_x}{\partial x} + \rho u_y\frac{\partial u_x}{\partial y} + \rho u_z\frac{\partial u_x}{\partial z} = -\frac{\partial P}{\partial x} + \mu\frac{\partial^2 u_x}{\partial x^2} + \mu\frac{\partial^2 u_x}{\partial y^2} + \mu\frac{\partial^2 u_x}{\partial z^2} + \rho g_x \tag{D.1-10}$$

$$\rho\frac{\partial u_y}{\partial t} + \rho u_x\frac{\partial u_y}{\partial x} + \rho u_y\frac{\partial u_y}{\partial y} + \rho u_z\frac{\partial u_y}{\partial z} = -\frac{\partial P}{\partial y} + \mu\frac{\partial^2 u_y}{\partial x^2} + \mu\frac{\partial^2 u_y}{\partial y^2} + \mu\frac{\partial^2 u_y}{\partial z^2} + \rho g_y \tag{D.1-11}$$

$$\rho\frac{\partial u_z}{\partial t} + \rho u_x\frac{\partial u_z}{\partial x} + \rho u_y\frac{\partial u_z}{\partial y} + \rho u_z\frac{\partial u_z}{\partial z} = -\frac{\partial P}{\partial z} + \mu\frac{\partial^2 u_z}{\partial x^2} + \mu\frac{\partial^2 u_z}{\partial y^2} + \mu\frac{\partial^2 u_z}{\partial z^2} + \rho g_z \tag{D.1-12}$$

D.2 CYLINDRICAL COORDINATES

The following forms of the r-, θ-, and z-components of the equations of motion in cylindrical coordinates allow for nonconstant physical properties, a body force due to a gravitational field, and unspecified viscous stress–tensor components τ_{ij}:

$$\rho\frac{\partial u_r}{\partial t} + \rho u_r\frac{\partial u_r}{\partial r} + \rho\frac{u_\theta}{r}\frac{\partial u_r}{\partial \theta} - \rho\frac{u_\theta^2}{r} + \rho u_z\frac{\partial u_r}{\partial z} = -\frac{\partial P}{\partial r} - \frac{1}{r}\frac{\partial}{\partial r}(r\tau_{rr}) - \frac{1}{r}\frac{\partial \tau_{r\theta}}{\partial \theta} + \frac{\tau_{\theta\theta}}{r} - \frac{\partial \tau_{rz}}{\partial z} + \rho g_r \tag{D.2-1}$$

$$\rho\frac{\partial u_\theta}{\partial t} + \rho u_r\frac{\partial u_\theta}{\partial r} + \rho\frac{u_\theta}{r}\frac{\partial u_\theta}{\partial \theta} + \rho\frac{u_r u_\theta}{r} + \rho u_z\frac{\partial u_\theta}{\partial z} = -\frac{1}{r}\frac{\partial P}{\partial \theta} - \frac{1}{r^2}\frac{\partial}{\partial r}(r^2\tau_{r\theta}) - \frac{1}{r}\frac{\partial \tau_{\theta\theta}}{\partial \theta} - \frac{\partial \tau_{\theta z}}{\partial z} + \rho g_\theta \tag{D.2-2}$$

$$\rho\frac{\partial u_z}{\partial t} + \rho u_r\frac{\partial u_z}{\partial r} + \rho\frac{u_\theta}{r}\frac{\partial u_z}{\partial \theta} + \rho u_z\frac{\partial u_z}{\partial z} = -\frac{\partial P}{\partial z} - \frac{1}{r}\frac{\partial}{\partial r}(r\tau_{rz}) - \frac{1}{r}\frac{\partial \tau_{\theta z}}{\partial \theta} - \frac{\partial \tau_{zz}}{\partial z} + \rho g_z \tag{D.2-3}$$

Equations (D.2-1) through (D.2-3) can be applied to non-Newtonian fluids if the appropriate constitutive equation relating the viscous stress to the rate of strain is known. The components of the viscous stress tensor τ_{ij} follow the sign convection for the force on a fluid particle discussed in Appendix A and for a Newtonian fluid are given by

$$\tau_{rr} = -\mu \left\{ 2\frac{\partial u_r}{\partial r} - \frac{2}{3}\left[\frac{1}{r}\frac{\partial}{\partial r}(ru_r) + \frac{1}{r}\frac{\partial u_\theta}{\partial \theta} + \frac{\partial u_z}{\partial z}\right]\right\} \tag{D.2-4}$$

$$\tau_{\theta\theta} = -\mu \left\{ 2\left(\frac{1}{r}\frac{\partial u_\theta}{\partial \theta} + \frac{u_r}{r}\right) - \frac{2}{3}\left[\frac{1}{r}\frac{\partial}{\partial r}(ru_r) + \frac{1}{r}\frac{\partial u_\theta}{\partial \theta} + \frac{\partial u_z}{\partial z}\right]\right\} \tag{D.2-5}$$

$$\tau_{zz} = -\mu \left\{ 2\frac{\partial u_z}{\partial z} - \frac{2}{3}\left[\frac{1}{r}\frac{\partial}{\partial r}(ru_r) + \frac{1}{r}\frac{\partial u_\theta}{\partial \theta} + \frac{\partial u_z}{\partial z}\right]\right\} \tag{D.2-6}$$

$$\tau_{r\theta} = \tau_{\theta r} = -\mu\left[r\frac{\partial}{\partial r}\left(\frac{u_\theta}{r}\right) + \frac{1}{r}\frac{\partial u_r}{\partial \theta}\right] \tag{D.2-7}$$

$$\tau_{\theta z} = \tau_{z\theta} = -\mu\left(\frac{\partial u_\theta}{\partial z} + \frac{1}{r}\frac{\partial u_z}{\partial \theta}\right) \tag{D.2-8}$$

$$\tau_{zr} = \tau_{rz} = -\mu\left(\frac{\partial u_z}{\partial r} + \frac{\partial u_r}{\partial z}\right) \tag{D.2-9}$$

For the special case of an incompressible Newtonian fluid with constant viscosity, equations (D.2.1) through (D.2-3) combined with equations (D.2-4) through (D.2-9) simplify to

$$\rho\frac{\partial u_r}{\partial t} + \rho u_r\frac{\partial u_r}{\partial r} + \rho\frac{u_\theta}{r}\frac{\partial u_r}{\partial \theta} - \rho\frac{u_\theta^2}{r} + \rho u_z\frac{\partial u_r}{\partial z}$$
$$= -\frac{\partial P}{\partial r} + \mu\frac{\partial}{\partial r}\left[\frac{1}{r}\frac{\partial}{\partial r}(ru_r)\right] + \mu\frac{1}{r^2}\frac{\partial^2 u_r}{\partial \theta^2} - \mu\frac{2}{r^2}\frac{\partial u_\theta}{\partial \theta} + \mu\frac{\partial^2 u_r}{\partial z^2} + \rho g_r \tag{D.2-10}$$

$$\rho\frac{\partial u_\theta}{\partial t} + \rho u_r\frac{\partial u_\theta}{\partial r} + \rho\frac{u_\theta}{r}\frac{\partial u_\theta}{\partial \theta} + \rho\frac{u_r u_\theta}{r} + \rho u_z\frac{\partial u_\theta}{\partial z}$$
$$= -\frac{1}{r}\frac{\partial P}{\partial \theta} + \mu\frac{\partial}{\partial r}\left[\frac{1}{r}\frac{\partial}{\partial r}(ru_\theta)\right] + \mu\frac{1}{r^2}\frac{\partial^2 u_\theta}{\partial \theta^2} + \mu\frac{2}{r^2}\frac{\partial u_r}{\partial \theta} + \mu\frac{\partial^2 u_\theta}{\partial z^2} + \rho g_\theta \tag{D.2-11}$$

$$\rho\frac{\partial u_z}{\partial t} + \rho u_r\frac{\partial u_z}{\partial r} + \rho\frac{u_\theta}{r}\frac{\partial u_z}{\partial \theta} + \rho u_z\frac{\partial u_z}{\partial z}$$
$$= -\frac{\partial P}{\partial z} + \mu\frac{1}{r}\frac{\partial}{\partial r}\left(r\frac{\partial u_z}{\partial r}\right) + \mu\frac{1}{r^2}\frac{\partial^2 u_z}{\partial \theta^2} + \mu\frac{\partial^2 u_z}{\partial z^2} + \rho g_z \tag{D.2-12}$$

D.3 SPHERICAL COORDINATES

The following forms of the r-, θ-, and ϕ-components of the equations of motion in spherical coordinates allow for nonconstant physical properties, a body force due to a gravitational field, and unspecified viscous stress–tensor components τ_{ij}:

$$\rho\frac{\partial u_r}{\partial t}+\rho u_r\frac{\partial u_r}{\partial r}+\rho\frac{u_\theta}{r}\frac{\partial u_r}{\partial \theta}+\rho\frac{u_\phi}{r\sin\theta}\frac{\partial u_r}{\partial \phi}-\rho\frac{u_\theta^2+u_\phi^2}{r}$$
$$=-\frac{\partial P}{\partial r}-\frac{1}{r^2}\frac{\partial}{\partial r}(r^2\tau_{rr})-\frac{1}{r\sin\theta}\frac{\partial}{\partial\theta}(\tau_{r\theta}\sin\theta)$$
$$-\frac{1}{r\sin\theta}\frac{\partial\tau_{r\phi}}{\partial\phi}+\frac{\tau_{\theta\theta}+\tau_{\phi\phi}}{r}+\rho g_r \tag{D.3-1}$$

$$\rho\frac{\partial u_\theta}{\partial t}+\rho u_r\frac{\partial u_\theta}{\partial r}+\rho\frac{u_\theta}{r}\frac{\partial u_\theta}{\partial \theta}+\rho\frac{u_\phi}{r\sin\theta}\frac{\partial u_\theta}{\partial \phi}+\rho\frac{u_r u_\theta}{r}-\rho\frac{u_\phi^2\cot\theta}{r}$$
$$=-\frac{1}{r}\frac{\partial P}{\partial \theta}-\frac{1}{r^2}\frac{\partial}{\partial r}(r^2\tau_{r\theta})-\frac{1}{r\sin\theta}\frac{\partial}{\partial\theta}(\tau_{\theta\theta}\sin\theta)-\frac{1}{r\sin\theta}\frac{\partial\tau_{\theta\phi}}{\partial\phi}$$
$$-\frac{\tau_{r\theta}}{r}+\frac{\cot\theta}{r}\tau_{\phi\phi}+\rho g_\theta \tag{D.3-2}$$

$$\rho\frac{\partial u_\phi}{\partial t}+\rho u_r\frac{\partial u_\phi}{\partial r}+\rho\frac{u_\theta}{r}\frac{\partial u_\phi}{\partial \theta}+\rho\frac{u_\phi}{r\sin\theta}\frac{\partial u_\phi}{\partial \phi}+\rho\frac{u_\phi u_r}{r}-\frac{u_\theta u_\phi}{r}\cot\theta$$
$$=-\frac{1}{r\sin\theta}\frac{\partial P}{\partial \phi}-\frac{1}{r^2}\frac{\partial}{\partial r}(r^2\tau_{r\phi})-\frac{1}{r}\frac{\partial\tau_{\theta\phi}}{\partial\theta}-\frac{1}{r\sin\theta}\frac{\partial\tau_{\phi\phi}}{\partial\phi}$$
$$-\frac{\tau_{r\phi}}{r}-\frac{2\cot\theta}{r}\tau_{\theta\phi}+\rho g_\phi \tag{D.3-3}$$

Equations (D.3-1) through (D.3-3) can be applied to non-Newtonian fluids if the appropriate constitutive equation relating the viscous stress to the rate of strain is known. The components of the viscous stress tensor τ_{ij} follow the sign convection for the force on a fluid particle discussed in Appendix A and for a Newtonian fluid are given by

$$\tau_{rr}=-\mu\left\{2\frac{\partial u_r}{\partial r}-\frac{2}{3}\left[\frac{1}{r^2}\frac{\partial}{\partial r}(r^2u_r)+\frac{1}{r\sin\theta}\frac{\partial}{\partial\theta}(u_\theta\sin\theta)+\frac{1}{r\sin\theta}\frac{\partial u_\phi}{\partial\phi}\right]\right\} \tag{D.3-4}$$

$$\tau_{\theta\theta}=-\mu\left\{2\left(\frac{1}{r}\frac{\partial u_\theta}{\partial\theta}+\frac{u_r}{r}\right)-\frac{2}{3}\left[\frac{1}{r^2}\frac{\partial}{\partial r}(r^2u_r)+\frac{1}{r\sin\theta}\frac{\partial}{\partial\theta}(u_\theta\sin\theta)\right.\right.$$
$$\left.\left.+\frac{1}{r\sin\theta}\frac{\partial u_\phi}{\partial\phi}\right]\right\} \tag{D.3-5}$$

$$\tau_{\phi\phi} = -\mu\left\{2\left(\frac{1}{r\sin\theta}\frac{\partial u_\phi}{\partial\phi} + \frac{u_r}{r} + \frac{u_\theta}{r}\cot\theta\right)\right.$$
$$\left. - \frac{2}{3}\left[\frac{1}{r^2}\frac{\partial}{\partial r}(r^2 u_r) + \frac{1}{r\sin\theta}\frac{\partial}{\partial\theta}(u_\theta\sin\theta) + \frac{1}{r\sin\theta}\frac{\partial u_\phi}{\partial\phi}\right]\right\} \tag{D.3-6}$$

$$\tau_{r\theta} = \tau_{\theta r} = -\mu\left[r\frac{\partial}{\partial r}\left(\frac{u_\theta}{r}\right) + \frac{1}{r}\frac{\partial u_r}{\partial\theta}\right] \tag{D.3-7}$$

$$\tau_{\theta\phi} = \tau_{\phi\theta} = -\mu\left[\frac{\sin\theta}{r}\frac{\partial}{\partial\theta}\left(\frac{u_\phi}{\sin\theta}\right) + \frac{1}{r\sin\theta}\frac{\partial u_\theta}{\partial\phi}\right] \tag{D.3-8}$$

$$\tau_{\phi r} = \tau_{r\phi} = -\mu\left[\frac{1}{r\sin\theta}\frac{\partial u_r}{\partial\phi} + r\frac{\partial}{\partial r}\left(\frac{u_\phi}{r}\right)\right] \tag{D.3-9}$$

For the special case of an incompressible Newtonian fluid with constant viscosity, equations (D.3-1) through (D.3-3) combined with equations (D.3-4) through (D.3-9) simplify to

$$\rho\frac{\partial u_r}{\partial t} + \rho u_r\frac{\partial u_r}{\partial r} + \rho\frac{u_\theta}{r}\frac{\partial u_r}{\partial\theta} + \rho\frac{u_\phi}{r\sin\theta}\frac{\partial u_r}{\partial\phi} - \rho\frac{u_\theta^2 + u_\phi^2}{r}$$
$$= -\frac{\partial P}{\partial r} + \mu\frac{1}{r^2}\frac{\partial^2}{\partial r^2}(r^2 u_r) + \mu\frac{1}{r^2\sin\theta}\frac{\partial}{\partial\theta}\left(\sin\theta\frac{\partial u_r}{\partial\theta}\right)$$
$$+ \mu\frac{1}{r^2\sin^2\theta}\frac{\partial^2 u_r}{\partial\phi^2} + \rho g_r \tag{D.3-10}$$

$$\rho\frac{\partial u_\theta}{\partial t} + \rho u_r\frac{\partial u_\theta}{\partial r} + \rho\frac{u_\theta}{r}\frac{\partial u_\theta}{\partial\theta} + \rho\frac{u_\phi}{r\sin\theta}\frac{\partial u_\theta}{\partial\phi} + \rho\frac{u_r u_\theta}{r} - \rho\frac{u_\phi^2\cot\theta}{r}$$
$$= -\frac{1}{r}\frac{\partial P}{\partial\theta} + \mu\frac{1}{r^2}\frac{\partial}{\partial r}\left(r^2\frac{\partial u_\theta}{\partial r}\right) + \mu\frac{1}{r^2}\frac{\partial}{\partial\theta}\left[\frac{1}{\sin\theta}\frac{\partial}{\partial\theta}(u_\theta\sin\theta)\right]$$
$$+ \mu\frac{1}{r^2\sin^2\theta}\frac{\partial^2 u_\theta}{\partial\phi^2} + \mu\frac{2}{r^2}\frac{\partial u_r}{\partial\theta} - \mu\frac{2\cos\theta}{r^2\sin^2\theta}\frac{\partial u_\phi}{\partial\phi} + \rho g_\theta \tag{D.3-11}$$

$$\rho\frac{\partial u_\phi}{\partial t} + \rho u_r\frac{\partial u_\phi}{\partial r} + \rho\frac{u_\theta}{r}\frac{\partial u_\phi}{\partial\theta} + \rho\frac{u_\phi}{r\sin\theta}\frac{\partial u_\phi}{\partial\phi} + \rho\frac{u_\phi u_r}{r} + \rho\frac{u_\theta u_\phi}{r}\cot\theta$$
$$= -\frac{1}{r\sin\theta}\frac{\partial P}{\partial\phi} + \mu\frac{1}{r^2}\frac{\partial}{\partial r}\left(r^2\frac{\partial u_\phi}{\partial r}\right) + \mu\frac{1}{r^2}\frac{\partial}{\partial\theta}\left[\frac{1}{\sin\theta}\frac{\partial}{\partial\theta}(u_\phi\sin\theta)\right]$$
$$+ \mu\frac{1}{r^2\sin^2\theta}\frac{\partial^2 u_\phi}{\partial\phi^2} + \mu\frac{2}{r^2\sin\theta}\frac{\partial u_r}{\partial\phi} + \mu\frac{2\cos\theta}{r^2\sin^2\theta}\frac{\partial u_\theta}{\partial\phi} + \rho g_\phi \tag{D.3-12}$$

APPENDIX E
Equations of Motion for Porous Media

E.1 RECTANGULAR COORDINATES

The following forms of the x-, y-, and z-components of the equations of motion in rectangular coordinates for flow through porous media are based on Brinkman's empirical modification of Darcy's law and assume a body force due to a gravitational field, and an incompressible fluid having constant viscosity μ and permeability k_p; the $\widehat{u}_i$ denote the components of the superficial velocity based on considering a porous medium to be homogeneous[1]:

$$0 = -\frac{\partial P}{\partial x} - \frac{\mu}{k_p}\widehat{u}_x + \mu\frac{\partial^2 \widehat{u}_x}{\partial x^2} + \mu\frac{\partial^2 \widehat{u}_x}{\partial y^2} + \mu\frac{\partial^2 \widehat{u}_x}{\partial z^2} + \rho g_x \tag{E.1-1}$$

$$0 = -\frac{\partial P}{\partial y} - \frac{\mu}{k_p}\widehat{u}_y + \mu\frac{\partial^2 \widehat{u}_y}{\partial x^2} + \mu\frac{\partial^2 \widehat{u}_y}{\partial y^2} + \mu\frac{\partial^2 \widehat{u}_y}{\partial z^2} + \rho g_y \tag{E.1-2}$$

$$0 = -\frac{\partial P}{\partial z} - \frac{\mu}{k_p}\widehat{u}_z + \mu\frac{\partial^2 \widehat{u}_z}{\partial x^2} + \mu\frac{\partial^2 \widehat{u}_z}{\partial y^2} + \mu\frac{\partial^2 \widehat{u}_z}{\partial z^2} + \rho g_z \tag{E.1-3}$$

E.2 CYLINDRICAL COORDINATES

The following forms of the r-, θ-, and z-components of the equations of motion in cylindrical coordinates for flow through porous media are based on Brinkman's empirical modification of Darcy's law and assume a body force due to a gravitational field, and an incompressible fluid having constant viscosity μ and permeability k_p; the $\widehat{u}_i$ denote the components of the superficial velocity based on considering a porous medium to be homogeneous[2]:

$$0 = -\frac{\partial P}{\partial r} - \frac{\mu}{k_p}\widehat{u}_r + \mu\frac{\partial}{\partial r}\left[\frac{1}{r}\frac{\partial}{\partial r}(r\,\widehat{u}_r)\right] + \mu\frac{1}{r^2}\frac{\partial^2 \widehat{u}_r}{\partial \theta^2} - \mu\frac{2}{r^2}\frac{\partial \widehat{u}_\theta}{\partial \theta} + \mu\frac{\partial^2 \widehat{u}_r}{\partial z^2} + \rho g_r \tag{E.2-1}$$

[1] H. C. Brinkman, *Appl. Sci. Res.*, **A1**, 27–34, 81–86 (1947).
[2] Ibid.

Scaling Analysis in Modeling Transport and Reaction Processes: A Systematic Approach to Model Building and the Art of Approximation, By William B. Krantz

$$0 = -\frac{1}{r}\frac{\partial P}{\partial \theta} - \frac{\mu}{k_p}\widehat{u}_\theta + \mu\frac{\partial}{\partial r}\left[\frac{1}{r}\frac{\partial}{\partial r}(r\,\widehat{u}_\theta)\right] + \mu\frac{1}{r^2}\frac{\partial^2 \widehat{u}_\theta}{\partial \theta^2}$$

$$+ \mu\frac{2}{r^2}\frac{\partial \widehat{u}_r}{\partial \theta} + \mu\frac{\partial^2 \widehat{u}_\theta}{\partial z^2} + \rho g_\theta \tag{E.2-2}$$

$$0 = -\frac{\partial P}{\partial z} - \frac{\mu}{k_p}\widehat{u}_z + \mu\frac{1}{r}\frac{\partial}{\partial r}\left(r\frac{\partial \widehat{u}_z}{\partial r}\right) + \mu\frac{1}{r^2}\frac{\partial^2 \widehat{u}_z}{\partial \theta^2} + \mu\frac{\partial^2 \widehat{u}_z}{\partial z^2} + \rho g_z \tag{E.2-3}$$

E.3 SPHERICAL COORDINATES

The following forms of the r-, θ-, and ϕ-components of the equations of motion in spherical coordinates for flow through porous media are based on Brinkman's empirical modification of Darcy's law and assume a body force due to a gravitational field, and an incompressible fluid having constant viscosity μ and permeability k_p; the $\widehat{u}_i$ denote the components of the superficial velocity based on considering a porous medium to be homogeneous[3]:

$$0 = -\frac{\partial P}{\partial r} - \frac{\mu}{k_p}\widehat{u}_r + \mu\frac{1}{r^2}\frac{\partial^2}{\partial r^2}(r^2\,\widehat{u}_r) + \mu\frac{1}{r^2 \sin\theta}\frac{\partial}{\partial \theta}\left(\sin\theta\frac{\partial \widehat{u}_r}{\partial \theta}\right)$$

$$+ \mu\frac{1}{r^2 \sin^2\theta}\frac{\partial^2 \widehat{u}_r}{\partial \phi^2} + \rho g_r \tag{E.3-1}$$

$$0 = -\frac{1}{r}\frac{\partial P}{\partial \theta} - \frac{\mu}{k_p}\widehat{u}_\theta + \mu\frac{1}{r^2}\frac{\partial}{\partial r}\left(r^2\frac{\partial \widehat{u}_\theta}{\partial r}\right) + \mu\frac{1}{r^2}\frac{\partial}{\partial \theta}\left[\frac{1}{\sin\theta}\frac{\partial}{\partial \theta}(\widehat{u}_\theta \sin\theta)\right]$$

$$+ \mu\frac{1}{r^2 \sin^2\theta}\frac{\partial^2 \widehat{u}_\theta}{\partial \phi^2} + \mu\frac{2}{r^2}\frac{\partial \widehat{u}_r}{\partial \theta} - \mu\frac{2\cos\theta}{r^2 \sin^2\theta}\frac{\partial \widehat{u}_\phi}{\partial \phi} + \rho g_\theta \tag{E.3-2}$$

$$0 = -\frac{1}{r\sin\theta}\frac{\partial P}{\partial \phi} - \frac{\mu}{k_p}\widehat{u}_\phi + \mu\frac{1}{r^2}\frac{\partial}{\partial r}\left(r^2\frac{\partial \widehat{u}_\phi}{\partial r}\right) + \mu\frac{1}{r^2}\frac{\partial}{\partial \theta}\left[\frac{1}{\sin\theta}\frac{\partial}{\partial \theta}(\widehat{u}_\phi \sin\theta)\right]$$

$$+ \mu\frac{1}{r^2 \sin^2\theta}\frac{\partial^2 \widehat{u}_\phi}{\partial \phi^2} + \mu\frac{2}{r^2 \sin\theta}\frac{\partial \widehat{u}_r}{\partial \phi} + \mu\frac{2\cos\theta}{r^2 \sin^2\theta}\frac{\partial \widehat{u}_\theta}{\partial \phi} + \rho g_\phi \tag{E.3-3}$$

[3] Ibid.

APPENDIX F
Thermal Energy Equation

F.1 RECTANGULAR COORDINATES

The following form of the thermal energy equation in rectangular coordinates allows for nonconstant physical properties, energy generation, and conversion of mechanical to internal energy through viscous dissipation, which is expressed in terms of unspecified viscous stress–tensor components τ_{ij}:

$$\rho C_v \frac{\partial T}{\partial t} + \rho C_v u_x \frac{\partial T}{\partial x} + \rho C_v u_y \frac{\partial T}{\partial y} + \rho C_v u_z \frac{\partial T}{\partial z} = \frac{\partial}{\partial x}\left(k \frac{\partial T}{\partial x}\right) + \frac{\partial}{\partial y}\left(k \frac{\partial T}{\partial y}\right)$$
$$+ \frac{\partial}{\partial z}\left(k \frac{\partial T}{\partial z}\right) - T \left.\frac{\partial P}{\partial T}\right|_{\rho} \left(\frac{\partial u_x}{\partial x} + \frac{\partial u_y}{\partial y} + \frac{\partial u_z}{\partial z}\right) - \tau_{xx} \frac{\partial u_x}{\partial x} - \tau_{yy} \frac{\partial u_y}{\partial y} - \tau_{zz} \frac{\partial u_z}{\partial z}$$
$$- \tau_{xy} \left(\frac{\partial u_x}{\partial y} + \frac{\partial u_y}{\partial x}\right) - \tau_{xz} \left(\frac{\partial u_x}{\partial z} + \frac{\partial u_z}{\partial x}\right) - \tau_{yz} \left(\frac{\partial u_y}{\partial z} + \frac{\partial u_z}{\partial y}\right) + G_e \tag{F.1-1}$$

where C_v is the heat capacity at constant volume, k the thermal conductivity, and G_e the energy generation rate per unit volume. Equation (F.1-1) can be applied to non-Newtonian fluids if the appropriate constitutive equation relating the viscous stress to the rate of strain is known. For the special case of an incompressible Newtonian fluid with constant thermal conductivity for which the components of the viscous stress tensor are given by equations (D.1-4) through (D.1-9), equation (F.1-1) simplifies to

$$\rho C_p \frac{\partial T}{\partial t} + \rho C_p u_x \frac{\partial T}{\partial x} + \rho C_p u_y \frac{\partial T}{\partial y} + \rho C_p u_z \frac{\partial T}{\partial z} = k \frac{\partial^2 T}{\partial x^2} + k \frac{\partial^2 T}{\partial y^2}$$
$$+ k \frac{\partial^2 T}{\partial z^2} + 2\mu \left(\frac{\partial u_x}{\partial x}\right)^2 + 2\mu \left(\frac{\partial u_y}{\partial y}\right)^2 + 2\mu \left(\frac{\partial u_z}{\partial z}\right)^2 + \mu \left(\frac{\partial u_x}{\partial y} + \frac{\partial u_y}{\partial x}\right)^2$$
$$+ \mu \left(\frac{\partial u_x}{\partial z} + \frac{\partial u_z}{\partial x}\right)^2 + \mu \left(\frac{\partial u_y}{\partial z} + \frac{\partial u_z}{\partial y}\right)^2 + G_e \tag{F.1-2}$$

where C_p is the heat capacity at constant pressure.

Scaling Analysis in Modeling Transport and Reaction Processes: A Systematic Approach to Model Building and the Art of Approximation, By William B. Krantz

F.2 CYLINDRICAL COORDINATES

The following form of the thermal energy equation in cylindrical coordinates allows for nonconstant physical properties, energy generation, and conversion of mechanical to internal energy using viscous dissipation, which is expressed in terms of unspecified viscous stress–tensor components τ_{ij}:

$$\rho C_v \frac{\partial T}{\partial t} + \rho C_v u_r \frac{\partial T}{\partial r} + \rho C_v \frac{u_\theta}{r}\frac{\partial T}{\partial \theta} + \rho C_v u_z \frac{\partial T}{\partial z} = \frac{1}{r}\frac{\partial}{\partial r}\left(rk\frac{\partial T}{\partial r}\right)$$
$$+ \frac{1}{r^2}\frac{\partial}{\partial \theta}\left(k\frac{\partial T}{\partial \theta}\right) + \frac{\partial}{\partial z}\left(k\frac{\partial T}{\partial z}\right) - T\left.\frac{\partial P}{\partial T}\right|_\rho \left[\frac{1}{r}\frac{\partial}{\partial r}(ru_r) + \frac{1}{r}\frac{\partial u_\theta}{\partial \theta} + \frac{\partial u_z}{\partial z}\right]$$
$$- \tau_{rr}\frac{\partial u_r}{\partial r} - \tau_{\theta\theta}\frac{1}{r}\left(\frac{\partial u_\theta}{\partial \theta} + u_r\right) - \tau_{zz}\frac{\partial u_z}{\partial z} - \tau_{r\theta}\left[r\frac{\partial}{\partial r}\left(\frac{u_\theta}{r}\right) + \frac{1}{r}\frac{\partial u_r}{\partial \theta}\right]$$
$$- \tau_{rz}\left(\frac{\partial u_z}{\partial r} + \frac{\partial u_r}{\partial z}\right) - \tau_{\theta z}\left(\frac{1}{r}\frac{\partial u_z}{\partial \theta} + \frac{\partial u_\theta}{\partial z}\right) + G_e \qquad \text{(F.2-1)}$$

where C_v is the heat capacity at constant volume, k the thermal conductivity, and G_e the energy generation rate per unit volume. Equation (F.2-1) can be applied to non-Newtonian fluids if the appropriate constitutive equation relating the viscous stress to the rate of strain is known. For the special case of an incompressible Newtonian fluid with constant thermal conductivity for which the components of the viscous stress tensor are given by equations (D.2-4) through (D.2-9), equation (F.2.1) simplifies to

$$\rho C_p \frac{\partial T}{\partial t} + \rho C_p u_r \frac{\partial T}{\partial r} + \rho C_p \frac{u_\theta}{r}\frac{\partial T}{\partial \theta} + \rho C_p u_z \frac{\partial T}{\partial z} = \frac{k}{r}\frac{\partial}{\partial r}\left(r\frac{\partial T}{\partial r}\right) + \frac{k}{r^2}\frac{\partial^2 T}{\partial \theta^2}$$
$$+ k\frac{\partial^2 T}{\partial z^2} + 2\mu\left(\frac{\partial u_r}{\partial r}\right)^2 + \frac{2\mu}{r^2}\left(\frac{\partial u_\theta}{\partial \theta} + u_r\right)^2 + 2\mu\left(\frac{\partial u_z}{\partial z}\right)^2 \qquad \text{(F.2-2)}$$
$$+ \mu\left(\frac{\partial u_\theta}{\partial z} + \frac{1}{r}\frac{\partial u_z}{\partial \theta}\right)^2 + \mu\left(\frac{\partial u_z}{\partial r} + \frac{\partial u_r}{\partial z}\right)^2 + \mu\left[\frac{1}{r}\frac{\partial u_r}{\partial \theta} + r\frac{\partial}{\partial r}\left(\frac{u_\theta}{r}\right)\right]^2 + G_e$$

where C_p is the heat capacity at constant pressure.

F.3 SPHERICAL COORDINATES

The following form of the thermal energy equation in spherical coordinates allows for nonconstant physical properties, energy generation, and conversion of mechanical to internal energy through viscous dissipation, which is expressed in terms of unspecified viscous stress–tensor components τ_{ij}:

$$
\begin{aligned}
\rho C_v \frac{\partial T}{\partial t} + \rho C_v u_r \frac{\partial T}{\partial r} + \rho C_v \frac{u_\theta}{r}\frac{\partial T}{\partial \theta} + \rho C_v \frac{u_\phi}{r\sin\theta}\frac{\partial T}{\partial \phi} &= \frac{1}{r^2}\frac{\partial}{\partial r}\left(r^2 k \frac{\partial T}{\partial r}\right) \\
&+ \frac{1}{r^2 \sin\theta}\frac{\partial}{\partial \theta}\left(k\sin\theta \frac{\partial T}{\partial \theta}\right) + \frac{1}{r^2\sin^2\theta}\frac{\partial}{\partial \phi}\left(k\frac{\partial T}{\partial \phi}\right) \\
&- T\left.\frac{\partial P}{\partial T}\right|_\rho \left[\frac{1}{r^2}\frac{\partial}{\partial r}(r^2 u_r) + \frac{1}{r\sin\theta}\frac{\partial}{\partial \theta}(u_\theta \sin\theta) + \frac{1}{r\sin\theta}\frac{\partial u_\phi}{\partial \phi}\right] \\
&- \tau_{rr}\frac{\partial u_r}{\partial r} - \tau_{\theta\theta}\left(\frac{1}{r}\frac{\partial u_\theta}{\partial \theta} + \frac{u_r}{r}\right) - \tau_{\phi\phi}\left(\frac{1}{r\sin\theta}\frac{\partial u_\phi}{\partial \phi} + \frac{u_r}{r} + \frac{u_\theta \cot\theta}{r}\right) \\
&- \tau_{r\theta}\left(\frac{\partial u_\theta}{\partial r} + \frac{1}{r}\frac{\partial u_r}{\partial \theta} - \frac{u_\theta}{r}\right) - \tau_{r\phi}\left(\frac{\partial u_\phi}{\partial r} + \frac{1}{r\sin\theta}\frac{\partial u_r}{\partial \phi} - \frac{u_\phi}{r}\right) \\
&- \tau_{\theta\phi}\left(\frac{1}{r}\frac{\partial u_\phi}{\partial \theta} + \frac{1}{r\sin\theta}\frac{\partial u_\theta}{\partial \phi} - \frac{\cot\theta}{r}u_\phi\right) + G_e \qquad \text{(F.3-1)}
\end{aligned}
$$

where C_v is the heat capacity at constant volume, k the thermal conductivity, and G_e the energy generation rate per unit volume. Equation (F.3-1) can be applied to non-Newtonian fluids if the appropriate constitutive equation relating the viscous stress to the rate of strain is known. For the special case of an incompressible Newtonian fluid with constant thermal conductivity for which the components of the viscous stress tensor are given by equations (D.3-4) through (D.3-9), equation (F.3-1) simplifies to

$$
\begin{aligned}
\rho C_p \frac{\partial T}{\partial t} + \rho C_p u_r \frac{\partial T}{\partial r} + \rho C_p \frac{u_\theta}{r}\frac{\partial T}{\partial \theta} + \rho C_p \frac{u_\phi}{r\sin\theta}\frac{\partial T}{\partial \phi} &= \frac{k}{r^2}\frac{\partial}{\partial r}\left(r^2 \frac{\partial T}{\partial r}\right) \\
&+ \frac{k}{r^2 \sin\theta}\frac{\partial}{\partial \theta}\left(\sin\theta \frac{\partial T}{\partial \theta}\right) + \frac{k}{r^2\sin^2\theta}\frac{\partial^2 T}{\partial \phi^2} \\
&+ 2\mu\left(\frac{\partial u_r}{\partial r}\right)^2 + 2\mu\left(\frac{1}{r}\frac{\partial u_\theta}{\partial \theta} + \frac{u_r}{r}\right)^2 \\
&+ 2\mu\left(\frac{1}{r\sin\theta}\frac{\partial u_\phi}{\partial \phi} + \frac{u_r}{r} + \frac{u_\theta\cot\theta}{r}\right)^2 + \mu\left[r\frac{\partial}{\partial r}\left(\frac{u_\theta}{r}\right) + \frac{1}{r}\frac{\partial u_r}{\partial \theta}\right]^2 \\
&+ \mu\left[\frac{1}{r\sin\theta}\frac{\partial u_r}{\partial \phi} + r\frac{\partial}{\partial r}\left(\frac{u_\phi}{r}\right)\right]^2 + \mu\left[\frac{\sin\theta}{r}\frac{\partial}{\partial \theta}\left(\frac{u_\phi}{\sin\theta}\right) + \frac{1}{r\sin\theta}\frac{\partial u_\theta}{\partial \phi}\right]^2 + G_e \\
&\qquad \text{(F.3-2)}
\end{aligned}
$$

where C_p is the heat capacity at constant pressure.

APPENDIX G
Equation of Continuity for a Binary Mixture

G.1 RECTANGULAR COORDINATES

The following form of the equation of continuity or species balance in rectangular coordinates for component A in a binary system allows for nonconstant physical properties and is expressed in terms of the mass concentration ρ_A and the mass flux components n_{Ai}:

$$\frac{\partial \rho_A}{\partial t} + \frac{\partial n_{Ax}}{\partial x} + \frac{\partial n_{Ay}}{\partial y} + \frac{\partial n_{Az}}{\partial z} = G_A \tag{G.1-1}$$

where G_A is the mass generation rate of component A per unit volume. The components of the mass flux are given for a binary system by Fick's law of diffusion in the form

$$n_{Ax} = \omega_A(n_{Ax} + n_{Bx}) - \rho D_{AB}\frac{\partial \omega_A}{\partial x} = \rho_A u_x - \rho D_{AB}\frac{\partial \omega_A}{\partial x} \tag{G.1-2}$$

$$n_{Ay} = \omega_A(n_{Ay} + n_{By}) - \rho D_{AB}\frac{\partial \omega_A}{\partial y} = \rho_A u_y - \rho D_{AB}\frac{\partial \omega_A}{\partial y} \tag{G.1-3}$$

$$n_{Az} = \omega_A(n_{Az} + n_{Bz}) - \rho D_{AB}\frac{\partial \omega_A}{\partial z} = \rho_A u_z - \rho D_{AB}\frac{\partial \omega_A}{\partial z} \tag{G.1-4}$$

in which ω_A is the mass fraction of component A, D_{AB} the binary diffusion coefficient, and u_i the mass-average velocity component in the i-direction. The form of the equation of continuity for a binary mixture or species balance given by equation (G.1-1) is particularly useful for describing mass transfer in incompressible liquid and solid systems for which the mass density ρ is constant. For the special case of an incompressible fluid or fluid having a constant mass density and a constant binary diffusion coefficient, equation (G.1-1) when combined with equations (G.1-2) through (G.1-4) simplifies to

$$\frac{\partial \rho_A}{\partial t} + u_x\frac{\partial \rho_A}{\partial x} + u_y\frac{\partial \rho_A}{\partial y} + u_z\frac{\partial \rho_A}{\partial z} = D_{AB}\frac{\partial^2 \rho_A}{\partial x^2} + D_{AB}\frac{\partial^2 \rho_A}{\partial y^2} + D_{AB}\frac{\partial^2 \rho_A}{\partial z^2} + G_A \tag{G.1-5}$$

Scaling Analysis in Modeling Transport and Reaction Processes: A Systematic Approach to Model Building and the Art of Approximation, By William B. Krantz

The following form of the equation of continuity or species balance in rectangular coordinates for component A in a binary system allows for nonconstant physical properties and is expressed in terms of the molar concentration c_A and the molar flux components N_{Ai}:

$$\frac{\partial c_A}{\partial t} + \frac{\partial N_{Ax}}{\partial x} + \frac{\partial N_{Ay}}{\partial y} + \frac{\partial N_{Az}}{\partial z} = \hat{G}_A \tag{G.1-6}$$

where $\hat{G}_A$ is the molar generation rate of component A per unit volume. The components of the molar flux are given for a binary system by Fick's law of diffusion in the form

$$N_{Ax} = x_A(N_{Ax} + N_{Bx}) - cD_{AB}\frac{\partial x_A}{\partial x} = c_A\hat{u}_x - cD_{AB}\frac{\partial x_A}{\partial x} \tag{G.1-7}$$

$$N_{Ay} = x_A(N_{Ay} + N_{By}) - cD_{AB}\frac{\partial x_A}{\partial y} = c_A\hat{u}_y - cD_{AB}\frac{\partial x_A}{\partial y} \tag{G.1-8}$$

$$N_{Az} = x_A(N_{Az} + N_{Bz}) - cD_{AB}\frac{\partial x_A}{\partial z} = c_A\hat{u} - cD_{AB}\frac{\partial x_A}{\partial z} \tag{G.1-9}$$

in which x_A is the mole fraction of component A. The form of the equation of continuity for a binary mixture or species balance given by equation (G.1-6) is particularly useful for describing mass transfer in gas systems for which the molar density c is constant at a fixed temperature and pressure. This equation is also used to describe reacting systems for which the generation rate of species is dictated by the reaction stoichiometry in terms of molar concentrations. For the special case of a fluid having a constant molar density and a constant binary diffusion coefficient, equation (G.1-6) combined with equations (G.1-7) through (G.1-9) simplifies to

$$\frac{\partial c_A}{\partial t} + \hat{u}_x\frac{\partial c_A}{\partial x} + \hat{u}_y\frac{\partial c_A}{\partial y} + \hat{u}_z\frac{\partial c_A}{\partial z} = D_{AB}\frac{\partial^2 c_A}{\partial x^2} + D_{AB}\frac{\partial^2 c_A}{\partial y^2} + D_{AB}\frac{\partial^2 c_A}{\partial z^2} + \hat{G}_A \tag{G.1-10}$$

G.2 CYLINDRICAL COORDINATES

The following form of the equation of continuity or species balance in cylindrical coordinates for component A in a binary system allows for nonconstant physical properties and is expressed in terms of the mass concentration ρ_A and the mass flux components n_{Ai}:

$$\frac{\partial \rho_A}{\partial t} + \frac{1}{r}\frac{\partial}{\partial r}(rn_{Ar}) + \frac{1}{r}\frac{\partial n_{A\theta}}{\partial \theta} + \frac{\partial n_{Az}}{\partial z} = G_A \tag{G.2-1}$$

where G_A is the mass generation rate of component A per unit volume. The components of the mass flux are given for a binary system by Fick's law of diffusion in the form

$$n_{Ar} = \omega_A(n_{Ar} + n_{Br}) - \rho D_{AB}\frac{\partial \omega_A}{\partial r} = \rho_A u_r - \rho D_{AB}\frac{\partial \omega_A}{\partial r} \tag{G.2-2}$$

$$n_{A\theta} = \omega_A(n_{A\theta} + n_{B\theta}) - \rho D_{AB}\frac{1}{r}\frac{\partial \omega_A}{\partial \theta} = \rho_A u_\theta - \rho D_{AB}\frac{1}{r}\frac{\partial \omega_A}{\partial \theta} \tag{G.2-3}$$

$$n_{Az} = \omega_A(n_{Az} + n_{Bz}) - \rho D_{AB}\frac{\partial x_A}{\partial z} = \rho_A u_z - \rho D_{AB}\frac{\partial x_A}{\partial z} \tag{G.2-4}$$

in which ω_A is the mass fraction of component A, D_{AB} the binary diffusion coefficient, and u_i the mass-average velocity component in the i-direction. The form of the equation of continuity for a binary mixture or species balance given by equation (G.2-1) is particularly useful for describing mass transfer in incompressible liquid and solid systems for which the mass density ρ is constant. For the special case of an incompressible fluid or fluid having a constant mass density and a constant binary diffusion coefficient, equation (G.2-1) combined with equations (G.2-3) through (G.2-4) simplifies to

$$\begin{aligned}\frac{\partial \rho_A}{\partial t} + u_r\frac{\partial \rho_A}{\partial r} + u_\theta\frac{1}{r}\frac{\partial \rho_A}{\partial \theta} + u_z\frac{\partial \rho_A}{\partial z} &= D_B\frac{1}{r}\frac{\partial}{\partial r}\left(r\frac{\partial \rho_A}{\partial r}\right)\\ &\quad + D_B\frac{1}{r^2}\frac{\partial^2 \rho}{\partial \theta^2} + D_{AB}\frac{\partial^2 \rho_A}{\partial z^2} + G_A\end{aligned} \tag{G.2-5}$$

The following form of the equation of continuity or species balance in cylindrical coordinates for component A in a binary system allows for nonconstant physical properties and is expressed in terms of the molar concentration c_A and the molar flux components N_{Ai}:

$$\frac{\partial c_A}{\partial t} + \frac{1}{r}\frac{\partial}{\partial r}(rN_{Ar}) + \frac{1}{r}\frac{\partial N_{A\theta}}{\partial \theta} + \frac{\partial N_{Az}}{\partial z} = \hat{G}_A \tag{G.2-6}$$

where $\hat{G}_A$ is the molar generation rate of component A per unit volume. The components of the molar flux are given for a binary system by Fick's law of diffusion in the form

$$N_{Ar} = x_A(N_{Ar} + N_{Br}) - cD_{AB}\frac{\partial x_A}{\partial r} = c_A\hat{u}_r - cD_{AB}\frac{\partial x_A}{\partial r} \tag{G.2-7}$$

$$N_{A\theta} = x_A(N_{A\theta} + N_{B\theta}) - cD_{AB}\frac{1}{r}\frac{\partial x_A}{\partial \theta} = c_A\hat{u}_\theta - cD_{AB}\frac{1}{r}\frac{\partial x_A}{\partial \theta} \tag{G.2-8}$$

$$N_{Az} = x_A(N_{Az} + N_{Bz}) - cD_{AB}\frac{\partial x_A}{\partial z} = c_A\hat{u}_z - cD_{AB}\frac{\partial x_A}{\partial z} \tag{G.2-9}$$

in which x_A is the mole fraction of component A and $\hat{u}_i$ is the molar-average velocity component in the i-direction. The form of the equation of continuity for a binary mixture or species balance given by equation (G.2-6) is particularly useful for describing mass transfer in gas systems for which the molar density c

is constant at a fixed temperature and pressure. This equation is also used to describe reacting systems for which the generation rate of species is dictated by the reaction stoichiometry in terms of molar concentrations. For the special case of a fluid having a constant molar density and a constant binary diffusion coefficient, equation (G.2-6) combined with equations (G.2-7) through (G.2-9) simplifies to

$$\frac{\partial c_A}{\partial t} + \hat{u}_r \frac{\partial c_A}{\partial r} + \hat{u}_\theta \frac{1}{r}\frac{\partial c_A}{\partial \theta} + \hat{u}_z \frac{\partial c_A}{\partial z} = D_{AB}\frac{1}{r}\frac{\partial}{\partial r}\left(r\frac{\partial c_A}{\partial r}\right) + D_{AB}\frac{1}{r^2}\frac{\partial^2 c_A}{\partial \theta^2} + D_{AB}\frac{\partial^2 c_A}{\partial z^2} + \hat{G}_A \tag{G.2-10}$$

G.3 SPHERICAL COORDINATES

The following form of the equation of continuity or species balance in spherical coordinates for component A in a binary system allows for nonconstant physical properties and is expressed in terms of the mass concentration ρ_A and the mass flux components n_{Ai}:

$$\frac{\partial \rho_A}{\partial t} + \frac{1}{r^2}\frac{\partial}{\partial r}(r^2 n_{Ar}) + \frac{1}{r\sin\theta}\frac{\partial}{\partial \theta}(n_{A\theta}\sin\theta) + \frac{1}{r\sin\theta}\frac{\partial n_{A\phi}}{\partial \phi} = G_A \tag{G.3-1}$$

where G_A is the mass generation rate of component A per unit volume. The components of the mass flux are given for a binary system by Fick's law of diffusion in the form

$$n_{Ar} = \omega_A(n_{Ar} + n_{Br}) - \rho D_{AB}\frac{\partial \omega_A}{\partial r} = \rho_A u_r - \rho D_{AB}\frac{\partial \omega_A}{\partial r} \tag{G.3-2}$$

$$n_{A\theta} = \omega_A(n_{A\theta} + n_{B\theta}) - \rho D_{AB}\frac{1}{r}\frac{\partial \omega_A}{\partial \theta} = \rho_A u_\theta - \rho D_{AB}\frac{1}{r}\frac{\partial \omega_A}{\partial \theta} \tag{G.3-3}$$

$$n_{A\phi} = \omega_A(n_{A\phi} + n_{B\phi}) - \rho D_{AB}\frac{1}{r\sin\theta}\frac{\partial \omega_A}{\partial \phi} = \rho_A u_\phi - \rho D_{AB}\frac{1}{r\sin\theta}\frac{\partial \omega_A}{\partial \phi} \tag{G.3-4}$$

in which ω_A is the mass fraction of component A, D_{AB} the binary diffusion coefficient, and u_i the mass-average velocity component in the i-direction. The form of the equation of continuity for a binary mixture or species balance given by equation (G.3-1) is particularly useful for describing mass transfer in incompressible liquid and solid systems for which the mass density ρ is constant. For the special case of an incompressible fluid or fluid having a constant mass density and a constant binary diffusion coefficient, equation (G.3-1) combined with equations (G.3-2) through (G.3-4) simplifies to

$$\frac{\partial \rho_A}{\partial t} + u_r \frac{\partial \rho_A}{\partial r} + u_\theta \frac{1}{r}\frac{\partial \rho_A}{\partial \theta} + u_\phi \frac{1}{r\sin\theta}\frac{\partial \rho_A}{\partial \phi} = D_{AB}\frac{1}{r^2}\frac{\partial}{\partial r}\left(r^2 \frac{\partial \rho_A}{\partial r}\right) + D_{AB}\frac{1}{r^2 \sin\theta}\frac{\partial}{\partial \theta}\left(\sin\theta \frac{\partial \rho_A}{\partial \theta}\right) + D_{AB}\frac{1}{r^2 \sin^2\theta}\frac{\partial^2 \rho_A}{\partial \phi^2} + G_A \quad \text{(G.3-5)}$$

The following form of the equation of continuity or species balance in spherical coordinates for component A in a binary system allows for nonconstant physical properties and is expressed in terms of the molar concentration c_A and the molar flux components N_{Ai}:

$$\frac{\partial c_A}{\partial t} + \frac{1}{r^2}\frac{\partial}{\partial r}(r^2 N_{Ar}) + \frac{1}{r\sin\theta}\frac{\partial}{\partial \theta}(N_{A\theta}\sin\theta) + \frac{1}{r\sin\theta}\frac{\partial N_{A\phi}}{\partial \phi} = \hat{G}_A \quad \text{(G.3-6)}$$

where $\hat{G}_A$ is the molar generation rate of component A per unit volume. The components of the molar flux are given for a binary system by Fick's law of diffusion in the form

$$N_{Ar} = x_A(N_{Ar} + N_{Br}) - cD_{AB}\frac{\partial x_A}{\partial r} = c_A\hat{u}_r - cD_{AB}\frac{\partial x_A}{\partial r} \quad \text{(G.3-7)}$$

$$N_{A\theta} = x_A(N_{A\theta} + N_{B\theta}) - cD_{AB}\frac{1}{r}\frac{\partial x_A}{\partial \theta} = c_A\hat{u}_\theta - cD_{AB}\frac{1}{r}\frac{\partial x_A}{\partial \theta} \quad \text{(G.3-8)}$$

$$N_{A\phi} = x_A(N_{A\phi} + N_{B\phi}) - cD_{AB}\frac{1}{r\sin\theta}\frac{\partial x_A}{\partial \phi} = c_A\hat{u}_\phi - cD_{AB}\frac{1}{r\sin\theta}\frac{\partial x_A}{\partial \phi} \quad \text{(G.3-9)}$$

in which x_A is the mole fraction of component A and $\hat{u}_i$ is the molar-average velocity component in the i-direction. The form of the equation of continuity for a binary mixture or species balance given by equation (G.3-6) is particularly useful for describing mass transfer in gas systems for which the molar density c is constant at a fixed temperature and pressure. This equation is also used to describe reacting systems for which the generation rate of species is dictated by the reaction stoichiometry in terms of molar concentrations. For the special case of a fluid having a constant molar density and a constant binary diffusion coefficient, equation (G.3-6) combined with equations (G.3-7) through (G.3-9) simplifies to

$$\frac{\partial c_A}{\partial t} + \hat{u}_r \frac{\partial c_A}{\partial r} + \hat{u}_\theta \frac{1}{r}\frac{\partial c_A}{\partial \theta} + \hat{u}_\phi \frac{1}{r\sin\theta}\frac{\partial c_A}{\partial \phi} = D_{AB}\frac{1}{r^2}\frac{\partial}{\partial r}\left(r^2 \frac{\partial c_A}{\partial r}\right) + D_{AB}\frac{1}{r^2 \sin\theta}\frac{\partial}{\partial \theta}\left(\sin\theta \frac{\partial c_A}{\partial \theta}\right) + D_{AB}\frac{1}{r^2 \sin^2\theta}\frac{\partial^2 c_A}{\partial \phi^2} + \hat{G}_A \quad \text{(G.3-10)}$$

APPENDIX H
Integral Relationships

H.1 LEIBNITZ FORMULA FOR DIFFERENTIATING AN INTEGRAL

Moving boundary problems usually involve determining the time derivative of an integral of a scalar quantity φ over a system whose volume $V(t)$ can be changing in time. The scalar quantity φ can be a function of time as well as the spatial coordinates. Leibnitz's formula permits recasting this volume integral into a more convenient form given by

$$\frac{d}{dt}\iiint_V \varphi \, dV = \iiint_V \frac{\partial \varphi}{\partial t} dV + \iint_S \varphi(\vec{u}_B \cdot \vec{n}) \, dS \tag{H.1-1}$$

where S is the time-dependent area enclosing the volume, $\vec{u}_B$ the local velocity of the bounding surface enclosing the volume, and $\vec{n}$ a unit normal vector to this surface. For a system having whose surfaces move in only one direction, equation (H.1-1) simplifies to

$$\frac{d}{dt}\int_{L_1}^{L_2} \varphi \, ds = \int_{L_1}^{L_2} \frac{\partial \varphi}{\partial t} ds + \varphi(L_2, t)\frac{dL_2}{dt} - \varphi(L_1, t)\frac{dL_1}{dt} \tag{H.1-2}$$

where s denotes an unspecified spatial coordinate for which L_1 and L_2 are lower and upper limits that can be time-dependent.

H.2 GAUSS–OSTROGRADSKII DIVERGENCE THEOREM

For a closed volume V surrounded by a surface S, the Gauss–Ostrogradskii or divergence theorem permits converting a volume integral into a surface integral. This theorem can be applied to scalars as follows:

$$\iiint_V \nabla \varphi \, dV = \iint_S \vec{n}\varphi \, dS \tag{H.2-1}$$

Scaling Analysis in Modeling Transport and Reaction Processes: A Systematic Approach to Model Building and the Art of Approximation, By William B. Krantz

where φ denotes an unspecified scalar quantity and $\vec{n}$ is a unit normal vector to the surface S. The divergence theorem can be applied to vectors as follows:

$$\iiint_V (\nabla \cdot \vec{u})\, dV = \iint_S (\vec{n} \cdot \vec{u})\, dS \qquad \text{(H.2-2)}$$

where $\vec{u}$ denotes an unspecified vector quantity. The divergence theorem can be applied to tensors as follows:

$$\iiint_V (\nabla \cdot \underline{\underline{\sigma}})\, dV = \iint_S (\vec{n} \cdot \underline{\underline{\sigma}})\, dS \qquad \text{(H.2-3)}$$

where $\underline{\underline{\sigma}}$ denotes an unspecified second-order tensor quantity.

NOTATION

Dimensions are given in terms of mass (M), moles (mol), length (L), time (t), force (F), and temperature (T).

A	unspecified constant, variable dimensions
A	amplitude of axial oscillation, L
a	integer or fractional quantity, dimensionless
a	surface area per unit volume of contacting device, L^{-1}
a_C	surface area per unit volume of carbon particles, L^{-1}
B	unspecified constant, variable dimensions
b	integer or fractional quantity, dimensionless
C_n	coefficients defined by equations (4P.12-2), (4.P.13-2), and (4.P.14-2)
C_p	heat capacity at constant pressure, $L^2/t^2 \cdot T$
C_{pi}	heat capacity at constant pressure of region, medium, or component i, $L^2/t^2 \cdot T$
C_v	heat capacity at constant volume, $L^2/t^2 \cdot T$
c	integer or fractional quantity, dimensionless
c	molar density or molar concentration, mol/L^3
c	speed of sound, L/t
c_{Ai}	molar concentration of component A at a point or plane defined by i, mol/L^3
c_i	molar concentration of component i, mol/L^3
$\hat{c}_i$	molar concentration of component i in the bulk fluid on the macroscale, mol/L^3
c_i^E	equilibrium or maximum molar concentration of component i, mol/L^3
c_i°	molar concentration of component i at the interface, mol/L^3
c_{ir}°	reaction equilibrium molar concentration of component i, mol/L^3
c_{is}	scale factor for the derivative of the concentration with respect to the i-coordinate, mol/L^4
c_{lm}	log-mean concentration, mol/L^3
c_w	molar concentration at the wall, mol/L^3
c_0	initial or upstream molar concentration, mol/L^3
D_i	binary diffusion coefficient in medium i, L^2/t
D_{ij}	binary diffusion coefficient for components i and j, L^2/t
D_L	axial dispersion coefficient, L^2/t
D_0	binary diffusion coefficient at infinite dilution, L^2/t
$\vec{F}_f$	force vector exerted by fluid particle on surrounding fluid, $M \cdot L/t^2$
$\vec{F}_p$	force vector exerted by surrounding fluid on a fluid particle, $M \cdot L/t^2$
f	frequency in cycles per unit time, t^{-1}

Scaling Analysis in Modeling Transport and Reaction Processes: A Systematic Approach to Model Building and the Art of Approximation, By William B. Krantz

$f_i(\cdots)$ denotes a function of the variables or parameters in parentheses, variable dimensions

G_e energy generation rate per unit volume, $M/L \cdot t^2$

G_i mass generation rate of component i per unit volume, $M/L^3 \cdot t$

G_0 amplitude of periodic energy generation rate per unit volume, $M/L \cdot t^2$

$\tilde{G}$ heating rate, $M \cdot L^2/t^3$

$\hat{G}$ total molar generation rate per unit volume, mol/$L^3 \cdot t$

$\hat{G}_i$ molar generation rate of component i per unit volume by homogeneous chemical reaction, mol/$L^3 \cdot t$

$\hat{G}_i^m$ maximum molar generation rate of component i by homogeneous chemical reaction, mol/$L^3 \cdot t$

g gravitational acceleration, L/t^2

g_c dimensional constant in Newton's law of motion, $M \cdot L/t^2 \cdot F$

g_i component of the gravitational acceleration in the i-direction, L/t^2

$g_i(\cdots)$ denotes a function of the arguments in parentheses, variable dimensions

H thickness, depth, or location of fixed boundary, L

H Henry's law constant, T^2/L^2

H_i enthalpy per mole of component i, $M \cdot L^2/t^2 \cdot$ mol

H_i thickness of layer i, L

h heat-transfer coefficient, $M/t^3 \cdot T$

h_P heat-transfer coefficient between particles and gas phase, $M/t^3 \cdot T$

h_W heat-transfer coefficient between wall and gas phase, $M/t^3 \cdot T$

h_1, h_2 thickness of layers 1 and 2, L

K integration rate constant, variable dimensions

K crystallization rate constant, t^{-n}

K_i distribution coefficient of component i between two phases, variable units

k thermal conductivity, $M \cdot L/t^3 \cdot T$

k an integer, dimensionless

k_G thermal conductivity of gas phase, $M \cdot L/t^3 \cdot T$

$k_G^{\bullet}$ gas-phase mass-transfer coefficient based on the molar flux relative to the molar average velocity, L/t

$k_{G0}^{\bullet}$ gas-phase mass-transfer coefficient in the absence of chemical reaction based on the molar flux relative to the molar average velocity, L/t

$k_i^{\bullet}$ mass-transfer coefficient of region, medium, or component i, L/t

k_L liquid-phase mass-transfer coefficient based on the molar flux relative to a stationary observer, L/t

$k_L^{\bullet}$ liquid-phase mass-transfer coefficient based on the molar flux relative to the molar average velocity, L/t

$k_{L0}^{\bullet}$ liquid-phase mass-transfer coefficient in the absence of chemical reaction based on the molar flux relative to the molar average velocity, L/t

k_m membrane permeability, $L^2 \cdot t/M$

k_n nth-order homogeneous reaction-rate parameter, mol$^{(1-n)} \cdot L^{3(n-1)}/t$

$\hat{k}_n$ nth-order heterogeneous reaction-rate parameter, mol$^{(1-n)} \cdot L^{(3n-2)}/t$

k_p permeability coefficient in Darcy's law for flow through porous media, L^2

L length in a particular coordinate direction, L

L location of fixed or moving boundary, L

L_d final depth corresponding to draining time t_d, L

L_e entrance length required for flow to become fully developed, L

L_i initial width of pulse of injected particles in field-flow fractionation, L

L_m	maximum thaw depth, L
L_s	unspecified length scale, L
L_0	initial depth corresponding to $t = 0$, L
$\dot{L}$	time rate of change of the thickness L, L/t
l_i	equilibrium adsorption distribution coefficient for component i, $L \cdot t^2/M$
M	mass, M
M_i	molecular weight of component i, M/mol
m	an integer, dimensionless
m_{ad}	total mass of adsorbent, M
m_{O_2}	total mass of oxygen produced per day in the pulsed PSA process, M
N_{Ai}	molar flux of component A in the i-direction or at surface i, mol/$L^2 \cdot t$
N_i	molar flux of component i, mol/$L^2 \cdot t$
N_{iw}	molar flux of component i at the wall, mol/$L^2 \cdot t$
$\overline{N}_{iw}$	time-averaged molar flux of component i at the wall, mol/$L^2 \cdot t$
n	an integer, reaction order, or Avrami exponent, dimensionless
n_{Ai}	mass flux of component A in the i-direction, $M/L^2 \cdot t$
n_G	mass flux in the gas phase, $M/L^2 \cdot t$
n_i	mass flux of component i, $M/L^2 \cdot t$
$\vec{n}$	unit normal vector, dimensionless
P	pressure, $M/L \cdot t^2$
$\dot{P}$	time rate of change of the pressure, $M/L \cdot t^3$
P_d	dynamic contribution to the total pressure, $M/L \cdot t^2$
P_h	hydrostatic contribution to the total pressure, $M/L \cdot t^2$
p	an integer, dimensionless
$p(\cdots)$	denotes a function describing the pressure, $M/L \cdot t^2$
p_i°	vapor pressure of component i, $M/L \cdot t^2$
$\overline{p}_i$	partial pressure of component i, $M/L \cdot t^2$
Q	volumetric flow rate, L^3/t
Q_L	volumetric flow rate of liquid, L^3/t
q	heat-transfer flux, M/t^3
q_i	heat-transfer flux at surface i, M/t^3
q_i	moles of component i per unit volume of adsorbent, mol/L^3
q_i^e	moles of component i per unit volume of adsorbent at equilibrium, mol/L^3
q_i^{∞}	value of q_i^e at infinite partial pressure of component i, mol/L^3
$\overline{q}$	heat flux averaged over the surface, M/t^3
$\dot{q}_i$	rate of molar adsorption of component i per unit volume, mol/$L^3 \cdot t$
$\dot{q}_i^e$	equilibrium-controlled rate of molar adsorption of component i per unit volume, mol/$L^3 \cdot t$
R	radius, L
R	gas constant, $M \cdot L^2/t^2 \cdot T \cdot$ mol
R	regression coefficient, dimensionless
R_i	rate of homogeneous reaction of component i per unit volume, mol/$L^3 \cdot t$
R_P	particle or pore radius, L
R_1, R_2	radii at surfaces defined by 1 and 2, L
r	radial coordinate, L
S	surface area, L^2
S_c	cross-sectional area, L^2
S_t	energy generation rate per unit volume, $M/L \cdot t^2$
$\dot{S}$	rate of entropy generation per unit volume, $M/L \cdot t^2 \cdot T$

T	temperature, T
T_c	crystallization temperature, T
T_f	freezing or front temperature, T
T_G	temperature of gas phase, T
T_i	temperature evaluated at surface i, T
T_{is}	scale factor for derivative of temperature in the i-direction, T/L
T°	reference temperature, T
$\overline{T}$	average temperature, T
$\overline{T}_m$	maximum value of the spatially averaged temperature, T
$\dot{T}$	time rate of change of temperature, T/t
t	time, t
$\vec{t}$	unit tangential vector, dimensionless
t_a	time at which rate of phase separation is a maximum, t
t_{ad}	characteristic adsorption time, t
t_c	contact or cooking time, t
t_d	characteristic axial dispersion, draining, or depressurization time, t
t_g	time for growth of dispersed phase, t
t_H	time required for the temperature to penetrate thickness H, t
t_o	observation time, t
t_p	pressurization time, t
t_{sp}	time scale corresponding to the periodic motion, t
t_{st}	time scale corresponding to the observation time, t
t_{sv}	time scale corresponding to the viscous penetration, t
U	superficial velocity in porous media, L/t
U_G	linear velocity of gas, L/t
U_m	maximum velocity, L/t
U_t	terminal velocity, L/t
U_w	axial velocity at the wall, L/t
U_0	specified constant velocity, L/t
$\overline{U}$	average velocity, L/t
$\overline{U}_w$	average velocity in the wall region, L/t
$\vec{u}$	mass-average velocity vector, L/t
$\vec{\hat{u}}$	molar-average velocity vector, L/t
$\vec{u}_B$	local mass-average velocity of a surface enclosing a volume, L/t
u_i	mass-average velocity component in the i-direction, L/t
$\hat{u}_i$	molar-average velocity component in the i-direction, L/t
${}^1\widehat{u}_i$	mass-average velocity component in the i-direction in a porous media, L/t
V	volume, L^3
V_{-H}, V_{+H}	velocity at which fluid is injected and withdrawn at the surfaces at $\pm H$, L/t
V_s	scale for velocity of a moving boundary, L/t
V_T	total volume, L^3
V_u	volume of upper chamber, L^3
V_0	specified blowing or suction velocity, L/t
W	width, L
W_G	molar flow rate of gas, mol/t
$\hat{W}_H$	molar injection rate of hydrogen per unit length, mol/$t \cdot L$
W_i	molar flow rate of component i, mol/t
W_i^0	molar flow rate of component i in the feed stream, mol/t
$\tilde{W}_m$	mass injection or removal rate per unit area, $M/t \cdot L^2$

w	gap width, L
X_c	fractional dissociation or conversion, dimensionless
x	spatial coordinate, L
x_i	mole fraction of component i, dimensionless
y	spatial coordinate, L
y_{BL}	product purity measured in terms of the mole fraction of component B, dimensionless
y_i	gas-phase mole fraction of component i, dimensionless
$\tilde{y}$	translated spatial coordinate, L
y_m^*	dimensionless y-coordinate in the equations of motion
y_t^*	dimensionless y-coordinate in the energy equation
z	spatial coordinate, L

GREEK SYMBOLS

α	thermal diffusivity, L^2/t
α	undetermined constant, variable units
α	arbitrary quantity considered in dimensional analysis, variable units
α_i	thermal diffusivity of medium i, L^2/t
β	arbitrary quantity considered in dimensional analysis, variable units
β	unspecified time constant, t^{-1}
β	constant coefficient for temperature dependence of vapor pressure, $M/L \cdot t^2 \cdot T$
β	constant coefficient for temperature dependence of thermal conductivity, $M \cdot L/t^2 \cdot T^2$
β_s	scale factor for the spatial derivative of the velocity, L/t
β_s	coefficient of solutal volume expansion, L^3/mol
β_t	coefficient of thermal volume expansion, T^{-1}
γ	ratio of heat capacity at constant pressure to that at constant volume, dimensionless
γ	arbitrary quantity considered in dimensional analysis, variable units
γ_t	second-order coefficient of thermal volume expansion, T^{-2}
Δc_A	$c_{A0} - c_{A\infty}$, concentration difference between control surfaces defined by 0 and ∞, mol/L^3
Δc_{lm}	log-mean concentration driving force, mol/L^3
ΔE	activation energy for reaction, $M \cdot L^2/t^2 \cdot$ mol
ΔH_f	latent heat of fusion per unit mass, L^2/t^2
ΔH_v	latent heat of vaporization per unit mass, L^2/t^2
ΔP	$P_2 - P_1$, pressure difference between control surfaces defined by 1 and 2, $M/L \cdot t^2$
$\Delta\rho_{AB}^0$	$\rho_A^0 - \rho_B^0$, difference between pure component mass densities, M/L^3
δ	region of influence or boundary-layer thickness, L
δ	arbitrary quantity considered in dimensional analysis, variable units
$\underline{\underline{\delta}}$	identity tensor, dimensionless
δ_c	characteristic film thickness, L
$\vec{\delta}_i$	unit vector in the i-direction, dimensionless
δ_m	momentum boundary-layer thickness, L
δ_p	thickness of region of influence in porous media, L
δ_r	thickness of reaction boundary layer, L
δ_s	solutal boundary-layer thickness, L

δ_t	thermal boundary-layer thickness, L
ε	fractional void space or porosity, dimensionless
ε	small distance, L
ε	radiative emissivity of surface, dimensionless
ς	dimensional constant, M/L^2
η	local film thickness, L
θ	angular coordinate, radians
θ	angle of inclination or angle between horizontal and unit tangential vector, dimensionless
κ	bulk viscosity, $M/L \cdot t$
κ	ratio of the mass fluxes for binary diffusion, dimensionless
κ	stoichiometric coefficient, dimensionless
λ_n	roots of equation (4.P.12-3), (4.P.13-3), or (4.P.14-3), dimensionless
μ	shear viscosity, $M/L \cdot t$
ν	kinematic viscosity, L^2/t
ξ	a dimensionless constant having a value of either 0 or 1, depending on the local concentration relative to that at reaction equilibrium
Π_i	ith dimensionless group whose definition depends on the particular problem
ρ	mass density, M/L^3
$\overline{\rho}$	average mass density, M/L^3
$\tilde{\rho}$	mass density of the gas phase, M/L^3
ρ_i	mass density of region, medium, or component i, M/L^3
ρ_i^0	mass density of pure component i, M/L^3
ρ_p	mass density of pure polymer, M/L^3
ρ_w	mass density of liquid water, M/L^3
σ	Stefan–Boltzmann constant, $M/t^3 \cdot T^4$
$\underline{\underline{\sigma}}$	total stress tensor, $M/L \cdot t^2$
σ_{AB}	collision diameter for molecules A and B, L
σ_{ij}	ij-component of the total stress tensor, $M/L \cdot t^2$
$\underline{\underline{\tau}}$	viscous stress tensor, $M/L \cdot t^2$
τ_{ij}	ij-component of the viscous stress tensor, $M/L \cdot t^2$
ϕ	angular coordinate, radians
ϕ	unspecified constant, dimensionless
ϕ	volume of liquid per unit interfacial area, L
φ	unspecified scalar having variable units
φ	dimensionless derivative of the concentration
χ	ratio of the total absorption rate relative to its maximum value for purely physical absorption, dimensionless
Ψ_r	reaction-rate parameter, dimensionless
ψ	liquid volume fraction in a multiphase system, dimensionless
$\dot{\psi}$	time rate of change of the liquid volume fraction, t^{-1}
ψ_{bf}	bed-size factor equal to the ratio of the total mass of adsorbent to the total mass of oxygen produced per day, dimensionless
ψ_r	product recovery equal to the ratio of the volumetric flow of the product to that of the feed, dimensionless
Ω	collision integral, dimensionless
ω	angular frequency or velocity, radians/t
ω_i	mass fraction of component i, dimensionless
$\tilde{\omega}_i$	mass fraction of component i in the gas phase, dimensionless

SUPERSCRIPTS

d depressurization or diluent
E equilibrium value
p pressurization
0 initial value
$\circ$ saturation or reference value
$-$ average value
$\hat{}$ variable evaluated in the bulk fluid when using microscale-macroscale modeling
$\frown$ variable evaluated in a porous medium
$*$ dimensionless variable
$\dagger$ transpose of a second-order tensor
$\cdot$ time derivative
$\tilde{}$ dummy integration variable or a translated spatial coordinate
$\rightarrow$ vector quantity

SUBSCRIPTS

A component A
atm atmospheric conditions
B component B
b evaluated at the boundary
C component C or carbon
c variable in the species-balance equation
D property of turkey dressing or the diluent phase
d dynamic pressure contribution
f frozen region or feed concentration
G property of the gas phase
h hydrostatic pressure contribution
I iron or iron phase
L evaluated at L, downsteam, or in the liquid
l evaluated in the lumen of a hollow fiber membrane
m maximum value, denotes a variable or parameter associated with the fluid or a with a membrane
o observation time
P property of the polymer or particulate phase
R rust or rust phase
r reference factor or component in the r-direction
s scale factor or denotes a solutal quantity
T property of turkey
t variable for heat transfer or terminal velocity
t time derivative
u unfrozen region
W water or water phase
w evaluated at the wall or in the wall region
x component in the x-direction
y component in the y-direction
z component in the z-direction

θ component in the θ-direction
ϕ component in the ϕ-direction
0 initial, upstream, reference, or infinite dilution value
∞ evaluated as the spatial or temporal variable approaches infinity
1 property of phase 1 or denotes a particular function
2 property of phase 2 or denotes a particular function
$=$ tenor quantity

COMMONLY USED DIMENSIONLESS GROUPS

Bi_m Biot number for mass transfer or solutal Biot number, ratio of the total mass transfer external to the system to the diffusive mass transfer within the system
Bi_t Biot number for heat transfer or thermal Biot number, ratio of the total heat transfer external to the system to the conductive heat transfer within the system
Da^{II} second Damkohler number, ratio of the time scale for radial diffusion to that for heterogeneous reaction
Fo_m Fourier number for mass transfer, ratio of the contact time to the characteristic time for species diffusion
Fo_t Fourier number for heat transfer, ratio of the contact time to the characteristic time for heat conduction
Fr Froude number, ratio of the kinetic energy to the gravitational potential energy of the flow
Gr_m Grashof number for mass transfer, ratio of the solutally induced convective to viscous transport of momentum
Gr_t Grashof number for heat transfer, ratio of the thermally induced convective to viscous transport of momentum
Gz Graetz number, a measure of the relative convection to diffusional transport of species
Le Lewis number, ratio of heat conduction to species diffusion
Ma Mach number, ratio of the characteristic fluid velocity divided by the speed of sound in the medium
Nu Nusselt number, a measure of the overall heat transfer to that by conduction alone
Pe_m Peclet number for mass transfer or solutal Peclet number, ratio of convective to diffusive transport of mass
Pe_t Peclet number for heat transfer or thermal Peclet number, ratio of convective to conductive transport of heat
Pe_V Peclet number for mass transfer or solutal Peclet number based on the transverse velocity
Pr Prandtl number, ratio of heat conduction to viscous transport of momentum
Ra_m Rayleigh number for mass transfer, ratio of solutally induced convective to diffusive transport of mass
Ra_t Rayleigh number for heat transfer, ratio of thermally induced convective to conductive transport of heat
Re Reynolds number, ratio of the kinetic energy per unit mass of the flow to the principal viscous stress or equivalently, ratio of convective to viscous transport of momentum
Re_V Reynolds number based on the transverse velocity
Sc Schmidt number, ratio of viscous transport of momentum to species diffusion

Sh Sherwood number, a measure of the overall mass transfer to that by diffusion alone

Sh_v Sherwood number in the presence of axial oscillations or vibrations

Th Thiele modulus, ratio of the characteristic time for diffusion relative to that for homogeneous reaction

MATHEMATICAL OPERATIONS

$\frac{D}{Dt}$ substantial time derivative or time derivative in a coordinate system convected at the fluid velocity

$\frac{d}{dt}$ total time derivative allowing for the change in the quantity in time as well as with spatial position

$\frac{\partial}{\partial t}$ partial time derivative allowing for the change in the quantity in time at a fixed spatial position

erf(φ) error function of z, $= \frac{2}{\sqrt{\pi}} \int_0^{\varphi} e^{-\tilde{\varphi}^2}\, d\tilde{\varphi}$

exp (φ) exponential function of φ, $= e^{\varphi}$

ln (φ) logarithm of φ to the base e

∇ del or nabla operator

INDEX

Scaling Analysis in Modeling Transport and Reaction Processes: A Systematic Approach to Model Building and the Art of Approximation, By William B. Krantz